BIOTECHNOLOGY: BRIDGING RESEARCH AND APPLICATIONS

Biotechnology: Bridging Research and Applications

Proceedings of the U.S.-Israel Research Conference on Advances in Applied Biotechnology June 24–30, 1990; Haifa, Israel

Edited by

DAPHNE KAMELY

ANANDA M. CHAKRABARTY

STEVEN E. KORNGUTH

KLUWER ACADEMIC PUBLISHERS

BOSTON / DORDRECHT / LONDON

Library of Congress Cataloging-in-Publication Data

Biotechnology : bridging research and applications / edited by Daphne Kamely, Ananda M. Chakrabarty, Steven E. Kornguth.
p. cm.
Includes index.
ISBN 0-7923-1144-2
1. Biotechnology. I. Kamely, Daphne. II. Chakrabarty, Ananda M., 1938- . III. Kornguth, Steven E.
TP248.2.B5515 1991
660'.6--dc20 91-6985

ISBN 0-7923-1144-2

Published by Kluwer Academic Publishers

Kluwer Academic Publishers incorporates
the publishing programmes of
D. Reidel, Martinus Nijhoff, Dr W. Junk and MTP Press.

Sold and distributed in the U.S.A. and Canada
by Kluwer Academic Publishers,
101 Philip Drive, Norwell, MA 02061, U.S.A.

In all other countries, sold and distributed
by Kluwer Academic Publishers,
P.O. Box 322, 3300 AH Dordrecht, The Netherlands.

Printed on acid-free paper

Table of Contents

Foreword ix

Acknowledgments xiii

KEYNOTE ADDRESS

Biological Applications of Synthetic Polypeptides 1
Ruth Arnon

SESSION I: BIOLOGICAL DETECTORS

Manufacturing and Biotechnology Synergisms for the Early 2000's 15
Christopher Canfield Green

Biosensors Based on Solvated Bilayers Attached to Electrodes 43
Carlos Gitler, Yechiel Shai, Tuvia Bercovoci, Diana Bach, Haim Garty, Israel Rubinstein, Suzi Steinberg and Itzhak Yuli

Alternatives to Radioimmunoassay 63
G. Barnard, J. De Boever and Fortune Kohen

The Use of Intact Luminous Bacteria in Analytical Chemistry, Clinical Microbiology and Acute Toxicity 79
Shimon Ulitzur

Deposition of Enzymes by Electropolymerization 89
Nicholas F. Almeida and Lemuel B. Wingard Jr.

Computerized Biospecific Electrodes 95
Judith Rishpon

NMR-Structural Studies of Membrane Bound Peptides and Proteins 109
Ki-Joon Shon, Patricia Schrader, Yongae Kim, Burkhard Bechinger, Michael Zasloff and Stanley Opella

SESSION II: MOLECULAR BIOLOGY IN FIELD DETECTION

DNA as a Biosensor for Environmental Agents 127
Lawrence Grossman and William Athas

Detection of Pathogens with Nucleic Acid Probes 141
David H. Gillespie

Structure-Function Relationships, In Vivo Mutability and Gene Amplification in Human Cholinesterases, Targets for Organophosphorus Poisons 153
Hermona Soreq, Lewis Neville, Averell Gnatt, Revital Ben-Aziz, Yaron Lapidot-Lifson, Gal Ehrlich, Shlomo Seidman, Efrat Lev-Lehman, Rachel Beeri, Dalia Ginzberg and Haim Zakut

Assessment of Genetic Damage in Human Tissue by Immunoassay 167
Paul T. Strickland, Nathaniel Rothman and Miriam C. Poirier

Field Operable Devices for Immunological, Molecular and Toxicological Diagnosis – A Review on a Unified Approach 179
Falk Fish

Design and Properties of Enzymes Immobilized in Sol-Gel Glass Matrices 205
Sergei Braun, S. Rappoport, S. Shtelzer, R. Zusman, S. Druckmann, D. Avnir and M. Ottolenghi

SESSION III: BIOLOGICAL DECONTAMINATION

Microbial Degradation of Agent Orange and Mustard Related Compounds 221
Steven Harvey, Joseph J. DeFrank, Daphne Kamely, James J. Valdes and Ananda M. Chakrabarty

Surface Active and Drag-Reducing Bacterial Polymers 231
Eugene Rosenberg

Towards Catalytic Antibodies for the Degradation of Toxic Agents 249
Bernard S. Green and Michael Glikson

Bacterial Heavy Metal Resistance Systems and Possibility of Bioremediation 265
Simon Silver

The Cholinergic Binding Site: Structure and Application 289
Bella Ohana, Rachel Mosckovitz and Jonathan M. Gershoni

SESSION IV: MATERIAL BIOTECHNOLOGY

Outdoor Cultivation of Microalgae in a Closed System for the Production of Valuable Biochemicals 301
Shoshana (Malis) Arad and Ephraim Cohen

Chemical Sensing Applications of Semiconductor Photoluminescence 317
Elizabeth R.M. Luebker, Larry K. Leung, Catherine J. Murphy, George C. Lisensky and Arthur B. Ellis

High Speed DNA Sequencing and the Analysis of the Human Genome 333
Lloyd M. Smith, Jon D'Cunha and John A. Luckey

Interacting Light Energy with Biocatalytic Assemblies: Biocatalysed Photosynthesis and Photoregulation of Enzymes 341
Itamar Willner, Noa Lapidot, Shai Rubin, Azalia Riklin and Bilha Willner

Modifying Specificities of Restriction Enzymes 371
Waclaw Szybalski

Pulsed Laser Fluorometry for Environmental Monitoring 377
George C. Saunders, John C. Martin, James H. Jett, Mark E. Wilder, Adelmo Martinez, Bill F. Bentley, John Lopez and Lara Hutson

SESSION V: PROTEIN ENGINEERING

Recognition of the Human CD4 Receptor by the AIDS Virus: A Molecular Target for AIDS Therapeutics 401
Raymond Sweet and Martin Rosenberg

Antibody-Catalyzed Concerted Chemical Reactions 413
Donald Hilvert

Engineering Altered Substrate Specificity into Subtilisin BPN′ 429
Richard Bott, Thomas Graycar and David Estell

Summary of Conference 441
Steven E. Kornguth

Concluding Discussion 445

List of Participants 451

Index 457

Foreword

Biotechnology is advancing at a rapid pace with numerous applications in medicine, industry, agriculture and environmental remediation. Recognizing this, government, industrial and academic research and development investment in biotechnology has expanded rapidly. The past decade has seen the emergence of applications of this technology with a dual-use potential. Military applications focus on four major areas: biomedical technology, such as vaccine development and medical diagnostics; detection of toxins, chemicals and pathogens; material biotechnology; and biological decontamination, including biodegradation and bioremediation.

This conference emphasizes the non-medical applications of biotechnology. The first two sessions focus on the synthesis and properties of molecules that may be used in detectors. The traditional approach to detection of chemical and biological agents relied on the development of specific assays or analyses for known agents. Advances in molecular biology have made possible the production of large quantities of toxins which were previously available in minute quantities, and the molecular engineering of toxins and pathogens with specific pharmacologic and physical-chemical properties.

In addition to the traditional approaches to detection of specific known compounds, biotechnology now offers generic approaches to detection. Physiological targets, known as receptors, are primary targets for many drugs and toxins. Similarly, pathogens rely on receptors to gain access to cells. These receptors function as sensitive detectors, generating signals which are transduced and amplified.

Biosensors are devices which use biological recognition elements such as receptors, antibodies, enzymes and gene probes to recognize an agent, and electronic devices to transduce and amplify the resultant signal. Biosensors are currently being developed for both continuous monitoring and one-time, immediate testing of aqueous and air samples. The successful development of biosensors will require further advances in molecular biology, such as cloning receptor genes, fermentation, scale-up production of antibodies and receptors, protein engineering to create proteins with enhanced stability or unique properties, and signal transduction.

The development of an array of biosensors has enormous potential. Aside from the obvious military applications, biosensors can be used to monitor industrial effluents and hazardous wastes, control fermentation processes in the food industry, and detect minute levels of hormones and drugs in body fluids. This conference addresses many issues related to the development of biosensors, with particular emphasis on the biological components.

Considerable emphasis is also placed on solutions to environmental pollution. Environmental pollution is a major problem not only in the United States, but throughout the industrialized world. There are unique pollution problems associated with the manufacture and storage of military hardware, hazardous chemicals, and other pollutants. With the lessening of tension throughout the world and increased cooperation within Europe, the question of how hazardous materials can be disposed of in an economically safe and ecologically acceptable manner has assumed considerable importance. Biotechnology has the potential to play a major role in solving some of the pollution problems that confront the defense industry, and a number of presentations in this conference address these problems with potential biological treatment strategies for the reduction of waste volume as well as actual detoxification of the chemicals. These include genetic manipulation strategies and construction of novel microbial strains for the decomposition of various toxic chemicals, the detoxification and potential recovery of heavy metals, the isolation and characterization of novel microbial polymers with surface-active and drag-reducing properties, as well as the synthesis of catalytic antibodies for the detection and degradation of toxic agents.

Material biotechnology is still in the early stages of development. Using biotechnology, it is now possible to engineer new and improved materials of higher strength, lighter weight, and increased elasticity. Higher quality materials can be achieved in the production of ceramics, bioadhesives, and antifouling agents. Exciting developments in this area, to name two discussed in the conference, include semiconductor photoluminescence and the construction of automated sequencing machines for rapid sequencing of large segments of DNA, introduced as a prelude to human genome sequencing.

The above described applications in biotechnology are based on advances in basic molecular biology and biochemistry. Understanding the mechanisms of DNA replication and gene regulation and expression, enables scientists to genetically engineer novel materials and analyze protein structure and function. Thus, through the use of modified proteins, light-induced catalysis and electron transfer reactions can be studied. Similarly, modification of restriction enzymes allows more efficient and altered substrate specificities, resulting in improved methods of cloning and sequencing. Crystallography and protein modelling and manipulation of enzyme specificity using oligonucleotide directed mutagenesis of cloned genes constitute the last session on protein engineering. Interesting reports on polysaccharide engineering and interaction between receptors and envelope proteins of viruses lead to better understanding of the molecular mechanism of disease.

While the subdisciplines of biotechnology are expanding rapidly, this conference focuses on a few more recent development which may also be relevant to defense research and applications. Through this conference and the advances demonstrated by the excellent variety of research topics, we hope that further collaborations between scientists and institutions in Israel and the United States can be established leading to fruitful and productive research programs.

Daphne Kamely and Ananda M. Chakrabarty

Acknowledgments

The organizing committee wishes to thank Dr. Robin Woo of Woo Wester Conference Consultants for organization and coordination of the conference and the University of Illinois for financial coordination. The organizers extend their appreciation to Dr. James Valdes and Mrs. Marcia Goforth for their technical and administrative assistance.

Many thanks are extended to Mrs. Daniella Dreman of Kenes and Mrs. Audrey Hakim of the Gil Travel Agency for local organization support and travel assistance.

The organizers wish to thank the following sponsors for their generous support:

United States Army
- Chemical Research, Development and Engineering Center
- Natick Research, Development and Engineering Center
- Army Research, Development and Engineering Center
- European Research Office
- Toxic and Hazardous Materials Agency

United States Navy
- Office of Naval Technology
- Office of Naval Research

United States Air Force
- Air Force Engineering and Services Center

State of Israel, Ministry of Defence

The Organizers:

Mr. Joseph Bar-Tov
Directorate of Defence Research and Development
Israel Ministry of Defence
Hakirya, Tel-Aviv, Israel

Daphne Kamely, Ph.D.
Director, Research and Laboratory Management
United States Army
SARD-TR
The Pentagon
Washington, D.C. 20310-0103

Ananda M. Chakrabarty, Ph.D.
Microbiology Department
University of Illinois College of Medicine
Chicago, Illinois 60612

Steven E. Kornguth, Ph.D.
Department of Neurology and Physiological Chemistry
University of Wisconsin
Madison, Wisconsin 53705

Biological Applications of Synthetic Polypeptides

Ruth Arnon
The Weizmann Institute of Science, Rehovot 76100, Israel

This paper discusses three immunological applications of synthetic polypeptides – the use of synthetic polypeptides in the development of synthetic vaccines, our studies on a synthetic copolymer of amino acids with potential therapeutic value for multiple sclerosis, and the use of synthetic polypeptides as drug carriers.

Development of Synthetic Vaccines

Our desire to develop synthetic vaccines stems from the fact that conventional vaccines, although being very successful in the sense that they have reduced both incidence and morbidity of many infectious diseases, still have some drawbacks. There are now various approaches in the development of new vaccines. Conventional vaccines are based on the entire disease-inducing organism which is either attenuated or killed and thus, while not causing the disease, they can still immunize and protect against it. The drawbacks that we envisage are 1) it is not always possible to culture the organism in amounts sufficient to produce a vaccine (e.g., some viruses and parasites) and 2) there can be serious biohazards associated with culturing disease-causing organisms, especially in large amounts (e.g., the AIDS virus).

New approaches for vaccine development include recombinant DNA genetic engineering techniques, which are being used to (1) express a relevant disease-specific protein either in yeast, bacteria, or mammalian cells, and isolate the protein for the purpose of vaccination, (2) express this protein in *Vaccinia* or avirulent strains of *Salmonella* and produce a "live" vaccine against the inserted protein by immunization with these recombinant organisms, (3) produce an anti-idiotype vaccine that mimics the relevant epitope, and (4) design a vaccine based on synthetic peptides, which I will discuss here.

We have been studying this latter approach for many years, since we were using synthetic polypeptide antigens to produce antibodies against any desired specificity, thus characterizing the structural requirements for rel-

D. Kamely et al. (eds), Biotechnology: Bridging Research and Applications, 1–12.

evant epitopes and carriers. We showed that this same approach could be used with a completely synthetic antigen carrying a peptide corresponding to a given epitope, to produce antibodies that are reactive with native proteins.

To summarize our approach, we have a carrier that is based on either a synthetic, in this case branched, polymer, or a natural protein, to which we attach a peptide with a relevant epitope, a fragment of a protein, or a fragment of a virus. In addition, a synthetic adjuvant can be attached to this synthetic antigen for augmenting the immune response, thus leading to a synthetic vaccine with built-in adjuvanticity.

Lysozyme

Working with the protein lysozyme, we identified a fragment, looped by internal disulfide bonds, which appeared to be an important epitope of the molecule according to its three-dimensional structure. Using this fragment, as isolated from lysozyme, and attaching it to our synthetic carrier, we produced antibodies which reacted with whole lysozyme (Arnon and Sela, 1969). We then synthesized this same fragment and showed that it was possible to prepare a completely synthetic molecule which, when used for immunization, induced antibodies that were reactive with the native protein, lysozyme (Arnon, et al., 1971).

MS2 Bacteriophage

Could this approach work for viral proteins as well? For our model system we used the MS2 bacteriophage, which has in its capsis 180 identical subunits of the coat protein whose sequence was known. We identified a fragment of 20 amino acid residues, synthesized it, attached it to our synthetic carrier, and immunized rabbits with this conjugate. The resultant antibodies completely inactivated and neutralized the virus (Langbeheim, Arnon and Sela, 1976). Furthermore, if we immunized with the same molecule, conjugated with the synthetic adjuvant muramyl dipeptide (MDP), instead of using Complete Freund's Adjuvant, we observed only slightly lower neutralization activity against the virus. This completely synthetic molecule has, therefore, "built-in adjuvanticity", and can be used for immunization without additional adjuvant: a solution in buffered saline can be used for effective immunization. Of course, no one is interested in being vaccinated against a bacteriophage, but the above results demonstrated the feasibility of the approach, that a synthetic molecule can be chemically engineered that has the capacity to induce virus-neutralizing antibodies. With this capability now in hand, we can pursue other systems that are of greater interest to humans.

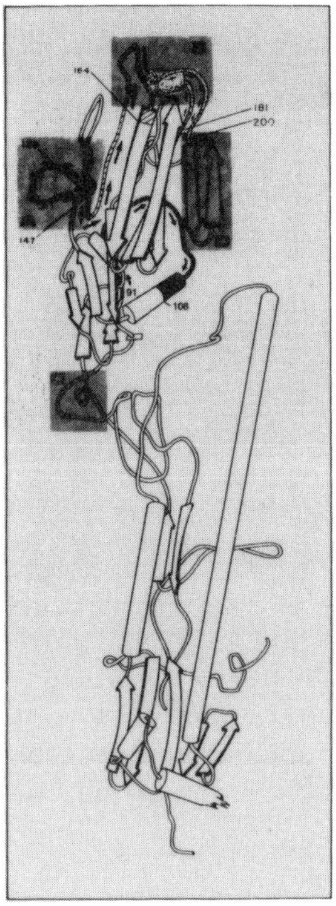

Figure 1. Three-dimensional structure of influenza hemagglutinin. The four proposed antigenic sites are shaded (from Wiley, Wilson and Skehel, 1981).

Influenza Virus

Our next model is the influenza virus, which has a membrane composed of several proteins and a core of RNA, surrounded by a nuclear protein. Two membrane proteins, a hemagglutinin (HA) and a neuraminidase, are serologically important. We selected the hemagglutinin as a candidate for our synthetic vaccine studies. As can be seen in the three-dimensional structure of the protein (Figure 1), Don Wiley and his collaborators (Wiley, Wilson and Skehel, 1981) proposed four regions as the major antigenic sites of HA. Changes in amino acid sequences in these regions are responsible for the serological differences in the different strains. One of the difficulties in preparing an efficient vaccine against influenza is due to these frequent changes (shifts and drifts), such that a vaccine that worked effectively against last year's strain is useless against this year's. If one could circumvent this problem and develop a vaccine that would be efficient against all strains of the virus, regardless of the changes that might occur in the sequence of the hemagglutinin, a great step forward would be realized.

We synthesized the peptide corresponding to the sequence 91–108, which is a conserved region in all hemagglutinins of the A-H3 type. We attached this peptide to a tetanus toxoid carrier and immunized against it (Muller, Shapira and Arnon, 1984). The antibodies we obtained reacted similarly with three different strains of influenza in a radioimmunoassay, and we surmise that they will similarly react to all other strains of influenza containing this same peptide sequence. The antibodies were also capable of neutralizing the biological activity of the hemagglutinin molecule.

The influenza virus, besides being a very interesting system for which a good vaccine is not yet available, is also a very good model system for study, since one can study the reactivity of the immune response against various epitopes on four different levels. At the lowest level, immunochemical reactivity, one can measure the capacity of antibodies, by radioimmunoassay for instance, to recognize the peptide or the intact virus, which is what we have shown here. In addition, at the second level, since the hemagglutinin has biological activity and can agglutinate chicken erythrocytes, one can measure the capacity of the antibodies to neutralize this activity. At the third level, since the virus can be grown in tissue culture, the capacity of the antibodies to inhibit plaque formation of the virus can be measured. Finally, at the fourth level, which is the most important one, it is possible to check whether immunization with the peptide induces protection against the disease. Fortunately, mice are sensitive to the same strains of influenza that are infectious in humans.

We studied the efficiency of our peptide on all four levels, with the following results: our synthetic peptide-induced antibodies inhibited up to 75% of the hemagglutination of chicken red blood cells. Viral plaque formation was inhibited up to 67%. When we immunized mice with the peptide, we achieved 50% protection against infection by three different strains of influenza: Texas, Port Chalmers, and England. When we looked at the titer of the virus in the lungs of infected mice, there was a difference of two orders of magnitude. So, at least in this case, it seems that a synthetic peptide vaccine is indeed capable of inducing partial protection against the disease. This is certainly not sufficient protection; we have tried a few other peptides of the hemagglutinin, but cannot augment this percentage. We are now trying a peptide of the nuclear protein that induces cytotoxic T cells; perhaps a combination of epitopes of two proteins will be more efficient.

Bacterial Toxins

Another system that we are studying is bacterial toxins, which I will illustrate with results obtained with Cholera toxin and Shiga toxin. These toxins have a similar structure, composed of two subunits, A and B. The B subunit, comprising a pentamer of B chains, is the most immunogenic part; antibodies against Cholera toxin react mainly with the B subunit and not with the A subunit, making the B chain a natural candidate for our studies. The sequence of the B chain is known, and shows high level of homology with the heat-

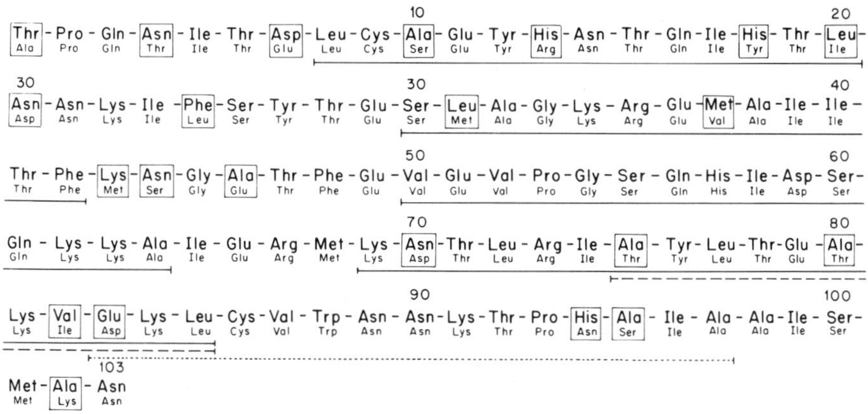

Figure 2. The amino acid sequence of the B chain of Cholera toxin (upper) and heat-labile toxin of E. coli (lower). The residues by which the two toxins differ are in boxes. The regions synthesized are underlined.

labile toxin of toxigenic *E. coli* (Figure 2). We synthesized several peptides indicated by the underlined sequences and used them for immunization (Jacob, Pines and Arnon, 1984). All elicited very efficient immune responses against the respective peptides, but only two (peptides 1 and 3) elicited immune responses against the intact toxin. We next checked the neutralizing capacity of these two peptides against the biological activity of the toxin, measured either *in vivo* by inhibiting the fluid secretion induced by the toxin into the ligated ileal loop in rat or rabbit, or biochemically by inhibition of cyclic AMP induction by the toxin. In both cases, with antibodies against peptide 1 and even more so with antibodies against peptide 3, we observed efficient (70%) inhibition of the biological activity of the toxin (Figure 3). In this case, as well as in the case of influenza, we can show that using a relatively short synthetic peptide (about 15 amino acids long), we can induce an immune response that is not only efficient in an immunochemical reaction, but also in neutralizing the biological activity of the toxin.

In the case of Shiga toxin, we again synthesized several peptides and have checked their capacity to induce inhibition of biological activity (Arnon, in press). We chose the peptides according to the folding of the B chain. Both the amino and carboxyl terminal regions have high hydrophilicity and high content of surface residues, whereas the middle part of the molecule has low hydrophilicity and no surface residues. We synthesized peptides covering the entire molecule, and conjugated them to various carriers. Table 1 shows neutralization of Shiga toxin by antibodies produced against the peptide conjugates. The peptides from both the amino and carboxyl terminal regions led to efficient neutralization of three different biological activities of the Shiga toxin – 1) cytotoxicity towards HeLa cells, 2) enterotoxicity measured by secretion of fluid into ligated ileal loops in rabbits, and 3) neurotoxicity, namely lethal effect in mice. In contrast, the peptides from the middle of

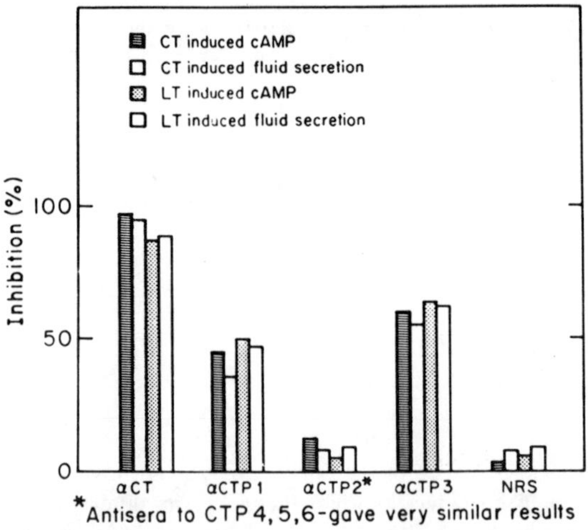

Figure 3. Inhibition of Cholera toxin (CT) and heat labile toxin of E. coli (LT) by rabbit antisera against the intact toxins, the various synthetic peptides, as well as by normal rabbit serum.

the molecule (residues 34–47) had no neutralizing activity. Since the two termini of the B chain are linked by a disulfide bridge, we have thus shown that the three-dimensional structure of the molecule is thus an important feature in deciding which part of the molecule is more likely to be a relevant epitope for induction of neutralizing immune response. Furthermore, immunizing mice with conjugates of the appropriate peptides protected them against neurotoxicity, and immunizing rats protects them against enterotoxicity of the Shiga toxin.

To summarize, we have shown that synthetic peptides can indeed serve as the basis for vaccine preparation. Conjugates of such peptides may induce antibodies with neutralizing capacity of either antiviral or antitoxin activity. Immunization with these synthetic vaccines leads to protection against the toxin's biological damage, as well as against viral infection. If the peptide is

Table 1. Neutralization of biological activities of Shiga toxin by anti-peptides sera

Peptide	Inhibition of the Various Biological Activities (%)		
Immunogen	Cytotoxic	Enterotoxic	Neurotoxic
None (normal)	23	19	0[a]
5–18	78	83	n.d.
7–26	60	80	60
13–26	65	67	55
19–29	55	n.d.	40
34–47	n.d.	n.d.	0
54–67	85	72	58
57–67	65	75	40
Shiga Toxin	87	90	100[a]

[a]Defined conditions for the calculations of relative neutralization.

designed in conjunction with an appropriate synthetic adjuvant, it is feasible to develop a completely synthetic vaccine with built-in adjuvanticity that can be used as a soluble material without any external adjuvant.

Development of Synthetic Copolymer of Amino Acids that Has Potential Therapeutic Value for Multiple Sclerosis

Multiple sclerosis (MS) is a neurological disease, with all the features of an autoimmune disease. However, the specific nature of the antigen is still unknown. Since MS occurs only in humans, research on it is difficult because there is no animal model. There is, however, a related experimental animal model, denoted experimental allergic encephalomyelitis (EAE), which can be induced in various species by injecting brain or spinal cord homogenate (Paterson, 1971). Within two weeks, the animals develop paralysis, mainly of the hind legs. The protein that is responsible for this encephalitogenic activity has been identified as the myelin basic protein. Several regions in its sequence have been shown to be responsible for the encephalitogenic activity in different species. The whole protein can induce EAE in practically every species that has been tested.

We have synthesized a copolymer of amino acids, denoted Cop 1, that is immunologically similar to the myelin basic protein, and have demonstrated that it can inhibit the encephalitogenic effect of that protein. Over 80% of guinea pigs injected with myelin basic protein alone succumbed to EAE, but if they had been treated with Cop 1 before challenge with the intact protein (inhibition), or if treated after challenge (suppression), only 20–25% of the guinea pigs manifested the clinical symptoms of disease (Teitelbaum et al., 1971). This effect was completely reproducible in a second batch of the polymer, and with many batches that were prepared and tested thereafter. The suppressive activity is due to the immunological similarity to the basic protein, mainly on the cellular level, but also on the humoral level. This was proven by synthesizing a series of copolymers related to Cop 1, and showing that those that were active in suppression were also immunologically cross-reactive, and those that were not active in suppression were not cross-reactive with the myelin basic protein. Cop 1 is also active in several other animal species –rabbits, mice, and primates. In the control groups 70–100% of the animals succumbed to the disease, while in the treated groups only 10–20% manifested clinical symptoms.

There is yet another experimental system, chronic relapsing EAE, that is induced in guinea pigs and produces symptoms that mimic MS more closely than acute EAE. We have shown that Cop 1 leads to very efficient suppression in this model as well. In the control group some of the guinea pigs died immediately and the rest had severe relapses; but in the treated group these symptoms were greatly delayed and were much milder (Keith et al., 1979).

In view of these combined results we were interested in examining

Table 2. Summary of results of a double-blind clinical trial with Cop 1 in MS patients

Parameter	Arm of Treatment	
	COP 1	Placebo
No. of attack-free patients	14/25	6/23
No. of attacks in 2 years	16	62
Average no. of attacks in 2 years	0.6	2.7
2-year change in DSS	−0.12	+0.74

whether Cop 1 might be of therapeutic value not only in EAE, but also in MS. We collaborated in a small clinical trial with physicians at Hadassah Hospital. Cop 1 was not expected to elicit any beneficial effects in this trial since the four patients included in it had very advanced MS, but we could examine possible side effects. Six months of monitoring revealed no notable side effects. We then received permission from the FDA in the U.S. to perform a double-blind clinical trial in patients in New York (Bornstein, et al., 1987). We carefully matched placebo and treated groups with 25 patients in each. As shown in Table 2, the placebo group suffered many more attacks (average 2.7 in two years) than the treated group (average 0.6). There were more attack-free patients in the Cop 1 treated group. Neurological changes were also more severe after two years in the placebo group, whereas in the Cop 1 treated group there was even a small improvement. There were no significant side effects. We are now engaged in further clinical trials, but we feel that Cop 1 might indeed be a candidate drug for the treatment of MS.

Synthetic Peptides as Drug Carriers

Chemotherapy is one of the most important means of treatment for cancer, but it has drawbacks. Chemotherapeutic drugs are very toxic, and it is difficult to adjust the dosage so that they will not cause too much damage to the patient and still efficiently eradicate the tumor. An immunological solution would be to use an affinity compound such as an antibody that can differentiate between tumor and normal cells, attach the drug to this antibody, and use this system to deliver the drug locally (Arnon and Sela, 1983).

The drug we used in our studies is cis-diamine dichloroplatinum(II) (cis-DDP) or its derivative cis-diamine diaquo dichloroplatinum (cis-aq) which is even more toxic than cis-DDP (Figure 4). To augment the amount of cis-DDP or cis-aq that we could attach to the antibody molecule, we used a spacer arm. The resultant conjugates were active *in vitro* as measured by the inhibition of DNA synthesis, with little loss in activity when compared with the drug alone. More importantly, we obtained improved *in vivo* activity in

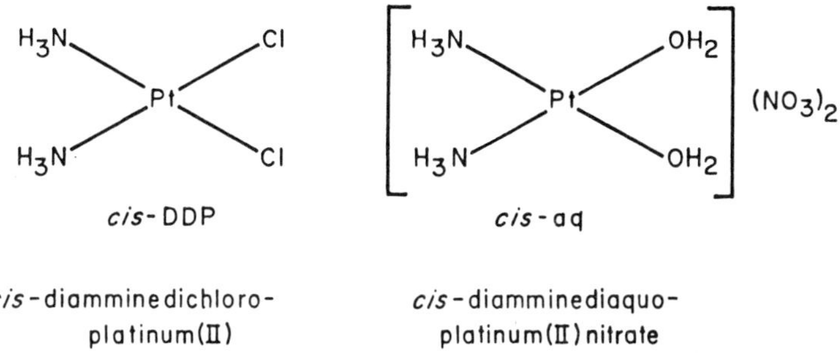

Figure 4. The chemical structure of cis-diamine dichloroplatinum(II) (Cis-DDP) and its cis-diaquo derivative (Cis-aq).

tumor-bearing mice. Using the free cis-DDP, a low concentration led to partial survival, but it was very difficult to adjust the dose so that all the animals would survive. If we increased that dose, they died of toxicity; if we gave too little, they died from the tumor. When we used the conjugate (with either low or high molecular weight polyglutamic acid) of cis-DDP or cis-aq, we could obtain 100% survival of the mice over a broad range of drug doses. Polyaspartic acid conjugates were also effective, showing an even broader range of non-toxic, tumor-killing concentrations (Figure 5).

Summary

This paper describes three aspects of our studies, all related to immunology, that utilize polypeptides, of either a particular sequence, or random poly-

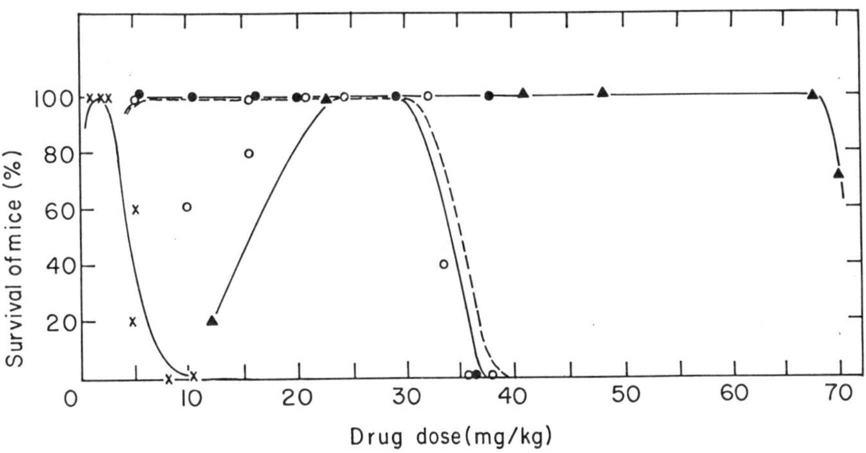

Figure 5. Antitumor activity and toxicity of free Cis-DDP (x) or Cis-Pt complexed to p-L-Glu (●), p-D-Glu (○) or p-L-Asp (▲). The data represent cumulative results of several experiments using the F9 embryonal carcinoma in Balb/C mice.

mers. Thus, polypeptides can be used for vaccine development, therapy of multiple sclerosis, and as spacer arms to enhance drug targeting in cancer chemotherapy.

Acknowledgments

For over 30 years I have collaborated with Michael Sela. Over the years the synthetic vaccine studies were done by Elchanan Maron, Harry Langbeheim; Gunhild Muller and Michal Shapira who worked on influenza; Chaim Jacob on Cholera toxin; and Ilana Harari on Shiga toxin. Dvora Teitelbaum collaborated in the studies on Cop 1. Oded Abramsky at Hadassah Hospital is still collaborating with us on clinical trials. Murray Bornstein performed the clinical trials at Albert Einstein in New York. The cis-DDP targeting work was done mainly in collaboration with Meir Wilchek and Bilha Schechter.

References

Arnon, R. Synthetic peptides as the basis for anti-shiga vaccines. In: New Generation Vaccines. Woodrow and Levine (eds.). Marcel Dekker Inc., New York. In press.

Arnon, R., Maron, E., Sela, M. and Anfinsen, C.B. 1971. Antibodies reactive with a native lysozyme elicited by a completely synthetic antigen. Proc. Nat. Acad. Sci. 68: 1450.

Arnon, R. and Sela, M. 1969. Antigens to a unique region in lysozyme provoked by a synthetic antigen conjugate. Proc. Nat. Acad. Sci. 62: 163.

Arnon, R. and Sela, M. 1983. Targeted chemotherapy: Drugs conjugated to anti-tumor antibodies. Cancer Surveys 1: 429.

Bornstein, M., Miller, A., Slagle, S., Weizman, N., Crystal, H., Drexler, E., Keilson, M., Merriam, A., Wassertheil-Smolle, S., Spada, V., Arnon, R., Jacobson, I., Teitlebaum, D. and Sela, M. 1987. A pilot trial of Cop 1 in exacerbating-remitting multiple sclerosis. New England J. Med. 317: 408.

Jacob, C.O., Pines, M. and Arnon, R. 1984. Neutralization of heat-labile toxin of E. coli by antibodies to synthetic peptides derived from the B subunit of cholera toxin. EMBO J. 3: 2889.

Keith, A.B., Arnon, R., Teitlebaum, D., Caspary, E.A. and Wisniewsky, M.M. 1979. The effect of Cop 1, a synthetic polypeptide on chronic relapsing EAE in guinea pigs. J. Neur. Sci. 42: 267.

Langbeheim, H., Arnon, R. and Sela, M. 1976. Antiviral effect on MS-2 coliphage obtained with a synthetic antigen. Proc. Nat. Acad. Sci. 73: 4636.

Muller, G.M., Shapira, M. and Arnon, R. 1982. Anti-influenza response achieved by immunization with a synthetic antigen. Proc. Natl. Acad. Sci. 79: 569.

Paterson, P.Y. 1971. Experimental allergic encephalomyelitis and autoimmune disease. Adv. Immunol. 5: 131.

Teitelbaum, D., Meshorer, A., Hirshfeld, T., Arnon, R., and Sela, M. 1971. Suppression of experimental allergic encephalomyelitis by a synthetic polypeptide. Eur. J. Immunol. 1: 242.

Wiley, D.C., Wilson, I.A. and Skehel, J.J. 1981. Structural identifications of the antibody binding sites of Hong Kong influenza hemagglutinin and their involvement in antigenic variation. Nature 289: 373.

Discussion (DR. ARNON)

Kornguth: I would like to thank Dr. Arnon for the beautiful presentation. The Weizmann Institute has been very active for over 35 years in the study of biological and chemical properties of synthetic polypeptides. These are three examples of excellent use of these compounds. Two questions come to mind. First, what are the tumors that have been used with cis-platinum polyglutamate in your studies? Second, the length of the polypeptides that you used for the vaccination probes are 10 to 15–mers. The problem with vaccination encountered with the shift in the amino acid sequence of the entire viral protein should also arise with the synthetic peptide antigen you are using. How will the peptide vaccines eliminate the shift in antigenic properties observed with complete viral immunogens?

Arnon: The tumors examined with cis-platinum polyglutamate complexes are mainly F-9 cells. The *in vivo* studies use mainly F-9 because it is very sensitive to cis-platinum. The lengths of the peptides we used are about 12–18 amino acid residues. There is not enough experience using synthetic peptides in sufficient numbers of generations of animals or even sufficient number of generations of the virus to see whether there is a resistance to a particular peptide which will be manifested in phenomena that are similar to the shifts and the drifts. As far as we know, the peptide that we are dealing with here is a conserved peptide and no change has been observed in this region. The immunogenic peptide used in our studies is a conserved protein in all strains of influenza. I suspect that the danger of shift and drift changes may be smaller here. Does this answer your question?

Kornguth: In part. I would have thought that a conserved, highly antigenic epitope would have expressed itself *in vivo* as a unique vaccine. That has not been found of course. How do you reconcile that with your observations?

Arnon: The region that we are using is not necessarily the most immunogenic part in the intact virus. It may be what we call a hidden epitope; in the intact virus it is not expressed as an important epitope. Maybe the immunized animal or the infected animal does not see that epitope as an immunizing epitope. However, if you present it to the animal artificially as an isolated epitope, the immune response is effective against the intact virus.

Eisenstadt: I think of epitopes as structural components of proteins. What is the conformation of your polypeptides?

Arnon: A small peptide of 15 amino acids in the isolated form definitely will not have exactly the same structure as it has in the intact protein. The peptide has different forms in solution. Some forms of it are more similar to the structure in the intact protein, especially if we are looking at regions that have spatial features. If they stem from regions in the molecule that have specific folding, then there is a higher probability that the antibodies formed against them will be reactive with the intact protein. I do not think that the peptide in the isolated form has exactly the same structure or folding that it has in the intact protein. But, antibodies formed against them recognize the intact protein.

Lotan: It is exciting to see that, after 30 years, polypeptides are still alive and kicking. There are a few questions that I would like to ask you. First, you mentioned the hydrophilicity of the Shiga protein. Can you elaborate on the index of hydrophilicity? Second, you presented data on response of patients to Cop I. It is clear that this peptide has no toxic effects. There were, however, one or two effects (one was dizziness) which were more expressed in the treated patients. I assume that was it something temporary. Is that so? Finally, can you elaborate on the difference between the effects of cis-platinum when bound to polyglutamic acid and polyaspartic acid?

Arnon: I will answer the third question first. We do not have enough results on the polyaspartic acid, but preliminary data suggest that it is more efficient than polyglutamic acid. In answer to your second question, the cop I results are not decisive at all. We interviewed patients each time they came in, and got these results. The hydrophilicity technique is not our work, but an established predictive model for protein structure. Each residue is assigned a value, with the most hydrophilic being glutamic acid and other polar amino acids. By consolidating the data we can create a continuous curve where whole regions of the protein can be defined as more hydrophilic and others more hydrophobic.

Key Words: Antibodies, Cholera toxin, Cis-diamine dichloroplatinum, Copolymer I (Cop I), Drug targeting, Experimental allergic encephalomyelitis (EAE), Hemagglutinin, Influenza virus, MS2 bacteriophage, Multiple sclerosis, Polyglutamic acid, Shiga toxin, Synthetic peptides, Synthetic vaccines

Session I

Biological Detectors

Manufacturing and Biotechnology Synergisms for the Early 2000's

Christopher Canfield Green
Head, Biomedical Science Department, General Motors Research Laboratories, 30500 Mound Road, Warren, MI 48090–9055

Introduction: Why Biotechnology in Large Industry?

Manufacturing industries ranging from the pharmaceutical and medical, commodity, industrial chemical, and energy-intensive or alternative fuel enterprises pace the biotechnological revolution for the next decade. The papers which follow in these proceedings make ample food for the argument that, indeed, absent a strong applied biotechnology research and development component each of these industries will languish or fail by early in the 21st Century. This paper will suggest that other manufacturing disciplines, such as those in the automotive and "medium machine-building" industries, have slowly begun to consider specialized process technologies, orphan products, and unique biotechnologically derived opportunities. And they will provide synergisms that will enhance their competitiveness. It is believed that although interactions between some of the problems and biotechnology prospects are daunting, failure to recognize them in some cases will certainly cause giants to fall and others to discover new strategies to race the competition and win.

The most important problems amenable to biotechnology solutions concern increased world attention on the Scylla and Charybdis of emerging national regulation, and public awareness of toxic materials that are required for or are products of manufacturing. We now know some of them must be remediated or recycled as a cost of doing business. Examples include products of combustion from both stationary and mobile sources, air toxics, polycyclic organic compounds and some complex aerosols. But at the other extreme are truly novel opportunities which biotechnology offers for the creation of biocomposites and materials with entirely new uses. Even in an apparently traditional discipline such as automotive engineering the 21st Century will find the creation of "smart" materials made of substances not yet invented, applied to the interiors for both civilian and military vehicles based on electrorheological fluids and composites that will be used for safer and real-

D. Kamely et al. (eds), Biotechnology: Bridging Research and Applications, 15–42.

time adaptive energy-absorbing instrument panels, knee-bolsters, side impact absorbing doors, and seats.

The purpose of this paper will be to suggest several areas of synergism between near and mid-term medium manufacturing requirements and biotechnology. Numerous other far-reaching industrial applications requiring more developed and mature approaches in the fields of energy conversion and storage, food processing, and medicinal chemistry and pharmaceuticals will be subjects of the following presentations by others.

The intention here is neither to second guess the cutting-edge science foreshadowed by the impressive list of titles and authors in this symposium, nor to serve as a comprehensive technical review of industrial opportunities. Several recent studies have completed that task in widely varying contexts (Kamely, 1990; Kamely, et al., 1989; Obringer and Tillinghast, 1989; Malinski, 1990; Green, et al., 1990). Instead, it is intended that these remarks be a broad apologia for newer and different investment strategies that include greater participation by the mature manufacturing sector in new forms of partnership with the (obviously) core emerging biotechnology companies. The argument is for diversification well beyond the current limited 1,000 or so biotechnology enterprises with their plethora of biotechnological products and promises to match industrial needs of the 21st Century. In the last two decades biotechnologist-entrepreneurs have matched promise of potential with hundreds of real products. The conservative business world that a few years ago worried aloud about biological hyperbole has become newly silent as low energy and low labor intensive biotechnology products of low cost are reported in both the technical and popular press daily. The established large industries have begun to recognize competitive advantage, and seek sensible strategies to invest. We need to help, and we need your help.

Biotechnology product lists are long in terms of applications for industry, so long that they have been viewed by non-biologists as being overdrawn and sometimes fanciful. The first has been a legitimate criticism because the lists have not usually been well thought out, and the "needs" analysis or research requirements for scaled-up process technology have not been done well except for some pharmaceutical applications. And in many of these cases the return on investment has been a reasonable (but often uninteresting) five years to reach commercialization, and then projected to top out at $10–100 million gross sales frequently considered by financial officers to be too long, too risky, and too small.

Definitions: What is Industrial Biotechnology?

Biotechnology for industry in the 21st Century should be defined in terms of choices rather than accidents of invention. The field is not now limited as much by missing discoveries as by investment resources. That is, unlike other technologies with industrial application, this set of related biochemical and chemical process engineering disciplines remains immature by definition in a business but not scientific sense: the rates of change of discovery and invention are high, are only now beginning an ascending limb of increase in knowledge, but lack adequate numbers of biologically trained scientists working with or even co-located at quality academic institutions together with excellent industrial engineering programs.

There are several important differences between biotechnology and the other engineering disciplines that have emerged in the industrial revolution. The differences are important because they dramatize, when taken together, how exceptionally small resource enhancements (either personnel or financial, occasionally both) can be leveraged enormously. The differences (opportunities, really) exist because biotechnology contains an emerging set of engineering applications based upon a curious mixture of known systems- and process-engineering sciences. The field is not unlike physics in the 1920's, electronics in the 1940's, and molecular biology in the 1960's. Biotechnology of the 1980's, indeed was emergent from these three.

The classic definitions of biotechnology include the application of scientific principles for clinical and industrial uses of biological systems to produce goods and services. Living organisms or their parts are used to make or modify products, or develop organisms for specific purposes. The subsystem technologies include molecular and cellular manipulation, enzyme definition, design and production and microbial techniques for growth and fermentation.

In setting the stage for this conference which is hoped to serve as a non-biological catalyst for industrial participation on broader scales than in the past, let me suggest that even the above definition is too restrictive and needs to be re-formulated in terms more meaningful and useful for industrialists.

The biotechnologies are:

- A rapidly maturing set of diverse engineering applications, math-based, which are emergent from the sciences of molecular biology and physical chemistry.
- A given biotechnology uses knowledge of living systems or their biochemical subsystems or cellular production processes including molecular genetics
- to make a commodity product,
- to modify genetically specific plants or animals to achieve a unique phenotypic result to enhance products qualitatively or quantitatively, or
- to mimic a biophysical effect through electrochemical means.

The important applications of biotechnology are:

PROCESS ENGINEERING in which efficiencies of process variables such as heat and pressures result in quantity amounts of manufactured biochemicals, commodity chemicals, or materials.

MOLECULAR DESIGN based upon structure-function relationships at cellular, molecular, and atomic levels of substances from living systems which can result in new products or devices, and

BIONICS whereby electrochemical signal processing capabilities of living neural systems are modeled to design novel information-intensive products.

Biotechnology can be defined by size of individual production units:

LARGE SCALE mega-ton products, in which biotechnology must compete with petroleum and coal, as a primary source of carbon compounds for fuels and high tonnage industrial products. Sales must be measured in the billions of dollars.

MEDIUM SCALE in which multi-ton products compete with petroleum technology (whether the carbon source is petroleum or vegetable) to produce either commodity chemicals currently in use or substitutes for them. In agriculture, biotechnologies can produce products such as proteins and lipids. Sales should be measured in terms of hundreds of millions of dollars per year.

SMALL SCALE "orphan" products in which specific substances are produced for which no other routes can currently be foreseen, or are as cost-effective, are research tools, or serve very special purposes. Specially competitive or possessing certain unique military advantages, the production costs and returns on investment are less important because possessing the technology confers a strategic business advantage.

Focus Areas Common to Industry: What are the Key Techniques and Methods?

From the field of biotechnology as a whole, there are six focus areas that share common methodologies as opposed to applications. Key investment strategies and choices can be better made if common resource-intensive developments can be found which serve several common and ongoing development efforts, even if they are typified by widely divergent applications. These areas and key synergisms are:

BIOMOLECULAR ENGINEERING: The specific *a priori* design and production of materials and composites using biological (biochemical) methods and starting material-stocks. Development depends heavily on CAD/-CAM/CAE technologies and are especially useful in the pharmaceutical and polymer chemistry industries.

BIOCOUPLING: The use of electronics together with biological molecules to detect physical-chemical substances, or conversely to use microelectronics to detect biological molecules. Particularly synergistic with VLSI (very large-scale integration), AI (artificial intelligence), and electro-physiological advances.

GENE TECHNOLOGIES: Includes the set of methods to "touch the genome" and modify it. Largely depends of molecular biology to identify, recognize, replace, remove, or permit replication at the cellular level of DNA, RNA, and regeneration or differentiation. Will expand much more rapidly than even biologists now suspect because of the current explosion in development of near-real-time protein engineering tools, e.g., liquid chromatography continuous flow fast atom bombardment tandem MS-MS. Especially synergistic for the following two focus areas.

BIOPRODUCTION TECHNOLOGIES: These are linkages of established process engineering methods especially found in current chemical engineering systems that manufacture kilogram to multi-ton quantities of both simple chemicals that are now cost limited (fuels and foods) or complex biochemicals limited by insufficiently understood synthetic mechanisms (vaccines and medicines). An especially critical enabling sub-technology will come from the invention of new real-time biological microprocessing systems and knowledge of how to couple them to low-temperature enzyme-catalyst reaction kinetics (e.g., safe production of explosives for construction industries, solvents, and surfactants).

TARGETED DELIVERY SYSTEMS: Very specific biochemical reactions can be enabled efficiently because they act only upon chosen parts of the genome. New drugs can be invented using biomolecular engineering, design for increased efficacy, for both prophylaxis and therapy. Synergistic with this focus will be advances in world-wide epidemiological models and basic research in virology.

BIONICS: The modeling of biological systems, especially of materials properties and neural-systems offers shortcuts to invention of devices, systems, and even organic and inorganic chemicals with diverse biokinetic attributes. Vehicle surfaces that are "adoptive" to electromagnetic energy and "adaptive" to kinetic energy are possible. Such "smart" and protective or environmentally adaptive surfaces will be possible with the synergistic development in expert systems, AI, and knowledge in neural networks and neurophysiology.

Biotechnology Applications in Industrial Settings: What Can Be Leveraged?

It is clear that the potential applications of the above outlined focus areas are highly segmented across civilian, military, and "hybrid" medium and large industries. Explication of the areas and the following rationale can be found elsewhere (Green, et al., 1990) but a summary here is useful. As asserted above there are several "econometric" differences in the developing biotechnologies compared to the more mature physical-science disciplines. They tend to be simple and obvious when examined in terms of the extreme capital and scientist-intensivity of the computing sciences, electronics, and other moderate-scale industries. Compared with these industries and the underlying physics and engineering-based manufacturing worlds, there are six fundamental differences (Green, 1989) that can be recast here in terms of "unfair advantages" that may be of value to the non-biologist financial officers of large corporations who recognize the importance of diversification:

○ The current rate of change in our understanding of the biology that underlies biotechnology is greater now than at any time before in history. Networking of scientists spanning numerous technical disciplines exists in hundreds of conferences and symposia yearly. Highly technical specialty conferences such as this one exist in which energies are concentrated on discovery and enabling of new applications. Our field, unlike the physical and mathematical sciences, does not wait for opportunity to discover us.

- The number of discoveries per scientist or laboratory by any measure (publications, patents, or other means) is higher by a factor of two than with any other area of science.
- The physical scale of laboratories and specific equipments required to develop new bioprocess systems or to conduct large production operations is small enough to be incorporated easily in existing facilities.
- Capital investment and sophistication for the construction of new production facilities is low. While the intellectual property required, is paradoxically both large and largely unavailable (*vide infra*), the field is not as much equipment or capital-intensive as found in other start-up enterprises. Opportunities for merging of established but small biotechnology concerns having limited ($1–$5 million) revenues with established manufacturing industries has never been better. This is particularly true for multinational corporations in which trade-offsets are important, and for companies in which tax-incentives still exist for IR&D in support of government contracting relationships and wherein research and development is taxpayer financed and doesn't impact negatively on shareholder equity. The prototypical model is found between numerous corporations in the United States and Israel.
- The production and process technology involved in batch processing and fermenting, chemical separation, and manufacture of biomaterials is relatively unsophisticated, becoming highly automated, and non-energy intensive. Within the next decade applications in microprocessor and biosensor-controlled nutrient and fermentation growth conditions will enhance and strengthen this difference.
- Modernization of existing facilities without requirements for enormous capital expenditures and heavy construction, plant rearrangements, or large space-occupying equipments is possible. Economies of scale for diversification are reasonable.

Biotechnology in Large Industry: A Paradox, or Contradiction in Terms?

Analysts such as this author, trained in the science but who operate at arms length from the biotechnologist's bench become awed when scanning lists of applications being tested, processes being patented, or even the agenda for this conference. What possible investment or research strategy could have comprehended the pace of change referred to above? How can corporations even today become convinced that biotechnological advances are critical for

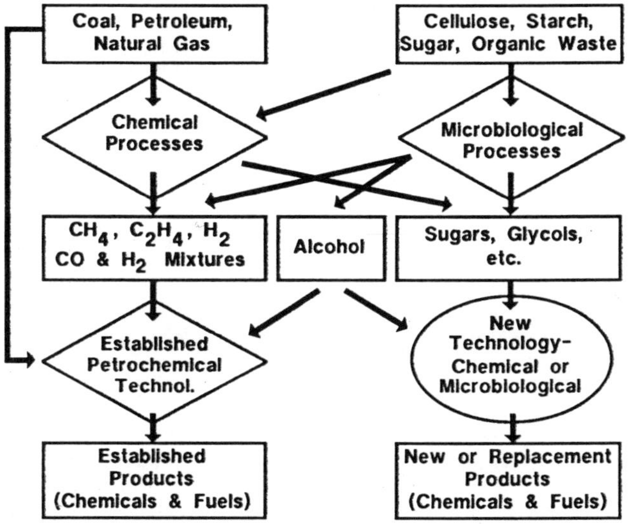

Figure 1. The competitive routes to large tonnage products.

their growth when the language and corporate cultures are removed from complex business of, say, producing millions of automobiles each year? And when the words of competition are "quality" and "cost" and "innovation" which really mean adapting inventions from last year, how can an industrialist be convinced by a biologist that an as yet uncloned gene producing a protein yet unidentified much less sequenced can be scaled-up to save millions of dollars within the next year? Such hyperbole is the stuff of proposed initiatives that are never given seed money. But in today's biotechnology business it frequently is not hyperbole. Our most pressing need is to learn how to perform accurate situational analyses of potential needs versus projects to be placed on laboratory benches, and then translate the analyses as engineering opportunities.

A few years ago several of us performed an internal assessment of chemical products then being considered for "manufacture through DNA technologies." We were embarrassed to place the ideas before our planning staffs because the traditional research/development/test/evaluation and engineering paradigms were so much a part of our planning process that Gantt charts timing the steps for production of reasonable products in interesting (to us capitalists) time-frames could not be constructed. All the potentials were then only in the research phase (Figure 1 and Tables 1–4). Let me touch briefly on just one or two examples from each of my four favorites: bioengineering, biocontrol, biosensing, and biomaterials.

Today, the costs of mineral and precious metal extraction are becoming interesting, as it is learned how to avoid self poisoning the cells that co-metabolize complex mixtures and concentrate metal ions through use of fluidized bed bioreactors. A decade ago industry had difficulty finding scien-

Table 1. Some chemical products for manufacture by DNA technologies

Products	Potential use
Mineral concentrating organisms	
• Uranium	For use in mining
• Gold	
• Iron	
• Metals	
• Copper	

tists willing to call themselves process engineers who felt knowledgeable enough about fermentation technology to even work on the problems. Now, the Solar Energy Research Institute in Colorado has pilot-plants that interest those of us working in alternative fuel vehicles. The industrial enzymes (Figure 2) include the ligninases, and products include the cellulosics (Table 2) that make possible competitive routes for alcohol production (Scott, 1989). Inexpensive cellulosic biomass is plentiful, feedstocks can be acquired economically, and genetic engineering coupled with simultaneous saccharification and fermentation (SSF) will permit the production of ethanol and oxidant enhancers at competitive rates in some world markets in less than five years (Hinman, et al., 1989, Wyman, C.E., pers. comm.).

Perhaps no single area of biotechnology that requires simultaneous recombinant DNA research together with process engineering is moving so fast, and needs to move fast, to solve the escalating problems in medium and heavy manufacturing industries as those relative to pollution and toxic/hazardous waste control. When we originally contemplated the problem in 1987 we had no idea of the scope of research then underway (Kamely, 1989). Having become better educated, our laboratory efforts have now gravitated toward enhancing currently available treatments that are already effective in dealing with high concentration contaminants which require nutrient supplements. Our goals press toward use of modified photosynthetic bacteria to degrade low concentration of PCBs and chlorinated hydrocarbons (Tables 5a and 5b), and (Siak, pers. comm.).

Table 2. Some chemical products for manufacture by DNA technologies

Products	Potential use
Industrial enzymes	
• Oxidases	Production of glycols
• Amylases	Production of sugars
• Lignases	Production of phenols from wood cellulose for plastics
• Cellulases	Production of fermentable sugars from wood cellulose
Fibers and plastics	For use in textiles and plastics
• Cellulosics	
• Silk	
• Polyhydroxybutyrate	
• Pollutants	

Table 3. Some chemical products for manufacture by DNA technologies

Products	Potential use
Commodity chemicals	Intermediates and bulk chemicals
• Amino acid/Acetic acid	
• Acetone	
• Acrylic acid/Adipic acid	
• Citric acid	
• Ethanol	
• Ethylene glycol	
• Ethylene oxide	
• Glycerol	
• Methanol	
• Propylene oxide	
• Salicylic acid	

At the General Motors Research Laboratories (GMR) Gibson and Abdul (1988) have been looking at the problems associated with PCBs and contaminated soil. A novel microbial strain identified as *Pseudomonas pickettii* PK01 was isolated from the native soil microorganisms in a shallow sandy aquifer. The bacteria were able to degrade benzene, toluene, and xylenes in water contaminated with gasoline, under aerobic and anoxic conditions within six days. Bacteria with the capability had not been reported previously. Figure 3 shows chromatograms with gasoline peaks in a simulated contaminated ground water mixture: (top) at the start and (bottom) after six days degradation by the microorganisms. The disappearance of the peaks representing benzene, toluene, xylenes and other aromatics indicates biodegradation has occurred. No decrease was seen in the peaks in killed controls. In Figure 4 a bacterial strain which grew on toluene vapor as the carbon source

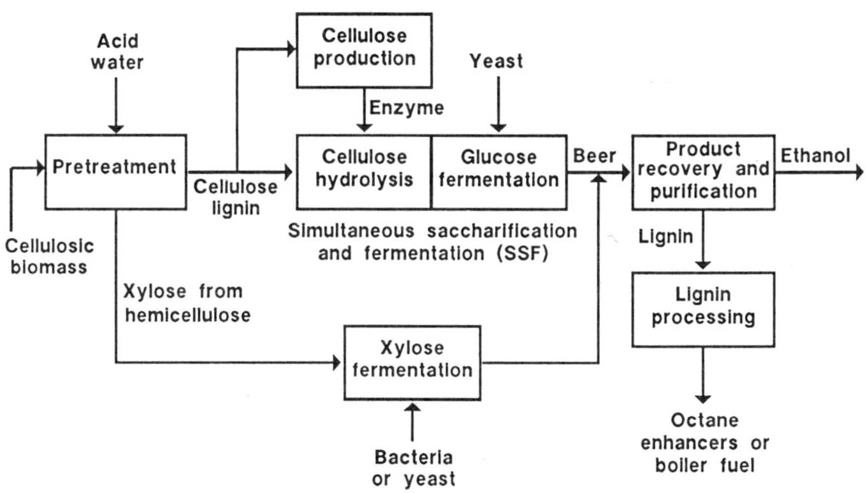

Figure 2. Improved xylose conversion for cellulosic biomass (SERI, from Scott, 1989).

Table 4. Some chemical products for manufacture by DNA technologies

Products	Potential use
Chemical degrading organisms • 2,4-D • 2,4,5-trichlorophenoxyacetic acid • Petroleum • Mercury • Agent orange (2,4,5-T/2,4-D) • Pesticides and phenoxy herbicides • Speciality chemicals • Oils • Lubricants • Surfactants	For use in pollution control systems

was isolated from the soil microorganisms using Agar plates in the laboratory. A seed culture of PK01 was added to sterile sandy aquifer material at a level higher than one million cells ml^{-1}. The new population degraded aromatics as the total native population had done. The reaction occurred under both aerobic or anoxic conditions.

One of the most pressing needs in any medium manufacturing industry including ours is the treatment of waste water contaminated with cutting

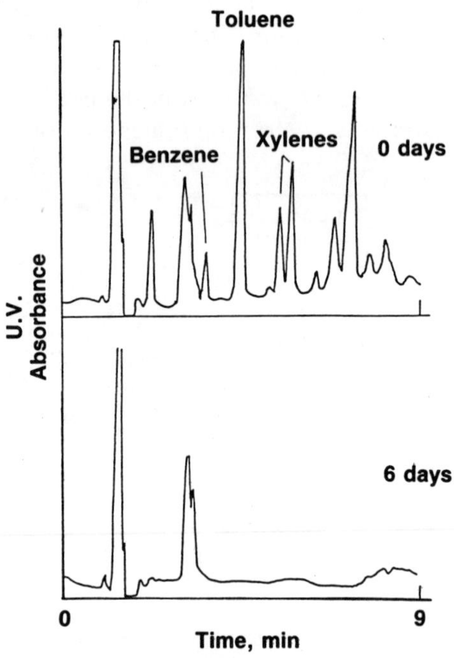

Figure 3. High performance liquid chromatography of gasoline hydrocarbons before and after degradation by native soil microbes (Gibson and Abdul, 1988).

Table 5a. Biodegradation as waste treatment

Problem:
Bulk waste with high organic content
- Paint sludge
- Waste water with high biological oxygen demand

Approach:
Study the microbial population involved in the degradation procedure and develop strategy to improve the efficiency of the system. Selection/sorting of fauna, examine metabolic pathways, develop process technology.

Source: Siak, 1990.

fluids. Laboratory testing at GMR has demonstrated aerobic and anaerobic biodegradation also is applicable for this problem. Bacteria consumed the organic chemicals in the influent, thereby reducing the oxidant demand of the effluent waste water.

Byung Kim in the GMR Environmental Science Department has developed an anaerobic, granular activated carbon (GAC), fluidized-bed reactor (Figure 5) to treat the waste water from cutting fluids containing fatty acids, alkanolamines, amino acids, alcohols, glycols, and surfactants. Figure 6 shows GAC used as a medium to support growth of microorganisms when operated at 36°C. A dilute mixture of eight metal cutting fluids used at a GM plant was used in the simulation. Approximately 64% of the organics, measured as chemical oxygen demand (COD) was degraded while the rest appeared to be non-biodegradable. But when the effluent from the anaerobic reactor was treated aerobically, approximately 65% of the anaerobically non-biodegradable COD was degraded aerobically (Kim, et al., 1989). In Figure 7 a series of adsorption isotopic experiments using the simulated waste water revealed that biological treatment enhances adsorption capacity probably due to selective biodegradation of poorly adsorbable organics, and organics in the effluent displayed adsorption characteristics. The results show the synergistic relationship between biological treatment and adsorption when used together (Kim, et al., 1990).

We have found some surprises concerning the scope of the biotechnology market place and multiple niches that would surely develop, even in our own

Table 5b. Biodegradation as waste treatment

Problem:
Low concentration, high volume waterborne pollutants (such as halogenated solvent and PCB)
- Water contamination of pollutants at parts per million or lower is difficult to decontaminate.

Approach:
Using DNA recombinant techniques introduce functional enzymes in the unicellular photosynthetic microbes as a way to become a self supporting system to treat low concentration waste. Specific "company-intensive" or manufacturing pollutants can be targeted.

Source: Siak, 1990.

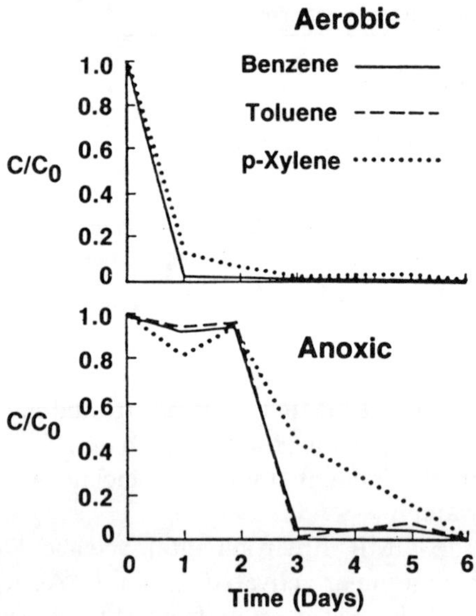

Figure 4. Reconstitution of the microbial population with *Pseudomonas pickettii* strain PK01 added to sterile aquifer material (Gibson and Abdul, 1988).

automotive industry (Table 6). Various studies in our laboratory by Peter Lee had identified the size and scope of the chemical laboratory market place as being about one billion dollars in 1980 and ranging to five billion dollars today. One third of these values were found in the area of immunoassay reagent sales, and these values have been consistent throughout the decade when viewed by other market analysts. Numerous observers have predicted

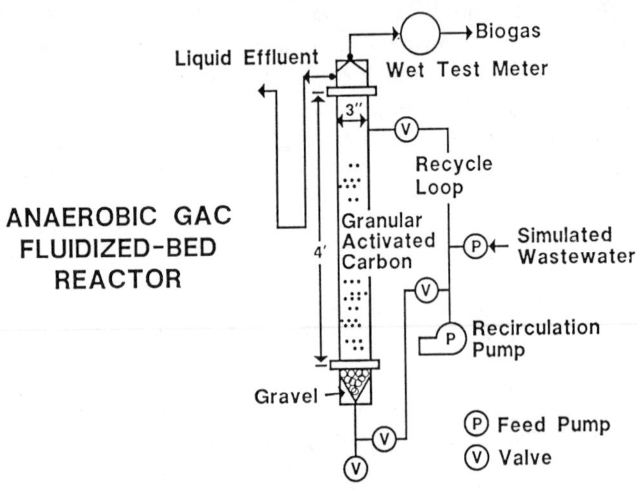

Figure 5. Anaerobic GAC fluidized-bed reactor (Kim, et al., 1989).

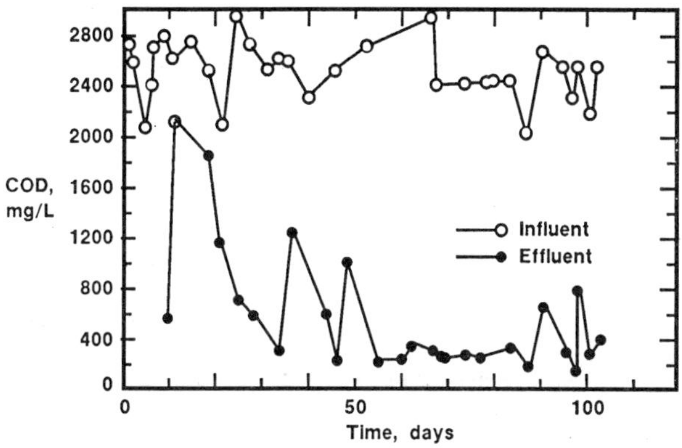

Figure 6. Wastewater treatment for Saltillo (Mexico) plant (Kim, et al., 1989).

that biotechnology in the development of biosensors as we defined them in the first part of this paper could obviate the radioactive liability issue, provide near real-time analysis and chemical applications for clinical triage diagnostics, and be less labor-intensive, expensive and time consuming than other options. There could be significant reductions of both material expense costs for consumables and the non-controllable overhead and personnel burdens. Thus, the billion or so dollars could be leveraged highly against the other costs generated around the several hundred billion immunoassays performed with the several hundred immunoassay kits each year!

In our laboratory, Lee, Heremans, and Perry have developed an immunosensor system that has great interest to us because it allows the measurement

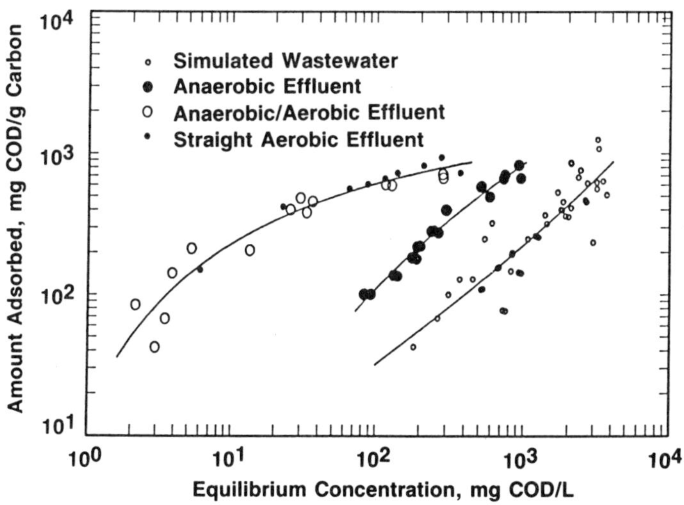

Figure 7. Effects of biological treatments on adsorption (Kim, et al., 1990).

Table 6. Biotechnology: Electrochemistry

Research area	Automotive objectives	Biotechnology connection
Electrochemical sensors	Real time analysis of oils, fuels, additives and gases.	Surface modified electrodes using enzymes; molecular engineering, microelectrodes.
Electrocatalysts	Advanced fuel cells for auxiliary power and vehicle propulsion.	Mechanisms similar to biological processes for oxygen reduction and fuel oxidation. Structure/property relationships for macromolecular catalysts, theoretical analysis.
Electrolytes	Membranes for fuel cells and solid state batteries.	Structure/property relationships, synthesis of new membranes, analysis of transport mechanisms, separation processes.

of a protein (α 2 μ-globulin) associated with gasoline vapor toxicity. Other sensors that detect molecules of alcohol in low concentrations have obvious automotive applications, and follow essentially the same technology.

In Figure 8 a dose response curve for an antigen-antibody interaction with a capacitance biosensor driven by a variable frequency oscillator is shown to be monitored through complex impedance measurements. The change in capacitance (defined as a dimensionless quantity delta C/C) increases as more antigen is bound to the antibody coated to the surface of a parallel plate air capacitor. The same antibody coated capacitor can be conditioned for reuse. In this figure, all the changes were referred to capacitance of the

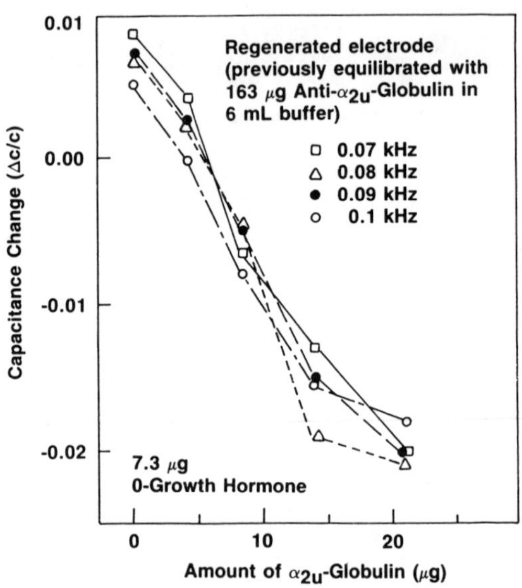

Figure 8. Measurement of σ_{2u}-globulin via capacitance biosensor.

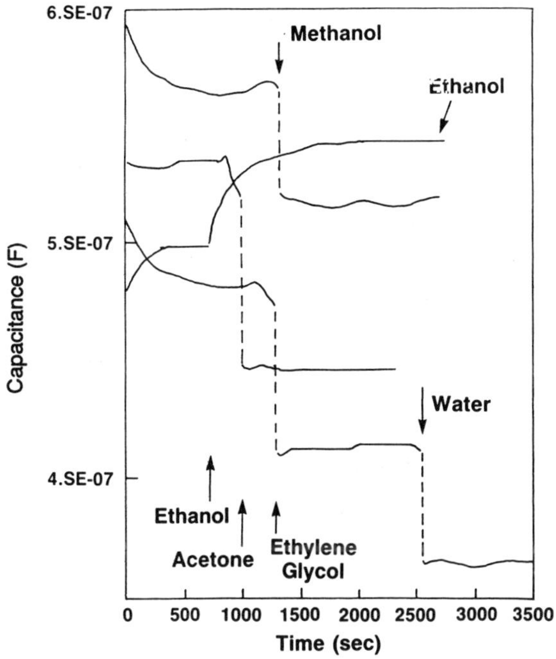

Figure 9.

buffer measured at 0.1 kHz. In another immunosensor system based on the same technology Lee showed that the capacitance or dielectric constant changed as NAD was transformed into NADH by ethyl alcohol in the presence of the enzymatic catalyst ADH. The measurements were conducted at 5 kHz using a parallel plate air capacitor. From Figure 9 it can be seen that the system is highly selective and discriminates between methanol, ethanol, acetone, ethylene glycol, and water.

Milestones: Can they be Identified?

Last year a study on critical technologies was performed for the United States Congress, and an important part of that study addressed issues of biotechnology milestones (Kamely, 1989). In that study the authors did a credible job identifying what would be needed to accomplish certain goals by the end of 1990, 1995, and 2000. Two examples are selected here (Tables 7a and 7b) of the many in the study, which have some bearing on the topics just covered. From the past summary and the papers to be presented at this conference it is surprising that most of the 1995 and some of the year 2000 targets have already been met, or will be long before even the best judgments of those of us who two years ago considered ourselves biotechnology futur-

Table 7a. Milestones – Biotechnology materials and processing

	1990	1995	2000
Bioprocessing			
• Bioremediation	• Restoration of soil contaminated with fuels, lubricants, & nitroaromatics	• In situ treatment of PCBs & TCEs in ground water	• End stream bioreactors using engineered organisms
• Biopaintstripping	• Degradative enzymes characterized	• Scale-up for batch processing	• Complete self-contained biostripping system
• Biomining	• Microbial strain selection	• Batch recovery of gallium from low grade ore	• Immobilized biochelators for industrial end stream recovery
• Energetics synthesis	• Room temp synthesis of some nitrate esters & aliphatic nitrocompounds	• Full range of nitrifying & denitrifying biotech products	• Novel, high-energy propellants & explosives

Table 7b. Milestones – Biotechnology materials and processing

	1990	1995	2000
Biomaterials			
• Fibers	• Cloned synthesis fibers	• Solid state synthesis fiber assembly	• High strength material
• Polymers	• Intermediates identified • Cotratus identified	• Intermediate synthesis pathway cloned solid state synthesis	• Composite matrix
• Elastomers	• Producers identified • Polymer characterization • Cloned synthesis	• Production optimization • Material optimization • Solid state synthesis	• Gaskets, coatings
• Adhesives	• Cloned variants	• Cloned synthesis	• General use adhesive
	• Product variants	• Variants synthesis	• Tissued adhesive repair
• Bioelectronlcs	• Tubule fabrication	• EMI coating & ceramics	• Protected equipment
	• Fiber processing	• Structural microwave composites	• Microwave devices
	• High resolution deposition	• 0.1 micron devices	• 3-D circuitry

Source: Ref. 7.

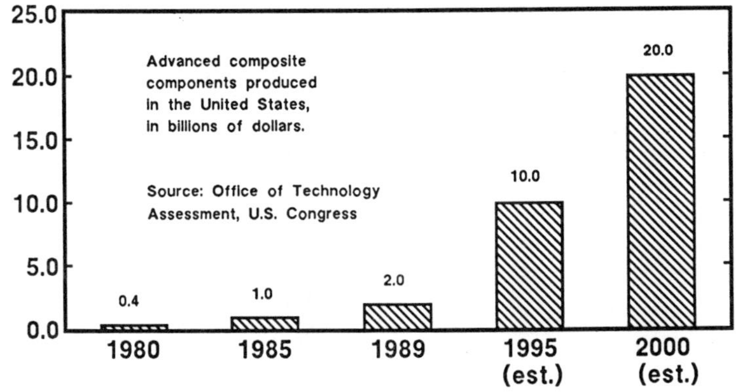

Figure 10. The growing use of composite materials.

ists. Probably no area in our field as much as that of biomaterials is likely to surprise us to the extent that past advances have done.

Composites, a significant portion of which can be either biologically designed or biologically derived, is a significant portion of the materials science future. Figures 10–12 show the scope of the potential market place. (Incidentally, about 10% of the weight of an automobile consists of composites). The Congressional Office of Technology Assessment indicates that growth will reach $20 billion by the year 2000. "Advanced composites" even in 1989 shipments included 5% industrial use (much of which was biological composites) and cars and transportation accounted for one-quarter of basic composite shipments. The biokinetic applications of composites engineered from

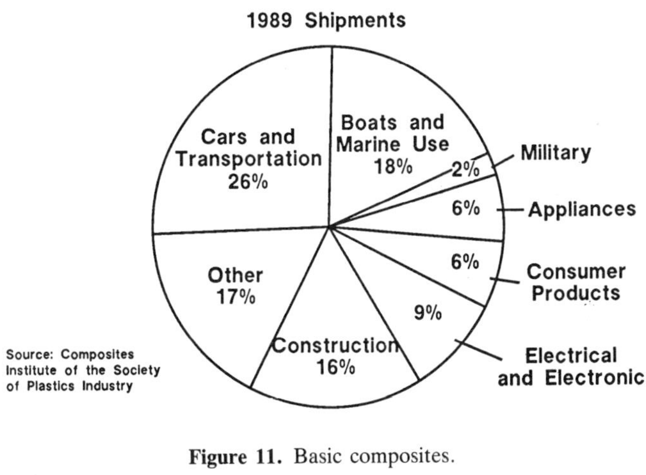

Figure 11. Basic composites.

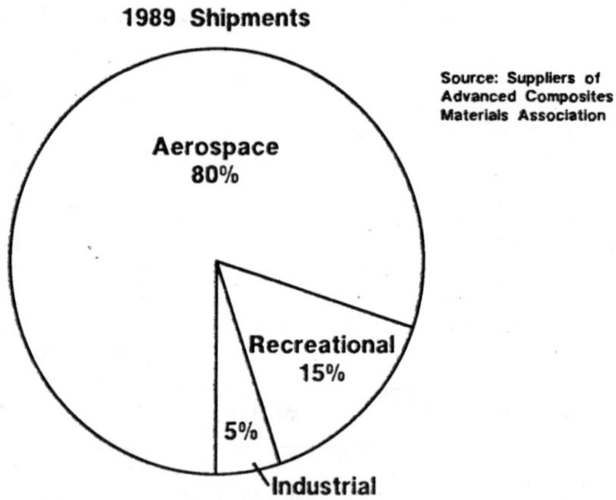

Figure 12. Advanced composites.

bioceramics produced through engineering proteins and carbohydrates, and significant cost reductions through the decreased energy requirements and process improvements were mentioned earlier.

In 1989 the Gorham Advanced Material Institute predicted that the global market for all biomaterials would be in excess of $5 billion in 1988, and had been growing between 12%–14% per year throughout the decade. Although half of the large growth has been accounted for in the medical and dental professions associated with orthopedics and implants, structural applications for large bodies (like automobiles) will occupy an increasing percentage of the growth potential. Finally, it can be predicted that the $5 billion or so inorganic biomaterials market of today will grow at an enhanced rate if it becomes possible to produce bioceramics together with traditional metals or possibly aluminize them to permit increased use of hybrid coatings. Research in a number of laboratories is now targeted toward the production of tailored and controlled ceramic particle formation through improved understanding of mineralization processes at a molecular level (Kamely, 1989, Nyce, 1989, and Cooper, et al., 1989).

Finally, the Institute for Future Technology in Japan has recently conducted its fourth technology forecast survey to the year 2015. The study utilizes an emerging technique called "probability encoding." This is a decision-making analysis technique employed in the presence of significant uncertainty, but which uses Delphic approaches. The first three surveys were conducted between 1970 and 1982, and were used by the Japanese Science and Technology Policy Bureau for determination of R&D support to many projects. The life sciences in general were given the highest ratings in terms of "degree of importance" and advances in biological materials and biocomposites were viewed as occupying the broadest range of applications for the highest degree of financial potential (Kato, 1988).

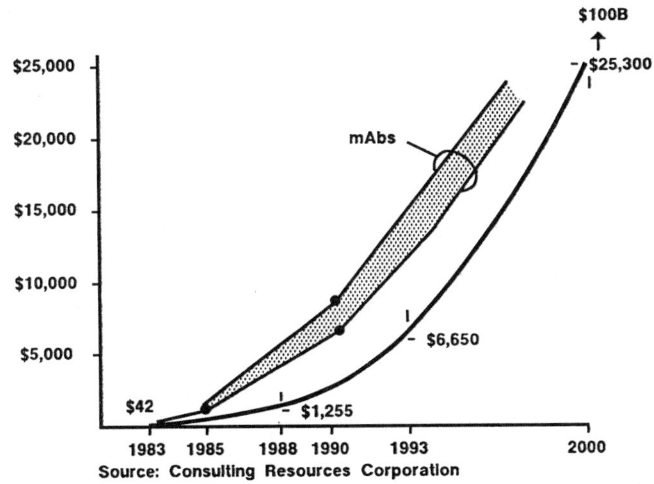

Figure 13. U.S. sales forecast for biotechnology (million dollars).

The Industrial Biotechnology Business Environment Today: Promise or Problems?

Early in the decade, overdrawn promises by the venture capital markets, penny stock values for over the counter biotechnology start-ups, and some key business successes in biotechnology public offerings confused us all. Today, with projections as high as the biomaterials segment of the discipline just reviewed, and as high as the 1990 CRC projections (Figure 13) in the face of a bare handful of profitable companies remaining of the sometime 1,000 or so in the United States the situation is no less schizophrenic, (Morris 1990, in press). Even as early as 1986, more conservative projections placed the industry sales volume by the year 2000 at from $10–$50 billion (Rathman, 1986). And this was after tax law changes made the incentives for limited R&D partnerships less important, especially for new pharmaceutical partnerships held at $200–$250 million through 1986 in the face of market capitalization of over $4 billion.

By 1987, federal and private sector spending together on biotechnology R&D exceeded $4.5 billion. Of this, about $2.7 billion was a result of federal investment while most within the private sector was performed in pharmaceutical and agricultural areas. The National Institutes of Health provided almost $2.5 billion of the federal spending, and the Department of Defense was the second largest spender with a pecuniary $120 million. In 1990 the total Department of Defense spending for biotechnology was only $100 million (Green, 1989).

So what can we glean from this information?

- The "ultimate" market by the year 2000, if everything works, will be less than $100 billion per annum. To date there has been a total of $3 billion invested in research and engineering (cf. above 4.5 including capitalization and advertizing).

- There are about 1,000 development-stage biotech companies worldwide, 700 in the United States. Most of the successful corporations do 70%–90% contract research, and having $200 million in the bank is rare. The average capitalization is about $10 million. By 1990 there were less than five with valuation of $500 million.
- Obtaining FDA-approved products remains slow – only a handful today, with the first in 1985 five years ago!
- Product patents are holding in court for the most part, but process technology patents are failing.
- Cumulative valuation of all companies may be $4–$5 billion but the number is soft and based on assumptions of near total success.

Even from all these accounts, things look a little better in terms of projections for itemized sales. A conservative arithmetic would include the following numbers, based upon the author's readings from recent annual reports and financial pages taking into account corporate projections, conservatively drawn for the next 20 years:

- Pharmaceuticals, $15–$20 billion per year.
- Health care/biomedical products, $10 billion.
- Speciality chemicals, $10 billion.
- Commodity chemicals, $20 billion.
- Agriculture, $100 billion.
- Energy related industry, $?
- Environmental/toxicology control, $?

So it is clear that although the pace has slowed over projections from 10–15 years ago, the future is far from uninteresting. This is especially true in the last two areas which appear to this observer to hold the most promise.

What do we want from biotechnology in medium industry? Not surprisingly there are only three possibilities: a product, a process to sell, or a process to use from which we can gain a competitive advantage.

Why are we interested now in biotechnology, when the "hit-rate" from ventures is so low, the returns on investment are almost non-existent, a few percent of all initiatives succeed, and tax incentives are missing? There are five fundamental reasons for business interest:

- Biopolitics: it is clear that renewable resources can economically provide raw materials for products that are approaching competition with those now based on non-renewable ones (e.g., for some areas of the world, oxygenated fuels).
- Dollars make cents (or sense!): processes that produce commodity or speciality chemicals especially compared with today's chemical processing of vegetable materials appear to be economical. Scaling up bioreactors and fermentors, using numerically controlled systems to modulate conditions in the fermentation stream are being accomplished as process engineers are brought to the biologist's bench.

- Innovation: an incredibly large number of possibly valuable products are being defined by biological methods that appear to now be possible with what will become routine genetic manipulation in the next two decades. It is no longer considered fanciful to develop crops that can with enhanced nitrogen fixation and metabolism at temperature extremes produce food and fodder in record quantities.
- Patents equal licenses: the discoveries are coming faster, and while patents are not directly resulting in protection per se, they are being turned increasingly into licensing arrangements with existing firms selling products similar or identical to ones being designed biotechnologically.
- Secrets are as slow as publishing: the proprietary nature of the laboratory is affording its own protection as science in general has been unable to discover reasonably fast ways to communicate new findings. Thus, there is a built in protection for corporations to do biotechnology.

On the down side, we are finding it increasingly difficult to find qualified biologists who can understand enough about engineering to be worthy of the title "biotechnologist" and thereby become involved in research programs. Beginning in 1985 through the current day the number of scientists and engineers employed primary in R&D programs in industry that could be called biotechnology ranged from about 5,000–7,000 today (Stewart, 1987 and Morris, pers. comm.). About half of the scientists are engaged in recombinant DNA/RNA research, and about one third in bioprocess technology. Ten to fifteen percent of those employed are trained in biochemistry, genetics, cell biology, immunology, or microbiology with approximately 8% in bioprocess engineering. A healthy quarter of the scientists are being attracted from other fields. Two thirds of the dollars being spent by industry are in the health care field, but we have indicated that this is likely to change in the next decade dramatically. Thus, we are manpower limited, somewhat scarce on engineering expertise and interest, but poised to increase capital resources. The major problem will remain finding the intellectual property to accompany the dollars.

Conclusions: How Can Large Industry Participate in Biotechnology?

A year or so ago Bert Maidment and I attempted a quality function deployment (QFD) approach to see if we could glean from the literature and experience of the start-up companies to which we had access a reasonable strategy for investment. The approach we took was to arrange nine process engineering topics in increasing order of complexity, from gene splicing to automated hardware and biocomputers and intersect materials resources required for each against six cells representing product families in order of increasing time required to reach the market from pharmaceutical products and research tools toward bioremediation and waste management appli-

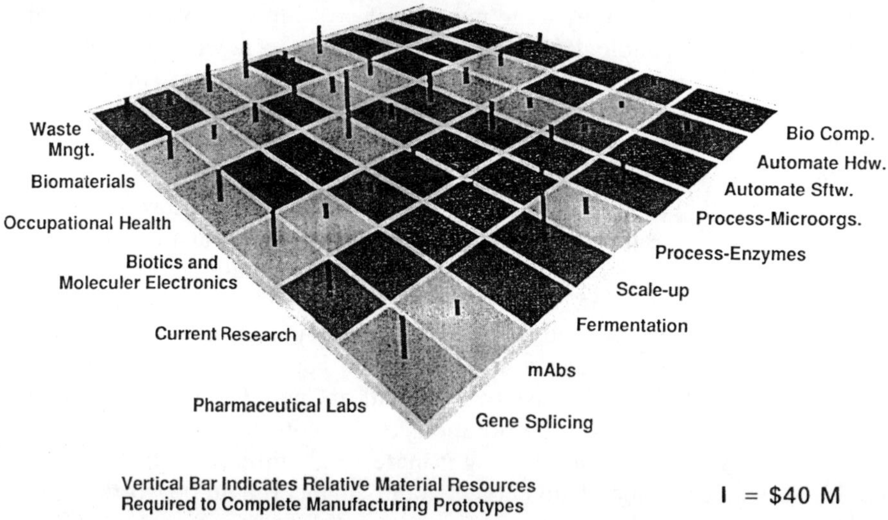

Figure 14. Econo-political forcing functions vs. biotechnology applications engineering.

cations (Figure 14). The result was as confusing to us as the graphic indicates. While we still believe the relative numbers are accurate, ranging from small units of investment ($20 million) to produce, for example, an automated hardware device for research application, to $200 million for an occupational health application/scale-up system, the level of information served no more purpose for decision making than good intuition. The device does serve, however, to display to business associates how the "hot-spots" such as scaling up for a pharmaceutical laboratory require maximum material resources. On the other hand, applications for the cluster of cells involving biomaterial and waste management products are seen to eventually not be rate-limited by automated software and hardware investment. The exercise did predict cor-

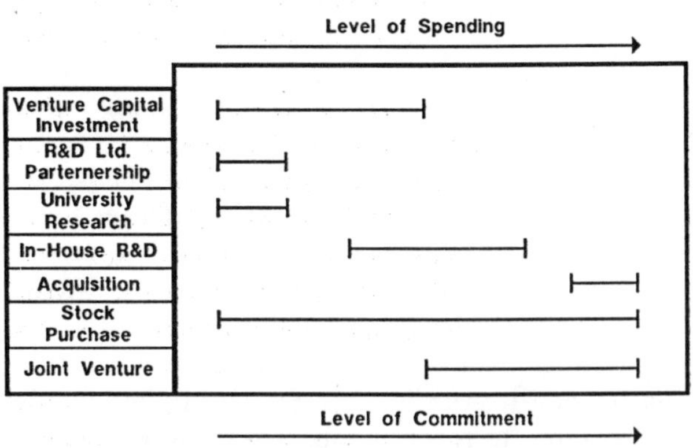

Figure 15. Biotechnology: methods of entry.

Table 8. Biotechnology: entering the new frontier

'Passive' entry	'Active' entry
• Venture capital investment	• Stock purchase and acquisition
• R&D limited partnership investment	• Joint ventures
• University research investment	• In-house research

rectly that finding functional process related enzymes for specific waste management tasks would be critical and yet would require modest ($40 million) investment for applications engineering.

Figure 15 tabulates the key advantages and disadvantages for new ventures in biotechnology. In brief, the advantages are the same as with any new venture, opportunities to reach into new competitive arenas and diversify abound. On the other hand, there are still problems with technology catching up with the real world of profit-taking. Many industries are not used to dealing with the biomedical regulatory environment and the number of years required to show a profit. This is particularly true in the United States in which a five year plan is generally thought to be long-term for the manufacturing industries. All of this serves to support our earlier conclusions, namely that industry must be sold on individual products or processes as opposed to the idea of "doing biotechnology." Those of us who are familiar with the field recognize that the panoply of opportunities must certainly afford profit for the corporation that stays the course, but we need to find more traditional ways to enter the marketplace.

There are three passive and three active entry strategies (Table 8). Arranged in terms of level of spending required or level of commitment versus the method of entry, there is a wide range of either dollars required or intellectual activity needed within a corporation to be successful (Figure 15). In the case of stock purchase and acquisition, considerable differences exist because of the time required for legal costs and personal involvement. Without quite understanding why, this is probably the reason that corporations are becoming more involved in direct acquisitions after products appear from start-up companies as opposed to early participation. R&D partnerships and support of university research is seen to be relatively inexpensive, but lack of commitment from within usually results in inability to sell programs to

Table 9. Key biotechnology problems and opportunities

Advantages	Disadvantages
• Embryonic industry-ground floor	• Not for the risk-averse
• Out-perform traditional industry segments	• May require years to show profit
• Use B/T techniques to enter existing markets	• Will new methods be acceptable?
• Diversify company	• Possible over-diversification
• No established market oligopolists	• Still many problems with technology
• Virgin markets	• Regulatory requirements

Table 10. Entering the biotechnology business via in-house R&D

Advantages	Disadvantages
• Benefits remain in-house	• Financial risk remains in-house
• Full control of scope/ direction of R&D	• Small in-house staff may not keep up with rapid technological advance
• R&D can be highly targeted for corporate strategy	• Corporate culture may inhibit creative R&D

Table 11. Large industry "buy-in" option: vehicle to enter the biotechnology business (investors' viewpoint)

Advantages	Disadvantages
• "Window" on technology	• Technology "window" can be very limited
• Exposure to biotechnology markets and key personnel	• Large financial commitment to become fund partners
• Inside track for licensing and joint ventures	• Too many dollars chasing too few good deals?
• Share costs and risks with other investors	• Fund goals and corporate goals may differ

Table 12. Acquisitions: vehicle to entering the biotechnology business

Advantages	Disadvantages
• Synergy from research staff integration	• May lose key people
• Retain all technologies and profits	• Large corporate environment may stifle responsiveness
• Managerial and financial synergy	• May lose entrepreneurial spirit

Table 13. Direct stock purchase vehicle to entering the biotechnology business

Advantages	Disadvantages
• Window on opportunity	• Limited control over project focus
• Window on technology	• No direct interface with technical staff
• Opportunity for capital appreciation	• Susceptible to possible stock depreciation

senior management. In-house R&D and joint ventures offer reasonable middle ground investment strategies and are nicely tailored to a corporation that has in-house research capability or, if not, finds reasonable satisfaction in joint ventures.

Tables 9–14 show the advantages and disadvantages of these strategies. Although there may be as many orders of ranking as there are individuals

Table 14. Entering the biotechnology business through joint ventures

Advantages	Disadvantages
• Synergy through combined strengths	• Tend to dissolve when there is no longer a fit
• Control of scope/direction of project	• Partner goals may differ
• Project can be highly targeted for corporate synergy	• Technology window limited to one project
• Exposure to key technologies and personnel	• May require large financial commitment

Table 15. A summary of entry vehicle characteristics

	Venture capital invest.	R&D Ltd. Partner-ship	Univ. research	In-house R&D	Acqui-sition	Stock purchase	Joint venture
"Window" on technology	*	○	●	*	○	○	○
Exposure to markets & key personnel	*	○	*	–	●	●	●
Captial appreciation	●	●	*	*	○	●	●
Synergy from staff interaction	●	●	●	*	○	○	*
Targeted project focus	●	●	●	*	○	○	*

* = Excellent; ● = Good; ○ = Satisfactory.

who view these tables, this author finds that, on balance, entering the biotechnology business via in-house R&D is by far the most saleable in terms of limitation of investment, maintenance of active championship from internal R&D staff, and ability to target the product or process toward a specific corporate need. The disadvantages are far from trivial, particularly that of corporate culture inhibiting creative research in times of budget constraints. On the other hand, the financial risk remains in-house while a disadvantage can be turned in these cases to serve as a control. A final QFD matrix (Table 15) that summarizes the entry characteristics versus the five most desired attributes of a research program also supports the argument that participating in an in-house R&D program is the most likely to succeed. No other set of strategies in the "active" category is satisfactory. Venture capital investment is reasonable, but there is little opportunity or "taste" for large industry to play in that arena. R&D limited partnerships are becoming less advantageous for tax purposes as mentioned earlier. University research satisfies the need for acquiring outside intellect, but serves little chance in the business world to offer substantial reward.

Summary

The arguments in this paper have been intended to stimulate increased attention by individuals in academic or small industrial environments to reconsider attempts to penetrate large manufacturing industry. The opportunities are important in terms of affecting millions of lives because of the problems that biotechnology addresses head-on: toxic material handling, alternative energy sources, and repair of broken lives along with genes. Compared with the scale of investments required for new initiatives in physics, electronics, and conventional material science, the competitive prospects in biotechnology are encouraging. By and large business predictions are being shown to have general validity, and earlier predictions of $100 billion per year total sales by the year 2000 may yet be realized; important profits are attainable. Biologists are finding themselves forced to deal with the engineering community, not because they now understand it is required for development of process technologies and scale-up, but because there are no other scientific manpower pools available. For whatever reason, the results are synergistic to industrial needs. Similarly, the budget-consciousness of medium and heavy industry is serving as a forcing-function to require that biotechnologies with reasonable application in the mid-term (5–10 years) are those that receive resources. This cannot be good news for those who prefer the taste of pure research, but for the rest who believe that working on applied problems is exciting and requires no less a search for fundamental scientific principles and is challenging to the extreme, the view ahead looks promising. The lack of reasonable information exchange on a timely basis through the traditional publication route is clearly obviated through the positive outcomes from networking in venues such as this conference.

Acknowledgments

The author is grateful for the interest and access to the research programs of Wayne France, Head of Environmental Science at General Motors Research Laboratories. Together with his scientists in the biotechnology programs of Environmental Science, others in the author's department have contributed information for these thoughts without reservation as cited in the text. Special appreciation is also due to Ann Stark and the executives responsible for the NRC/BAST/STAR Biotechnology Study chaired by the author, and the members of the Biotechnology Experts Group of that study who contributed immensely to the specifics contained within this paper.

References

Chirichiello, J.R. 1987. Biotechnology research and development activities in industry. 1–27. NSF. pp. 87–311.

Cooper, G.J., Pearce, B.P., Cater, S.R., Kenward, C.E., and Townend, D. 1989. Augmentation by foam materials of lung injury. Proceedings of International Research Council on Biokinetics of Impact (IRCOBI), Stockholm.

Gibson, T.L., and Abdul, A.S. 1988. Microbial degradation of aromatic hydrocarbons in hydrogeological materials: Microcosm studies. Proceedings of Second National Outdoor Action Conference on Aquifer Restoration. pp. 53–70.

Green, C.C. 1989. Biotechnology and arms control negotiations. In: European arms control: Problems and prospects. Hatchett, R.L., ed., Mosher Institute and Texas A&M, University Press, College Station. pp. 168–177.

Green, C.C., Soukup, J.F., Maidment, B.W., Petsko, G.A., and Kamely, D. 1990. Biotechnology: from diversity toward opportunity. NAS/NRC/STAR Press, Washington. In press.

Hinman, N.D., Wright, J.D., Hoagland, W., and Wyman, C.E. 1989. Xylose fermentation: an economic analysis. Applied Biochemistry and Biotechnology 20/21: 391–401.

Kamely, D. 1990. Biotechnology materials and processes. In: The Department of Defense critical technologies plan. Committees on Armed Services, United States Congress, AO–A219300. 22: A-225–232.

Kamely, D., Chakrabarty, A., and Omenn, G.S. 1989. Advances in biotechnology IV. Gulf Publishing, Houston. p. 504.

Kato, S. 1988. Future technology in Japan: forecast to the year 2015. Institute of Future Technology Science and Technology Policy Bureau. pp. 5–17, 73–89, 144–146, 157–158, 247–258.

Kim, B.R., Anderson, S.G., and Zemla, J.F. 1990. Effect of biological treatments on COD absorption. Water Research. In press.

Kim, B.R., Matz, M.J., and Lipari, F. 1989. Treatment of a metal-cutting-fluid wastewater using an anaerobic, GAC fluidized-bed reactor. J. Water Pollution Control Fed. 61: 1430.

Malinski, T. 1990. Industrial applications of biotechnology. Biotechnology Forum Workshop I. Oakland University Press, Rochester. 8: 7.

Morris, D.W. 1990. Biotechnology: the science talent, and concerns – 2020. Green, et al., op. cit.

Nyce, A.C. 1989. The global markets for inorganic biomaterials: market analysis and forecasts. Gorham Advanced Materials Institute Press, Gorham.

Obringer, J.W. and Tillinghast, H.S. 1989. Advances in applied biotechnology III. Gulf Publishing, Houston. p. 272.

Rathman, G.B. 1986. Meeting the global challenge in biotechnology. Research Management. Sept–Oct: pp. 36–42.

Scott, Charles D. 1989. Advanced bioprocessing concepts for the production of fuels and chemicals. In: Biotechnology for Aerospace Applications. Op. cit. pp. 49–58.

Discussion (DR. C. GREEN)

Kamely: You mentioned your biosensors program. What kind of advantages do biosensors provide over conventional sensors? What in your industry can be applied with biosensors?

C. Green: There are three reasons that we are interested in biosensors. First, biosensors are a research tool. We are learning that numerous industry and manufacturing problems indeed intersect with biological systems. Human beings have to be monitored in exquisitely sensitive ways. We have found in our research in toxicity of our products and processes – gasoline vapors, hydrocarbons, PCB's – we cannot detect at the parts per billion sensitivity that we need for some of our research without looking at something like immunological field effect transistor technology, for one example.

Second, other than sensitivity, there are different classes of compounds that we are interested in studying. You saw the importance of using an immunologically based biosensor in discriminating ketones, alcohols and various other alcohol fractions. There may come the day when one will not be able to operate a vehicle while under the influence of alcohol as determined by breath and/or neurochemical indicators. The extent to which that will be true will not only be determined by a sensor system which does not exist today, but also by law which does not exist today. The ability to discriminate compounds in terms of off gassing of breath is one important biological strategy. There will be about $8,000 of electrical content in your vehicle ten years from now. Today there is $1,500. A large portion of the electrical content will include sensing strategies and systems, of which a good portion will be reconstituted biosensor systems.

Third, for some special real-time experiments we cannot measure products of combustion in our laboratories to the degree that we need with systems based on spectroscopy alone. Biosensors offer a potential, and we are exploring some novel approaches using synthetic porphyrins and protoporphyrins in similar constructs, as I discussed earlier.

Key Words: Biotechnology, Manufacturing, Bioremediation, Biomaterials, Biocomposites, Bioenergy, Biosensors, Industry, Biodegradation

Biosensors Based on Solvated Bilayers Attached to Electrodes

Carlos Gitler, Yechiel Shai, Tuvia Bercovoci, Diana Bach, Haim Garty, Israel Rubinstein[1], Suzi Steinberg[1] and Itzhak Yuli
Departments of Membrane and Material [1]Research, Weizmann Institute of Science, Rehovot, 76100, Israel

Introduction

Need for Biosensors

Next Generation of Chemical Monitors
In our world, influenced by microelectronics and science fiction, if we have to imagine the next generation of diagnostic biosensors, it is clear that removal of blood from the body will not be required to monitor the health of an individual. Rather, sensing devices will be inserted into the body, where, after detecting the desired chemical, they will relay by continuous telemetry, the levels of the chemical species being monitored.

What Has To Be in a Sensor?

In order to achieve this end, the biosensors will have to meet several requirements:

Biocompatibility. The continued exposure of the sensor to biological fluids in the body will elicit responses that will be directed against the device as a foreign body. Unless taken into account, these will tend to sequester by encapsulation the irritating device. Long term implantation therefore requires that the device be completely inert.

Selectivity. The sensor will have to detect the analyte with high selectivity; interference should approach nil. This requirement implies that sensors based on class reactions such as oxidation/reduction (Gregg and Heller, 1990) will

D. Kamely et al. (eds), Biotechnology: Bridging Research and Applications, 43–61.

not be sufficiently selective because of the plethora of chemicals present in biological fluids.

Sensitivity. The fluctuations in the concentrations of analytes generally vary within a narrow range about a given median. Therefore, the sensors have to have sufficient sensitivity to detect the fluctuations required for clinical analyses.

Reproducibility. This represents one of the most difficult problems in the design of a biosensor. High selectivity means that the biosensor must have a recognition step that involves high affinity binding of the analyte to a component of the sensor device. This means that the off-rates have to be low and that the recognition portion of the sensor will remain occupied with the analyte, and thus, would considerably delay the next sensing step. Therefore, the sensor must include a secondary reaction that makes the off-rates increase transiently (Yuli and Snyderman, 1986). Alternatively, the analyte levels have to fall in an oscillatory manner. These aspects will be further discussed below.

Nature of Biological Sensing

The initial interaction of molecules with biological sensing systems takes place through high affinity binding to a protein that resides in a membrane. Three types of proteins are involved in this recognition:

Receptor Proteins. These are proteins that, upon binding of small molecules, can induce an intracellular signal which can be due to the activation of an intracellular enzyme that is part of the receptor protein. Examples include the epidermal growth factor receptor, the insulin receptor, etc., where the receptor acts as a cytoplasmic tyrosine kinase (Roth and Casell, 1983). Alternatively, receptor binding to the agonist can modify a separate protein that acts as a regulatory element, altering the activity of an enzyme. An example is the activation of adenylate cyclase or of phospholipases mediated by the receptor activation of G-proteins (Stryer and Baurne, 1986; Burch, et al., 1986).

Transport Proteins. These are proteins that bind small molecules with intermediate affinities and then undergo conformational changes that induce the translocation of the bound molecule across the membrane (Klingenberg, 1989).

Ion Channels. These are proteins that, upon binding to a small molecule, undergo a conformational change that creates a pathway for the flow of ions

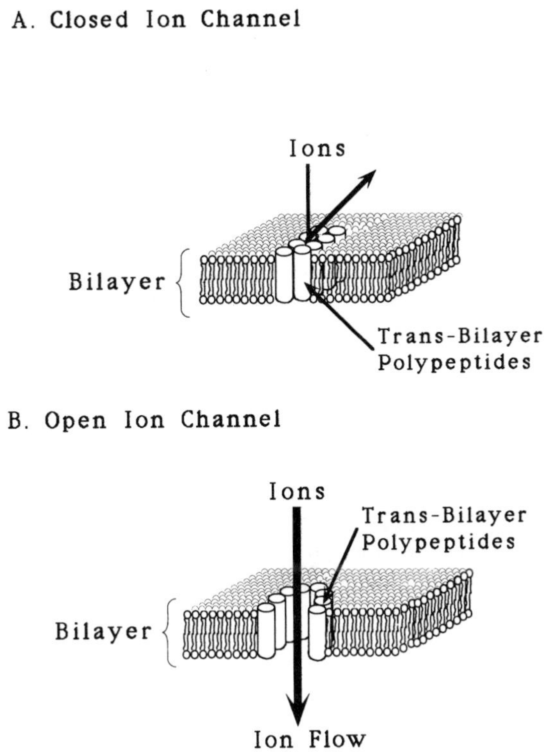

Figure 1. Schematic representation of a closed and open ion channel.

across the membrane. Figure 1 shows a diagrammatic representation of the two states of an ion channel according to the current paradigm (Numa, 1986; Montal, 1987; Tanabe, et al., 1987). The trans-bilayer helical segments of the ion-channel protein can undergo a conformational change from a closed state (A) to an open state (B) in which an aqueous pathway is opened for the flow of ions through the ion-impermeable bilayer.

Each of these proteins could be used as the basis for the construction of biosensors. However, the first two categories are problematic since, if the receptors are used alone, they have a low amplification of the signal. To use them as they function in the cells themselves requires the participation of coupled reactions that increase exponentially the complexity of the system. For this reason, the sensor in our development program is based on the use of ion channels.

Detection Involving Ion Channels

Biological systems perceive extracellular signals such as light, smell, nerve-nerve stimuli, etc., by the initiation of coupled, cascade-like amplification reactions (Mooribroke and Wang, 1988). In many of these, the initial or intermediate steps in the cascade involve the opening of membrane-associated ion-channels (Fox, 1987). In ligand-activated channels, the process is initiated by the binding of the small effector molecule to the receptive part of the channel protein. This induces conformational changes that lead to the opening of a pore across the lipid bilayer, causing a step increase in electrical conductance. Biological channels may be functionally reconstituted in artificial lipid bilayers, leading to effector-induced currents essentially similar to those occurring in the biological membranes. By means of the electronic amplification available today, a single channel opening event can be detected (Corronado and Latorre, 1983; Corronado, 1986).

Ion channels are either high molecular weight proteins (in the range of >1000 kDa) (Hediger, et al., 1987; Shull and Greeb, 1988) made of a single chain or an association of several polypeptides. In addition, there are small molecules such as alamethicin (Boheim and Kolb, 1978; Fox and Richards, 1982; Hall, et al., 1984), melittin (Habermann, 1972; Terwillinger and Eisenberg, 1982), or pardaxin (Thompson, et al., 1986; Shai, et al., 1990), which have molecular weights in the range of 2–3 kDa. These proteins associate in water and also in the membrane where they form pores that induce ion fluxes.

Why Use of Bilayers in a Sensing Device?

Biological systems, as mentioned above, represent a wide reservoir of ion channels. They could readily be used as high amplification sensing devices, if an adequate medium could be created for their functional incorporation into the sensing device. However, their normal function requires that the medium in which they are incorporated retain the properties of fluidity, high impedance and adequate width to allow the membrane-embedded apolar and polar segments of the ion channels to be oriented properly. Furthermore, the medium into which they are incorporated must allow the ion channels to undergo conformational changes or association between polypeptides to create the conductive pathways. In addition, ion-channel proteins have only 10–15% of their polypeptide chains in contact with or immersed in the lipid bilayer (see National Biomedical Research Foundation – protein and nucleic acid databases). The rest protrudes into the aqueous layers on both sides of the lipid membrane.

These requirements can best be met by use of lipid bilayers formed by natural or artificial lipids. Most attempts to achieve this goal have been based on the use of Langmuir-Blodgett mono- or multiple bilayers deposited on the surface of an electrode (Sygawara, et al., 1987; Stelzle and Sackmann, 1989). However, these bilayers suffer from lack of hydration and limited mobility. Furthermore, they are limited because there is little water between the bilayers and between the bilayer and the electrode.

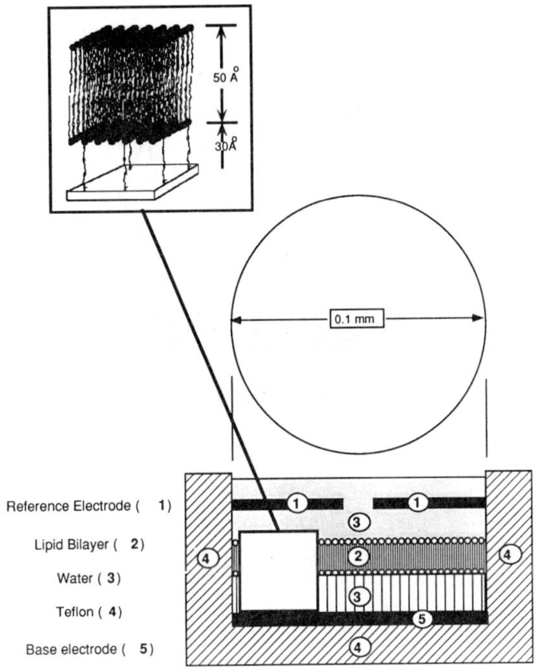

Figure 2. Design of the biosensor based on a solvated bilayer attached to a gold electrode.

Nature of the Sensor that is Being Developed

General Description

In general terms, the device consists of an electrode-attached lipid bilayer, doped with either synthetic or biological ion channels. Opening of the channels is induced by interaction with the molecule that is to be detected. This causes a step increase in the bilayer conductance which is sensed by the electrodes.

A teflon block contains a circular well at the bottom of which the working electrode is placed (Figure 2). A lipid bilayer is attached to this electrode (see insert) in such a way that there is a bulk aqueous layer between the bilayer and the electrode. The bilayer boundary is sealed by an apolar contact between the hydrophobic component of phospholipids and the teflon. The bilayer would contain, in addition, the ion channel forming polypeptides (not shown). A second reference electrode is placed above the bilayer (see section on Current Status for details).

Attachment of a Lipid Bilayer to an Electrode

The concept for attachment of a fluid and hydrated phospholipid bilayer to an electrode (at present, a gold electrode) is shown schematically in Figure

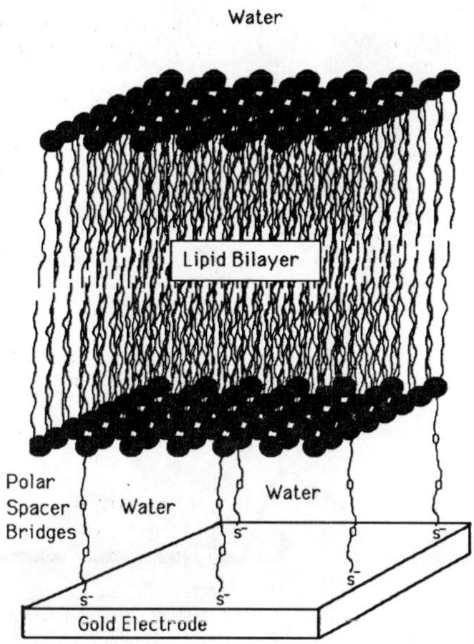

Figure 3. Method of attachment of the lipid bilayer to the electrode.

3. The idea is based on the attachment of the bilayer to the electrode through a hydrophilic spacer arm joined chemically to a phospholipid molecule. The arm in current use is constructed of a string of oxyethylene groups. This arm is similar to neutral detergents such as the Tritons (Helenius and Simons, 1975), in which the hydrophilic portion is made up of oxyethylene chains that have strong interactions with water. The use of nonionic polar groups reduces charge repulsion. However, since the distance between the arms is expected to be rather high, other charged groups may be used to increase the interaction with bulk water. The end of the arm not attached to the phospholipid is substituted with a thiol or thiomethyl group, which is responsible for the high affinity interaction with the gold or silver of the electrode. This design represents an extension of the techniques developed by Rubinstein, Steinberg, Tor, Shanzer and Sagiv at the Weizmann Institute of Science (Rubinstein, et al., 1988).

Our approach for attaching the bilayer to an electrode is unique in that:

i) It allows bulk water to be present at the electrode surface and on both sides of the bilayer. This allows the incorporation of large ion-channel proteins. Furthermore, conductivity or impedance can be used to register channel opening.
ii) It permits motion of the non-anchored lipids and of the polypeptides incorporated into the bilayer.
iii) It gives the bilayer stability and a fixed geometry with regards to the electrode surface.

Sequence:

1 5 10 15
NH2-Gly-Ile-Gly-Ala-Val-Leu-**Lys**-Val-Leu-**Thr**-**Thr**-Gly-Leu-**Pro**-Ala-

Leu-Ile-**Ser**-Trp-Ile-**Lys-Arg-Lys-Arg-Gln-Gln-amide.**
20 25

Helical rod:

side PROJECTION **Schiffer-Edmundson PROJECTION**

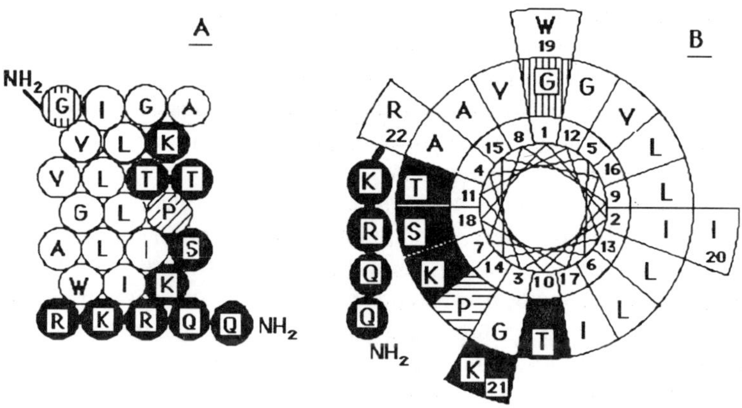

Figure 4. Structure of bee venom melittin.

iv) It minimizes the requirements for a smooth electrode surface.

The bilayer deposition on the electrode surface is achieved by the introduction of mixed micelle population consisting of the adequate phospholipids and a small, water-soluble detergent. Following the spontaneous attachment of the mixed micelles to the electrode surface by the hydrophilic arms, slow removal of the detergent results in the formation of a continuous bilayer structure.

Formation of Ion-channels by Small Polypeptides

The principles underlying formation of ion-channels by small (20–25 amino acid) peptides are not completely understood. It has been suggested that peptides such as alamethicin and melittin (50% of the dry weight of the honey bee venom) form ion-channels because they behave as amphipathic-helical-peptides (AHP). That is, the polar amino acids are present in the polypeptide chain with 3,4 periodicities. On formation of a 3.6 residues/turn α-helix, the polar side-chains are directed towards the same side of the helical rod. This is exemplified by the sequence of melittin and the side and top projections of the α-helical rod it forms (see Figure 4).

Melittin has been chosen as the starting point for peptide design in this project. It is water soluble and exists, most likely, as a tetramer (Terwillinger and Eisenberg, 1982). Yet, it transfers spontaneously into lipid bilayers probably due to the presence of an apolar 6 amino acid segment at the NH_2-terminal end of the polypeptide. Once incorporated into the bilayer, it assumes an α-helix secondary structure and forms anion-selective channels. A channel is made of an aggregate of several helices in the plane of the membrane with their hydrophobic side-chains in contact with the surrounding lipids. Their hydrophilic side-chains are directed away from the lipids and form the aqueous ion conductive pore. However, with time and/or additional incorporation of melittin into the bilayer, relatively large pores are generated and cell lysis becomes dominant. Upon examination of the sequence of melittin (Figure 4), it was observed that it has a cluster of six basic amino acids (bold and underlined) at the carboxy-terminal end. We reasoned that this cluster of basic amino acids could, like poly-L-lysine, be responsible for the lytic properties of melittin. Melittin, like alamethicin, contains a proline in position 14. Interestingly, proline is abundant within the sequence of the ion-conductive segment of most native channels (Brandl and Deber, 1986). Proline is an amino-acid; its presence results in the destruction of the regular α-helix backbone-hydrogen bonds. One or two amide-carbonyl moieties will not be hydrogen bonded, and their non-bonding electrons will contribute to the overall polarity and amphipathic configuration of the α-helix.

Design of CH-1

Based on the above considerations, we devised a homologue of melittin that we refer to as CH-1. Several significant modifications were introduced into CH-1. First, three of the basic amino acids and one glutamine of the carboxy-terminal segment of melittin were removed to eliminate lytic properties from CH-1. Second, residues 2 and 19 of melittin were substituted: tryptophan for isoleucine and cysteine for tryptophan, respectively. The rest of the CH-1 molecule is identical with melittin. The cysteine at position 19 was introduced as a reactive center. It permits substitution, by direct attachment or disulfide formation, of fluorescent probes, haptens or peptides containing desirable epitopes, without altering the helical structure. It was protected after synthesis as the thiomethyl mixed disulfide (see Figure 5).

Nature of the Bilayer Aggregates

The relative hydrophilic angular sector of the CH-1 α-helix seems to imply that a stable ionic channel is constructed of a tetramer assembly, as shown in Figure 6.

It is likely that other amphipathic peptides with a different ratio of hydrophilic to hydrophobic residues would aggregate with different assembly numbers (Lear, et al., 1988; DeGrado, et al., 1989). The aggregates that can be formed are not well defined. It is likely that, if the proline bend of the helix is not very extensive, dimers can exist. It is also apparent from Figure 7,

Sequence:

1 **5** **10** **15**
NH2-Gly-Trp-Gly-Ala-Val-Leu-**Lys**-Val-Leu-**Thr**-**Thr**-Gly-Leu-**Pro**-Ala-

Leu-Ile-**Ser**-Cys-Ile-**Lys**-**Gln**-**amide.**
20

Helical rod:

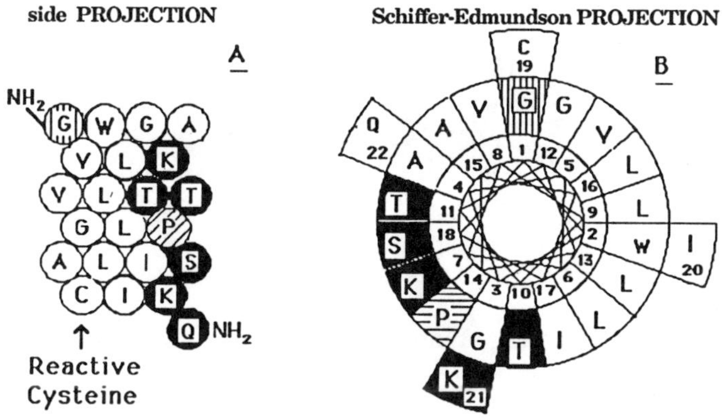

Figure 5. Structure of the synthetic peptide CH-1.

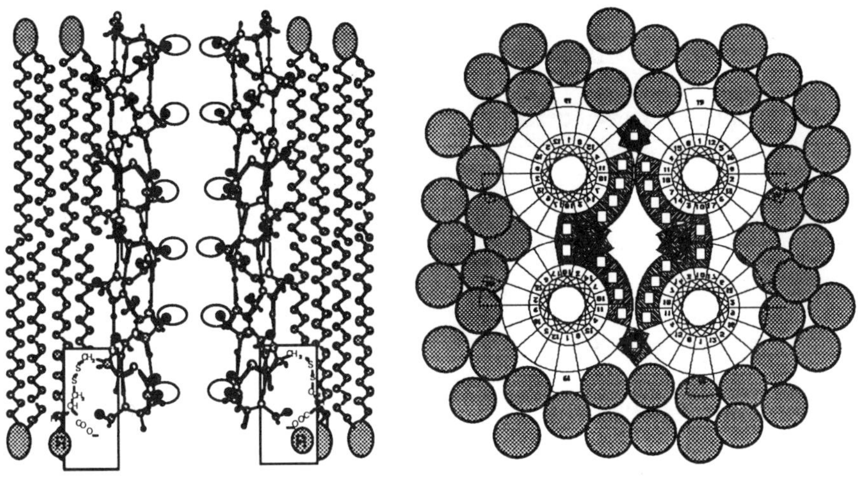

A B

Figure 6. Schematic representation of CH-1 as a helical rod aggregate immersed in the lipid bilayer (A). The polar side-chain groups (open circles not drawn to scale) are directed towards the center of the aggregate. Schiffer-Edmunson helical projection of the aggregate (B), showing the polar central aqueous pore region. Chemical group attachment site is shown by a rectangle.

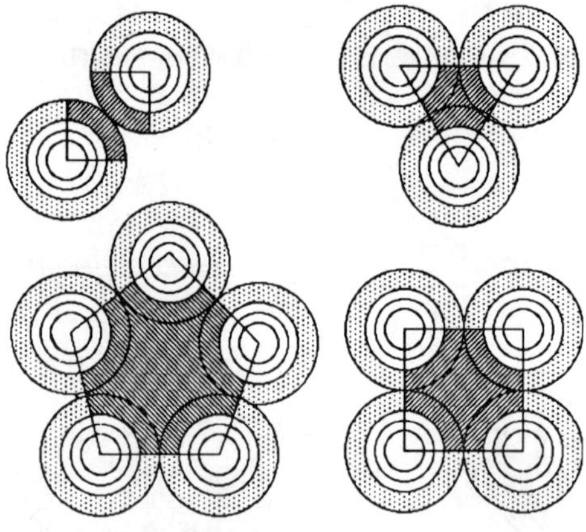

Figure 7. Normal view of the helical-rod aggregates.

that the larger the assembly number, the larger the aqueous pore obtained. This feature has already been used, to design channels differing in their selectivity to ions, by control of pore-diameter. For the purpose of the present discussion, it is likely that factors that perturb the aggregation will decrease the probability that open channels will be formed. They will also affect the open times of those channels that are formed. The fact that channel opening is due to the transient formation of aggregates due to spontaneous lateral diffusion in the bilayer plane implies also that any perturbation to the diffusion will decrease the probability of formation of ion-channels.

The Role of Proline in the Peptide Channels

In an attempt to better understand the role of proline in peptide aggregation and the regulation of the ion-conductivity, we designed a second polypeptide CH-2. It differs from CH-1 only in residue 14 where the proline of CH-1 is substituted by serine. In this way, CH-2 retains all structural features of melittin and CH-1 except for the absence of the proline-induced break in the α-helix rod.

Two Ways of Inducing Channel Opening

Depending on the component protein or polypeptide present in the device, opening of the channel can be induced by two alternative mechanisms, (1) a ligand induced opening of pre-existing fixed or assembled channels and (2) specific removal of a perturbation that prevents amphipathic peptides, immersed in the bilayer plane, from spontaneously aggregating to form a channel assembly.

In the first category we refer to natural channels present in biological membranes, such as the acetylcholine or GABA receptors. These channels can be either highly purified from suitable cells or can be genetically engineered. Then, they can be incorporated into the electrode-supported artificial bilayer. The bilayer should be adjusted to support the receptor-channels in their closed state when unoccupied. The binding of an agonist to its specific receptor site is expected, as in the native systems, to induce the conformational change that would trigger the opening of the channel. This approach will allow the formation of a device which mimics directly the agonist perception by the in vivo receptors. In the case of the acetylcholine receptor, the further addition of acetylcholine esterase to the sensor assembly would allow simulation of the susceptibility of the nervous system to both agonists and antagonists.

In the second triggering category, we envisage detectors based on the use of an adequate group attached to the ion-channel peptide as the receptive part of the element. Attachment of a chemical moiety to the end of the CH-1 should allow interaction with small or large molecules that will alter its channel function. The thiol of cysteine-19 of CH-1 was designed as the site of group attachment.

The procedure envisaged for the attachment of the chemical groups to CH-1 is to treat the peptide, after removal of the thiol-protective group, with active disulfides. Thiol-disulfide exchange should result in the formation of chemical moiety-SS-CH-1. The disulfide allows the formation of derivatives that do not alter the helical packing of the channel-peptide.

The behavior of the modified peptides can be studied initially in liposomes using the liposome-potential collapse method we developed (Loew, et al., 1983; Garty, et al., 1983). This will allow us to determine the range and the working conditions of the system under study, prior to its incorporation into the sensor.

Reversibility of the Sensing Reaction

In order to allow the sensor ion channels to return to their closed state, certain relaxation mechanisms are envisaged. Receptor bound agonists will be removed by an adequate enzyme that will be incorporated in the bilayer or, if necessary, will be attached to the wall of the cell in the aqueous compartment above the bilayer. The enzyme activity will be adjusted in such a way that, upon exposure of the device to the substance to be detected, the concentration will transiently rise to open the channels. The miniature dimension of the proposed device offers a short response time due to diffusion. Due to the enzymatic breakdown, the substance to be detected will fall to low levels and allow re-initiation of the detection cycle. Since the channels respond very fast, the adequate combination of channels to enzyme should be easily established by experimentation.

Restoration of the electrolyte medium, in the aqueous layer between the electrode and the bilayer, is designed to be made by reversed ion flux. The electric field would be generated between the detection and the reference electrodes. The same mechanism is considered for arresting the ion flux once a sufficient signal is perceived, so that the sensor lifetime would be prolonged.

The susceptibility of the ion channel opening/closing to externally imposed electrical fields depends on the dipole moment of the peptide, its spatial orientation and mechanical coupling to the helix backbone. These properties will be studied with the aid of specifically designed peptides.

Advantages Foreseen for the Device Under Development

High Sensitivity

Because of the typically high single channel conductance (between pS and nS), opening of one channel induced by the binding of few molecules should be a readily detectable event. The recorded sensitivity of these systems is in the nanomolar range.

Versatility

The device to be developed is designed to be part of a electronic system that can readily be adapted to different needs. It is suitable for the detection of a broad spectrum of chemical signals. By control of the lipid components, the system can be made to function at temperatures from 4–40°C. Clearly, a microchip of the type to be developed can easily be heated or cooled, for instance by Pelttier devices, so that its use may be extended to temperatures down to −40°C and above 40°C. A device consisting of a series of different sensors may be used to obtain a fingerprint of a given environment.

Generic

Once a lipid bilayer can be attached in a stable functional form to an electrode, the detector can be made to detect any desired molecule for which an antibody can be made or for which there exists a ligand-activated natural channel.

The use of the actual receptors present in the nervous systems of humans allows fidelity to the biological requirements because it mimics the biological detectors themselves. This means that the detectors may be used to determine the adequacy of design of inhibitors of the receptor function and of molecules that may protect against inhibition.

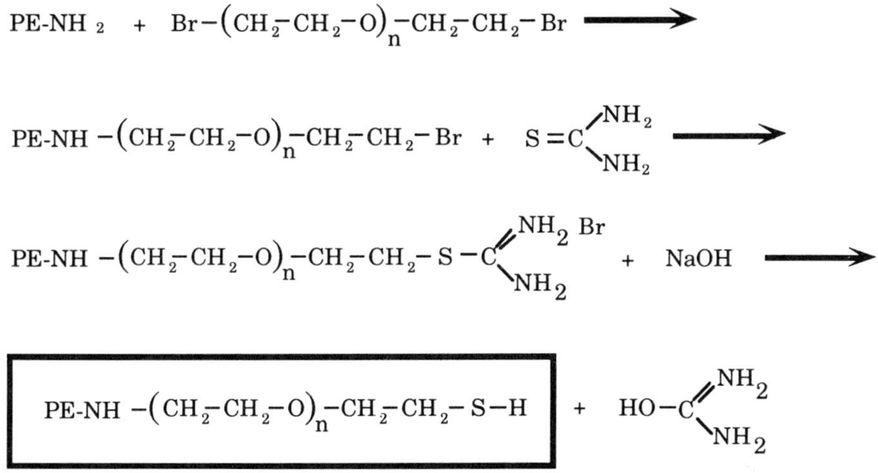

Figure 8. Scheme for the synthesis of the long-chain omega-thiol bridge designed to anchor the bilayer to the electrode.

Current State of the Project

Synthesis of the Hydrophilic Linker Arm

The first of a homologous series of phosphatidylethanolamines derivatized with thio-oxyethylene polymer (PE-(OE)n-SH) has been prepared and tested. The synthesis was made by a general procedure developed in our laboratory that is shown in Figure 8.

In the synthesis, the length of the polymer was n = 12. We expect this arm to support an aqueous spacing of approximately 30 Å. If necessary, the distance from the bilayer can be varied by choosing the adequate poly(oxyethylene) length.

Attachment of Liposomes and Mixed Micelles to the Electrode Surface

In these preliminary studies, the gold electrode was deposited on glass and has been made hydrophilic by argon sputtering plus heating. This procedure yielded highly uniform surfaces that are regularly monitored by ellipsometry. A lacquer seal around the gold was formed (see Figure 9, top) so as to provide hydrophobic support for the boundaries of the phospholipid bilayer. Then, a mixed large-unilamellar-liposome suspension, made of purified soybean asolectin (phosphatidylcholine, phosphatidylethanolamine, phosphatidylserine) and PE-(OE)12SH (50:1 molar ratio), was applied to the electrode (Figurc 9; middle). The liposome electrode interaction was carried

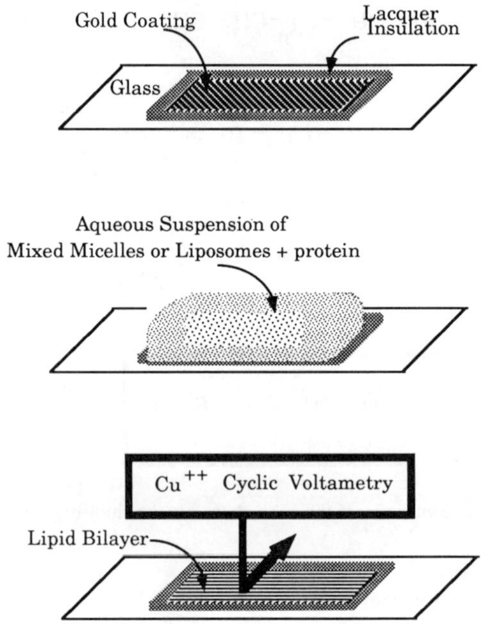

Figure 9. Schematic representation of the method used to cover the electrode with a lipid bilayer.

out in a 100% humidified chamber for 24–48 h (in different experiments). The excess liposomes were removed by extensive washing. Then, the access of Cu^{++} to the electrode was studied by cyclic voltametry (Figure 9; bottom).

The results indicate a marked reduction in the access of the copper ions to the electrode surface. The highly significant 40 mV shift in the Cu+ + deposition potential implied that an ordered access obscuring layer has been formed at the electrode surface. In addition, water was found by H+ reduction to be present adjacent to the electrode.

These results are preliminary. However, they suggest that we can, within a reasonable time, explore the attachment and properties of the layers formed. We are currently testing the use of mixed micelles of soybean asolectin (50 parts) and PE-(OE)12–SH (1 part) in mixed micelles with a short-chain detergent to obtain improved attachment and bilayer formation by regulating the removal of the detergent.

Synthesis of CH-1 *and CH*-2

Both peptides were prepared a unique peptide synthesizer that is based on conventional solid phase synthesizer with improved coupling and removal methods. Overall yields of 80% and 65% were obtained for the 22-step synthesis of CH-1 and CH-2, respectively. The peptides were then purified to 99 + % yielding single symmetric peaks by HPLC analysis.

Channel Activity of CH-1 and CH-2

The channel activity of the two peptides has been examined both in a macroscopic and a microscopic scale. Initially, the peptides were monitored for their capacity to depolarize liposomes made of soybean asolectin supplemented with cholesterol and polarized by valinomycin-mediated $K+$ diffusion potential (Loew, et al., 1983). The experimental results showed that both peptides induce ion fluxes that collapse the $K+$ valinomycin potentials. However, CH-1 was approximately two orders of magnitude more potent than CH-2.

At the single ion channel level, we found that both peptides are active. However, the lack of proline was found to decrease significantly the capacity of the polypeptide to form ion-channels. CH-1 opens channels at a higher frequency and for longer periods than CH-2. Furthermore, CH-1 tends to open several channel levels, with incubation time and concentration, while CH-2 displays primarily a single level with a lower unit conductance than that induced by CH-1.

The single channel activity is currently being analyzed by different models in an attempt to correlate the single channel data with the ion-fluxes observed at the macroscopic level.

References

Boheim, G. and Kolb, H.A. 1978. Analysis of the multi-pore system of alamethicin in a lipid membrane. I. Voltage-jump current-relaxation measurements. J. Membr. Biol. 38: 99–150.

Brandl, C.J. and Deber, C.M. 1986. Hypothesis about the function of membrane-buried proline residues in transport proteins. Proc. Natl. Acad. Sci. 83: 917–921.

Burch, R.M., Luini, A. and Axelrod, J. 1986. Phospholipase A2 and phospholipase C are activated by distinct GTP-binding proteins in response to alpha 1-adrenergic stimulation in FRTL5 thyroid cells. Proc. Natl. Acad. Sci. 83: 7201–7205.

Coronado, R.R. and Latorre R. 1983. Phospholipid bilayers made from monolayers on patch-clamp pipettes. Biophys. J. 43: 231–236.

Coronado, R.R. 1986. Recent advances in planar phospholipid bilayer techniques for monitoring ion channels. Ann. Rev. Biophys. and Biophys. Chem. 15: 259–277.

DeGrado, W.F., Wasserman, Z.R. and Lear, J.D. 1989. Protein design, a minimalist approach. Science 243: 622–628.

Fox, J.A. 1987. Ion channel subconductance states. J. Membr. Biol. 97: 1–8.

Fox, R.O. and Richards, F.M. 1982. A voltage-gated ion channel model inferred from the crystal structure of alamethicin at 1.5–Å resolution. Nature 300: 325–330.

Garty, H., Rudy, B. and Karlish, S.J.D. 1983. A simple and sensitive procedure for measuring isotope fluxes through ion-specific channels in heterogenous populations of membrane vesicles. J. Biol. Chem. 258: 13094–13099.

Gregg, B.A. and Heller, A. 1990. Cross-linked redox gels containing glucose oxidase for amperometric biosensor applications. Anal. Chem. 62: 258–263.

Habermann, E. 1972. Bee and wasp venoms. Science 177: 314–322.

Hall, J.E., Vodyanoy, I., Balasubramanian, T.M. and Marshall, G.R. 1984. Alamethicin: A rich model for channel behavior. Biophys. J. 45: 233–247.

Hediger, M.A. Coady, M.J., Ikeda, T.S. and Wright, E.M. 1987. Expression cloning and CDNA sequencing of the Na^+/glucose co-transporter. Nature 330: 379–381.

Helenius, A. and Simons, K. 1975. Solubilization of membranes by detergents. Biochim. Biophys. Acta 415: 29–79.

Klingenberg, M., 1989. Survey of carrier methodology: Strategy for identification, isolation, and characterization of transport systems. Methods in Enzymology 171: 12–23.

Lear, J.D., Wasserman, Z.R. and DeGrado, W.F. 1988. Synthetic amphiphilic peptide models for protein ion channels. Science 240: 1177–1181.

Loew, L.M., Rosenberg, I., Bridge, M. and Gitler, C. 1983. Diffusion potential cascade. Convenient detection of transferable membrane pores. Biochemistry 22: 837–844.

Montal, M. 1987. Reconstitution of ion channel proteins from excitable cells in planar lipid bilayer membranes. J. Membr. Biol. 98: 101–115.

Mooribroke, M.J. and Wang, J.H. 1988. Integration of signal-transducing processes. Biochem. Cell Biol. Canada 66: 557–566.

Numa, S. 1986. Molecular basis for the function of ionic channels. Biochem. Soc. Symp. England 52: 119–143.

Roth, R.A. and Casell, D.J. 1983. Insulin receptor: Evidence that it is a protein kanase. Science 219: 299–301.

Rubinstein, I., Steinberg, S., Tor, Y., Shanzer, A. and Sagiv, J. 1988. Ionic recognition and selective response in self assembling monolayer membranes on electrodes. Nature 332: 426–429.

Shai, Y., Bach, D. and Yanovsky, A. 1990. Channel formation properties of synthetic pardaxin analogues. J. Biol. Chem. In press.

Shull, G.E. and Greeb, J. 1988. Molecular cloning of two isoforms of the plasma membrane Ca^{2+}-transporting ATPase from rat brain. Structural and functional domains exhibit similarity to NA^+, K^+- and other cation transport ATPases. J. Biol. Chem. 263: 8646–8657.

Stelzle, M. and Sackmann, E. 1989. Sensitive detection of protein adsorption to supported lipid bilayers by frequency-dependent capacitance measurements and microelectrophoresis. Biochim. Biophys. Acta 981: 135–142.

Stryer, L., and Bourne, H.R. 1986. G proteins: A family of signal transducers. Ann. Rev. Cell Biol. 2: 391–419.

Sugawara, M., Kojima, K., Sazawa, H. and Umezawa, Y. 1987. Ion channel sensors. Anal. Chem. 59: 2842–2846.

Tanabe, T., Takeshima, H., Mikami, A., Flockerzi, V., Takahashi, H., Kangawa, K., Kojima, M., Matsuo, H., Hirose, T. and Numa, S. 1987. Primary structure of the receptor for calcium channel blockers from skeletal muscle. Nature 328: 313–318.

Terwilliger, T.C. and Eisenberg, D. 1982. The structure of melittin. II. Interpretation of the structure. J. Biol. Chem. 257: 6016–6022.

Thompson, S.A., Tachibana, K., Nakanishi, K. and Kubota, I. 1986. Melittin-like peptides from the shark-repelling defense secretion of the sole *Pardachirus pavoninus*. Science 233: 341–343.

Yuli, I. and Snyderman, R. 1986. Extensive hydrolysis of N- formyl-L-methionyl-L-leucyl-L-[^{3}H] phenylalanine by human polymorphonuclear leukocytes. A potential mechanism for modulation of the chemoattractant signal. J. Biol. Chem. 261: 4902–4908.

Discussion (DR. GITLER)

Ellis: Could you comment on the temperature control needed for these systems? Do you have to worry that at higher temperatures you will create channels and have signal-to-noise problems?

Gitler: Yes, it is obvious from the behavior of the lipid bilayer that it is very sensitive to temperature. We would have to incorporate some system that keeps it at constant temperature. Alternatively, if we knew the uses we could design the lipids in such a way that they would be fluid at the temperature that we would require. There are bacteria that grow at 80° C; they must make a bilayer which can function at that temperature. That is done by making phospholipids which contain branch chains or different degrees of unsaturation. Given the temperature, we could design the lipid bilayer which would be functional. Given large changes, we would have to incorporate electronic systems that would keep the temperature constant. This is not such a difficult problem at the level of energy exchange needed. We only have 60 Å to keep warm. So we are not really keeping very large quantities of bulk fluid warm; we are keeping very minute surfaces at constant temperature.

Ward: I am interested in the mechanism whereby you make your bilayers. Is it necessary that the micelles be unilamellar? It seems to me that if you have multilamellar micelles you are not going to get a simple bilayer formed eventually in the way that you showed. Is that critical in the design of your apparatus? Can you say something about the length of the alkyl chains that accounts for the gap between the gold and the bottom end of the bilayer.

Gitler: When I normally reconstitute the acetylcholine receptor, I extract it from the membrane by means of detergents, usually a small dialyzable detergent. When I add the detergent to the membrane, it intercalates into the bilayer and disperses it together with the receptor. When I begin to dialyze out the detergent, unilamellar liposomes are formed. When you dialyze our detergent you usually get formation of monolamellar liposome structures; you do not get multibilayer structures. We would prefer to have the simplest type of design which would be one bilayer attached through a polar group directly to the electrode. Prior to dialysis of the detergent, we are allowing the micelles first to attach. Then we can take away the excess lipid by exchanging the surface, by removing it at the same time as you add it. Then we begin to dialyze out the detergent, so there is no more lipid, or very little, left to attach to other places after the first attachment has occurred. There is thus no possibility of making several layers, because you have already begun with one. We have shown that the liposomes having the long polar omega thiol arm attach quite nicely to a gold surface.

Ward: Is there experimental evidence that it is the single bilayer?

Gitler: We are getting there. For the time being, we have done it with liposomes and now we are doing it with dialysis to show that it is actually one bilayer. We can use oxyethylene chains of almost any length that we want, and we are developing very nice chemistry of activating the two extremes. The one we are using now is eleven oxyethylene groups, about 33–35 Å.

Lotan: What is the mechanism which allows you to dialyze the detergent only and not the phospholipids? Their structure is similar. Second, you mentioned that you can change thermal stability of the bilayer by using different lipids with different hydrophobic groups? On the other hand, melittin when fully helical is about 30 Å long, just the right size to cross the bilayer. However, if you use a different hydrophobic group, the bilayer will no longer be 30 Å deep, perhaps more. Accordingly, melittin will not span it any more. Finally, do you have any information you can share with us on the thermal stability of helical melittin?

Gitler: First, concerning dialysis, we deal with this every day in the laboratory. The detergent is in equilibrium between the part that is incorporated in the lipid bilayer and the micelle and free monomers. The detergent micelle in turn is in equilibrium with the monomer. The phospholipid, because it has two chains of 16–18 carbons, 32–36 carbons total, is completely insoluble in water for all practical purposes. What you are dialyzing out is first the free monomeric detergent, and you never dialyze phospholipid because it can never reach monomer concentrations. That has been done so many times that I, even if my description is wrong, the experiment works. You get no phospholipid coming out. In answer to your other question, we do not change the apolar quality of the lipid by changing the length only. We can easily change it by the number of double bonds, and so we do not change the net thickness of the bilayer. We can change it by introducing side chain groups, like methyl groups, which makes it very rigid. For example, in bacterial purple membranes, there are phytic acid groups which have side chain methyl groups. So that is no problem; we can control that. We can also add more cholesterol, less cholesterol, and change the rigidity of the bilayer. With regards to the thermal stability, there are many systems which work at 80° C so we should be able to get systems which have thermal stability.

Lotan: Melittin seems to be particularly suited for such studies. When fully helical this molecule will assume the shape of a rod about 30 Å long. This is exactly what is needed in order for it to cross the lipid bilayer. In view of this requirement, do you have any experimental evidence regarding the actual structure, particularly the helicity, of melittin when it is embedded in the bilayer? Also, is there any indication as to the thermal stability of helical melittin?

Gitler: Once you introduce a protein or polypeptide into an apolar environment, the hydrogen bonding is so stable that there is almost no alternative for it to be changed. Doty showed a long time ago that helical segments in solvents have a tremendous stability.

Lotan: Is melittin helical when it is incorporated in the bilayer?

Gitler: Prior to incorporation, it is not all helical. The only thing we do is we put melittin monomer into solution, and it spontaneously incorporates into the lipid bilayer. The surprising thing is that then it does not come out almost at all.

Lotan: But are you sure that it is going into the bilayer, and at the same time it is undergoing a conformational transition to the helical form?

Gitler: Melittin is soluble in water and it transfers into the bilayer spontaneously. There is an extensive literature dealing with initial polar attachment followed by insertion.

Kornguth: Perhaps it should be pointed out that polycations have been shown many times to adsorb on surfaces, not to necessarily traverse or enter into the bilayer. You pointed out clearly how a polycationic pentapeptide is most likely to adsorb on, not enter in, to the unit.

Gitler: That is part of the process of melittin penetration. However, we removed the pentabasic group present in melittin when we designed CH1 and it forms beautiful ion channels. CH1 does not have then the basic groups of melittin, yet it forms very nice ion channels. We did not want attachment of the type you describe; that is why we took the basic groups away.

Key Words: Biosensors, Hydrophilic linker arms, Ion channels, Lipid bilayer, Melittin

Alternatives to Radioimmunoassay

G. Barnard, J. De Boever[1] and Fortune Kohen[2*]
Department of Chemistry, City University, London, U.K.;
[1]*Akademisch Ziekenhuis Vrouwenkliniek/Poli III, De Pintelaan 185, B-9000, Belgium;*
[2]*Department of Hormone Research, The Weizmann Institute of Science, Rehovot, Israel*
[*]*Corresponding author*

Introduction

The most significant developments in recent years in immunoassay technology have been: (1) the use of monoclonal antibodies; (2) the introduction of simpler techniques to separate the antibody-bound and free fractions; (3) the tagging of an antibody instead of the antigen in a move from limited to excess reagent systems; and (4) the emergence of several methods utilizing non-radioactive labels as an alternative to radioactive labeling. To date, at least ten labels have been advocated (bacteria, erythrocytes, latex particles, gold sols, enzymes, substrates and cofactors; dyes, fluorescent, phosphorescent and chemiluminescent compounds).

At present, four types of label have emerged as major competitors with radiolabels. These include particles, enzymes, fluorophores and chemiluminophores. The sensitivity range of these probes is shown in Table 1. The choice of a non-isotopic label depends largely upon the specific activity which can be achieved. ^{125}I is actually rather inefficient as a label since only a minute fraction of the total signal is available for detection at any time (1 disintegration/second for every 7.5×10^6 atoms present). In the case of enzyme labels, the specific activity will depend on the amplification factor of the signal generating system. In a chemiluminescent reaction we would expect a maximum yield of up to 1 detectable signal/molecule, while in a fluorometric system by repeated excitation several detectable signals/labeled molecule can be derived. While such considerations are sound in theory, it is the practical constants which determine the relative success of each system. The main consideration in practice is to amplify the signal as much as possible while reducing background to a minimum since it is actually the signal/noise ratio which determines the detectability of the labeled molecule.

In our laboratory, we investigated luminescence procedures as viable alternatives to radioimmunoassay (RIA) and immunoradiometric assay

D. Kamely et al. (eds), Biotechnology: Bridging Research and Applications, 63–78.

Table 1. The sensitivity range of various probes

Probes	Detection limit (mole/L)
Enzyme	
colorimetric end-point	10^{-12}–10^{-13}
luminometric end-point	10^{-14}–10^{-16}
Fluorophore	
FITC	10^{-12}
rare-earth chelates	10^{-14}–10^{-16}
Radioactive	
^{3}H	10^{-12}
^{125}I	10^{-14}
Chemiluminophore	10^{-14}–10^{-16}
Sol Particles	
colorimetric end-point	10^{-12}

Modified from: L. Dibbelt and E. Kuss, Endocrinologie Informationen, **11**, 18–20 (1987), by permission.

(IRMA). Two different approaches were explored for labeling antigens or antibodies. In the direct approach antigens or antibodies have been coupled covalently to luminescent compounds or rare-earth chelates, and the resulting conjugates served as labels in immunological reactions (Kohen, et al., 1988). In the indirect approach we have explored the affinity of avidin (KD 10^{-15}M) for biotin to amplify the sensitivity of the assay of haptens and polypeptide hormones. In this system biotin covalently coupled to antibodies serves as the indirect probe, and labeled streptavidin as the secondary probe (Kohen, et al., 1988; Strasburger, et al., 1988). After the immunological reaction the end-point is measured by chemiluminescence or fluorescence depending on the marker on the streptavidin. Table 2 shows the streptavidin conjugated probes and the corresponding end-points.

These conjugated probes were used in the development of competitive type immunoassays for haptens, drugs and polypeptide hormones and in two-site assays for polypeptide hormones. Several formats were investigated for the development of competitive type assays for haptens. These include non-separation assays based on antibody-enhanced hydrolysis (Kohen, et al., 1980), and antibody enhanced chemiluminescence (Kohen, et al., 1979) and conventional separation assays based on first or second antibody on solid phase or on magnetic particles (Kohen, et al., 1987), and on solid-phase antigen mediated by the biotin-avidin reaction (Strasburger, et al., 1988). In our most recent work, we studied two novel approaches in the development of robust, sensitive assays for haptens, (1) non-separation assays based on quenching of the fluorescence of the labeled antigen when the latter is bound to the antibody and (2) separation assays using the idiotypic-anti-idiotypic format monitored by time-resolved fluorescence.

Table 2. Streptavidin conjugated probes for use in immunoassay procedures

Probe	End-point
Enzymes	
Penicillinase	Disappearance of color
β-Galactosidase	Colorimetry, fluorometry
Alkaline phosphatase	Colorimetry
Horseradish peroxidase	Colorimetry, chemiluminescence
Glucose-6-phosphate dehydrogenase	Bioluminescence
Chemiluminescent labels	
Isoluminol	Chemiluminescence
Acridinium esters	Chemiluminescence
Flourescent labels	
FITC	Fluorometry
Texas Red	Fluorometry
Phycobiliproteins	Fluorometry
Metals	
Europium	Time-resolved fluorescence
Colloidal Gold	Colorimetry
Radioactive Markers	
^{125}I	Counting of radioactivity

Currently, we are investigating in detail two-site time-resolved immunofluorometric assays for peptide hormones mediated by the streptavidin-biotin interaction. Although these two-site assays are potentially sensitive and precise, several two-site assays have been shown to be negatively biased when results were compared with the corresponding values obtained from conventional competitive radioimmunoassays (RIA) (Reiter, et al., 1988). In an attempt to select the optimum pair of monoclonal antibody (Mab) combination for a given assay that will give numerical values comparable to those obtained by RIA, we evaluated in detail pairs of Mabs combinations for capture Mab and labeled Mab.

We report here the recent results obtained using some of the approaches just described.

Materials and Methods

Carboxy derivatives of steroids and drugs, urinary steroid metabolites and prostaglandins were linked covalently to amino alkyl derivatives of isoluminol (6–amino-2,3–dihydrophthalazine-1,4–dione) to yield the corresponding hapten-aminohexyl ethyl isoluminol, and hapten-aminobutyl ethyl isoluminol (ABEI) (see Figure 1 for representative structures of labeled steroids), essentially as previously described (Kohen, et al., 1986). Purified IgG fractions of monoclonal antibodies were reacted with an N-hydroxysuccinimide ester of biotin (Strasburger and Kohen, 1990) or with an EDTA-chelated europium,

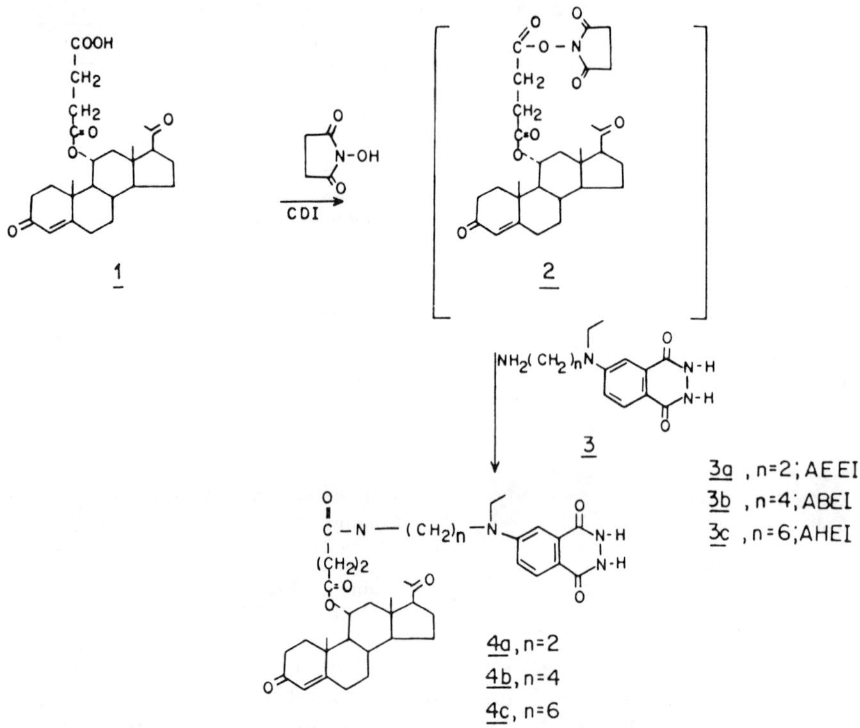

Figure 1. Representative synthesis and structures of steroid-chemiluminescent marker conjugates. 1 = progesterone-11α-hemisuccinate; 3a: aminoethyl ethyl isoluminol (AEEI); 3b: aminobutyl ethyl isoluminol (ABEI); 3c: aminohexyl ethyl lisoluminol (AHEI).

as previously described (Hemmila, et al., 1984). The resulting conjugates were purified by gel filtration on Sephadex G-25, and stored at -20°C until use. Streptavidin labeled with the EDTA europium chelate (13 moles of europium per mole of streptavidin), the fluorescent europium chelate covalently linked to estrone-3–glucuronide and enhancement solution were donated by T. Lovgren and I. Hemilla (Wallac Oy, Turku, Finland). All other reagents and assay procedures were as previously described (Strasburger, et al., 1989a,b), unless specified otherwise.

Light emission was measured with a Luminometer Model 2080 (Lumac Systems, Basel) or Arcus time resolved fluorimeter (Wallac, Finland).

Results

Chemiluminescence-based Immunoassays for Haptens

In these assays, haptens, specific antibodies or streptavidin are linked covalently to derivatives of isoluminol (Strasburger and Kohen, 1990). These compounds, upon oxidation, release energy in the form of light at approximately 430 nm and the light yield is measured with a simple Luminometer. Using haptens conjugated to isoluminol derivatives, we have developed two types of formats for chemiluminescence monitored immunoassays for steroids and drugs: (a) assays that do not require physical separation of bound and free hormone, so called "homogeneous assays", as exemplified by antibody-enhanced chemiluminescence immunoassay of progesterone, estradiol, cortisol and total estrogens (Kohen, et al., 1985); (b) assays that require a separation step, so called "heterogeneous assays". The homogeneous immunoassays, based on antibody-enhanced chemiluminescence, proved satisfactory with respect to sensitivity and precision. However, these assays were affected by nonspecific interference from luminescent compounds present in extracts of plasma or diluted urine samples. To overcome these problems, assays requiring separation of bound and free ligand were developed.

Separation of bound and free antibody fractions was achieved either in the liquid or solid phase (e.g. microtiter plates, polystyrene balls, tubes, magnetic particles, etc.) (Kohen, et al., 1987). These approaches were adopted for the measurement of plasma, urinary and salivary steroids, and drugs. Figure 2 shows salivary progesterone levels during pregnancy, using second-antibody coated microtiter plates, progesterone-11α-hemisuccinate-aminobutylethyl isoluminol as the label and monoclonal antibodies to progesterone as the specific antibody (De Boever, et al., 1989).

More recently, we have used the biotin-avidin interaction as an amplification system for the measurement of haptens from biological fluids (Strasburger, et al., 1988). In this format a limited amount of antigen is immobilized on a solid phase, and the specific homologous antibody labeled with biotin serves as the primary probe. After the immunological reaction, labeled streptavidin is added, and the end-point is determined by fluorescence or luminometry, depending on the marker attached to streptavidin. Using chemiluminescent labeled streptavidin we have developed assays for cortisol, progesterone and estradiol (Strasburger, et al., 1988). The sensitivity of these assays was similar to those achieved by conventional RIA procedures. The advantages and disadvantages of this system are summarized in Table 3.

Time-resolved Fluoroimmunoassay: Principles

Recently, methods involving the use of time-resolved fluorometry have been introduced for the measurement of haptens and protein hormones. While the fluorescence decay time of molecules such as fluorescein is similar to that of the background (~10 nsec), certain rare earth chelates [e.g., chelates of europium (Eu) and terbium (Tb)] possess extended decay times (~ μsec). Consequently, the emission from europium chelates can be distinguished from the background fluorescence (which has a short decay time) using a time-resolved fluorimeter with appropriate delay, counting and cycling times

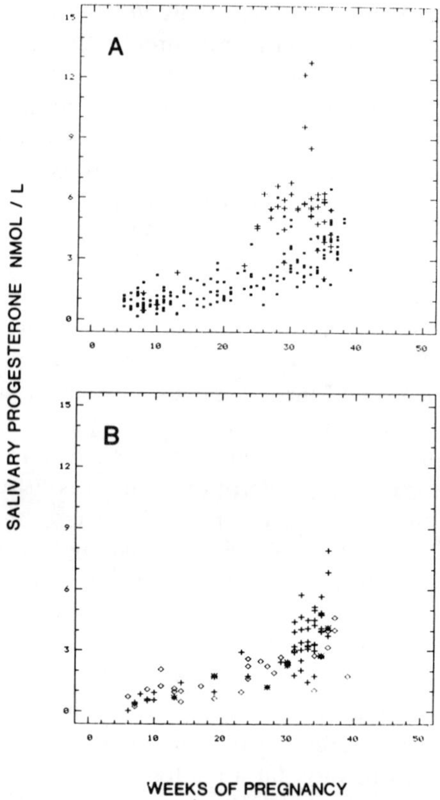

Figure 2. Salivary progesterone levels during pregnancy using chemiluminescence as an end-point. (A) ■ Normal; +, twin pregnancy. (B) ◇ Intra-uterine growth retardation; +, pre-term delivery; *, toxemia.

Table 3. Advantages and disadvantages of chemiluminescence immunoassays

Advantages of Isoluminol Tagged Labels:

a. Shelf life of labels at least two years as compared to ^{125}I-tracers.
b. An increase in sensitivity in immunometric assays as compared to immunoradiometric assays.
c. The cost per tube is relatively low.
d. Hapten or peptide assays can be easily set up in a research laboratory.

Disadvantages of Isoluminol Tagged Labels:

a. The measurement of chemiluminescence is a more complicated step than the measurement of radioactivity.
b. The measurement is a one-off procedure and cannot be repeated.

Table 4. Advantages of using rare-earth chelates as labels in immunoassay procedures

a. Small, stable, non-radioactive label
b. High-specific activity
c. Stable signal
d. High signal/noise ratio
e. Rapid and reproducible measurements
f. Easy disposal of reagents
g. Non-hazardous and non-toxic

(Barnard, 1988). The advantages of using rare-earth chelates in immunoassay procedures is shown in Table 4.

Competitive Fluoroimmunoassays for Haptens

In these assays, haptens, specific antibodies or streptavidin are linked covalently to EDTA-chelated europium. The fluorescence of europium is measured in an Arcus time-resolved fluorimeter. Using haptens conjugated to a novel europium chelate or to polylysine-EDTA Eu we have developed non-separation assays for steroid glucuronides (Barnard, et al., 1989) and a conventional solid-phase separation assay for steroids and drugs.

Recently, the specific anti-idiotypic antibodies were labeled directly with EDTA-chelated europium, and an idiotypic-anti-idiotypic approach was developed. This assay enabled the direct measurement of plasma estradiol.

Non-separation Fluoroimmunoassays for Haptens: Principles

The principle of this assay is based upon the fact that the fluorescence of the hapten-Eu conjugate is quenched when the labeled ligand is bound to specific antibodies. Consequently, after the antibody-antigen binding reaction, the fluorescence measured mainly emanates from the antibody-free label (see Figure 3). An increase in the concentration of competing analyte results in an increase in the antibody-free labeled derivative and a concomitant increase in the fluorescence. Therefore, the need to separate antibody-bound and free fraction is obviated.

A Non-separation Time-resolved Fluoroimmunoassay for Estrone-3-glucuronide

In the procedure developed in our laboratory, estrone-3–glucuronide coupled to europium chelate (Barnard, et al., 1989) serves as the labeled antigen and monoclonal anti-estrone-3–glucuronide (clone $3F_{11}$) as the antibody (Kohen and Lichter, 1986). A representative dose-response curve is shown in Figure

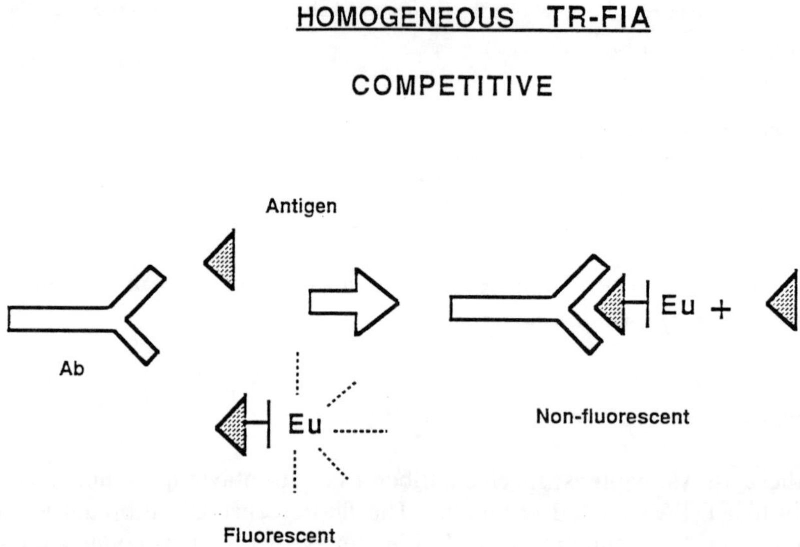

Figure 3. Principles of competitive non-separation assays for haptens monitored by time-resolved fluorescence. The hapten-europium conjugate is fluorescent in aqueous solution. However, when the fluorescent labeled ligand is bound to the specific antibody, quenching of fluorescence is observed. TR-FIA: time-resolved fluorescence immunoassay.

4. The sensitivity, specificity and accuracy of this assay are similar to those for a conventional separation fluoroimmunoassay (FIA) or RIA, but the speed, convenience, precision, reliability and clinical utility of the non-separation assay are more advantageous. The whole assay can be performed in

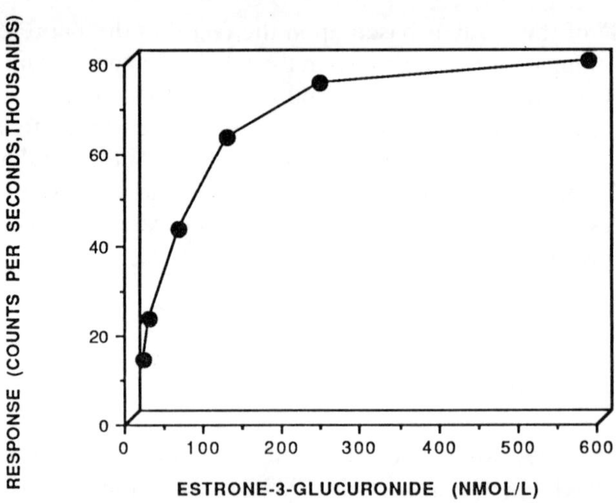

Figure 4. A calibration curve for a non-separation time-resolved fluoroimmunoassay for estrone-3-glucuronide.

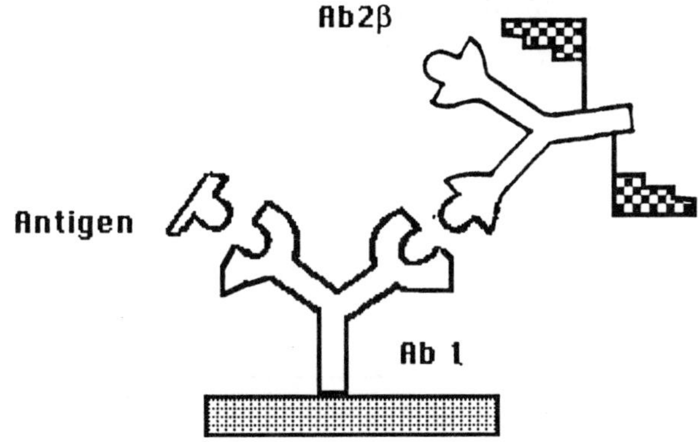

Figure 5. Principles of a separation assay for haptens based on the idiotypic-anti-idiotypic approach. The anti-idiotype, $Ab_{2\beta}$, bears on its Fab region a structural counterpart of the hormone binding site which is shared by the idiotype Ab_1 and by the hormone. Thus $Ab_{2\beta}$ can compete with the antigen (hormone) for the binding sites of Ab_1.

15 minutes using 10 μl of undiluted urine and is useful for monitoring ovarian function in women.

Separation Time-resolved Fluorometric Assays for Haptens Based on the Idiotypic-anti-idiotypic Approach: Principles

In this assay, the primary idiotypic antibody (Ab_1, Figure 5) is captured, and the anti-idiotypic antibody ($Ab_{2\beta}$) generated against the paratypic region of Ab_1 is labeled with europium. Since $Ab_{2\beta}$ has the capacity to mimic the original antigen used to generate Ab_1, it can compete with the original antigen for the binding sites of Ab_1 (Bona, et al., 1986). The principles of this assay are shown in Figure 5.

Separation Assay for Estradiol Based on the Anti-idiotypic-idiotypic Approach

In this system, the primary antibody (Ab_1) to estradiol [(clone # $2F_9$ (Kohen and Lichter, 1986)] serves as the solid phase, and europium labeled anti-idiotypic antibody (clone # $1D_5$, $Ab_{2\beta}$) directed against the paratypic region of the primary anti-estradiol antibody is the marker. Representative dose-response curves are shown in Figure 6. The assay is suitable for the direct measurement of estradiol from plasma. The sensitivity of the assay is 2 pg/well. The values of estradiol as determined by this method were compared to a conventional liquid-phase radioimmunoassay using tritiated estradiol and an extraction procedure. The FIA gave numerical values that agreed well with those obtained by RIA ($r = 0.96$, $n = 70$). In this approach, the

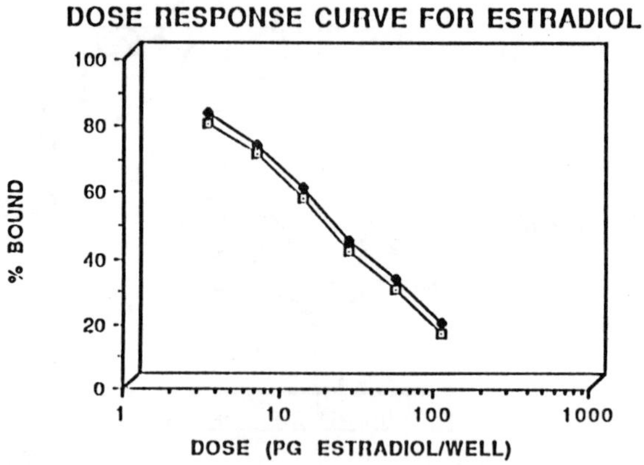

Figure 6. Calibration curves for estradiol based on the idiotypic-anti-idiotypic approach; □———□ in buffer; -■———■ in the presence of 25 μl of serum.

anti idiotypic antibody serves as the marker. Since tagging of antibodies is easier than labeling of haptens, the assay is easy to set up in small laboratories.

Time-resolved Immunofluorometric Assays for Polypeptide Hormones

Currently, we are using monoclonal antibodies to polypeptide hormones (e.g. FSH, hGH, LH, etc.) in the development and evaluation of two-site time-resolved immunofluorometric assays (IFMAs) based on the streptavidin-biotin interaction. The assays involve the use of one Mab immobilized to the wells of polystyrene microtiter plates and a second biotinylated Mab directed against a separate epitope of the polypeptide hormone. After the binding reaction and the removal of the reaction mixture, streptavidin labeled with EDTA chelated europium is added. The end-point is determined by time-resolved fluorometry (Strasburger, et al., 1989a). We have investigated the effect of various combinations of monoclonal antibodies to polypeptide hormones on the performance of alternative immunofluorometric assays (IFMAs) for the measurement of polypeptide hormones in serum. As an example, we describe here the evaluation of various IFMAs for the measurement of hGH in serum.

Human Growth Hormone as Measured by a Two-site Immunofluorometric assay

To date, many of the current criteria for diagnosis of human growth hormone (GH, somatotropin) deficiency have been based upon measurement of this hormone by competitive RIA with use of polyclonal antibodies. In recent

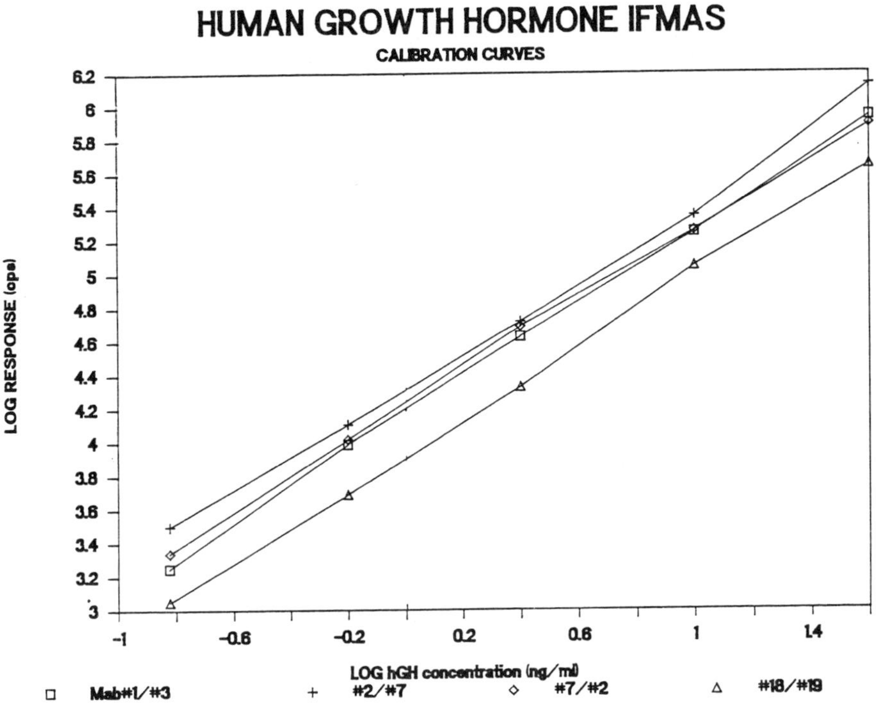

Figure 7. Calibration curves for GH using different pairs of monoclonal antibody combinations.

years, however, the development of hybridoma technology has led to the generation of various Mabs to GH with different affinities and epitope specificities. Subsequently, these reagents have been used in the development of non-competitive two-site immunometric assays (e.g., immunoradiometric assay, IRMA). In general, the values obtained for serum GH by IRMA have been lower than those obtained by RIA because of the epitope specificity profile of the Mabs in the IRMA. Attempting to obtain GH values numerically similar to those by RIA, we used combinations of six Mabs to GH in developing and evaluating two-site IFMAs based on the streptavidin-biotin interaction.

For this purpose six well-characterized Mabs to GH (clones #1, #2, #3, #7, #18 and #19) (Strasburger, et al., 1989b) with different specificity profiles were selected, and evaluated in the alternative IFMAs. Calibration curves obtained for the measurement of GH by the alternative IFMAs are shown in Figure 7.

The specificities of the alternative IFMAs were investigated by measuring two concentrations (1 ng/ml and 10 ng/ml) of structurally related compounds [22K GH, 20K GH, MetLeu^{8} GH, Met^{14}GH and human placental lactogen (hPL)] using the different Mab combinations. The results are shown in Table 5.

Table 5. Estimates of specificity for immunofluorometric assays (IFMAs) for GH

METHOD	IFMA 1	IFMA 2	IFMA 3	IFMA 4	RIA
CAPTURE Mab:	#1	#2	#7	#18	-
LABELED Mab:	#3	#7	#2	#19	-
		hGH (ng/ml)			
PATIENT #1	2.9	2.7	3.1	3.3	2.9
PATIENT #2	3.0	3.0	3.5	3.5	3.3
PATIENT #3	53.4	50.0	95.1	33.7	81.1
PATIENT #4	10.0	6.4	8.3	7.1	8.0
PATIENT #5	3.9	3.3	6.0	4.2	5.6

The different Mab combinations were used to measure GH in serum obtained from GH-deficient and acromegalic patients. The results obtained with the acromegalic patients are shown in Table 6.

Of the four combinations investigated, one pair (capture Mab: Clone #7; labeled Mab; clone #2) gave numerical results that were very similar to those obtained by the reference RIA (see Table 6). It may be speculated that the two epitopes recognized by these antibodies are available for binding in the alternative forms of the circulating GH (e.g., 20K variant, proteolytic fragments, etc.). The other combinations gave results which were negatively biased, although all the IFMAs had good sensitivity and precision.

Our findings are similar to those reported for the measurement of other analytes by two-site immunometric assays. This has resulted with a re-assessment of the value of the "sandwich" assay and in some cases has lead to a return to the older reference RIAs. This is unfortunate since the advantages of the non-competitive assays are significant. In many cases, careful selection of the appropriate Mabs can obviate most of these difficulties and can facilitate the development of more sensitive, precise and yet still clinically useful diagnostic procedures.

Conclusions

It is anticipated that in many countries the use of radioactivity in immunoassays will be resisted. There is now increasing pressure to find alternative labels to replace the use of radioactivity in immunoassay procedures. A wide variety of labels and instrumentation have become available commercially, and the choice of the label depends on the desired sensitivity, the instrumentation available and the location of the test (out-patient clinic, home, or hospital).

It is evident that some of these labels (e.g., enzymes, chemiluminescent

Table 6. Measurement of hGH in serum obtained from acromegalic patients by IFMAs and RIA

METHOD	IFMA 1	IFMA 2	IFMA 3	IFMA 4
CAPTURE Mab:	#1	#2	#7	#18
LABELED Mab:	#3	#7	#2	#19
		% CROSS-REACTION		
22K GH	100	100	100	100
20K GH	>100	>100	>100	1.1
LeuMet^{8}GH	0.6	3.2	0.2	75
Met^{14}GH	0.4	0.5	-	62
hPL	31.1	27.1	1.2	3.7

markers) require further processing (e.g., chemical reactions) for their detection. This extra step may introduce pipetting errors. In the case of a chemiluminescent label, using isoluminol as a marker, the generation of the light signal cannot be repeated. On the other hand, chemiluminescent and fluorescent labels are capable of yielding high specific activities and high speed of detection as compared to ^{125}I-tracers, thus enabling the endocrinologist to measure very low levels of circulating hormones.

In this review, we have described various types of competitive assays for haptens and non-competitive two-site assays for polypeptide hormones based on monitoring chemiluminescence or fluorescence. Two novel approaches for the measurement of haptens have been investigated, namely non-separation assays based on quenching of fluorescence and separation assays based on the idiotypic-anti-idiotypic approach. The non-separation assay requires the availability of a hapten coupled to a europium chelate that is fluorescent in aqueous solution. When this fluorescent labeled hapten is bound to the homologous antibody, the fluorescence of the conjugate is quenched. Since a separation step is not required, the assay is simple, fast, precise and has the potential for total automation.

The separation assay based on the idiotypic-anti-idiotypic approach requires the idiotypic antibody (Ab_1) and the labeled anti-idiotypic antibody ($Ab_{2\beta}$) directed against the paratypic region of the primary antibody (Ab_1). This method has great potential since the antibody instead of the hapten is labeled. It is generally easier to label an antibody with a non-isotopic marker (e.g., enzyme, rare-earth chelate, etc.) than the hapten. The conjugation step can thus be performed in the investigator's laboratory. In addition, the use of two antibodies in the system provides a sensitive, precise and robust assay.

In addition, we have described here two-site assays, based on streptavidin-biotin interaction. We have attempted to show the drawbacks of employing monoclonal antibodies in two-site assays, using measurements of GH in serum as examples. The method of the two-site assay which has been described here can be applied to the measurement of other polypeptide hormones (e.g., LH, hCG, TSH, prolactin) since use of some of the reagents (e.g., labeled streptavidin) is universally applicable. The method requires only a high-affinity capture antibody, and a biotin-labeled antibody directed against a different epitope of the antigen. Thus, the use of the avidin-biotin amplification system and of labeled streptavidin as a marker for peptide hormones unifies immunoassay methodology.

Acknowledgments

We are grateful to Drs. C. Strasburger, Z. Zadik, P. Pinzani, Mr. L. Toldo and Ms. A. Altamirano-Bustamante for permission to quote collaborative work; to Mrs. M. Kopelowitz for excellent secretarial assistance; to Mr. B. Resch of Corning, Germany, for the loan of a Berthold Clini-Lumat; to Dr. T. Lovgren of Wallac, Finland, for the loan of an "Arcus"; to Dr. R.D. Mackintosh and Mr. J. Vossen of Lumac bv, Landgraaf, The Netherlands, for the loan of a Biocounter M2000 luminometer; to the Special Programme of Research, Development and Research Training in Human Reproduction, the World Health Organization (to F.K.), the Belgian National Fund for Scientific Research (NFWO grants to JDB), and to the Institute for International Studies in Natural Family Planning, Department of Obstetrics and Gynecology, Georgetown University Medical School, Washington, D.C. (to G.B. and F.K., grant No. BR-IS-002) for financial support.

References

Barnard, G. 1988. The development of fluorescence immunoassays. In: Non-radiometric Assays: Technology and Application in Polypeptide and Steroid Hormone Detection. Alan R. Liss, Inc., New York. pp. 15–37.

Barnard, G., Lichter, S., Kohen, F., Mikola, H. and Lovgren, T. 1989. The measurement of estrone-3–glucuronide in urine by a rapid non-separation time-resolved fluoroimmunoassay. Clin. Chem. 35: 555–559.

Bona, C.A., Kang, C-Y., Kohler, H. and Monestier, M. 1986. Epibody: The image of the network created by a single antibody. Immunol. Rev. 90: 116–171.

De Boever, J., Vanderkerckhove, D. and Kohen, F. 1989. Development of a chemiluminescence immunoassay for salivary progesterone using microtitre plates as a solid phase. Analytica Chimica Acta. 227: 119–127.

Hemmila, I., Dakubu, S., Mukkala, V-M., Siitari, H. and Lovgren, T. 1984. Europium as a label in time-resolved immunofluorometric assay. Anal. Biochem. 137: 335–343.

Kohen, F., Pazzagli, M., Kim, J.B., Lindner, H.R. and Boguslaski, R.C. 1979. An assay procedure for plasma a progesterone based on antibody enhanced chemiluminescence. FEBS Lett. 104: 201–205.

Kohen, F., Kim, J.B., Barnard, G. and Lindner, H.R. 1980. Antibody-enhanced hydrolysis of steroid esters. Biochim. Biophys. Acta. 629: 328–337.

Kohen, F., Bayer, E.A., Wilchek, M., Barnard, G., Kim, J.B., Collins, W.P., Beheshti, I., Richardson, A. and McCapra, F. 1984. Development of luminescence-based immunoassays for haptens and for peptide hormones. In: Analytical Applications of Bioluminescence and Chemiluminescence. Kricka, L. and Whitehead, T.P., eds. Academic Press, Inc. pp. 149–158.

Kohen, F., Pazzagli, M., Serio, M., de Boever, J. and Vanderckhove, D. 1985. Chemiluminescence and bioluminescence immunoassays. In: Alternative Immunoassays. Collins, W.P., ed. John Wiley and Sons. pp. 103–121.

Kohen, F. and Lichter, S. 1986. Monoclonal antibodies to steroid hormones. In: Monoclonal Antibodies: Basic Principles, Experimental and Clinical Applications in Endocrinology. Forti, G., Lipsett, M.B. and Serio, M., eds. Raven Press. pp. 87–95.

Kohen, F., De Boever, J. and Kim, J.B. 1986. Surface chemiluminescent immunoassays (steroids). In: Methods in Enzymology on Bioluminescence and Chemiluminescence. De Luca, M.A. and Mc Elroy, W.D., eds. Academic Press, Inc. pp. 133: 387–403.

Kohen, F., de Boever, J. and Kim, J.B. 1987. Recent advances in chemiluminescence-based immunoassays for steroid hormones. J. Steroid Biochem. 27: 71–82.

Kohen, F., Amir-Zaltsman, Y., Strasburger, C.J., Bayer, E.A. and Wilchek, M. 1988. The avidin-biotin reaction in immunoassay. In: Complementary Immunoassays. Collins, W.P., ed. John Wiley & Sons, Ltd. pp. 57–70.

Reiter, E.O., Morris, A.H., MacGillivray, M.H. and Weber, D. 1988. Variable estimates of serum growth hormone concentrations by different radioassay systems. J. Clin. Endocrin. Metab. 66: 68–71.

Strasburger, C.J. and Kohen, F. 1990. Biotinylated probes in immunoassay. In: Methods in Enzymology: Avidin-biotin Technology. Wilchek, M. and Bayer, E., eds. Academic Press Inc. 184: 481–496.

Strasburger, C.J., Amir-Zaltsman, Y. and Kohen, F. 1988. The avidin-biotin reaction as a universal amplification system in immunoassays. In: Non-radiometric Assays: Technology and Application in Polypeptide and Steroid Hormone Detection. Albertson, B. and Hazeltine, F., eds. Alan R. Liss. pp. 79–100.

Strasburger, C.J., Kostyo, J., Vogel, T., Barnard, G.J. and Kohen, F. 1989a. The antigenic epitopes of human growth hormone as mapped by monoclonal antibodies. Endocrinology. 124: 1548–1557.

Strasburger, C., Barnard, G., Toldo, L., Zarmi, B., Zadik, Z., Kowarski, A. and Kohen, F. 1989b. Somatotropin as measured by a two-site time-resolved immunofluorometric assay. Clin. Chem. 35: 913–917.

Discussion (DR. KOHEN)

Gillespie: The sensitivity of a detection process depends critically on the component interaction with the lowest binding constant. Because antibodies bind with a $Kd = 10^{-7}$–10^{-9}, the sensitivity of immunoassays is unlikely to be improved by employing more readily detected probes. Assuming an antibody with a binding constant of 10^{-12} for its antigen, a very avid antibody, and employing a noise level of 0.01%–0.1% of the input reporter, immunoassays cannot detect under 10^5 target molecules. Sensitivity can best be improved by engineering superavid antibodies and by reducing assay noise.

Ellis: In the Eu chelate luminescence method, there is also the opportunity to see whether more than one species is present: a single exponential decay would be expected for a single species, whereas more complex (multiexponential) decays would be expected if several Eu complexes contribute. In the Eu and Tb mixed systems, determination of concentrations requires determining whether the two complexes interact with one another by energy transfer processes.

Key Words: Immunoassay, Fluoroimmunoassay, Isoluminol, Europium, Streptavidin-biotin, Immunofluorometric assays

The Use of Intact Luminous Bacteria in Analytical Chemistry, Clinical Microbiology and Acute Toxicity

Shimon Ulitzur
Department of Food Engineering and Biotechnology
Technion-Israel Institute of Technology, Haifa 32000, Israel
and
Biolume Ltd., Centre for Industrial Research, Technion City, Haifa 32000, Israel

Introduction

Luminous bacteria are widely distributed in the sea, either as free-living and saprophytic form or as symbionts to marine animals. The luminescence (lux) system of different marine bacteria has been cloned by us into different Gram negative and Gram positive bacteria, as well as into strains of *Streptomyces*, *Mycobacteria* and into yeasts (unpublished results).

The bacterial luminescence system can be viewed as a branch of the electron transport system, in which electrons from reduced substrates are shunted via reduced flavin to oxygen. The enzyme luciferase, a mixed function oxidase, simultaneously oxidizes the reduced form of FMN and a long chain aldehyde with the aid of molecular oxygen. The products of the reaction are: Light; FMN; H_2O and a long chain fatty acid. Although different long chain aldehydes (C_8-C_{16}) are active in vitro, only myristic (C_{14}) aldehyde is recycled in the cells, and this aldehyde is assumed to be the intrinsic aldehyde in luminous bacteria (Ulitzur and Hastings, 1978b, 1979a). The biochemistry and physiology of luminous bacteria have recently been reviewed by Meighen (1988).

The luciferase and other enzymes which take part in the luminescence reaction are inducible enzymes. In most of the luminous bacteria these enzymes are not formed together with other cellular proteins but are synthesized de novo only late in the exponential phase of growth, upon the accumulation in the growth medium of a species-specific metabolite (autoinducer).

Recently, Ulitzur and Kuhn (1988) and Ulitzur (1989) have shown that other cellular factors are involved in the regulation of the luminescence system including Sigma 32 protein and *lexA* protein.

The analytical applications of the bacterial luminescence system have been reviewed by several authors (Gorus and Schram, 1979; Thore, 1979; Whitehead, et al., 1979; Stewart, 1990). Almost all the analytical applications

D. Kamely et al. (eds), Biotechnology: Bridging Research and Applications, 79–87.

described so far are based on the luminescence system of the cell-free extracts. These enzymatic assays are basically performed as a two-step procedure. At the first stage NADH or NADPH, which can be formed in a separate reaction, react continuously with NAD(P)H-FMN oxidoreductase. At the second stage, the $FMNH_2$ formed and long chain aldehydes are simultaneously oxidized by the bacterial luciferase with the participation of molecular oxygen.

$$NAD(P)H + FMN + H^+ \xrightarrow{\text{oxidoreductase}} NAD(P)^+ + FMNH2 \qquad (1)$$

$$FMNH_2 + O_2 + RCHO \xrightarrow{\text{luciferase}} FMN + RCOCHH + O_2 + \text{light} \qquad (2)$$

This assay is used for direct determination of FMN, NADH and NADPH, as well as for the determination of numerous substrates and enzymes involved in NAD(P)H production or consumption (Gorus and Schram, 1979; Thore, 1979; Whitehead, et al., 1979). The availability of purified enzyme preparations and the use of immobilized enzymes have made these bioluminescent tests highly attractive for analytical purposes.

While the potential use of the bacterial luciferase system has been extensively developed, the use of whole, intact luminous bacteria for analytical purposes has been almost completely ignored. This field has only been explored in the last decade (mainly by my colleagues and myself) and will be reviewed briefly in this communication.

Analytical Applications of Intact Luminous Bacteria

The use of whole intact luminous bacteria for analytical purposes has some clear advantages. Intact luminous bacteria are self-maintaining luminescence units, and the *in vivo* luminescence of a single bacterium can be detected with the aid of a sensitive photometer. Luminous bacteria may be lyophilized and stored for many years in a cold room. Such lyophilized cultures promptly resume their full level of *in vivo* luminescence upon hydration.

The intensity of the *in vivo* luminescence emitted by luminous bacteria reflects both their cellular metabolic activity and their viability. Chemophysical and biological factors that affect the synthesis of the luminescence system, its activity or the integrity of the luminous bacterial cell can be determined under appropriate conditions. These applications can be further expanded by cloning the luminescence system into different microorganisms and by using genetic vectors carrying luminescence genes. The light emitting reaction in bacteria involves mixed function oxidation of $FMNH_2$ and long chain fatty aldehyde by molecular oxygen catalyzed by bacterial luciferase. The precursor of the natural aldehyde is myristic acid. Both aldehyde and myristic acid, as well as molecular oxygen, can be determined in nM range by intact luminous bacteria under appropriate conditions.

Tests utilizing luminous bacteria have been applied by us in the fields of analytical chemistry, food technology, medical microbiology and acute toxicity. The different approaches and some of the tests that have been successfully applied will be described briefly. The basic principles of the Bioluminescence Test are given below:

1. Tests for fatty acids, aldehydes and lypolytic enzymes.
Certain dark mutants of luminous bacteria, as well as *E. coli* cells in which the luciferase gene was cloned, emit light in the presence of very low concentrations of certain long chain aldehydes, while other mutants also respond to long chain fatty acids. These bacterial strains have been used to determine the enzymatic activity of lipase, phospholipase and monoamineoxidase or to test oil rancidity. Lyophilized cultures of luminous bacteria can be applied for all tests (Rottem, et al., 1979; Tenne, et al., 1985a, 1985b; Ulitzur, 1979, 1982; Ulitzur and Hastings, 1978a, 1978b, 1979a, 1979b, 1980; Ulitzur and Heller, 1978, 1981; Ulitzur, et al., 1979).

2. Determination of NAD + , $NADH_2$, dehydrogenases and their substrates with intact luminous bacteria.
Using aldehyde-requiring mutant of luminous bacteria or *E. coli* cells harboring the luciferase genes, one can determine different factors and enzymes that affect aldehyde concentration in the assay medium. One of these assays uses the horse liver alcohol dehydrogenase and 2% ethanol (in excess) with limited concentration of long chain aldehyde. In the presence of NAD+, alcohol dehydrogenase generates $NADH_2$ by ethanol oxidation and uses it to reduce the long chain aldehyde to the corresponding alcohol. Neither acetaldehyde nor the long chain alcohol are active in the production of light. Such coupled system allows detection of 10^{-11}M NAD + within a few minutes'. The same system permits detection of ethanol or alcohol dehydrogenase (Ulitzur, 1982).

3. Determination of cAMP.
The development of ın vivo luminescence in *Vibrio harveyi* cells is dependent on cAMP. By utilizing the cAMP-requiring mutant of this strain, we were able to determine nanogram range of cAMP in a short test (Ulitzur and Yashphe, 1975). The bioluminescence is also affected by cGMP (Ulitzur, et al., 1976).

4. Use of luminous bacteria for determination of oxygen.
The Km of bacterial luciferase for O_2 is about 3×10^{-8}M. Actually, the oxygen concentration determined with the aid of luminous bacteria is below 10^{-8}M (Lloyd, et al., 1982; Oshino, et al., 1972). (For comparison, a platinum electrode does not allow detection of oxygen below 10^{-6}M). Using

optic fiber technology we were able to determine oxygen concentrations in air or liquid. For this purpose we trapped a high concentration luminous bacteria in a metal chamber connected at one side to the optic fiber and at the other side covered by light nontransmissable oxygen permeable film (paper in preparation). A somewhat similar technology has been developed by us for rapid determination of oxygen transmissibility through different plastic films (Miltz and Ulitzur, 1980).

5. Tests for phagocytosis and serum bactericidal activity.
Vibrio cholerae, non-marine luminous bacteria, are extremely sensitive to phagocytosis and bactericidal activity of human serum. These bacteria have been used to determine the phagocytic and opsonic abilities of P.M.N. phagocytes and to test deficiencies in C′3 and C′4 components (Barak, et al., 1983a, 1983b, 1983c, 1985).

6. Bacterial bioluminescence as an early indication of marine fish spoilage.
Luminous bacteria are widely distributed in the sea mainly as free-living bacteria or as saphrophytes of marine animals. During growth on fish the *in vivo* luminescence is drastically increased. Thus, the *in situ* luminescence of the fish can be used as a probe for bacterial biomass load of the fish and as an early indication of its spoilage. This method has been successfully applied for the determination of fresh and frozen marine fish spoilage. The luminescence level of the bacterial suspensions that were taken from fish skin increased during storage at 20–25°C and correlated well with the total bacterial count. The increase in the bioluminescence level appeared earlier than the increase in trimethylamine level and occurred approximately at the same time as the increase in the hypoxanthine concentration (Barak and Ulitzur, 1980).

7. Tests for antibiotics.
Bioluminescence assays for antibiotics are based on the following principles: the "Lysis Test" uses a sensitive, highly luminescent culture of Gram-negative or Gram-positive bacteria and determines the activity of the antibiotics that affect the integrity of bacterial cell wall or membrane. Using this test, the presence of penicillin or polymyxin B in milk can be detected in the ng range within 25 min.

The "Induced Test" is based on the ability of the tested antibiotic to inhibit the de novo formation of luciferase (proteins) in certain dark mutants of luminous bacteria or in preinduced cultures of *E. coli* cells. Another application of this test is to use bacteriophage-carrying luciferase genes and to determine the degree of inhibition of luminescence development in the infected culture.

DNA synthesis inhibitors can be determined by their ability to cure bacteriophage-infected luminous bacteria.

Using a dark, nonsense mutant of luminous bacteria, one can specifically determine the activity of aminoglycoside antibiotics which act as miscoding agents. All of these tests require only a few microliters and results are obtained within 15–20 min (Ulitzur and Goldberg, 1977; Ulitzur and Weiser, 1981; Weiser, et al., 1981; Naveh, et al., 1983; Naveh and Ulitzur, 1986; Ulitzur, 1986b).

Recently, we have developed a new concept for the determination of antibiotics in biological fluids based on mobile genetic elements that remove the dependence on an easily measurable biological function and allow tests to be created in organisms most sensitive and/or relevant to the said antibiotics (Ulitzur and Kuhn, Patent Application 1989). Using lux carrying plasmids or phages, we were able to introduce, for example, the lux system into *Bacillus* for the determination of penicillins in milk. This assay allows detection in the nanogram range of ampicillin in milk in a procedure lasting approximately 15 min.

8. Metabolic activity of bacteria under stress conditions – luminous bacteria as a model.

Determination of the metabolic activity of bacteria in foodstuffs is very problematic. The use of cloned luminous *E. coli*, *Salmonella* or *Bacillus* allows the determination of luminescence as an indicator of the bacterial metabolic activity under different stress conditions. Luminescence was detected in bacteria suspended in very low water activity (AW = 0.6–0.9) or at low temperatures (−20°C), and the effect of the thermal process on milk and egg powder containing luminous *Salmonella* was studied (Reinhertz, et al., 1989; see also Stewart, 1990).

9. Determination of water toxicity.

Lyophilized marine luminous bacteria are being used for the determination of water toxicity. Freeze-dried bacteria are hydrated and added at low concentration to the water in question. The water toxicity is reflected by the decrease in luminescence after 5 min of incubation. A selective sensitive strain of luminous bacteria with stable luminescence was chosen; 0.02 ppm of different organic phosphorous compounds and 1 ppb of sodium fluoreacetate were detected. A similar test has been developed by Microbics Corporation, Carlsbad, CA 92008, USA.

10. Tests for growth factors and vitamins.

The luminescence of auxotrophic mutants of *E. coli* carrying the whole lux system of *V. fischeri* is dependent on the presence of the required growth factor. The use of Vitamin B12–requiring mutant of *E. coli* allows detection of 1 pg B12 in blood or foodstuff. The test takes less than one hour. Similarly, tests for different vitamins and amino acids were developed. Using these assay systems one can determine very low activities of proteases or nucleases in a continuous manner. The release of the requiring amino acid or nucleotide from a more complex molecule induces light in the proper auxotrophic mutant of luminous bacteria (in preparation).

11. Using the luminescence system as a sensor for mass transfer during the process of fermentation.
Determination of mass transfer of oxygen during fermentation is very complex, especially in high-viscosity media. A novel method which uses bioluminescence for mass transfer measurements of oxygen and substrate components in biochemical reactions has recently been developed by us utilizing a cloned luminescent strain of *Xanthomonas campestris* (Vashitz, et al., 1988, 1989).

12. Test for genotoxic and carcinogenic agents.
Certain dark mutants of luminous bacteria restore their *in vivo* luminescence by very low concentrations of direct mutagens, DNA synthesis inhibitors and DNA damaging agents. All the tested agents that respond positively in the "Ames Test" as well as many other carcinogenic agents gave positive results with the Bioluminescence Test. The test was applied successfully to determine genotoxic activity in biological fluids, foodstuffs, water and air samples. The Bioluminescence Test requires only a few hours of incubation, and hundreds of samples can be tested in one day (Levi and Ulitzur, 1983; Ulitzur, 1981, 1986a; Ulitzur and Weiser, 1981; Ulitzur, et al., 1980, 1981; Weiser, Ulitzur and Yannai, 1981). Lyophilized cultures of the dark mutant can be obtained from Microbics Corporation, Carlsbad, CA 92008, USA.

13. Identification of bacteria and their antibiotic susceptibility.
Enumeration and identification of microorganisms concomitant with their antibiotic sensitivity represent the primary goal of diagnostic medical and food microbiology. The numerous techniques, tests and media that have been developed to enumerate and identify microorganisms testify to the importance of this task. Nevertheless, despite substantial diversity, none of the currently applied tests allows the determination of less than 10^4 bacterial cells per ml in a 1 h-long procedure. Achievement of this goal becomes possible with the aid of a luciferase gene.

The basic element of this test is a DNA fragment coding for all or part of the luminescence (lux) operon from luminous marine bacteria. Transient or stable transfer of the lux operon into phenotypically dark bacteria by transformation, conjugation or transduction results in a rapid emission of light. The susceptibility to antibiotics of the bacteria in question can be determined from the degree of inhibition of luminescence developed by the tested bacteria. Transformation of lux genes by bacteriophages with a specific host range offers the most promising tool for clinical microbiology. Using this technology we were able to determine the presence of 10–100 bacteria of *Salmonella* or *E. coli* per ml together with their antibiotic susceptibility in a procedure lasting 45 min (Ulitzur and Kuhn, 1987; Stewart, 1990).

References

Barak, M., Merzbach, D. and Ulitzur, S. 1985. Opsonization increases permeability of the outer membrane of the luminous bacteria. J. Appl. Bacteriol. 59: 57–61.

Barak, M. and Ulitzur, S. 1980. Bacterial bioluminescence as an early indication of marine fish spoilage. Eur. J. Appl. Microbiol. Biotechnol. 10: 155–165.

Barak, M., Ulitzur, S. and Merzbach, M. 1983a. Determination of serum bactericidal activity with the aid of luminous bacteria. J. Clin. Microbiol. 18: 248–253.

Barak, M., Ulitzur, S. and Merzbach, D. 1983b. Phagocytosis-induced mutagenesis in bacteria. Mut. Res. 121: 7–16.

Barak, M., Ulitzur, S. and Merzbach, D. 1983c. The use of luminous bacteria for determination of phagocytosis. J. Immunol. Meth. 64: 353–363.

Gorus, F. and Schram, E. 1979. Applications of bio- and chemiluminescence in the clinical laboratory. Clin. Chem. 25: 512–519.

Levi, B-Z. and Ulitzur, S. 1983. Proflavin and norharman induce luciferase synthesis under anaerobiosis. Arch. Mikrobiol. 134: 281–284.

Lloyd, D., Williams, J., Yarlett, N. and Williams, A.G. 1982. Oxygen affinities of the hydrogenosome-containing protozoa *Tritrichomonas foetus* and *Dasytricha ruminantium*, and two aerobic protozoa, determined by bacterial bioluminescence. Gen. Microbiol. 128: 1019–1022.

Meighen, E.A. 1988. Enzymes and genes from the lux operons of bioluminescent bacteria. Ann. Rev. Microbiol. 42: 151–176.

Miltz, J. and Ulitzur, S. 1980. A new and fast bioluminescence method for determination of oxygen transmission through plastic films used in food packaging. Food Technol. 15: 389–390.

Naveh, A., Potasman, I., Bassan, H. and Ulitzur, S. 1983. A new, rapid and sensitive bioluminescence of protein synthesis inhibitors – antibiotics. J. Appl. Bacteriol. 56: 457–462.

Naveh, A. and Ulitzur, S. 1986. Determination of misreading effect of antibiotics with the aid of luminous bacteria. J. Microbiol. Meth. 4: 241–249.

Oshino, R., Oshino, N., Tamura, M., Kobilinsky, L. and Chance, B. 1972. A sensitive bacterial luminescence probe for O_2 in biochemical systems. Biochim. Biophys. Acta. 273: 5–17.

Reinhertz, A., Kopelman, I.J. and Ulitzur, S. 1984. The metabolic activity of bacteria under conditions of low water activity – luminous bacteria as a model. In Analytical Applications of Bioluminescence and Chemiluminescence. Kricka, L.J., Stanley, P.E., Thorpe, G.H.G. and Whitehead, T.P., eds., Academic Press Inc. pp. 541–544.

Rottem, S., Ulitzur, S., Hasin, M. and Rasin, S. 1979. Susceptibility of phospholipids of *Proteus mirabilis* smooth and rough strains to endogenic phospholipase A activity. FEMS Lett. 5: 13–16.

Stewart, G.S.A.B. 1990. A review. *In vivo* bioluminescence: new potentials for microbiology. Letts. Appl. Microbiol. 10: 1–8.

Tenne, M., Finberg, J.P.M., Youdim, M.B. and Ulitzur, S. 1985a. A new, rapid and sensitive bioluminescence assay for monoamine oxidase activity. J. Neurochem. 44: 1378–1384.

Tenne, M. Youdim, M.B.H., Ulitzur, S. and Finberg, J.P.M. 1985b. Determination of aliphatic amines by monoamine oxidase A and B studied by using bioluminescence technique. J. Neurochem. 44: 1373–1377.

Thore, A. 1979. Luminescence in clinical analysis. Ann. Clin. Biochem. 16: 359–369.

Ulitzur, S. 1979. A fast and sensitive bioluminescence assay for lipase. Biochim. Biophys. Acta 572: 211–217.

Ulitzur, S. 1981. Bioluminescence test for genotoxic agents. Trends Anal. Chem. 1: 329–333.
Ulitzur, S. 1982. The use of intact luminous bacteria for analytical purposes. In: Luminescent assays: Perspectives in endocrinology and clinical chemistry. Serio, M. and Pazzagli, M., eds., Raven Press, New York. pp. 95–108. Ulitzur, S. 1986a. Bioluminescence test for genotoxic agents. Meth. in Enzymol. 133: 264–274.
Ulitzur, S. 1986b. Determination of antibiotic activities with aid of luminous bacteria. Meth. in Enzymol. 133: 275–284.
Ulitzur, S. 1989. The regulatory control of the bacterial luminescence system – a new view. J. Biolum. Chemilumn. 4: 317–325.
Ulitzur, S. and Goldberg, I. 1977. A sensitive, rapid and specific bioassay for the determination of antilipogenic compounds. Anti Microbiol. Chemot. 12: 308–313.
Ulitzur, S. and Hastings, J.W. 1978a. Bioassay for myristic acid and "long chain" aldehydes. Meth. in Enzymol. 57: 189–193.
Ulitzur, S. and Hastings, J.W. 1978b. Myristic acid stimulation of bacterial luminescence in "aldehyde" mutants. Proc. Natl. Acad Sci. USA 75: 266–269.
Ulitzur, S. and Hastings, J.W. 1979a. Evidence for tetradecanal as the natural aldehyde in bacterial bioluminescence. Proc. Natl. Acad. Sci. USA 76: 265–267.
Ulitzur, S. and Hastings, J.W. 1979b. The control of aldehyde synthesis in the luminous bacterium, *Beneckea harveyi*. J. Bacteriol. 137: 854–859.
Ulitzur, S. and Hastings, J.W. 1980. Reversible inhibition of bacterial bioluminescence by long chain fatty acids. Curr. Microbiol. 3: 295–300.
Ulitzur, S. and Heller, M. 1978. New, fast and very sensitive bioluminescence assay for phospholipase A and C. Anal. Biochem. 91: 421–431.
Ulitzur, S. and Heller, M. 1981. A sensitive bioassay for phospholipase A and phospholipase C using bacterial bioluminescence. Meth. in Enzymol. 72: 338–346.
Ulitzur, S. and Kuhn, J. 1987. Introduction of lux genes into bacteria, a new approach for specific determination of bacteria and their antibiotic susceptibility. In: Bioluminescence and chemiluminescence: New perspectives. Scholmerich, J., Andreesen, R., Kapp, A., Ernst, M. and Woods, W.G., eds., John Wiley, Chichester. pp. 463–472.
Ulitzur, S. and Kuhn, J. 1988. The transcription of bacterial luminescence is regulated by Sigma 32. J. Biolum. Chemilumn. 2: 81–93.
Ulitzur, S. and Weiser, I. 1981. Acridine dyes and other DNA-intercalating agents induce the luminescence in luminous bacteria and their dark variants. Proc. Natl. Acad. Sci. USA 78: 3338–3342.
Ulitzur, S. Weiser, I. and Yannai, S. 1980. A new, sensitive, fast and simple bioluminescence assay for mutagenic compounds. Mut. Res. 7: 113–124.
Ulitzur, S., Weiser, I. and Yannai, S. 1981. Bioluminescence test of mutagenic agents. In: Bioluminescence and chemiluminescence. DeLuca, M.A. and McElroy, W.D., eds., Academic Press, New York and London. pp. 139–145.
Ulitzur, S., Yagen, I. and Rottem, S. 1979. Determination of lipopolysaccharide by a bioluminescence technique. Env. Appl. Bacteriol. 37: 782–894.
Ulitzur, S. and Yashphe, J. 1975. An adenosine 3', 5'-cyclic monophosphate requiring mutant of the luminous bacteria *Beneckea harveyi*. Biochim. Biophys. Acta 404: 321–328.
Ulitzur, S., Yashphe, J. and Hastings, J.W. 1976. Inhibition and stimulation of the development of the bioluminescence system in *Beneckea harveyi* by cyclic GMP. Proc. Natl. Acad. Sci. USA 73: 4454–4456.
Vashitz, O., Sheintuch, M. and Ulitzur, S. 1989. Mass transfer studies using cloned-luminous strains of *Xanthomonas campestris*. Biotechnol. Bioengin. 34: 671–680.
Vashitz, O., Ulitzur, S. and Sheintuch, M. 1988. Mass transfer, batch and continuous kinetics, in a luminous strain of *Xanthomonas campestris*. Chem. Eng. Sci. 43: 1883–1890.
Weiser, I., Ulitzur, S. and Yannai, S. 1981. DNA damaging agents and DNA synthesis inhibitors induce the luminescence in dark variants of luminous bacteria. Mut. Res. 91: 443–450.
Whitehead, T.P., Kricka, L.J., Carter, T.J.N. and Thorpe, G.H.G. 1979. Analytical luminescence, its potential in the clinical laboratory. Clin. Chem. 25: 1531–1546.

Discussion (DR. ULITZUR)

Grossman: In terms of detecting the effects of benzo(a)pyrene, you might be able to clone in the genes involved in oxidation to activate this agent.

Ulitzur: You mean cytochrome P_{450}?

Grossman: That is right.

Ulitzur: It is true, for benzo(a)pyrene you indeed require activation. Interestingly, the luminous bacteria have cytochrome P_{450} that was found to be active under certain conditions. It is still being studied.

Szybalski: Which kind of luciferase-encoding gene constructs did you use – the two separate genes, luxA and luxB, or the fused product of these two genes? The luxA-luxB fusion which we tried to use was not very active. Do you have any better luxA-luxB fusion?

Ulitzur: In cloned bacteria we are using the natural non-fused luciferase. In eucaryotic cells we used a fused alpha-beta luciferase that indeed showed only about 10% of the activity. In addition, in eucaryotic cells the respiratory activity is located in the mitochondria. Thus, the luciferase is not fully expressed. We are now using a leader protein to clone the luciferase into the mitochondria where better expression of the luminescence system is expected.

Szybalski: When you use a photoluminescent bacteriophage, you have to be sure that the bacteria are not resistant to a given phage because then your test might fail. Is that correct?

Ulitzur: That is right. Therefore, you need a cocktail of phage that will cover those strains that are resistant to certain bacteriophages. We also found that many bacteria that are not showing plaques are still emitting light. It appears that the light formation mechanism is faster than the restriction enzymes.

Szybalski: You did not discuss the applications of the luxA-luxB system as a reporter gene. Do you have any comments and suggestions on such a use?

Ulitzur: About three years ago Dr. O. Carmi, Dr. J. Kuhn and I described the use of luxA-luxB as reporter genes for sporulation promoters in Bacillus strains (Carmi, A., Stewart, S.A.B., Ulitzur, S. and Kuhn, J. 1987. J. Bacteriol. 169: 2165–2170).

Brodman: How does your test correlate with mammalian cell testing; is it better than the Ames test?

Ulitzur: It is run now in the EPA laboratories. Indeed, they found that this test is better correlated than Ames test for teratogenicity.

Key Words: Bioluminescence applications, Analytical applications of bacterial bioluminescence, Luciferase

Deposition of Enzymes by Electropolymerization

Nicholas F. Almeida and Lemuel B. Wingard Jr.†
Department of Pharmacology, School of Medicine, University of Pittsburgh, Pittsburgh, PA 15261

Introduction

A significant problem in the large scale application of enzymes and other proteins for the development of microscale biosensors is the difficulty in fabricating uniform and reproducible coatings containing the active protein (Wingard, 1990b). With emphasis on biosensor miniaturization as well as the utilization of many types of biological materials in biosensor design (Rechnitz, 1987), a critical factor in the design is the ability to obtain functionally uniform and reproducible loadings of the protein on the transducer element.

In situ electropolymerization is a technique capable of producing apparently uniform and reproducible multilayer coatings over very small areas. The immobilization of protein within the polymer coating may be by entrapment within the polymer matrix during electropolymerization (Foulds and Lowe, 1986; Bartlett and Whitaker, 1987) or possibly by covalent attachment to functionalized monomers prior to electropolymerization. However, the large size of many proteins may cause steric problems during the electropolymerization, and such systems still need to be demonstrated experimentally. Entrapment during electropolymerization appears preferable to covalent attachment of the protein to the polymer film after the film is formed, especially if the polymer film occupies only a few μm^2 of area on the transducer surface.

We have been studying the effect of the conditions of electrochemical polymerization as they affect loading, spatial distribution, and activity of immobilized enzymes, using both conducting and non-conducting polymers and different sized enzymes. The results presented here show how some electropolymerization variables influence the apparent activity of glucose oxidase immobilized in poly(-phenylenediamine). Some of this work was summarized briefly at the 10th International Enzyme Engineering Confer-

† Deceased

D. Kamely et al. (eds), Biotechnology: Bridging Research and Applications, 89–93.

ence in Kashikojima, Japan in September 1989 (Almeida and Wingard, 1990).

Materials and Methods

Materials

Glucose oxidase Type VII from *Aspergillus niger*, horseradish peroxidase Type II, o-dianisidine, and β-D(+)-glucose all from Sigma, 1,3–phenylenediamine dihydrochloride from Aldrich and ammonium and potassium phosphates from Fisher were used as received. Glass microscope slides, vapor coated with 100 Å thick chromium followed by about 2000 Å thick gold were used as the working electrodes.

Electropolymerization

Electropolymerizations were carried out with a Model 273 Princeton Applied Research Potentiostat/Galvanostat, a jacketted glass cell thermostatted at 25 ± 1°C, a platinum gauze auxiliary electrode, a Ag/AgCl (1M KCl) reference electrode, and the working electrode. For the electropolymerization, glucose oxidase, 1,3–phenylenediamine, and 0.25 M buffered ammonium phosphate were placed in solution and deoxygenated. The working electrode potential was maintained at + 0.622 V [vs Ag/AgCl(1M KCl)] for 15 min. After electropolymerization, the glass slide working electrode with the attached polymer film was rinsed by immersion in gently stirred 0.1 M phosphate buffer (pH 6.0) for 15 min to remove loosely bound enzyme and unreacted monomer. The films were stored in 0.1 M phosphate buffer (pH 6.0) at 25°C, until assayed (within 2 hr) for immobilized enzyme activity.

Assay for Enzyme Activity

Enzyme activities were determined by a spectrophotometric assay (Wingard, et al., 1983) using a Perkin Elmer Model 320 UV-spectrophotometer. Soluble glucose oxidase was assayed at 25 °C by adding a sample of the enzyme solution to a cuvette preloaded with o-dianisidine, peroxidase and 18% glucose buffered at pH 6.0, and measuring the initial linear rate of change of absorbance at 460 nm. For the immobilized glucose oxidase, the glass slide coated with the enzyme-containing polymer film was incubated in 18% glucose, buffered at pH 6.0, at 25°C. Aliquots of the solution were taken at 2 min intervals and injected within 2 secs into cuvettes preloaded with pH 6.0 o-dianisidine and peroxidase. The linear rate of change of absorbance at 460 nm was used to determine the apparent activity of the immobilized enzyme (Wingard, et al., 1983). For both cases, one unit of enzyme activity was defined as the rate of change of absorbance equivalent to 1 μmole min^{-1} of hydrogen peroxide formed, at pH 6.0 and 25°C.

Results and Discussion

Results

The concentration of soluble enzyme in the starting solution prior to electropolymerization, the concentration of monomer (1,3–phenylenediamine) in the starting solution, and the pH of the monomer solution were three of the prime variables during the electropolymerization.

The maximum apparent activity of the immobilized glucose oxidase is observed at a soluble enzyme concentration of 0.1 mg ml^{-1} in the starting (pre-electropolymerization) solution. At higher soluble enzyme concentrations the ratio of soluble enzyme to monomer may be sufficient for enzyme interference with electropolymerization to produce defects in the polymer film through which the enzyme can leak out. Further explanation awaits additional experiments.

The concentration of 1,3–phenylenediamine dihydrochloride from which the film is electrochemically synthesized also influences the apparent activity of the immobilized enzyme. With a soluble enzyme concentration of 0.1 mg ml^{-1} and pH 7.0, the electropolymerization was carried out at different initial monomer concentrations. The results, summarized in Figure 1, show a maximum in apparent glucose oxidase activity at a starting monomer concentration of 0.04 M.

A similar series of experiments at fixed enzyme and monomer concentrations, but different pH values, resulted in maximum apparent enzyme activity at pH 7.0.

Work is underway to determine the loading and spatial distribution of the immobilized enzyme and how these are influenced by electropolymerization conditions. Studies are also planned to tell whether the differences in apparent enzyme activity result from different amounts of enzyme being immobilized, or whether some of the enzyme is inactivated or simply not accessible to substrate because of the morphology of the polymer film.

Discussion

Phenylenediamine also has been electropolymerized to form a thin film to cover glucose oxidase immobilized by covalent (glutaraldehyde or carbodiimide) coupling to platinized reticulated vitreous carbon for use as a biosensor detector for flow injection analysis (Sasso, et al., 1990). The monomer used was 1,2–phenylenediamine rather than the 1,3–isomer of the present study.

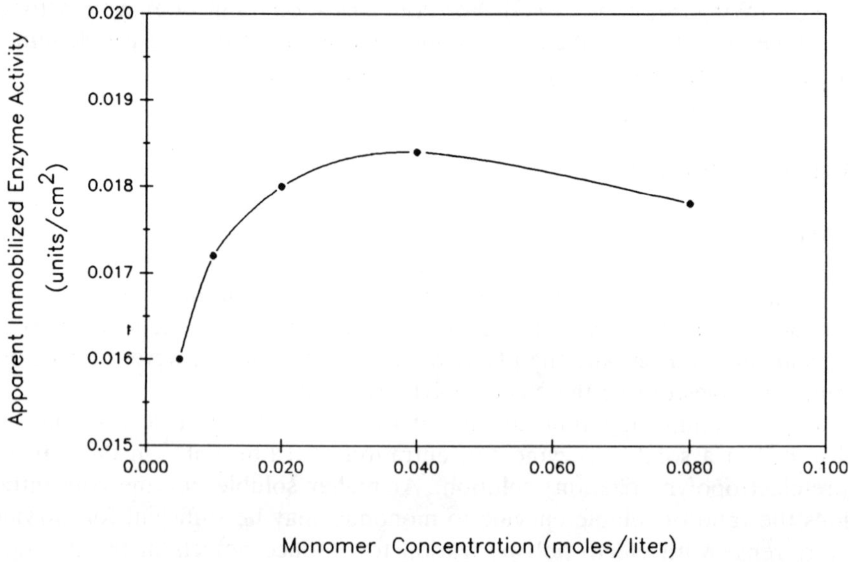

Figure 1. Influence of monomer concentration in starting solution prior to electropolymerization, for a fixed soluble glucose oxidase concentration of 0.1 mg ml^{-1} and pH 7.0.

In the 1,2–isomer work electropolymerization was not used for enzyme immobilization but only to form a protective coating over previously immobilized enzyme. The results indicate that the electropolymerized film from the 1,2–isomer permitted rapid passage of small molecular weight (< about 150 daltons) analytes but did not allow passage of large molecular weight materials such as albumin and other serum proteins (> about 50,000 daltons). Therefore, there is reason to believe that our films formed with the 1,3–isomer also are highly porous to small substrates such as glucose, but not to glucose oxidase enzyme with its molecular weight of about 153,000 daltons (Swoboda, 1969).

Considerable interest has developed recently in incorporation of neuroreceptors as the molecular recognition component of biosensors (Wingard, 1987). Since neuroreceptor proteins are not available in large supplies and miniature sensors are desirable, methodology is being explored to develop very small area immobilization techniques to give immobilized receptors that retain receptor binding specificity and possibly receptor functionality as well as provide uniformity and reproducibility of sensor manufacture. Several receptor biosensor designs have been described (Wingard, 1990a and 1990b), but none of the reports give convincing results concerning stability of the receptor immobilization or in some cases interpretation of the results. Thus, there is a need for considerable development of greatly improved receptor immobilization-stabilization-transduction techniques.

Electropolymerization should be a suitable technique for immobilizing neuroreceptors without loss of receptor binding specificity. However, retention of receptor ion channel or secondary messenger activity is questionable via any present method of protein immobilization, and much work remains to be carried out in this area.

Conclusions

The electropolymerization conditions that result in maximum apparent activity of the immobilized glucose oxidase are: 1,3–phenylenediamine monomer concentration 0.04 M (.025–0.06 M), enzyme concentration 0.1 mg ml^{-1}, and pH 7. The glucose oxidase results suggest that electropolymerization can also be used for immobilization of receptors with retention of receptor binding affinity.

References

Almeida, N.F. and Wingard, L.B. Jr. 1990. Immobilization of glucose oxidase by electropolymerization of monomers: Influence of polymerization conditions. Ann. N.Y. Acad. Sci. 855. In press.

Bartlett, P.N. and Whitaker, R.G. 1987. Electrochemical immobilization of enzymes. J. Electroanal. Chem. 224: 27–48.

Foulds, N.C. and Lowe C.R. 1986. Enzyme entrapment in electrically conducting polymers. J. Chem. Soc. Faraday Trans. 1. 82: 1259–1264.

Rechnitz, G.A. 1987. Biosensors: Challenges for the 1990s. In: Biosensors International Workshop 1987. Schmid, R.D., Gilbault, G.G., Karube, I., Schmidt, H.L. and Wingard, L.B., eds. VCH Publishers, Weinheim. pp. 3–12.

Sasso, S.V., Pierce, R.J., Walla, R. and Yacynych, A,M. 1990. Electropolymerized 1,2–diaminobenzene as a means to prevent interference and fouling and to stabilize immobilized enzyme in electrochemical biosensors. Anal. Chem. 62: 1111–1117.

Swoboda, B.E.P. 1969. The relationship between molecular conformation and the binding of flavin-adenine dinucleotide in glucose oxidase. Biochim. Biophys. Acta. 175: 365–379.

Wingard, L.B. Jr. 1987. Possibilities for biosensors based on neuroreceptors. In: Biosensors International Workshop 1987. Schmid, R.D., Gilbault, G.G., Karube, I., Schmidt, H.L. and Wingard, L.B., eds. VCH Publishers, Weinheim. pp. 133–137.

Wingard, L.B., Jr. 1990a. Amplification possibilities with neuroreceptor-based biosensors. In: Biosensor Technology: Fundamentals and Application. Bowden, E., Buck, R., Hatfield, W.E. and Umana, M., eds. Marcel Dekker, New York. In press.

Wingard, L.B., Jr. 1990b. Biosensor trends. Ann. N.Y. Acad. Sci. 855. In press.

Wingard, L.B., Jr., Cantin, L.A. and Castner, J.F. 1983. Effect of enzyme-matrix composition on potentiometric response to glucose oxidase immobilized on platinum. Biochim. Biophys. Acta. 748: 21–27.

Key Words: Glucose oxidase, 1,3–Phenylenediamine, Electropolymerization, Thin films, Enzyme immobilization, Biosensors, Conducting polymers

Computerized Biospecific Electrodes

Judith Rishpon
Department of Biotechnology, Tel-Aviv University, Ramat Aviv, 69978, Israel

Introduction

Electrochemical devices employing immobilized biomaterials combine the high selectivity of biological recognition with the high sensitivity and the relative simplicity of modern electroanalytical techniques. Such devices provide a convenient and direct measurement based on the generation of an electric signal by an electroactive biomaterial. The response is relatively quick, the measurement is nondestructive and enables repeated use of the electrode-immobilized biomaterial. Moreover, the measurements are applicable to colored and turbid samples and enable measurements in a continuous mode. These advantages have promoted extensive studies on the application of enzyme electrodes as biosensors in monitoring of physiological and biotechnological processes. Enzyme electrodes for measurement of blood components such as glucose and lactic acid are available commercially. However, in most cases the electrode is employed empirically requiring a predetermined calibration curve. This is mainly because the response of an immobilized bioactive material is a complex function of its activity and the transport processes to and within the bioactive layer. Both these parameters usually change with time.

The interfacing of electrochemical devices with computers is relatively easy. Such a combination, in the context of biosensors, is particularly advantageous for process monitoring and control. Furthermore, modern electrochemical techniques employ dynamic measurements which significantly increase the sensitivity (Bard and Faulkner, 1980) and enable a thorough investigation of processes occurring at and near an electrode. Computer analysis of the data will provide information on parameters relating to the processes occurring in the bioactive layer next to an electrode and, moreover, will enable an on-line evaluation of sensor performance, leading to the development of more reliable biosensors.

In this work we describe several biospecific electrodes connected to a computerized electrochemical system and employed in the determination of analytes at very low, as well as very high, concentration ranges. The examples include an enzyme electrode for the determination of glucose and its appli-

D. Kamely et al. (eds), Biotechnology: Bridging Research and Applications, 95–107.

cation in monitoring of yeast fermentation, an antibody electrode for the determination of isoenzymes, and a sensitive electrode system for detection of bacteria.

Materials and Methods

Glucose oxidase (GOX) from *Aspergillus niger* (E.C.1.1.3.4, Catalog No G 2133) was purchased from Sigma (St. Louis, Mo. USA). Lactate dehydrogenase (LDH) isoenzymes were purchased from Sigma; LDH 5 No L-5762 (M4) from porcine muscle type XXXII-S and LDH1 No L-9832 (H4) from porcine heart type VII-S. Alkaline phosphatase affinity pure rabbit anti-mouse IgG (H and L) was purchased from Jackson Immunoresearch Laboratories, West Grove, Pa. USA. Biotin-SP-Affinity Pure rabbit anti-mouse IgG (H and L) was purchased from Jackson. Mouse monoclonal antibodies to pLDH5 (mAB 2/66) were obtained from Z. Hollander (Hollander and Katchalski-Katzir, 1986) Avidin from egg white affinity purified, lyophilized No A-9275 was purchased from Sigma. N-hydroxysuccinimidobiotin (Catalog No H1759), (NHS-biotin) was purchased from Sigma. 1,6–Diaminohexane (hexamethylenediamine) (HMD) was purchased from Fluka Chemie AG, Buchs, Switzerland. 1–Ethyl-(3 dimethyl amino propyl) carbon diimide hydrochloride (EDC) and 1–cyclohexyl-3–(morpholino-ethyl)carbodimide metho-p-toluenesulphonate were purchased from Sigma. p-Aminophenyl phosphate (APP) was synthesized as was previously described (Boyland and Manson, 1957). Formalin fixed *Staphylococcus aureus* (S 2014) and protein A were obtained from Sigma.

The Electrochemical System

A three-electrode cell was used with a Pt gauze as a counter electrode and a KCl saturated calomel electrode as a reference electrode. A PAR 173 or PAR 273 potentiostat was employed coupled to a microcomputer (IBM PC or compatible) and the Tecmar Lab master interface. The electrochemical measurements were conducted employing the chronocoulometric technique previously described (Rishpon, 1987). The working electrode was connected and disconnected in a repetitive mode using a CMOS inverter MC 4069 and preprogrammed digital pulses. The resulting decay curves of the current (Figure 1) were sampled, averaged and analyzed by the computer. In order to increase the signal-to-noise ratio the results were integrated by the computer during the period T_f (see Figure 1) and are presented in some of the cases as charge per given period of time.

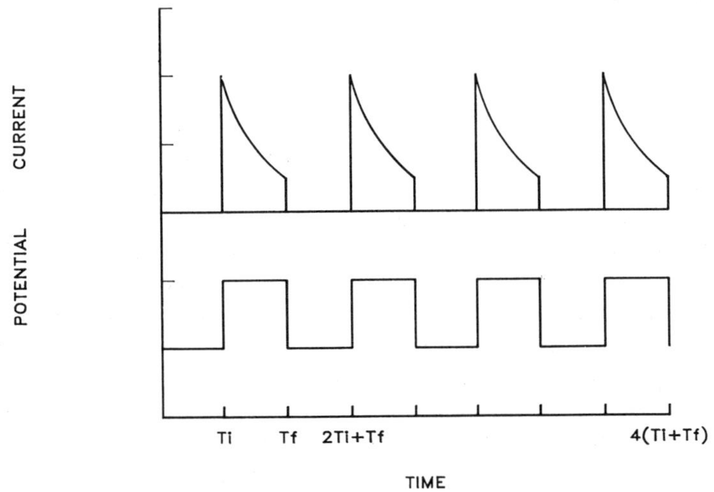

Figure 1. Schematic representation of applied electrode potential and resulting current as function of time. T_i is the time period at which no electron transfer takes place at the electrode. T_f time at which the electrode is connected and held in the diffusion-controlled region of the product.

Specific Electrodes

1. Glucose electrode: A glucose oxidase membrane was constructed on the surface of 1 × 1 cm platinum net (390 mesh). The enzyme was entrapped according to a previously described procedure (Rishpon, et al., 1989). Glucose concentration was determined by measuring the electro oxidation of H_2O_2 at 0.6V vs. SCE.

2. LDH5 electrode: The electrode consisted of a glassy carbon rod (High Performance Material Co., Inglewood, Ca. USA) embedded in a teflon housing exposing a single surface of a disc to the solution. The antibody was covalently bound to the carbon surface by the EDC method (Rishpon and Rosen, 1989).

3. Electrode system for the detection of bacteria: The electrodes were made of carbon felt (RVG 1000, Le Carbon Lorraine, France) cut into 5 mm diameter discs to which antibodies were bound via a biotin avidin bridge. The discs were exposed to 1M HMD and 500 mg ml^{-1} EDC in water, followed by incubation in NHS-biotin; The discs were incubated in avidin, and finally in biotinilated anti-mouse IgG antibodies. The detailed procedure is described elsewhere (Hadas, et al., 1990, manuscript in preparation). Prior to use the discs were assembled in a teflon housing connected from inner side to a Pt wire for electrical connection and exposing the other side for the immunological reactions.

Results and Discussion

A Glucose Sensor

A computerized chronocoulometric system for the characterization of a glucose electrochemical sensor has been recently developed in our laboratory (Rishpon, 1987). The chronocoulometric results were obtained at different glucose concentrations with the glucose oxidase crosslinked to different matrices. The experimental curves were compared with simulated data calculated for enzyme membrane having different permeability and activity. Parameter fitting was employed to estimate the activity of the enzyme in the membrane and the diffusion coefficient of the substrate in the enzyme membrane. The general shape of the curves could be correlated with the operating condition of the electrode, such as low enzyme activity, cosubstrate depletion and mass transfer limitations. In particular, we followed the changes of the diffusion coefficient of solutes in enzyme-membranes freshly prepared or aged (Rishpon and Rosen, 1987). The application of the glucose sensor was further extended to the measurement of glucose in very low and in very high concentrations.

Figure 2 presents results obtained with the Pt net glucose electrode in the ppb concentration range. In the upper curve the electrode was disconnected for 10 s and then connected and poised at 0.6V vs SCE and the current sampled for 0.5 s. The lower curve presents results of the same electrode operated in a steady state mode at 0.6V vs SCE. The data demonstrates that the application of the chronocoulometric technique significantly increases the sensitivity of the measurements. Disconnecting the electrode electrically while the enzymatic reaction is taking place leads to accumulation of the enzyme reaction product (H_2O_2) near the Pt. Hence, when the potential is applied on the electrode, the current is much higher than at steady state. This type of sensitivity enhancement was further exploited in the antibody electrode, as will be described below.

In situ monitoring of glucose during fermentation processes by means of glucose oxidase electrode faces a series of intrinsic difficulties. The device must probe a wide range of glucose levels from very low to relatively high concentrations, well above the Km of the enzyme. The local oxygen tension is unstable and inherently restricted by the low solubility of oxygen in the fermentation medium. Especially at high glucose concentration the bimolecular enzymatic reaction is impeded by oxygen depletion. In addition all the components of the enzyme electrode in contact with the fermentation broth must be sterile or aseptic. We have lately developed an electrode design which copes with these difficulties (Rishpon, et al., 1990). A schematic layout of the glucose electrode assembly is shown in Figure 3. A dialysis membrane separates the enzyme gel layer from the glucose solution. An outer oxygen permeable silicon membrane serves for the confinement of the enzyme chamber on the open-air side allowing unlimited transport of oxygen into the

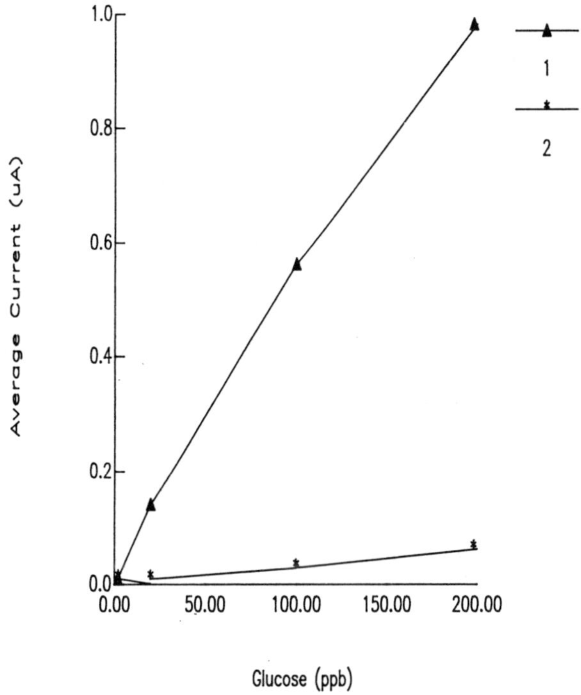

Figure 2. Calibration curve for glucose at low concentration (ppb = ng ml^{-1}). Upper curve – electrode was disconnected for 10 s and connected for 0.5 s repeatedly. Lower curve – electrode was connected continuously.

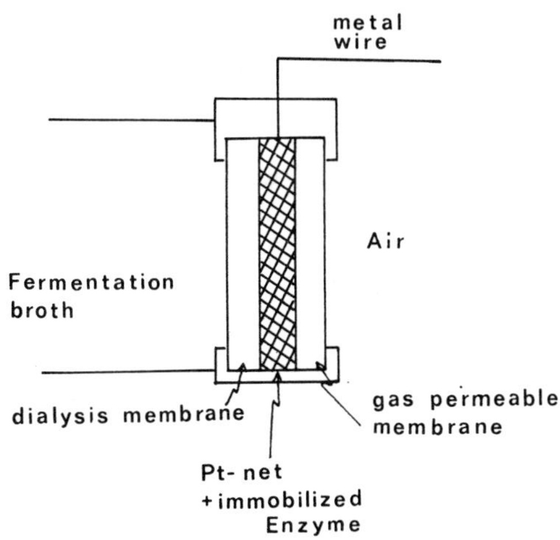

Figure 3. The schematic layout of the "sandwiched" glucose-oxidase electrode (SGE).

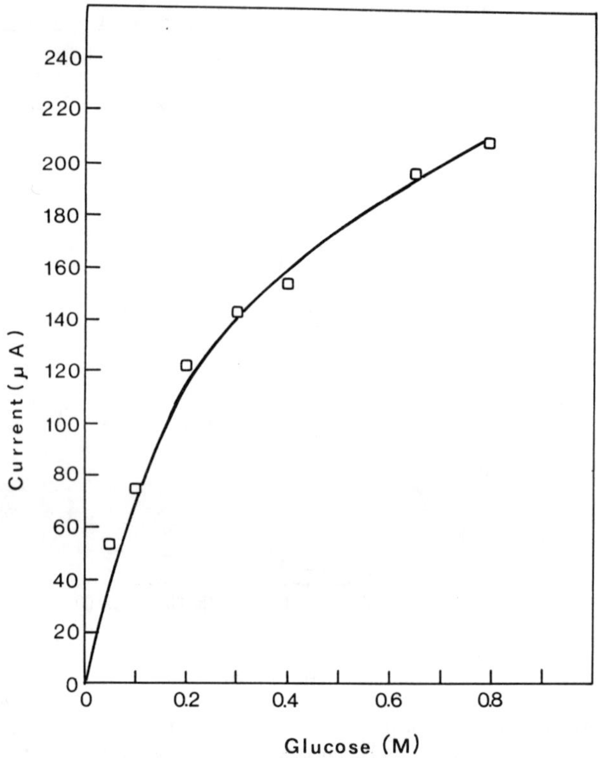

Figure 4. SGE response curve to 0–0.8M glucose concentration, using 0.2 μm filter and preboiled dialysis membrane with molecular-weight cutoff of 6000, as inner membrane.

enzyme gel layer. This electrode design was named SGE (sandwich glucose electrode). A calibration curve of this electrode at high glucose concentration is presented in Figure 4.

The ability to continuously monitor the changing glucose concentration throughout a fermentation process was demonstrated employing a yeast fermentation as a model. The results of simultaneous measurements of glucose by the above described glucose electrode are demonstrated in Figure 5. The glucose level was measured for 28 h. The fermentation process was started by the addition of glucose to the yeast culture to a final concentration of 200 mM. The data show continuous monitoring of glucose consumption by the growing yeast during the first 8 h. The concentration of glucose dropped to a steady level below 20 mM. Upon a second and third addition of glucose (final concentration 0.2M) the glucose signal was somewhat lower. This was possibly due to fouling of the inner membrane. However, the glucose consumption by the yeast was clearly followed and was found again to fall below 20 mM. The data thus show that the SGE is capable of closely

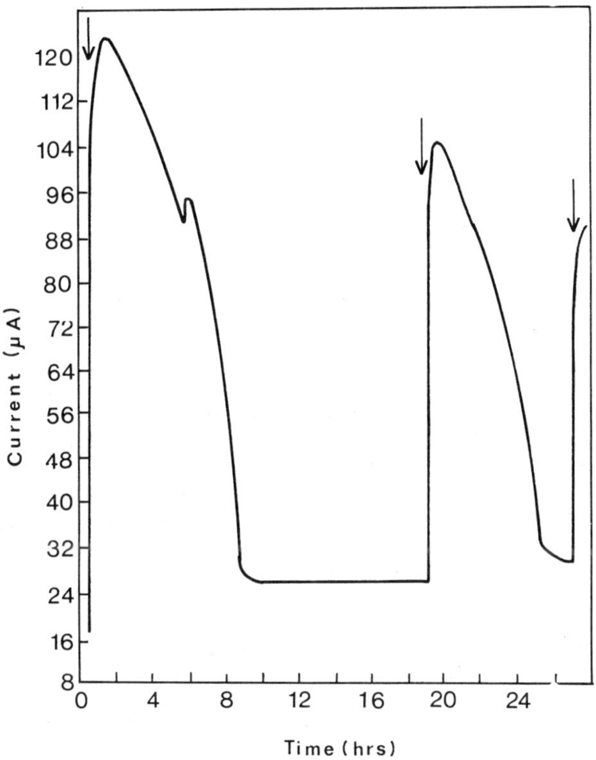

Figure 5. Long-term glucose monitoring. Measurements were carried out under the same experimental setup described in Figure 4. Glucose input (0.2M final concentration) was added at the points indicated.

following the kinetics of glucose consumption during a fermentation process for extended periods of time.

An Electrode for the Isoenzyme LDH5

Human sera contains five LDH isoenzymes and their relative properties change significantly under certain pathological conditions. The distribution of the isoenzymes varies among the tissues. The determination of the relative distribution of these isoenzyme in serum is useful in the diagnosis of diseases involving tissue damage. LDH1 is predominant in heart muscle and is used as a marker for myocardial infarction. LDH5 is predominant in liver; high levels of the isoenzyme LDH5 appear in the serum during liver diseases, such as hepatitis or malignant liver processes (Kolaric, et al., 1975). Currently, serum LDH isoenzymes are measured in clinics and hospital laboratories by electrophoretic methods (McKenzie and Henderson, 1983) which

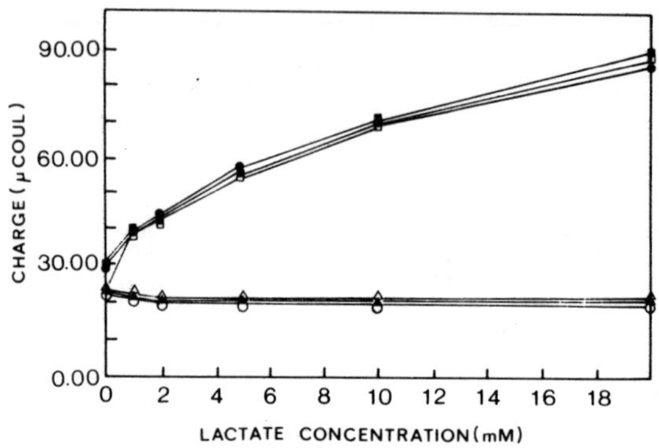

Figure 6. Response of gc/aLDH5 electrode to the addition of different concentrations of lactate. Measurement conditions: glycine-KC1 buffer 0.1 M, pH = 9.5 mM. NAD +, 0.70V vs. SCE, rotating disc electrode, 400 rpm. Upper curves: (1) filled squares, after first incubation with LDH5 (10 μg ml^{-1} h^{-1}); (2) filled circles, second measurement after the first incubation; (3) empty squares, after removal of the antigen and reincubation with LDH5 (10 μg ml^{-1} h^{-1}). Lower curves: (4) empty triangles, after removal of the antigen by washing at pH 2.2; (5) filled triangles, after incubation with LDH1 (10 μg ml^{-1}); (6) empty circles, after incubation with LDH5 (10 μg ml^{-1}) in the presence of aLDH5 (10 μg ml^{-1}).

are tedious and time consuming. We present here electrode for LDH5 based on monoclonal antibodies toward the isoenzyme.

Figure 6 shows the response of a glassy carbon electrode onto which the antibody 2/66 was covalently bound. The electrode was incubated in LDH5 solution and tested as a lactic acid electrode. in the presence of NAD^+. The electrochemical determination was based on the electrode oxidation (at 0.7V vs SCE) of NADH, generated by the enzyme reaction (Moiroux and Elving, 1978). The results are expressed as charge per 100 s, as explained in the experimental section. Curve 1 in Figure 5 demonstrates the activity of the LDH5 electrode prepared in this way after incubation in a solution of 10 μg ml^{-1} LDH5 for 1 h. Washing the electrode in a solution of low pH (glycine pH 2.2) completely removed the antigen: after washing, the electrode did not respond to the addition of lactic acid (curve 3). However, the enzyme activity of the electrode could be fully recovered by reincubation of the electrode in the LDH5 solution (curve 4). No activity was detected after incubating the electrode in a solution containing the isoenzyme LDH1 (curve 5). A competition effect could also be demonstrated: no enzyme electrode activity was observed after addition of the antibody anti-LDH5 (10 μg ml^{-1}) to the incubation solution. The covalently bound antibodies exhibited long operational and storage stability (i.e., more than 6 months).

In order to further improve the electrode sensitivity we have used the mediator PMS which reoxidizes the NADH formed by the enzymatic reaction. The resulting reduced $PMSH_2$ is oxidized on the glassy carbon electrode at much lower potential (0.13V vs. SCE). The high sensitivity of the improved system is presented in Figure 7.

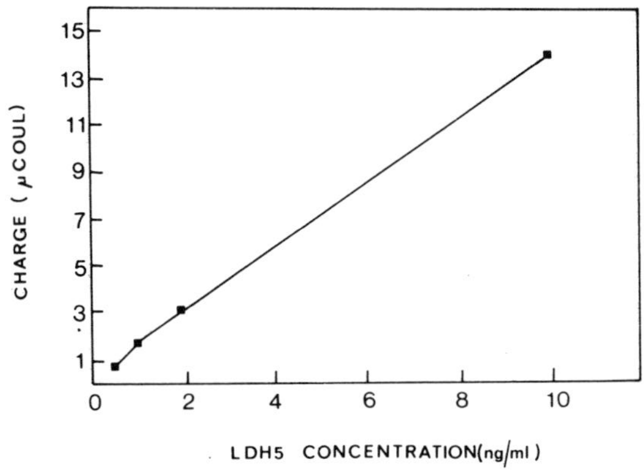

Figure 7. Calibration curve for LDH5. Gc/aLDH5 electrode was incubated with different concentrations of LDH5 for 2h at 37°C. Measurement conditions are as in Figure 6.

The applicability of the method developed for clinical detection of the LDH isoenzyme was demonstrated by measurements of human and dog sera in which the LDH5 activity was also estimated by electrophoresis in hospital laboratory (Rishpon and Rosen, 1989).

An Immunoelectrochemical System for the Detection of Bacteria

Enzyme immunoassays are generally performed with large number of samples including standards for the determination of the activity of the reagents. By the same approach we constructed a multielectrode device which takes advantage of the fact that in the computerized system described an electrode is electrically disconnected part of the time. This time interval is used for electrical connection and current measurements of another electrode. A schematic layout of the multielectrode device connected to the computerized electrochemical system is shown in Figure 8. Combination of the high sensitivity of the chronocoulometric technique with simultaneous measurements of many electrodes results in fast and sensitive system for the determination of antigens.

In a previous publication we have described an heterogeneous electrochemical enzyme immunoassay based on the detection of alkaline phosphatase (Rosen and Rishpon, 1989). Further developments of this enzyme-tagged immunooelectrochemical assay resulted in a highly sensitive system for determination of antigens. The antibodies are immobilized on disposable carbon felt discs via a biotin-avidin-biotin bridge, as described in the experimental part.

In a typical assay the antibody-bound carbon felt electrodes are incubated in the antigen solutions, washed and incubated in a solution of a second antibody conjugated to alkaline phosphatase. The alkaline phosphatase ac-

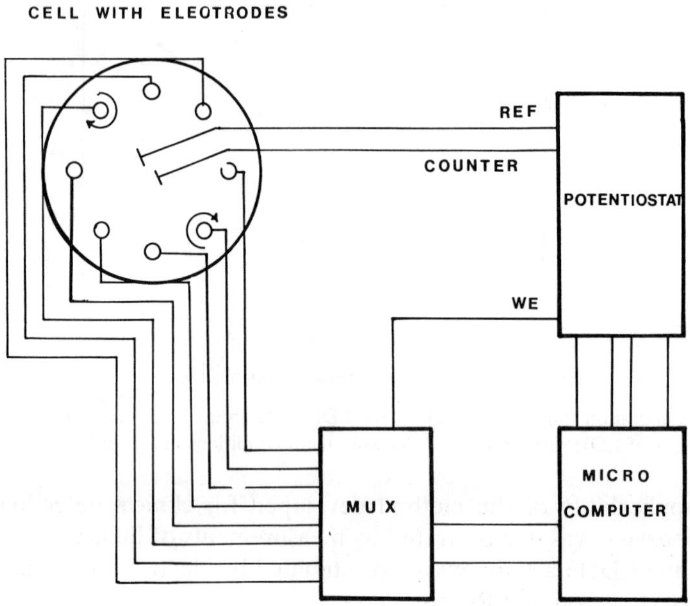

Figure 8. Schematic layout of the computerized multielectrode electrochemical system.

tivity is then measured. In the results presented in Figure 9 the antigen is *S. aureus*. The electrodes were incubated for 5 min in solutions containing different concentrations of the bacteria, washed and incubated for additional 5 min in a solution of rabbit anti-mouse IgG conjugated to alkaline phosphase. The results show that concentrations as low as 10 bacteria ml^{-1} can be detected.

The immunological reaction of *S. aureus* with rabbit IgG is based on binding of protein A abundant in this bacterium cell wall (Surolia, et al., 1982; Werner and Machleidt, 1978). Hence, in order to substantiate our results, we tested purified protein A as the antigen using the same immunoelectrochemical system. The results shown in Figure 10 indicate that the system is also very sensitive towards protein A and is capable of detecting protein A in the 0.1 pg ml^{-1} concentration range. It should be noted however that the reproducibility (as obtained from the results of two duplicates having the same concentration) is still not satisfactory, especially at high concentrations of bacteria. This is probably at least partially due to inhomogeneity of the carbon felt discs and is currently under investigation.

Concluding Remarks

This communication outlines some of the advantages of combining computerized electrochemical measurements with electrode-immobilized en-

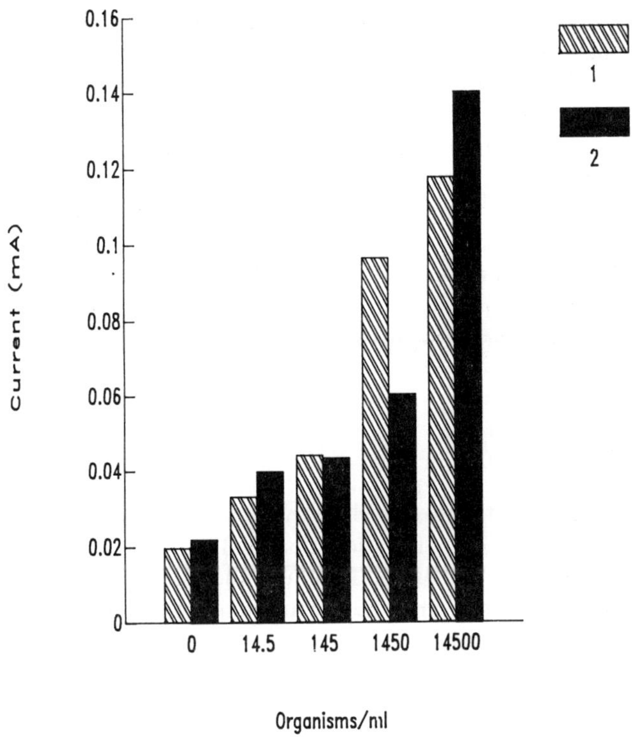

Figure 9. Response of the carbon felt disc electrodes after incubation with different concentrations of *S. aureus*. 1 and 2 denote two duplicates. See text for details.

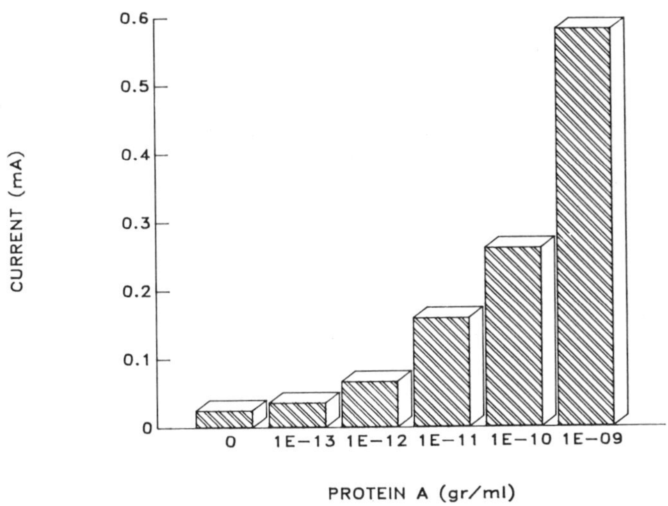

Figure 10. Response of the carbon felt disc electrodes after incubation with different concentrations of protein A.

zymes and antibodies. The examples shown represent the diverse aspects of the system aimed at the detection of analytes at concentrations above and below conventional ranges. The computerized glucose sensor has demonstrated the feasibility of continuous *in situ* monitoring of glucose in fermentation broth. The immunoelectrochemical system can be applied in clinical diagnostics and environmental control. The limits of detection of this method are at present being compared with recently reported heterogeneous assays using fluorescence (Kitamura, et al., 1989; 0.3 pg ml^{-1}) or luminescence (Bronstein, et al., 1989; 0.7 pg ml^{-1}). These reported immunassays are however more complicated and time consuming than the immunoelectrochemical assays described in this work.

Acknowledgments

The author wishes to thank the Ministry of Science and Technology, Israel, and Biovik K.B., Sweden, for their financial support. Thanks are also extended to I. Rosen, E. Hadas, A. Freeman and Y. Shabtai for their collaboration and to L. Soussan and Y. Gezundhajht for their assistance in the immunoelectrode research.

References

Bard, A.J. and Faulkner L.R. 1980. Electrochemical Methods Fundamentals and Applications. Wiley, New York.

Boyland, E. and Manson, D. 1957. The oxidation of aromatic amines. Part V. Oxidation by perphosphoric acids. J. Chem. Soc. 4689–4694.

Bronstein, I., Voyta, J.C. and Edwards, B. 1989. 1,2 Dioxetanes: Novel chemiluminescent enzyme substrate. Application to immunoassay. J. Biolumin. and Chemilumin. 4: 99–111.

Hollander, Z. and Katchalski-Katzir, E. 1986. Use of monoclonal antibodies to detect conformational alterations in lactate dehydrogenase isoenzyme 5 on heat denaturation and on adsorption to polystyrene plates. Mol. Imm. 23: 927–933.

Kitamura, K., Matsuda K., Ide, M., Tokunaga, T. and Honda, M. 1989. Fluorescence sandwich ELISA for detecting soluble and cell associated human interleukin-2. J. Imm. Meth. 121: 281–288.

Kolaric, K., Roguljic, A. and Perinovic, R. 1975. Comparative study of some enzymatic parameters and liver scanning in the early detection of the malignant liver processes. Clin. Chim. Acta. 60: 109–111.

McKenzie, D. and Henderson, A.R. 1983. Electrophoresis of lactate dehydrogenase isoenzymes. Clin. Chem. 29: 189–195.

Moiroux, J. and Elving, P.J. 1978. Effects of adsorption, electrode material and operational variables on the oxidation of dihydronicotinamide adenine dinucleotide at carbon electrodes. Anal. Chem. 50: 1056–1062.

Rishpon, J. and Rosen, I. 1987. Diffusion and activity in computerized enzyme electrode systems. Enzyme Engineering 8. 501: 292–297.

Rishpon, J. 1987. Chronocoulometric techniques in the study of computerized enzyme electrode systems. Biotech. and Bioeng. XXIX: 204–214.

Rishpon, J. and Rosen, I. 1989. The development of an immunosensor for the electrochemical determination of the isoenzyme LDH5. Biosensors. 4: 61–74.

Rishpon, J., Shabtai, Y., Rosen, I., Zibenberg, Y., Tor, R. and Freeman, A. 1990. In situ glucose monitoring in fermentation broth by "sandwiched" glucose-oxidase electrode (SGE). Biotech. and Bioeng. 35: 103–107.

Rosen, I. and Rishpon, J. 1989. Alkaline phosphatase as a label for a heterogeneous immunoelectrochemical sensor. J. Electroanal. Chem. 258: 27–39.

Surolia, A., Pain, D. and Islam Khan, M. 1982. Protein A: nature's universal anti-antibody. TIBS 0376: 74–76.

Tucker, D.F., Begent, R.H.J. and Hogg, N.M. 1978. Characterization of immune complexes in serum by adsorption on staphylococcal protein A: Model studies and application to sera of rats bearing a gross virus-induced lymphoma. J. of Immunol. 121: 1644–1651.

Usategui-Gomez, M., Wicks, M. and Warshaw, R.W. 1979. Immunochemical determination of the heart isoenzyme of lactate dehydrogenase (LDH-1) in human serum. Clin. Chem. 25: 729–734.

Werner, S. and Machleidt, W. 1978. Isolation of precursors of cytochrome Oxidase from *Neurospora crassa*: Application of subunit-specific antibodies and Protein A from *Staphylococcus aureus*. Eur. J. Biochem. 90: 99–105.

Discussion (DR. RISHPON)

Kornguth: Some of the reactions that you have shown evolve gas during the reaction process. I would expect the bubbles released to confound the results because of the non-uniform access of substrate to immobilized enzyme.

Rishpon: Bubbles will cause a change in the response of the electrode, but we can tell that it is an artifact. By employing this technique, we know that the rate in which glucose penetrates the enzyme layer is changed. This tells us that something is wrong, and we have to correct the response of the electrode. The current measured is the rate of glucose arriving. The technique described can tell us that it is changed, not because there is less glucose, but because the membrane is changed.

Key Words: Enzyme electrode, Antibody electrode, Immunoelectrochemical sensor, Computerized electrode, Chronocoulomteric, Glucose monitoring, Biosensors

NMR-Structural Studies of Membrane Bound Peptides and Proteins

Ki-Joon Shon[1], Patricia Schrader[1], Yongae Kim[1], Burkhard Bechinger[1], Michael Zasloff[2], and Stanley Opella[1]
[1]*Department of Chemistry, University of Pennsylvania, Philadelphia, PA 19104*
[2]*Departments of Pediatrics and Genetics, Division of Human Genetics and Molecular Biology, Children's Hospital of Philadelphia, Philadelphia, PA 19104*

Introduction

Membrane Proteins

Membrane proteins are responsible for many of the properties and functions of biological systems; they act as receptors and catalysts, transport ions and molecules, and have roles in the assembly, fusion, and structure of cells and viruses. More than half of all proteins are associated with membranes and, even though phospholipids in bilayers apparently account for many of the physical properties of membranes, biological membranes generally consist of greater than 50% protein by weight. Not only are many proteins tightly bound to membranes, numerous other proteins pass through membranes on their way to the locations where they carry out their functions.

In spite of their importance and prevalence, proteins that are either transiently or permanently associated with membranes are not nearly as well characterized as the water soluble globular proteins of the cytoplasm and periplasm. This reflects the difficulty in applying the modern biophysical methods of x-ray diffraction and NMR spectroscopy to membrane proteins. The multidimensional NMR methods that are so successful in describing the structures of small globular proteins in solution (Wuthrich, 1986) do not work very well with large, slowly reorienting aggregates. X-ray diffraction requires high quality single crystals, and membrane associated proteins are much more resistant to crystallization than globular proteins.

The structures of two membrane proteins, bacteriorhodopsin (Henderson, et al., 1990) and the photosynthetic reaction center (Deisenhofer, et al., 1985), have been determined because they form ordered arrays or crystals that have been amenable to structural analysis, albeit with extraordinary experimental efforts. The structures of these two proteins have many features in common. Just as the initial structures of myoglobin and lysozyme displayed

D. Kamely et al. (eds), Biotechnology: Bridging Research and Applications, 109–124.

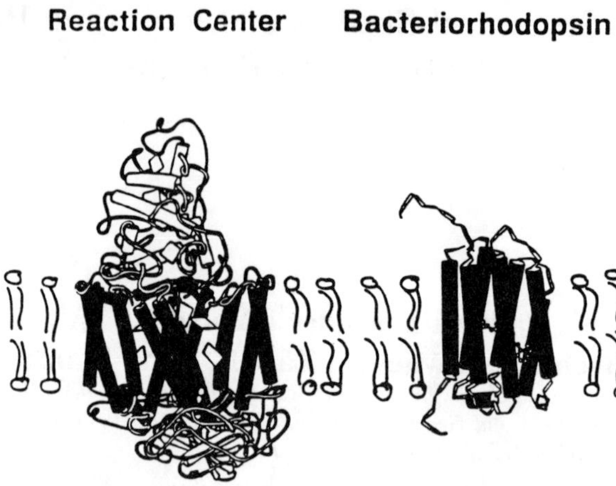

Figure 1. Structures of the photosynthetic reaction center (Deisenhofer, et. al.1985) and bacteriorhodopsin (Henderson, et. al., 1990).

most of the features found in subsequent studies of globular proteins (Dickerson and Geis, 1969), the structural features shown in Figure 1 are likely to turn up in various combinations in other membrane proteins. The structures of both proteins are dominated by long hydrophobic membrane spanning helices, approximately perpendicular to the plane of the bilayer. They also have amphipathic helices in the plane of the bilayer. The helices parallel and perpendicular to the plane of the bilayer are connected by loops or turns, at least some of which have been shown to be mobile and unstructured (Keniry, et al., 1984). The N- and C-terminal regions of these and other membrane proteins are mobile and unstructured as well (Leo, et al., 1987).

Some of the most interesting differences between membrane proteins and globular proteins arise from dynamics. Membrane proteins have highly mobile regions in distinctive locations, including the loops connecting the helices and both the N- and C- termini (Keniry, et al., 1984; Leo, et al., 1987). In contrast, globular proteins appear to be much more rigid with few mobile residues. Thorough investigations of all proteins must include their dynamics; however the strong correlation between structural features and mobility in membrane proteins means that dynamic and structural studies performed in parallel yield substantial dividends in characterizing and understanding their structures and functions.

NMR Spectroscopy

The development of NMR methods for describing membrane proteins will be facilitated by many of the same factors that contributed to the development of methods for globular proteins in solution. Foremost is the selection of tractable peptide and protein systems. This generally means relatively short polypeptide sequences that can be prepared by over-expression in bacteria or solid-phase peptide synthesis, enabling large amounts to be prepared and labeled specifically, selectively, and uniformly with stable isotopes (Shon, et al., 1989).

Both multidimensional solution NMR spectroscopy (Wuthrich, 1986) and solid-state NMR spectroscopy (Opella, et al., 1987) are independently capable of describing the structure and dynamics of proteins. However, at their present stages of development it is necessary to utilize a combination of these two methods to overcome the limitations inherent in applying either one of them alone to the molecular aggregates formed by proteins in membrane environments. This is feasible because there are two viable model membrane systems. Membrane proteins can be solubilized in detergent micelles; this is the type of preparation used to crystallize the reaction center protein for the x-ray diffraction studies (Deisenhofer, et al., 1985). Small membrane proteins in well behaved micelles are suitable samples for solution NMR studies; although the protein-micelle complexes do not reorient as rapidly as small globular proteins, they do reorient rapidly enough to give reasonably well resolved spectra. Membrane proteins can also be embedded in phospholipid bilayers, and these samples are appropriate for solid-state NMR studies because the proteins are immobilized by the lipid environment.

Several hundred measurements of short distances can be combined with distance geometry or other algorithms to determine the complete three-dimensional structures of proteins in solution. However, the secondary structure of a protein can be mapped out quite reliably by observing a few characteristic distances for each residue (Wuthrich, et al., 1984). Because of the limited range of possible secondary structures anticipated for membrane proteins, based on the results shown in Figure 1, knowledge of the secondary structure alone is much more valuable information in the case of membrane proteins than globular proteins. One of the initial goals of the NMR structural studies is to determine the secondary structure of a membrane protein.

The internuclear distances between ^{1}H sites are measured qualitatively using homonuclear nuclear Overhauser enhancements (NOEs). Although many different multidimensional experiments have been devised to enhance resolution, they are all designed to resolve and measure cross-peak intensities arising from $^{1}H/^{1}H$ NOEs from two nearby (<5 Å apart) ^{1}H sites. The interpretation of the observed NOEs can only be performed when the resonances are assigned to specific residues in the protein. Assignments can be made in number of ways; however, in the case of membrane proteins, many of the most direct experimental approaches do not work and isotopic labeling is essential for making many of the assignments.

The dynamics of individual sites in a protein in solution are described through nuclear relaxation measurements. In general, many different parameters must be measured as a function of field strength to give a detailed view of the intramolecular motions of a protein. This is especially difficult to accomplish for a membrane protein. Fortunately, the $^{1}H/^{15}N$ heteronuclear NOE, by itself, is a reliable indicator of local segmental backbone motions of a protein (Bogusky, et al., 1987). By labeling the backbone amide sites with ^{15}N, it is possible to measure heteronuclear NOEs for individual residues and, because of the negative gyromagnetic ratio of ^{15}N, these measurements can be interpreted in a straightforward manner. Sites that are rigidly held by the protein structure reorient with the correlation time of the protein-micelle complex and give resonances with small, positive intensity. In contrast, protein sites that undergo rapid motion give resonances with negative intensity of variable amplitude. The qualitative distinction between negative and positive peaks is so striking, even in protein spectra with limited resolution, that mobile sites can be readily identified. Although it would certainly be desirable to bring the full power of relaxation analysis to bear on this problem and describe internal motions in great detail, the observation of mobile sites in membrane proteins is extremely valuable at the present level of understanding these proteins.

Several different solid-state NMR experiments yield structural information, because both distances and angles can be measured with the various spectral parameters observable in these experiments. The approach we are developing for determining the structures of proteins by solid-state NMR spectroscopy takes advantage of the spectral characteristics of oriented and immobile protein samples, such as membrane proteins in phospholipids on glass plates, and measuring spectral parameters that yield the angles between bonds or chemical groups and the direction of orientation (and the applied magnetic field) (Opella, et al., 1987). In immobile samples, as in oriented proteins, the spectral parameters reflect the anisotropy of the various nuclear spin interactions that are present in natural abundance or isotopically labeled sites. Of course, exactly the same interactions are present in solution, but they are averaged to their isotropic values by the rapid overall reorientation of the protein. Because the heteronuclear dipole-dipole, chemical shift, and nuclear quadrupole interactions are highly anisotropic, the resonance frequencies and splittings directly reflect the orientations of the sites. The experimental goal is to assign and measure several angular parameters for each peptide plane (or sidechain) and to interpret these measurements to give the orientations of the peptide planes relative to the axis system defined by the orientation of the sample in the magnetic field. Once the orientations of all of the individual peptide planes are established, then it is possible to reconstruct the structure of the protein.

Solid-state NMR spectroscopy is well suited for describing the dynamics of proteins. In unoriented samples the powder pattern lineshapes from individual sites are strongly influenced by motions. The motionally averaged lineshapes can be interpreted in terms of specific models of motion by comparing calculated and experimental lineshapes (Opella, 1986). These methods are useful for both backbone and sidechain sites. They are particularly adept at distinguishing mobile sites in the presence of rigid sites, which is of considerable value in understanding membrane proteins.

Filamentous Bacteriophage Coat Protein

The coat protein from filamentous bacteriophage provides a tractable experimental system for investigating several events that occur at membranes (Marvin and Hohn, 1969; Makowski, 1984). Bacterial cells infected with filamentous bacteriophages synthesize the polypeptide coded for by the viral gene 8; for example, class I bacteriophages fd (or M13) produce many copies of the 73 residue procoat protein with its 23 residue N-terminal leader sequence (Chang, et al., 1978). The procoat protein is then inserted into the cell membrane where it is processed by the bacterial leader peptidase to its mature form which contains 50 residues (Ito, et al., 1980). Many copies of the coat protein are stored in the cell membrane prior to virus assembly, which occurs at the membrane. New virus particles are extruded through the cell membrane, making this a non-lytic infection.

NMR spectroscopy is essential for describing these proteins and the processes in which they participate. Although fiber diffraction has played an important role in describing the structure of the coat protein in the virus particles (Makowski, 1984), despite substantial efforts in several laboratories, none of the forms of coat protein have been crystallized. Even though some forms of the protein exist only transiently in vivo, all of the forms can be prepared in vitro as samples suitable for investigation by NMR spectroscopy. Procoat protein and coat protein can be reconstituted into detergent micelles for solution NMR studies or phospholipid bilayers for solid-state NMR studies of the membrane bound forms of the proteins. Comparisons among all of the various coat protein species are of interest, including between the structural form in the virus and the membrane bound form of the coat protein, between the procoat protein and the coat protein, and between the coat protein from class I bacteriophages (e.g., fd, M13) and class II bacteriophages (e.g., Pf1).

Bacteria infected with filamentous bacteriophages can be grown on defined media, enabling both uniform and selective isotopic labeling of the coat protein. The protein can also be prepared by automated solid-phase peptide synthesis, enabling specific labeling of sites. In both cases, sufficient quantities are available for NMR studies.

fd and M13 are essentially identical viruses. Their coat proteins differ by only a single residue and together they constitute the best studied example of a class I filamentous bacteriophage. The sequence of fd coat protein is:

AEGDDPAKAA$_{10}$FDSLQASATE$_{20}$YIGYAWAMVV$_{30}$
VIVGATIGIK$_{40}$LFKKFTSKAS$_{50}$.

The sequence is characterized by a long stretch of hydrophobic amino acids (residues 20–40). Globular proteins of this size can have their structures determined by multidimensional solution NMR methods in a straightforward manner without using isotopic labels. Because filamentous bacteriophage coat proteins are membrane proteins and are insoluble in aqueous solution, they present difficulties for conventional NMR investigations of their structures (Cross and Opella, 1981). However, it has been possible to reduce the impact of these difficulties on the structure determination through extensive isotopic labeling, optimal choice of micelles, heteronuclear NMR techniques, and the direction about structural motifs derived from bacteriorhodopsin and the photosynthetic reaction center. Isotopic labeling is the key and is essential for several purposes in the solution NMR experiments, including improving resolution, making resonance assignments, and controlling spin-diffusion (Shon and Opella, 1989).

Magainins

Magainins are a novel class of 20–25 residue peptides with potent antibacterial activity and high selectivity between animal and bacterial cells. Magainins were found initially in frog skin (Zasloff, 1987) and have since been identified in mammalian organs. As is the case for nearly all membrane associated peptides and proteins, little structural information is available. A conventional NMR study showed them to be alpha helical in trifluoroethanol-/water solution and unfolded in aqueous solution (Marion, et al., 1988). These peptides probably form amphipathic helices in lipid environments. Their high degree of selectivity between mammalian and bacterial cells means that peptide-lipid interactions are likely to have strong effects on their structures and functions. Like many other peptides, they are synthesized in the form of polyproteins with leader sequences and are subsequently processed by cellular enzymes. This means that there may be multiple mechanisms for selectivity, including the placement of the peptides in the proper membranes by the leader sequence.

The sequence of the Magainin peptide used in the NMR studies is:

GIGKFLHSAG$_{10}$KFGKAFVGEI$_{20}$MNS

It was prepared by solid-phase peptide synthesis. Samples can be prepared in detergent micelles for solution NMR studies and in phospholipid bilayers for solid-state NMR studies, enabling the determination of the secondary structure and orientation of these peptides as a function of lipid composition.

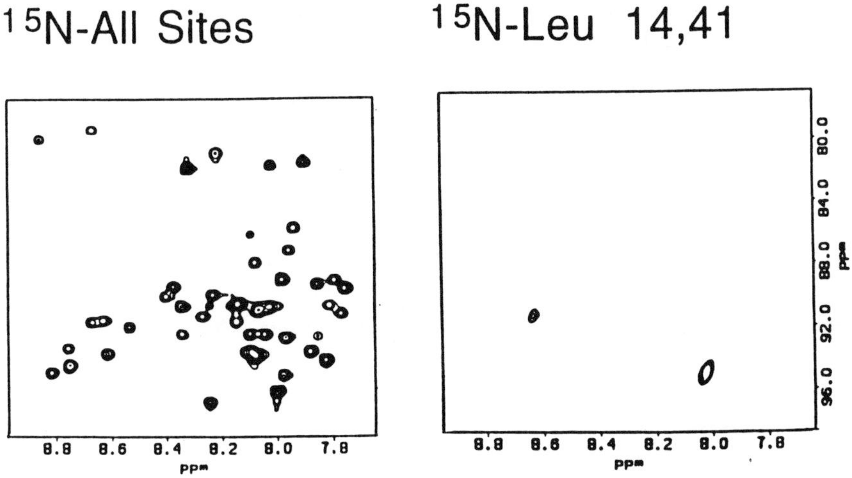

Figure 2. $^1H/^{15}N$ heteronuclear correlation spectra of fd coat protein in micelles in solution. The spectra are from a uniformly ^{15}N labeled sample (left) and a selectively ^{15}N-Leu labeled sample (right). Each cross-peak in these contour plots represents the resonance frequencies of the amide nitrogen and its bonded hydrogen.

Solution NMR Studies of Membrane Proteins in Micelles

Filamentous Bacteriophage Coat Protein

Even small proteins in micelles, such as the 50 residue fd coat protein, produce broad resonance lines because the protein-micelle complex has a large effective radius and reorients relatively slowly in solution. The broad lines combine with the limited chemical shift dispersion resulting from so many residues having the same helical secondary structure to give spectra with limited resolution. The most direct way around the resolution problem is to label the protein with ^{15}N and to perform heteronuclear correlation experiments. The two-dimensional $^1H/^{15}N$ heteronuclear correlation spectrum in Figure 2A was obtained on a uniformly ^{15}N labeled sample of fd coat protein in sodium dodecyl sulfate (SDS) micelles in H_2O solution. Each correlation-peak in this contour plot corresponds to a single amide site at the resonance frequencies associated with both the ^1H chemical shift and the ^{15}N chemical shift. Because these spectra are obtained by detecting the ^1H resonances with ^{15}N decoupling, the large 90 Hz $^1H/^{15}N$ spin-spin coupling is not observed. Even though the smaller $^1H/^1H$ spin-spin couplings are present, they are not observed because of the relatively broad linewidths of the ^1H resonances. This is an example of why it is more difficult to study membrane proteins than globular proteins by NMR spectroscopy, since these couplings are extremely useful for protein structure determination and yet are unavailable for membrane proteins in micelles.

All of the amide sites in fd coat protein contribute correlation peaks to the spectrum on the left in Figure 2, since all sites are labeled with ^{15}N and the protein sample is in H_2O. The spectrum on the right in Figure 2 was obtained under the same conditions as that on the left in Figure 2, except that the protein is selectively, rather than uniformly, labeled. The bacteria were grown on a defined medium where all of the amino acids were added, 19 of them with natural abundance and only leucine with ^{15}N. The spectrum on the right in Figure 2 was also was obtained in H_2O and shows two fully resolved correlation peaks, one for each of the two leucine residues in the protein. This clearly assigns two of the correlation peaks in the complete spectrum on the left in Figure 2 to leucine residues. Additional steps are needed to assign these peaks to individual leucines. fd coat protein has a limited population of amide hydrogens that are resistant to exchange with solvent deuterons (Cross and Opella, 1981). Only one of the leucine residues has an amide site that undergoes very slow exchange; based on previous detailed studies of hydrogen exchange kinetics (Henry, et al., 1987), Leu-41 is the residue that is slowly exchanging and gives the peak in D_2O solution. By difference, this also assigns the resonance for Leu-14.

In order to characterize the entire secondary structure of fd coat protein, rather than just the parts of it that give the best spectral results, the limitations resulting from the broad lines and efficient spin-diffusion that accompany the slow reorientation of proteins in micelles, as well as the extensive spectral overlap, had to be reduced. This was accomplished by labeling fd coat protein uniformly with ^2H (90%) on all carbon sites, with ^{15}N (99%) on all nitrogen sites, and with ^2H (50%) on all exchangeable sites and by combining $^1H/^{15}N$ heteronuclear correlation and $^1H/^1H$ NOE spectroscopy (Shon and Opella, 1988). The combined two-dimensional $^1H/^{15}N$ correlation and $^1H/^1H$ NOE spectrum of fd coat protein is shown twice in Figure 3A and 3B and has correlation peaks for all amide sites with NOE cross-peaks between all adjacent residues in alpha helices. The tracing of two continuous pathways of NOEs among the various amide ^1Hs directly maps out the two stretches of alpha helix in the protein; this is done in two separate spectra for clarity. The alpha helical segments that correspond to the paths of NOEs come from the long hydrophobic helix and the shorter amphipathic helix. The hydrophobic helix has turned out to be particularly difficult to characterize, largely because of the extensive spin diffusion among its resonances. This spin-diffusion causes loss of a number of expected cross-peaks in the spectrum in H_2O. However, key cross-peaks are present in the spectrum obtained in D_2O, probably because only the central part of the helix contributes the slowly exchanging amide resonances and they have more limited possibilities for spin-diffusion. The correlation peaks assigned to the amide sites in the N- and C- terminal regions and the region between the two helices do not have NOE cross-peaks to other amide sites.

One of the best characterized examples of a protein with a partially flexible backbone is the membrane bound form of fd coat protein (Leo, et al., 1987).

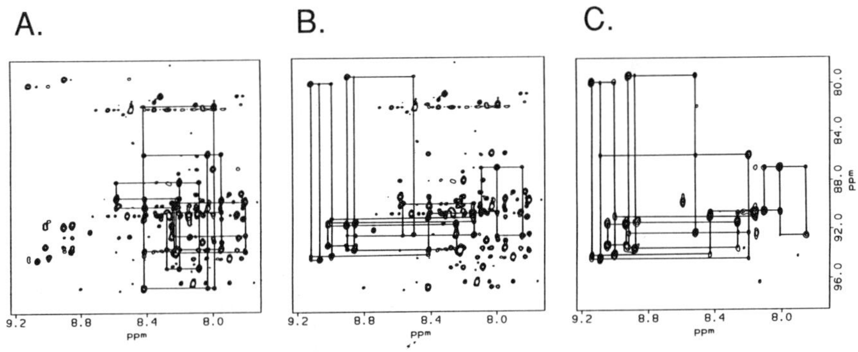

Figure 3. Combined $^1H/^{15}N$ heteronuclear correlation with $^1H/^1H$ NOE spectra of uniformly ^{15}N and C-^2H labeled fd coat protein (Shon and Opella, 1989). These spectra have NOE cross-peaks between correlation peaks from residues with amide hydrogen in close proximity, as in an alpha helix. A. and B. are the same spectrum, which was obtained in H_2O solution so that correlation peaks are present from all amide sites. The amphipathic helix between residues 7–19 is mapped out in A and the hydrophobic helix between residues 25–46 is mapped out in B. The spectrum in C. was obtained in D_2O solution, so there are only correlation peaks from residues with slowly exchanging amide hydrogens. Part of the hydrophobic helix is confirmed by these data with the helix between residues 28–42 mapped out in C.

Amide nitrogen sites in proteins have a single bonded proton, therefore ^{15}N spin-lattice relaxation is dominated by the heteronuclear dipole-dipole coupling between the ^1H and ^{15}N spins. Some of the protein backbone sites are highly mobile and give negative signal intensity in NOE experiments (Bogusky, et al., 1987). One of the mobile sites is Gly-3, as shown in Figure 4. The protein has four labeled glycine residues, three of which give small positive signals and one of which, Gly-3, gives a large negative signal. All results show that the N-terminal residues up to Pro-6 are mobile in both the membrane bound and viral forms of the protein. Similar experiments show that the C-terminal residues are mobile as well.

Magainin

Solid-phase peptide synthesis enables isotopes to be placed in specific sites. The 23 residue magainin sequence was synthesized with ^{15}N-Leu at residue 6. Figure 5 contains the $^1H/^{15}N$ heteronuclear correlation spectrum of specifically ^{15}N labeled magainin in micelles in solution. A single correlation peak is observed, consistent with a single conformation for the peptide in micelles. In separate experiments $^1H/^1H$ homonuclear NOEs are observed between amide NHs for many sites showing the presence of alpha helix in a substantial part of the peptide.

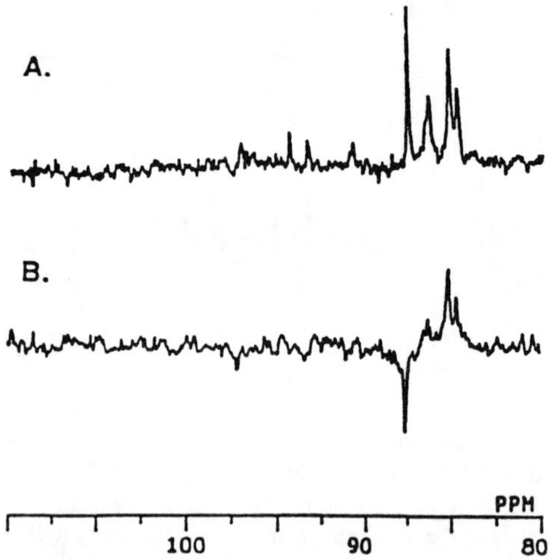

Figure 4. ^{15}N NMR spectra of ^{15}N-Gly labeled fd coat protein in SDS micelles (Bogusky, et al., 1987). A. is the conventional spectrum showing resolved resonances from the four glycine residues in the protein. B. is the spectrum obtained with the heteronuclear $^{1}H/^{15}N$ NOE expressed. The one mobile glycine gives a negative peak while the three rigidly held glycines give small positive peaks.

Studies of Protein in Bilayers

Filamentous Bacteriophage Coat Protein

fd coat protein can be transferred directly from the virus to phospholipids. The protein is immobilized by its interactions with the lipids, even though they are fully hydrated. The membrane bound form of the protein has mobile N- and C- terminal residues. An unoriented sample of uniformly ^{15}N labeled

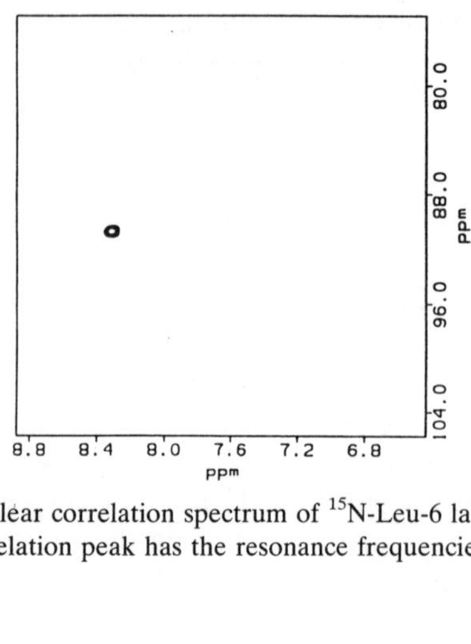

Figure 5. $^{1}H/^{15}N$ heteronuclear correlation spectrum of ^{15}N-Leu-6 labeled magainin in micelles in solution. The single correlation peak has the resonance frequencies of the ^{15}N amide and its bonded hydrogen.

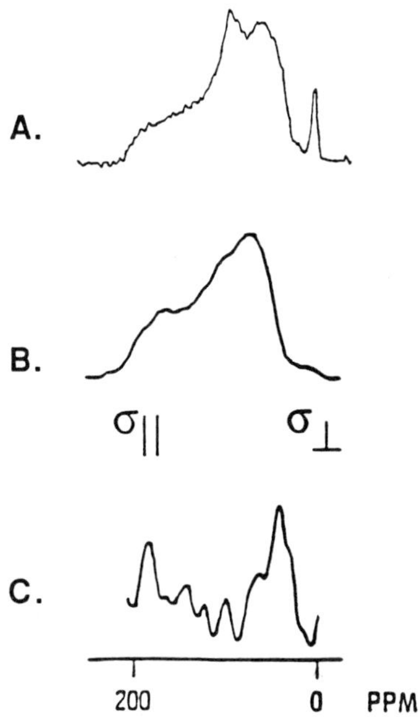

Figure 6. Solid-state ^{15}N NMR spectra of fd coat protein in hydrated phospholipid bilayers. A. is the spectrum of an unoriented uniformly 15N labeled sample. Isotropic intensity is superimposed on the powder pattern from the rigidly held amide sites. The narrow peak on the right is from the amino nitrogens. B. and C. are spectra of a ^{15}N-Leu labeled samples. B. is unoriented. C. is oriented between glass plates with the plane of the bilayers perpendicular to the direction of the applied magnetic field. The two peaks are from the Leu residues in different orientations relative to the direction of the magnetic field.

fd coat protein in phospholipids gives the solid-state NMR spectrum shown in Figure 6A (Leo, et al., 1987). There is a substantial intensity from narrow isotropic resonances from the mobile residues superimposed on the powder pattern from the rigidly held residues. The rigid and mobile residues can be identified by studying specifically and selectively labeled protein samples. The spectrum in Figure 6B is from a protein sample labeled selectively in the two leucine residues. Both leucines 14 and 41 are rigidly held in the coat protein structure, since they both contribute to the observed powder pattern. If one of the leucines were mobile, then isotropic intensity would be readily observed on top of the powder pattern.

Phospholipids containing membrane proteins can be oriented between glass plates. These proteins fulfill the two requirements for samples for solid-state NMR structural studies, since they are immobilized by their interactions with the lipids, and the direction of orientation can be arranged to be parallel to that of the applied magnetic field. This means that ^{15}N labeled sites give single line resonances rather than powder patterns, greatly enhancing resolution, and these single lines have resonance frequencies that reflect the orientation of the peptide group relative to the known direction of sample orientation.

Solid-state NMR spectroscopy can determine the complete structures of immobile proteins, such as those embedded in lipid bilayers, when several spectral parameters are measured to characterize the angles between each of the peptide planes and the direction of sample orientation (Opella, et al., 1987). With the secondary structure of the membrane bound form of fd coat protein determined on the basis of NOEs in solution, the ^{15}N chemical shift provides sufficient angular restrictions to determine the orientations of the helices relative to the direction of orientation and each other in the protein. The spectrum in Figure 6C presents solid-state ^{15}N NMR spectra of ^{15}N-Leu labeled fd coat protein oriented in phospholipid bilayers. These spectra can be simply interpreted by using the ^{15}N amide chemical shift tensor as a qualitative guide to the orientation of the peptide groups in the helices. An amide N-H bond approximately parallel to the direction of the applied magnetic field has a ^{15}N resonance frequency close to that of the principal element $\sigma_{\parallel}$ and an amide N-H bond perpendicular to the field has a ^{15}N resonance frequency close to that of the principal element $\sigma_{\perp}$(Harbison, et al., 1984). The spectrum of uniformly ^{15}N labeled protein has two bands of signal intensity, near the resonance frequencies associated with $\sigma_{\parallel}$ and $\sigma_{\perp}$. Spectral resolution is greatly improved by using selectively labeled protein samples. There are two leucine residues in fd coat protein and ^{15}N-Leu labeled fd coat protein has two resonances, one of which has a resonance frequency near $\sigma_{\perp}$. Therefore, the N-H bond of Leu-14 and hence the amphipathic helix that includes this residue must be approximately perpendicular to the direction of the magnetic field and parallel to the plane of the bilayer. One of the leucine residues in fd coat protein is in the hydrophobic helix, and the spectrum in Figure 6A has one resonance with intensity near $\sigma_{\parallel}$. This demonstrates that the N-H bonds of Leu-41 (and the hydrophobic helix) are approximately parallel to the direction of the magnetic field and perpendicular to the plane of the bilayer.

Magainin

In solution, magainin labeled at Leu-6 with ^{15}N shows a single line at its isotropic resonance position. The solid-state NMR spectral characteristics of this site can be used to determine the orientation of the peptide in the membrane. The powder pattern spectrum in Figure 7 has the frequency breadth and lineshape of a rigid polycrystalline peptide, demonstrating that the site and the peptide are immobilized by their interaction with the phospholipids.

Phospholipids containing magainin can be oriented between glass plates. The spectrum in Figure 7 is obtained from this sample. It has a single line near the position of $\sigma_{\perp}$ corresponding to an amide site with its N-H bond

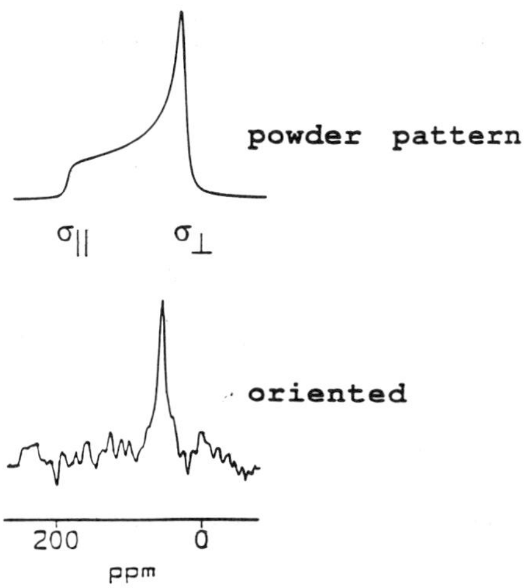

Figure 7. Solid state ^{15}N NMR spectra. Top: calculated ^{15}N amide powder pattern. Bottom: experimental spectrum of magainin labeled at Leu-6 in phospholipid bilayers; the spectrum was obtained after aligning the peptide containing bilayers between glass plates with the plane of the bilayer perpendicular to the magnetic field.

and the plane parallel to the plane of the bilayers. Since magainin forms an amphipathic helix, this indicates that the peptide is bound to the phospholipids parallel to the plane of the bilayer, most likely along the top of the membrane bilayer, enabling the hydrophobic side of the helix to interact with the hydrocarbon chains and the hydrophilic side of the helix to interact with the polar headgroups and the solvent.

Conclusions

The NMR spectra data presented here can be readily interpreted based on established NMR methods. The combination of several different experiments leads to a model for each protein. This is illustrated in Figure 8 where fd coat protein and magainin are shown in the presence of lipids. This figure should be compared to Figure 1. There are strong similarities in the basic structural features of all of the membrane proteins, with hydrophobic membrane spanning helices, amphipathic helices in the headgroups of the lipids, and mobile and rigid bridging loops.

The results presented here are among the first on membrane proteins. They benefit from the development of several spectroscopic techniques and their application to tractable systems. There are many peptide and protein systems that can have these methods applied to them in their present stage of development. Future developments, especially the implementation of high field magnets, three- and higher dimensional experiments, and 1H spectral

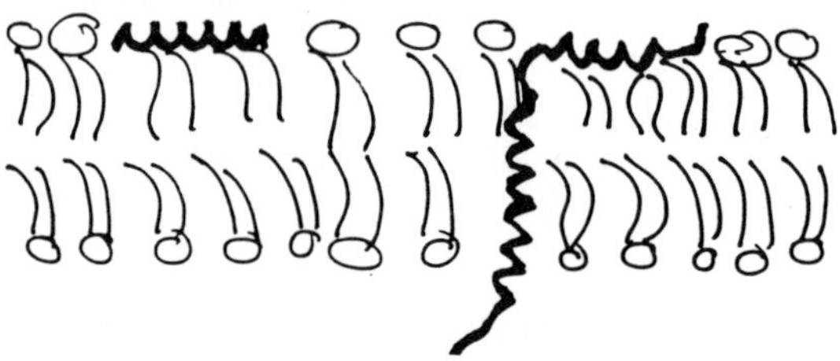

Figure 8. Structural models of fd coat protein and magainin in a lipid bilayer. fd coat protein has two helical segments; the amphipathic helix is parallel to the plane of the bilayer and the hydrophobic helix is perpendicular to the plane of the bilayer. The magainin helix is parallel to the plane of the bilayer.

parameters in the solid-state, will greatly enhance their applicability. These developments are all currently underway.

Acknowledgments

We thank Howard Eck for synthesizing the magainin peptide sample and Brian Sykes for helpful dicussions on coat protein. This research is supported by grants (GM-24266–14 and AI-20770–06) from the National Institutes of Health and the Benjamin Franklin Partnership and the Mathers Foundation. B.B. is supported by an EMBO long-term fellowship (ALTF 454–1989).

References

Bogusky, M.J., Schiksnis, R.A., Leo, G.C. and Opella, S.J. 1987. Protein backbone dynamics by solid state and solution ^{15}N NMR spectroscopy. J. Magn. Reson. 72: 186–190.

Chang, C., Blobel, G. and Model, P. 1978. Detection of prokariotic signal peptidase in an *E. coli* membrane fraction: Endoproteolytic cleavage of nascent f1 procoat protein. Proc. Natl. Acad. Sci. 75: 361–365.

Cross, T.A. and Opella, S.J. 1981. 'H and ^{13}C NMR of the aromatic residues of fd coat protein. Biochem. 20: 290–297.

Deisenhofer, J., Epp., O., Miki, K., Huber, R. and Michel, H. 1985. Structure of the protein subunits in the photosynthetic reaction center of *Rhodopseudomonas viridis* at 3 Å resolution. Nature 318: 618–624.

Dickerson, R.E. and Geis, I. 1969. The Structure and Action of Proteins. Benjamin/Cummings, Menlo Park, CA. pp. 1–120.

Harbison, G.S., Jelinski, L.W., Stark, R.E., Torchia, D.A., Hertzfeld, J. and Griffin R.G. 1984. ^{15}N chemical shift and ^{15}N-^{13}C dipolar tensors for the peptide bond in [1–13C] glycyl [15N] glycerine hydrochloride monohydrate. J. Magn. Reson. 60: 74–82.

Henderson, R., Baldwin, J.M., Ceska, T.A., Zemlin, N., Beckmann, E. and Downing, K.H.

1990. Model for the structure of Bacteriorhodopsin based on high resolution electron cryo-microscopy. J. Mol. Biol. 213: 899–929.
Henry, G.D., Werner, J.H. and Sykes, B.D. 1987. Backbone dynamics of a model membrane protein: Measurement of individual amide hydrogen-exchange rates in detergent solubilized m13 coat protein using ^{13}C NMR hydrogen/deuterium isotope shifts. Biochem. 26: 3626–3634.
Ito, K., Date, T. and Wickner, W. 1980. Synthesis, assembly into the cytoplasmic membrane, and proteolytic processing of the precursor of coliphage M13 coat protein. J. Biol. Chem. 255: 2123–2130.
Keniry, M.A., Gutowsky, H.S. and Oldfield, E. 1984. Surface dynamics of the integral membrane protein bacteriorhodopsin. Nature. 307: 383–386.
Leo, G.C., Colnago, L.A., Valentine, K.G. and Opella, S.J. 1987. Dynamics of fd coat protein in lipid bilayers. Biochem. 26: 854–862.
Makowski, L. 1984. Structural diversity in filamentous bacteriophages. In: Biological Macromolecules and Assemblies. McPherson, ed. Wiley, NY. p. 202.
Marion, D., Zasloff, M. and Bax, A. 1988. A two-dimensional NMR study of the antimicrobial peptide magainin 2. FEBS Lett. 227: 21–26.
Marvin, D. and Hohn, B. 1969. Filamentous bacterial viruses. Bacterial Rev. 33: 172–209.
Opella, S.J., Stewart, P.L. and Valentine, K.G. 1987. Protein structure by solid-state NMR spectroscopy. Q. Rev. Biophys. 19: 7–49.
Opella, S.J. 1985. Protein Dynamics by Solid State NMR. Methods Enzymol. 17: 327–361.
Schiksnis, R.A., Bogusky, M.J., Tsang, P. and Opella, S.J. 1987. Structure and dynamics of the Pf1 filamentous bacteriophage coat protein in micelles. Biochem. 26: 1373–1381.
Shon, K., Schrader, P., Opella, S.J., Richards, J. and Tomich, J. 1989. NMR spectra of synthetic membrane bound coat protein species. In: Frontiers of NMR in Molecular Biology, UCLA Symposia on Molecular and Cellular Biology, New Series, Vol. 109. Live, D., Armitage, I. and Patel, D., eds. Alan R. Liss, NY. 104–118.
Shon, K. and Opella, S.J. 1989. Detection of 'H homonuclear NOE between amide sites in proteins with $^{1}H/^{15}N$ heteronuclear correlation spectroscopy. J. Magn. Reson. 82: 193–197.
Wuthrich, K., Billeter, M. and Braun, W. 1984. Polypeptide secondary structure determination by NMR observation of short proton-proton distances. J. Mol. Biol. 180: 715–740.
Zasloff, M. 1987. Magainins, a class of antimicrobial peptides from *Xenopus* skin: Isolation, characterization of two active forms, and partial cDNA sequence of a precursor. Proc. Natl. Acad. Sci. 84: 5449–5453.

Discussion (DR. OPELLA)

Ward: I have a question about your migainin data and modelling. Do you have any experimental evidence that bears on the question of the stoichiometry when the helices form clusters perpendicular to the membrane?

Opella: No. That is not my modelling, but I took that picture just to illustrate a channel with a hole in the center.

Ward: Concerning your Pf1 model, I was somewhat confused. You showed us a lot of interesting data and ended up showing us some models at the end. It was not clear to me whether those models you showed were models based upon the break in the chain which you deduced by your mobility data, or whether the models also implied coordinates which you determined from your NMR data. Can you clarify that point for me?

Opella: At this stage, the models do not represent coordinates. They represent the secondary structure as determined, the dynamics as determined from heteronuclear NOEs from ^{1}H NOEs in solution and motionally averaged lineshapes, and the directions as determined for those pieces of secondary structure in solid-state NMR experiments on oriented samples. These experiments can give full three-dimensional coordinates, but we are not there yet in data analysis.

Kornguth: From the point of view of applications to various DoD needs, there are two interesting features that you have talked about. First, NMR spectroscopy is the only technique that we have now, as you pointed out, which is capable of providing atomic resolution, in solution as well as on a solid phase, in a non-crystalline state. There are biomaterials which are excellent organic electron conductors or phonon conductors – phthalocyanins, porphyrins. Can one utilize NMR spectroscopy to determine the distance in spacing between the face to face porphyrins in solution as well as in a matrix such as dimethyl formamide or polycinnamic acid materials? Second, can one distinguish between protein components that are adsorbed onto a surface by beta pleated sheet hydrogen bonding versus alpha-helical materials that are inserted in a non-solution state into a membrane? Does the powder pattern obtained from the former differ from the pattern of the latter structures?

Opella: I do not understand the last question very well. Are you trying to distinguish the effect of environment on the structure, or are you trying to distinguish how the proteins orient?

Kornguth: I am trying to distinguish how the proteins are interacting with each other. In the case of the alpha helix, we have multiple barrels, one unit interacting with the other, as compared to an adsorption on a surface of a protein not interacting by alpha helical features, or by side chain interactions perturbing the alpha helix.

Opella: I think what you want is a qualitative assay of something being inside or outside. I think the answer to that is no. We cannot sort that out by NMR spectroscopy. If you had a very good preparation though, where we could go ahead and determine secondary structure and orientation, then of course we could tell the difference between them. Regarding your first question on the interaction between the phthalocyanins or porphyrins in an oriented state as compared with the matrix around it, there is whole range of experiments in solid state NMR that will do those structural studies. It is a very well developed field.

Key Words: NMR spectroscopy, Membrane proteins, Peptides, Magainin, Filamentous bacteriophage coat protein, fd coat protein, Solid-state NMR, Two-dimensional NMR

Session II
Molecular Biology in Field Detection

DNA as a Biosensor for Environmental Agents

Lawrence Grossman and William Athas
Department of Biochemistry, The Johns Hopkins University School of Hygiene and Public Health, 615 North Wolfe Street, Baltimore, MD 21205

It is the thesis of this paper that DNA can be utilized as a target to detect environmental agents that can modify DNA directly or indirectly through metabolic activation. Expression of DNA modification will be achieved by using as a target a bacterial gene encoding an easily assessable enzyme. The modified gene will be physically introduced into DNA repair deficient lymphoid lines. The level of damage will be reflected as loss of gene activity manifested as reduction of the specific enzyme activity. This extremely sensitive and specific assay, in its essence, is a "Host Cell Reactivation" (HCR) in which damaged viruses are repaired by their hosts.

An introduction into the generic species of damaging agents and the fund of repair systems that normally accommodate to such changes, but can be mutated in the planned tester strains, will be presented.

The ability of cells to survive hostile environments is due in part to surveillance systems that recognize damaged sites in DNA and are capable of either reversing the damage or removing damaged bases or nucleotides, generating sites which lead to a cascade of events restoring DNA to its original structural and biological integrity.

Both endogenous and exogenous environmental agents can damage DNA. A number of repair systems are regulated by the stressful effects of such damage, affecting the levels of responsible enzymes or modifying their specificity. Repair enzymes appear to be the most highly conserved proteins, showing their important role throughout evolution.

The enzyme systems can either directly reverse the damage to form the normal purine or pyrimidine bases or the modified bases can be removed together with surrounding bases through a succession of events involving nucleases, DNA polymerizing enzymes and polynucleotide ligases which assist in restoring the biological and genetic integrity to DNA.

D. Kamely et al. (eds), Biotechnology: Bridging Research and Applications, 127–140.

Damage

As a target for damage, DNA possesses a multitude of sites which differ in their receptiveness to modification. On a stereochemical level, nucleotides in the major groove are more receptive to modification than those in the minor groove, the termini of DNA chains expose reactive groups, and some atoms of a purine or pyrimidine are more susceptible than others. As a consequence the structure of DNA represents a heterogeneous target in which certain nucleotide sequences also contribute to the susceptibility of DNA to genotoxic agents.

Endogenous Damage

Even at physiological pH and temperature in the absence of extraneous agents, the primary structure of DNA undergoes alterations. A number of specific reactions directly influence the informational content as well as the integrity of DNA. Although the rate constants for many reactions are inherently low it is because of the enormous size of DNA and its persistence in cellular life cycles that the accumulation of these changes can have significant long term effects.

Deamination

The hydrolytic conversion of adenine to hypoxanthine, guanine to xanthine, and cytosine to uracil-containing nucleotides is of sufficient magnitude to affect the informational content of DNA (Figure 1).

Depurination

The glycosylic bonds linking guanine in nucleotides are especially sensitive to hydrolysis – more than the adenine and pyrimidine glycosylic links. The end result, an apurinic (or apyrimidinic) site is recognized by surveillance systems and, as a consequence, is repaired.

Oxidative Damage

The production of oxygen radicals, superoxide or hydroxyl radicals, as a metabolic consequence as well as at inflammatory sites causes sugar destruction which eventually leads to strand breakage.

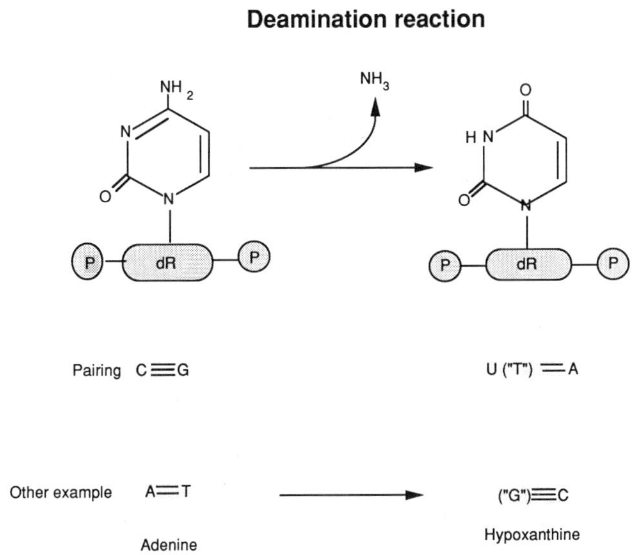

Figure 1. Deamination reactions have mutagenic consequences because the deaminated bases cause false recognition.

Exogenous Damage

The concept of DNA repair in biological systems arose from photobiologists and radiobiologists studying the viability and mutagenicity in biological systems exposed to either ionizing or ultraviolet irradiation. Target theories, derived from the random statistical nature of photon bombardment, led to the identification of DNA as the primary target for the cytotoxicity and mutagenicity of ultraviolet light. The photoproducts in DNA responsible for the effects of such irradiation have been attributed to 5,6–pyrimidine cyclobutane *cis*, *syn* dimers, 6,4–pyrimidine-pyrimidone dimers and 5,6–water addition products of cytosine (hydrates). In addition, most of the structural and regulatory genes controlling DNA repair in *E. coli* were identified facilitating the molecular characterization of the relevant enzymes.

Ionizing Radiation

The primary cellular effect of ionizing radiation is the radiolysis of water which mainly generates hydroxyl radicals ($HO^{\cdot}$). The hydroxyl radical is capable of abstracting protons from the $C^{4'}$ position of the deoxyribose moiety of DNA thereby labilizing the phosphodiester bonds and generating single and double strand breaks. The pyrimidine bases are also subject to $HO^{\cdot}$ addition reactions.

Ultraviolet Irradiation (UV)

Most of the UV photoproducts are chemically stable; their recognition provided direct biochemical evidence for DNA repair. The major photoproducts are 5,6–cyclobutane dimers of neighboring pyrimidines (intrastrand dimers), 6,4–pyrimidine-pyrimidone dimers (6–4 adducts), and 5,6–water addition products of cytosine (cytosine hydrates).

Alkylation Damage

This modification takes place on purine ring nitrogens (cytotoxic adducts), the O^6 position on guanine, the O^4 positions of the pyrimidines (mutagenic lesions), and the oxygen residues of the phosphodiester bonds of the DNA backbone (biologically silent). Alkylating agents are environmentally pervasive, arising indirectly from many foodstuffs and from automobile exhaust in which internal combustion of atmospheric nitrogen results in the formation of nitrate and nitrites.

Bulky Adducts

Large bulky polycyclic aromatic hydrocarbon modification occurs primarily on the N^2 and C^8 position of guanines invariably from the metabolic activation of these large hydrophobic uncharged macromolecules to their epoxide analogs. The major source of these substances is from the combustion of tobacco, petroleum products and food stuffs.

Direct Reversal or Removal Mechanisms

The simplest repair mechanisms involve the direct photoreversal of pyrimidine dimers to their normal homologs and the removal of O-alkyl groups from the O^6 methylguanine and from the phosphotriester backbone as a consequence of alkylation damage to DNA.

Photolyases – Photoreversal

The direct reversal of pyrimidine dimers to the monomeric pyrimidines is the simplest mechanism, and parenthetically it is chronologically the first mechanism described for the repair of photochemically damaged DNA. It is

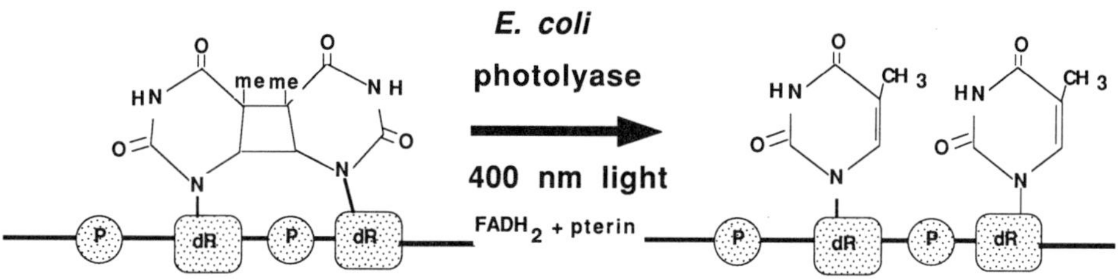

Figure 2. The direct photoreversal of pyrimidine dimers in the presence of visible light.

a unique mechanism characterized by a requirement for visible light as the sole source of energy for breaking two carbon-carbon bonds (Figure 2).

The enzyme protein has two associated light-absorbing molecules (chromophores) which can form an active light dependent enzyme. One of the chromophores is $FADH_2$ and other is either a pterin or a deazaflavin able to absorb the effective wavelengths of 365–400 nm required for photoreactivation of pyrimidine dimers. It is suggested that photoreversal involves energy transfer from the pterin molecule to $FADH_2$ with electron transfer to the pyrimidine dimer resulting in non-synchronous cleavage of the C^5 and C^6 cyclobutane bonds.

Enzymes that carry out photoreactivation have been identified in both prokaryotes and eukaryotes.

Alkyl Group Removal – Methyl Transferases

Cells pre-treated with less than cytotoxic or genotoxic levels of alkylating agents before lethal or mutagenic doses are more resistant. This is an adaptive phenomenon with anti-mutagenic and anti-cytotoxic significance. During this adaptive period a 39 kDa Ada protein is synthesized, which specifically removes a methyl group from a phosphotriester bond and from an O^6 methyl group of guanine (or from O^4 methyl thymine). The O^6 methyl group of guanine is not liberated as free O^6 methyl guanine during this process, but is transferred directly from the alkylated DNA to this protein; the Ada protein (methyl transferase) and an unmodified guanine are simultaneously generated. These alkyl groups specifically methylate cysteine 69 and cysteine 321 respectively in the protein (Figure 3).

The methyl transferase is used stoichiometrically (it does not turn over) and is permanently inactivated in the process. Nascent enzyme is, however, generated because the mono- or di-methylated transferase activates tran-

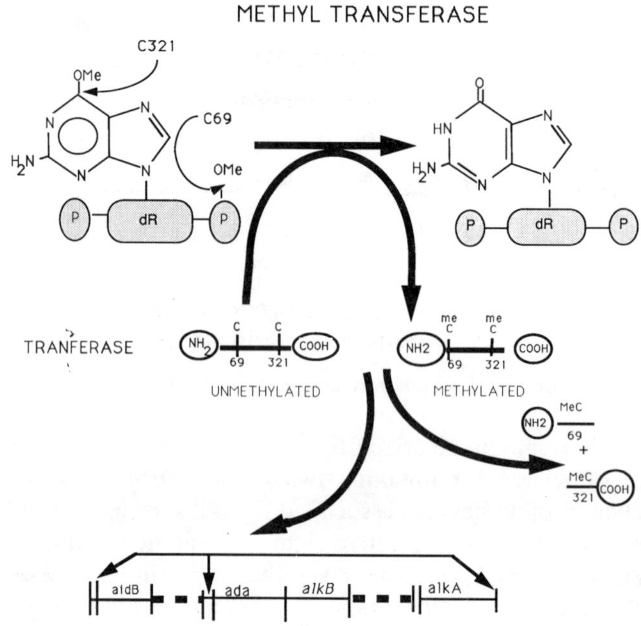

Figure 3. The direct reversal of alkylation damage removes such groups from the DNA backbone and the O-6 position of guanine. Such alkyl groups are transferred directly to specific cysteine residues on the transferase, the levels of which are influenced adaptively by levels of the alkylating agents. The methylation of the transferase inactivates the enzyme which is used up stoichiometrically in the reaction. The alkylated transferase acts as a positive transcriptive signal turning on the synthesis of unique mRNA. Regulation of transferase levels may be influenced by a unique protease.

scription of its own "regulon" which includes, in addition to the *ada* gene, the *alkB* gene of undefined activity and the *alkA* gene which controls a DNA glycosylase. The latter enzyme acts on 3–methyl adenine, 3–methyl guanine, O^2 methyl cytosine and O^2 methyl thymine. The methylated Ada protein can specifically bind to the operator of the *ada* gene acting as a positive regulator. Down regulation may be controlled by proteases acting at two hinge sites in the Ada protein.

Base-specific Responses

Base Excision Repair by Glycosylases and AP Endonucleases

Bases modified by deamination can be repaired by a group of enzymes called DNA glycosylases, which specifically break the N-glycosyl bond of that base

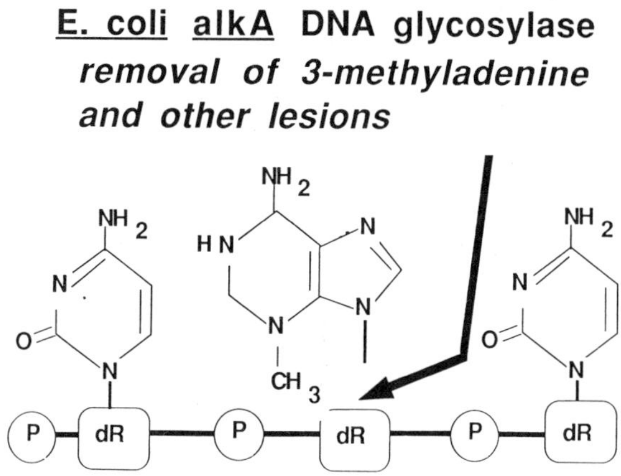

Figure 4. DNA glycosylases hydrolyze the N-glycosyl bond between damaged bases and deoxyribose generating an AP (apyrimidinic or apurinic) site.

and the deoxyribose of the DNA backbone, generating an apyrimidinic or an apurinic site (AP site)(Figures 4 and 5). These are rather small, highly specific enzymes that require no cofactor for functioning. They are the most highly conserved proteins, attesting to the evolutionary unity both structurally and mechanistically from bacteria to man.

As a consequence of DNA glycosylase action, the AP sites generated in the DNA are acted upon by a phosphodiesterase specific for such sites which can nick the DNA either 5′ and or 3′ to such damaged sites. If there is a sequential action of a 5′-acting and a 3′-acting AP endonuclease, the AP site is excised, generating a gap in the DNA strand.

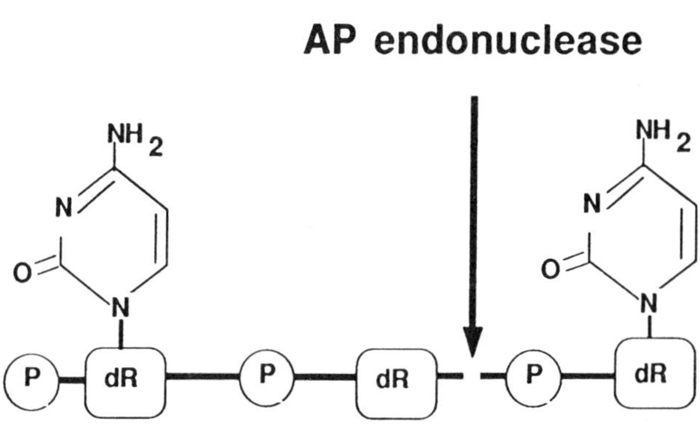

Figure 5. Endonucleases recognize AP sites and hydrolyze the phosphodiester bonds either 3', 5' or both sides of the deoxyribose moiety in damaged DNA.

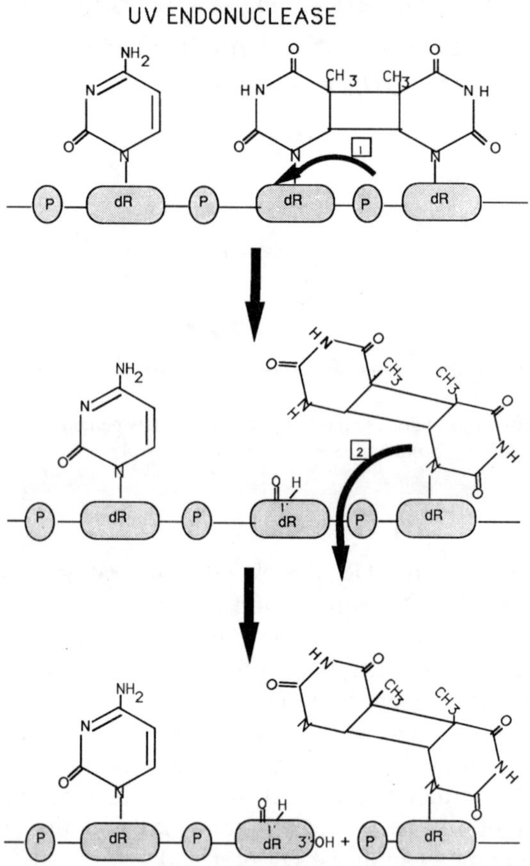

Figure 6. The same enzyme that can hydrolyze the N-glycosyl residue of a damaged nucleotide also hydrolyzes the phosphodiester bond linking the AP site generated in the first N-glycosylase reaction.

Glycosylase-associated AP Endonucleases

An enzyme from bacteria and phage-infected bacteria, encoded in the latter case by a single gene (*denV*), breaks the N-glycosyl bond of the 5′-thymine moiety of a pyrimidine dimer followed by hydrolysis of the phosphodiester bond between the two thymine residues of the dimer (Figure 6). This enzyme, referred to as the pyrimidine dimer DNA glycosylase, is found in *Micrococcus luteus* and phage T4 infected *E. coli*. This small uncomplicated enzyme does not require cofactors and is presumed to act by a series of linked beta-elimination reactions.

An enzyme behaving in a similar glycosylase-endonuclease fashion, but acting on the radiolysis product of thymine, thymine glycol, has been isolated from *E. coli* and is referred to as endonuclease III.

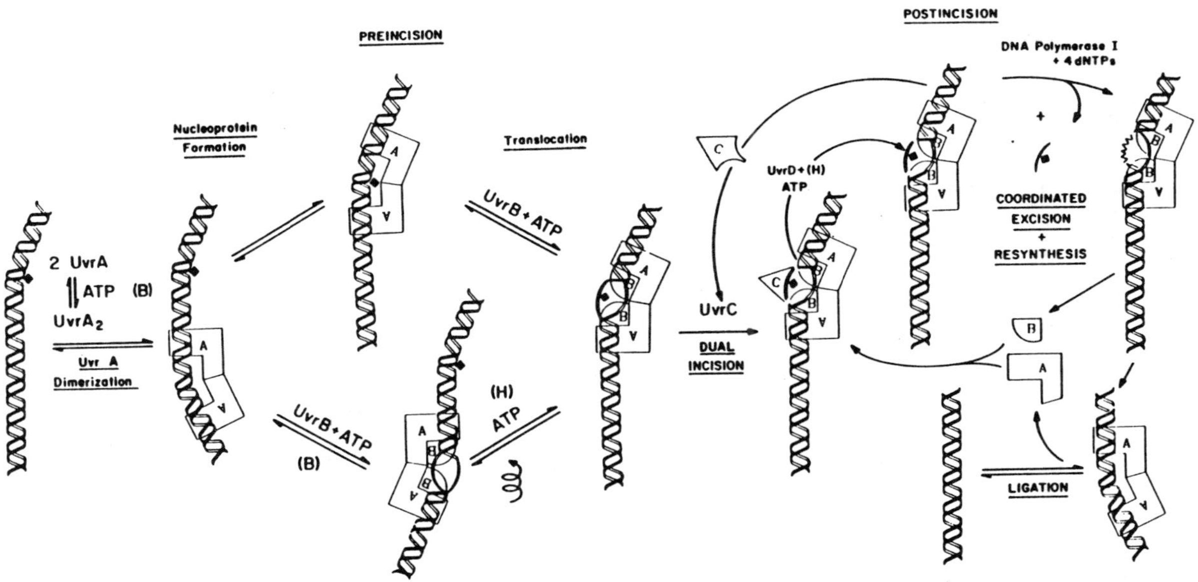

Figure 7. Nucleotide excision reactions. In this multi-protein enzyme system the UvrABC proteins catalyze a dual incision reaction 7 nucleotides 5′ and 3–4 nucleotides 3′ to a damaged site. The UvrA protein, as a dimer, binds to undamaged sites initially and in the presence of UvrB, whose cryptic ATPase provides the energy, is able to translocate to a damaged site. This pre-incision complex interacts with UvrC leading to the dual incision reaction. The incised DNA-UvrABC complex does not turn over and requires the coordinated participation of the UvrD and DNA polymerase reactions for damaged fragment release and turnover of the UvrABC proteins. Ligation, the final reaction, restores the integrity to the DNA strands.

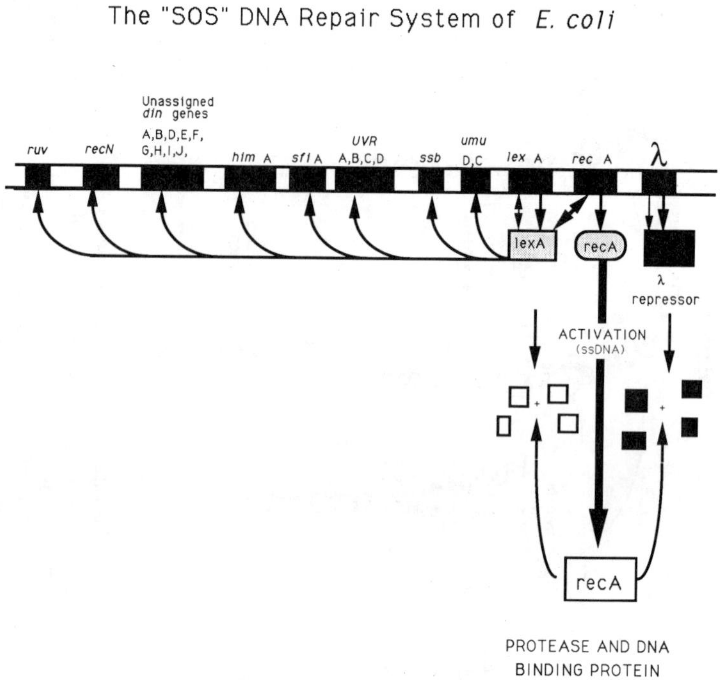

Figure 8. Regulation of the nucleotide excision pathway by the "SOS" system. The *lexA* and λ repressors negatively control a multitude of genes which are turned on when cells are damaged, leading to the over-production of the RecA protein which assists in the proteolysis of the LexA and lambda proteins, thereby derepressing the controlled gene systems. When DNA is fully repaired the level of RecA declines, restoring the "SOS" system to negative control.

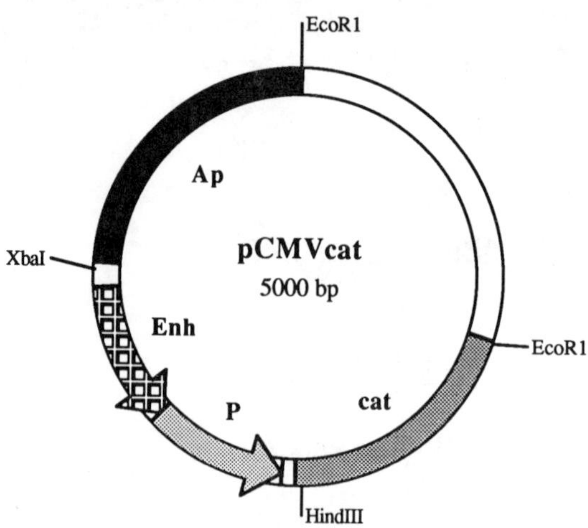

Figure 9. Structure of the pCV$_2$*cat*.

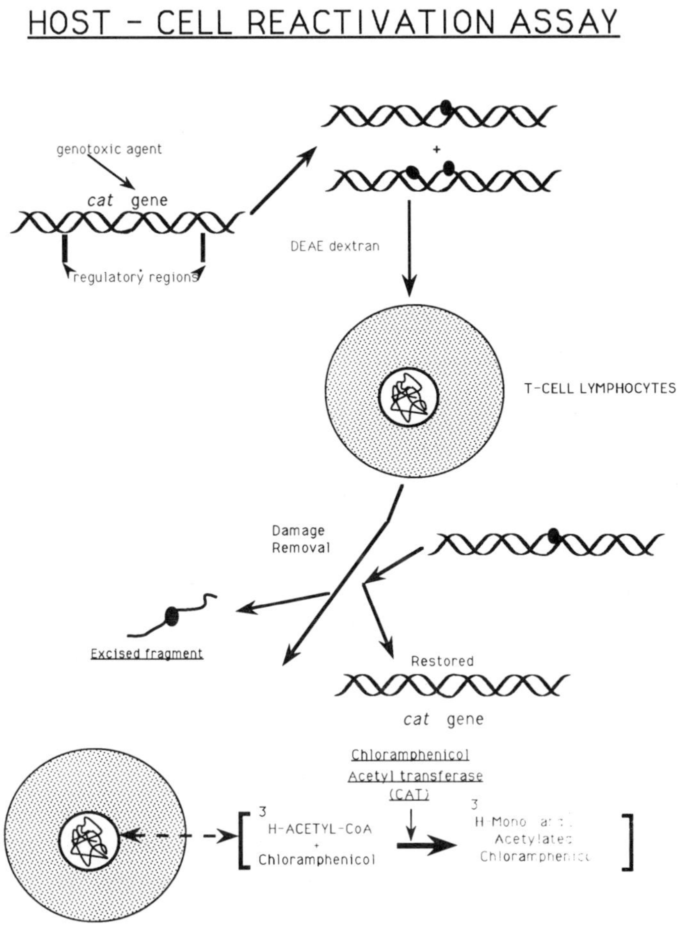

Figure 10. Host cell reactivation assay (HCR).

Nucleotide Excision Repair

Prokaryotes

The ideal repair system is one that is somewhat indiscriminate and which can respond to virtually any kind of damage. Such a repair system has been characterized in *E. coli* where it consists of at least six gene products of the *uvr* system (Figure 7). These proteins consist of the UvrA protein which binds as a dimer to DNA in the presence of ATP, followed by the UvrB protein which cannot bind DNA by itself. Translocation of the UvrAB complex from initial undamaged DNA sites to damaged sites is driven by a cryptic ATPase associated with UvrB which is activated by the formation of the UvrAB-undamaged DNA complex. This complex is now poised for

Table 1. List of repair-deficient lymphoblast lines available from the mutant cell repository

Indicator lymphoblasts*		
Cell lines	Generic damage	Specific damage
1. Xeroderma pigmentosum	Bulky adducts	Photochemical, 4-NQO Benzanthracene, AAF Aflatoxin, DMBA, Chrysene Nitrogen mustards, Bisulfan Psoralen + UV
2. SV40-Transformed cell lines *mer-*	Alkylation	MNU, MNNG and derivatives
3. Ataxia telangiectasia	Radiomimetic	Radiolysis products of water Bleomycin
4. Bloom's syndrome	SS breaks from: – Damage – Repair	Radio- and UV-mimetic
5. Fanconi's anemia	Interstrand cross-links	Mitomycin psoralen + UV

endonucleolytic activity catalyzed by the interaction of the UvrAB-damaged DNA complex with UvrC to generate two nicks in the DNA 7 nucleotides 5′ to the damaged site and 3–4 nucleotides 3′ to the same site. These sites of breakage are invariant regardless of the nature of the damage. In the presence of the UvrD (helicase III) and DNA polymerase I, the damaged fragment is released, accompanied by the turnover of the UvrA, UvrB and UvrC proteins. The continuity of the DNA helix is maintained, based on the sequence of the opposite strand. The final integrity of the interrupted strands is restored by the action of DNA polymerase I which copies the other strand and by polynucleotide ligase which seals the gap.

The levels of the Uvr proteins are regulated in *E. coli* by an "SOS" regulon monitoring a large number of genes which includes: the *uvrA*, *uvrB*, possibly *uvrC*, and *uvrD* as part of the excision repair system; the regulators of the SOS system, the LexA and RecA proteins, cell division genes *sulA* and *sulB*; recombination genes *recA*, *recN*, *recQ*, *uvrD*, and *ruv*; mutagenic by-pass mechanisms (*umuDC*, *recA*), damage inducible genes, and the lysogenic phage λ. The LexA protein negatively regulates these genes as a repressor by binding to unique operator regions. When the DNA is damaged, for example by ultraviolet light, a signal in the form of a DNA repair intermediate induces the synthesis of the recA protein. When induced, the recA protein acts as a protease assisting the LexA protein to degrade itself activating its own synthesis and that of the recA protein as well as some 20 other different genes. These genes permit the survival of the cell in the face of life-threatening environmental damage. Upon repair of the damaged DNA, the level of the signal subsides, reducing the level of recA, stabilizing the integrity of the intact LexA protein and its repressive properties on all the other genes (Figure 8). Then the cell returns to its normal state.

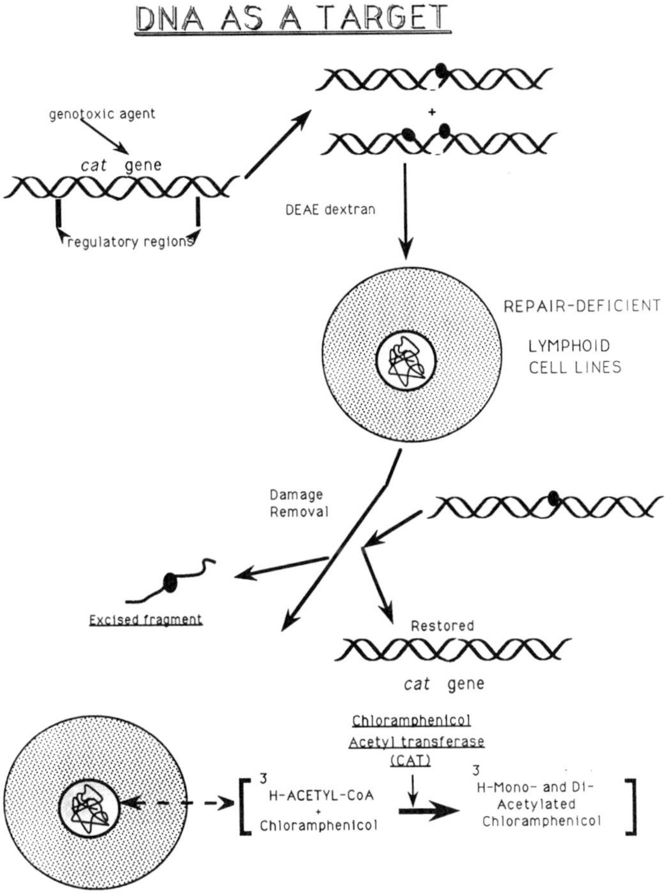

Figure 11. HCR assay for measuring environmental genotoxic damage.

Host Cell Reactivation Assay (HCR)

The assay depends on the use of a shuttle vector (Figure 9) which contains the *E. coli* cat gene encoding an enzyme, unique to bacteria, that is able to diacetylate chloramphenicol from labeled acetyl CoA. The vector contains viral enhancer sequences that allow this bacterial gene to be expressed in human cells and the elements of pBR322 for perpetuation in bacteria. Whole blood is fractionated using Ficoll gradients, and the buffy coat containing mostly lymphocytes is treated with mitogen to metabolically activate the T-cell lymphocytes. After 36 hours the T-cell lymphocytes are amenable to the physical introduction of the damaged or undamaged DNA facilitated by DEAE-dextran (Figure 10).

Thirty-six hours later the level of chloramphenicol acetyltransferase (CAT) is assessed. The level of expressed enzyme between undamaged and damaged states is a direct measure of the repair capacity of the individual donor. Introducing such dose-modified DNAs into human lymphoid lines derived from patients with DNA repair deficiencies (Table 1) permits an extremely sensitive estimate for the presence of a genotoxic modification of the target DNA (Figure 11). The target DNA can be used to act directly with the damaging agent in aqueous solution, with airborne materials soluble in aqueous medium, or indirectly with agents requiring the metabolic activation of lipophilic agents. In principle, this HCR assay is the mammalian equivalent of the Ames mutagenesis assay which measures mutations in repair-deficient bacterial cell lines.

References

Friedberg, E.C. 1984. DNA Repair. Freeman, New York.

Grossman, L., Caron, P.R., Mazur, S.J., and Oh, E.Y. 1988. Repair of DNA containing pyrimidine dimers. FASEB J. 2: 2696–2701.

Lindahl, T., Sedgwick, B., Sekiguchi, M., and Nakabeppu, Y. 1988. Regulation and expression of the adaptive response to alkylating agents. Ann. Rev. Biochem. 57: 133–157.

Modrich, P. 1989. Methyl-directed DNA mismatch correction. J. Biol. Chem. 264: 6597–6600.

Sancar, A. and Sancar, G.B. 1988. DNA repair enzymes. Ann. Rev. Biochem. 57: 29–67.

Walker, G.C. 1985. Inducible DNA repair systems. Ann. Rev. Biochem. 54: 425–457.

Weiss, B. and Grossman, L. 1987. Phosphodiesterases involved in DNA repair. Adv. in Enzymol. 60.

Discussion (DR. GROSSMAN)

Kamely: What other field applications could emerge from an assay such as this?

Grossman: The ones that would be of interest to us involve determining the presence of "unknown" environmental agents which are potentially genotoxic. Further, this assay can be used for prospective studies to determine whether individuals in the population may be at risk in their environment because of repair defects.

Key Words: DNA damage, DNA modification, DNA repair enzymes, Host cell reactivation (HCR)

Detection of Pathogens with Nucleic Acid Probes

David H. Gillespie
Hahnemann University, Broad and Vine, Philadelphia, PA 19102

Introduction

The presence of pathogenic microorganisms can represent a hazard in military and civilian situations. One approach to detecting pathogens is to probe for the presence of their genetic information using molecular hybridization with nucleic acid probes. This approach has the advantage of speed over culture procedures and the advantage of sensitivity over immunological procedures. The following discussion attempts to lay out the scope of the technical problems involved in pathogen detection with nucleic acid probes and provides possible solutions for these problems.

Molecular Hybridization

The process of nucleic acid probing, molecular hybridization, involves the interaction between a molecule of label-probe and a molecule of intended target to form a molecular hybrid, a duplex, ideally to the exclusion of any interaction between the label-probe and any other molecule in the specimen. A typical probing mixture consists of 10^9–10^{12} molecules of probe and the same number of nucleic acid molecules from the specimen. But with regard to the intended target, there can be as few as one or as many as $1–10 \times 10^6$ molecules. So the range of target molecules that needs to be detected within the specimen is great, whereas the number of probe molecules added to a hybridization reaction is always very large. Label-probe can hybridize with this intended target even if there is only one molecule present, and the hybridization can occur to the exclusion of stable interaction with any other nucleic acid molecule in the specimen. That sort of specificity can be obtained by adjusting conditions and by adjusting the nature of the probe as long as there is at least one nucleotide in the sequence of the intended target which distinguishes it from the sequence of every other nucleic acid molecule in the specimen.

The reason that a probe can hybridize extraordinarily and specifically with

D. Kamely et al. (eds), Biotechnology: Bridging Research and Applications, 141–151.

a target has to do with the fact that the chemistry of the interaction is relatively simple, relative for example, to an antibody-antigen interaction. Secondly, and again as opposed to an antibody-antigen interaction, the forces that maintain the hybrid of probe and target are extraordinarily stable to purification. A nucleic acid hybrid is about 10^9 times more stable than an antibody-antigen complex. It is possible with DNA probes to detect a single intended target molecule because the probe-target hybrid can be rigorously purified and will not fall apart.

The association of two strands of nucleic acid uses the specific base pairing rule of Watson and Crick, where adenine pairs with thymidine and guanidine pairs with cytidine. The requirement for Watson-Crick base pairing provides a uniqueness in the specificity of the interaction between the probe and the target. A given probe sequence 20 nucleotides in length will be represented once, on the average, in a random-sequence nucleic acid one trillion nucleotides long. Put another way, if a cell contains a billion nucleotides worth of genetic information, there is only a one in one thousand chance that the probe sequence will be found by chance in that cell's genome. A typical mammalian cell contains around one billion nucleotides in its genome. Therefore, when a 20 nucleotide long probe complementary to a known genome sequence is used and if all 20 nucleotides are required for hybridization, it is likely that that sequence will only be represented by the intended target. I emphasize the word likely. This is strictly a probability assessment. There is a probability that an unintended target could carry the same sequence by chance or by evolutionary design.

Accepting the fact that a probe sequence can be unique, will the probe uniquely interact with the intended target to the exclusion of all other targets? To approach this we need to understand the mechanism of nucleic acid interactions. The interaction between a probe and a target involves three steps. First, it involves a large number of more or less random collisions between nucleic acid molecules. Most of those are non-productive; that is, they are interactions between regions of nucleic acid molecules that have less than about five complementary nucleotides. These interactions are either not stable or are freely reversible; unless stabilized, they are non-productive and do not lead to the formation of a hybrid. The second step is the productive collision between a short region of the probe with a complementary region of the target, forming what Wetmur and Davidson (1968) defined as the nucleation complex. The nucleation complex is a small duplex region about six nucleotides long which causes the quasi-stable formation of a complex having a half-life long enough so that other parts of the probe and target that are complementary to each other now zip up rapidly.

The nucleation complex does not involve a length which provides sequence uniqueness. The vast majority of these quasi-stable interactions will occur with unintended targets. If there is no complementary flanking region, the nucleation complex will dissociate and the probe will collide with other targets until it collides with the intended target and the zipping up process takes place. That may sound as if the formation of a stable hybrid would take an agonizingly long time. The fact is that even with the large number of specimen nucleic acid molecules that have to be "scanned," the molecular hybridization process in practice can be driven to completion on the order of 10 min or so, regardless of how few intended targets exist. To do that, it is necessary to use an excess of probe. If a large amount of probe is used, two advantages are gained. First, the hybridization reaction will be "driven" by the probe so the reaction will go very quickly. And secondly, all of the available target molecules will become hybridized with probe, a condition known as "saturation" of target. This means that the amount of hybrid formed reflects exactly the amount of target present and provides a way to quantitate the number of targets present.

The most specific probes are short oligonucleotide DNA probes. Oligonucleotide probes are synthesized chemically in a totally automated format. Generally, such probes have ends which can be labeled. For a radioactively labeled probe which is relatively stable, a single radioactive phosphate can be added to the 5′ end of the oligonucleotide with gamma ^{32}p-ATP and polynucleotide kinase. Oligonucleotides or long probes can also be labeled nonradioactively, with affinity ligands which bind enzymes or with enzymes themselves.

The stability of a hybrid between various targets and a given probe varies with the degree of "mismatching" in the hybrids; unpaired bases and "non-Watson-Crick" base pairs. Furthermore, any environmental condition that destabilizes hydrogen bonds increases the specificity or "stringency" of the molecular hybridization reaction. For example, a mismatched hybrid may be stable at low temperatures, because a smaller number of hydrogen bonds per base are required to stabilize the hybrid. As the temperature is raised and more stress is put on the hybrid, more hydrogen bonds per base are required to stabilize the structure. For example, at about 5°C below the hybridization temperature rate optimum a hybrid will denature even though there is only a single mismatch, whereas the perfectly matched hybrid will be stable. Experimentally, it has been shown that oligonucleotide probes can distinguish between a wild type sequence and a mutant sequence where the mutant is a single nucleotide substitution (Wallace, et al., 1981).

Other conditions influence stringency. For example, low salt concentration or the addition of hydrogen bond breakers (e.g., formamide or chaotropic salts) increase stringency. For very stringent conditions, it is theoretically best to work at fairly low salt concentrations. However, high salt concentrations speed molecular hybridization, so there is a trade in deciding between conditions of specificity and rate of hybridization. The range of salts that are usually used for molecular hybridization are between 0.1 and 1.0 M (but see discussion below re: chaotropic salts). For accelerated hybridization that cannot be driven with an excess of probe, even higher salt concentrations can be used. Then the specificity will have to be adjusted by increasing temperature or including hydrogen bond breakers. The disadvantage of working at a high temperature is the instability of nucleic acids during hybridization. Accelerated hybridizations of high stringency at low temperatures can be performed with high salt and hydrogen bond breakers. This is one reason for formamide-based nucleic acid hybridization (Bonner, et al., 1967; Gillespie and Gillespie, 1971) and newer techniques based on guanidine thiocyanate, GuSCN (Thompson and Gillespie, 1987).

Sample Preparation and Solution Hybridization

One of the requirements for rapid diagnostics will be to perform hybridization in solution without purifying or immobilizing target nucleic acids. A major difficulty is to put the specimen in a state where the target nucleic acids are amenable to probing without having to purify nucleic acids. One approach is to utilize the combined action of detergents and enzymes and high salt concentrations. Proteolytic enzymes destroy proteins and detergents disrupt weak forces and high salt concentrations break electrostatic interactions. That logic is being used successfully to measure bacteria and mycoplasma. It remains to be seen how widely useful that approach is, because the hybridization conditions are not ideal and because of significant complex formation of lysate and target.

An alternative approach is the use of chaotropic salts to disrupt a specimen and present nucleic acids in a state where they can be probed directly in the same solution containing the chaotropic salt. Chaotropic salts were defined initially by Hamaguchi and Geiduschek (1962) as salts which destabilize DNA. Chaotropic ions include the iodide, thiocyanate, perchlorate and trichloracetate anions and probably the guanidinium cation. At high concentrations, chaotropic salts are thought to work by disrupting the water lattice in a solution and thereby destabilizing bonds which depend on water, such as nonpolar bonds. At high concentration, chaotropic salts also compete with electrostatic bonds. Chaotropic salts do not disrupt covalent or disulfide bonds.

Chaotropic salts have become popular during the last 10 years in membrane biochemistry for purification of membrane components that cannot be solubilized any other way. Chaotropic salts cause the disruption of cells and subcellular organelles basically by solubilizing the individual molecules of cellular structures; for example, they free RNA and DNA from proteins that they might have been associated with in the cell. Nucleic acids are presented in a way that is suitable for efficient molecular hybridization.

One chaotropic salt, guanidine thiocyanate (GuSCN) additionally promotes molecular hybridization and accelerates it some 100–fold (Thompson and Gillespie, 1987). In 2–5M GuSCN molecular hybridization occurs at temperatures ranging from room temperature to 37°C. GuSCN has other advantages – it denatures proteins by eliminating weak forces that are required for activity, it inactivates nucleases, so target nucleic acids are stable during the hybridization procedure, and it also inactivates microorganisms, providing a level of safety. All that is required for this format is a fine suspension of the specimen dissolved in about 5 M GuSCN/0.1 M EDTA, pH 7. For molecular hybridization, one simply needs to add probe, incubate for minutes or hours, depending on the concentration of probe and measure the amount of probe which has been hybridized to target. The principle to be emphasized with regard to clinical diagnostics is that the hybridization step will take place in solution. This not only increases the speed and simplicity of sample preparation and molecular hybridization, but in the longer run it will be much easier to automate than a system depending on the hybridization with a target fixed to a solid support.

Purification of Hybrids by Affinity Capture

Simple methods for sample preparation and molecular hybridization exist. The problem at this point is to purify the probe-target hybrids which are formed. This can be accomplished by affinity capture or sandwich hybridization (Virtanen, et al., 1983). The principle is to use a solid support with an attached affinity ligand as a matrix for purifying probe-target hybrid. In one version, this ligand is an oligonucleotide which is complementary to the target at a site near or adjacent to the site which hybridizes with the label-probe. The ligand acts as a hook to purify target, and in so doing purifies the label-probe which is hybridized with target. That hybrid is no longer capturable because the oligonucleotide site has been occupied by an oligonucleotide that has leached off the support. The efficiency of capture in this version depends critically upon the stability of the ligand-support bonds. A second disadvantage is the label-probe which will attach to solid support and give noise. And finally, the signal from hybridized label-probe is limited by the amount of target.

The solution for the leaching problem is the use of a second probe to hybridize with target. What is used here are two probes. One probe carries a label and the second probe, which is used to purify the hybrid, is called a capture-probe. The capture-probe is bifunctional. It contains a region which will hybridize with the intended target, and it also contains an adduct which can be captured onto the solid support by recognizing the affinity ligand. Essentially, all capture-probe molecules are captured by the solid support. Some of those capture-probes are also hybridized with target carrying label-probe. Again, some of the affinity ligand on the support can leach out. But now chances are very high that affinity ligand that leaches off the solid support will hybridize with a free capture-probe, rather than one which has hybridized with target, since probes are present in vast excess over target. This capture-probe version does not suffer from the efficiency disadvantage of the first affinity capture system. There still is a noise component from label-probe which interacts directly with the support and there still is a signal limitation because the amount of label is limited by target.

The solution to the noise problem is the following. Simply use a capture-probe which forms a reversible link with the affinity ligand on the membrane (Thompson, et al., 1989). Hybrids formed in solution are captured, then capture-probes and hybrids are released into solution. Some of the noise remains attached to the membrane, some of the noise is released along with the hybrid. Then this mixture is recaptured on fresh solid support. The same or lower percentage of unreacted label-probe binds to the solid support and that means that with every cycle of capture and release, noise is reduced by some factor which is characteristic of the support. Noise can be virtually eliminated by adjusting the number of cycles of capture and release.

Since the capture-probe can be an oligonucleotide, the specificity of hybridization can be controlled by defining the structure of the oligonucleotide capture-probe. It does not matter whether the label-probe is specific, because in order to get a signal, label-probe and capture-probe must both hybridize with the target. This version of solution hybridization has the high specificity of oligonucleotides as well as the relatively high signal of long label-probes.

Preferred label-probes are RNA probes, also called "riboprobes." They are synthesized enzymatically, beginning with basically the same sort of recombinant DNA that was used to generate a nick translation probe, except that in addition to an inserted sequence, which is the probe sequence, at one or both ends of the insert that are promoter sequences which recognize DNA dependant RNA polymerase and cause the initiation of transcription at that site in a particular direction (Green, et al., 1983). RNA copies made by in vitro transcription have a defined starting point. If template DNA is cleaved with a restriction endonuclease in a strategic position, RNA transcripts can also have a defined endpoint. Finally, since RNA transcription is asymmetric, RNA probes are single stranded and cannot reanneal with themselves during hybridization in solution. Hybridization can be scores by measuring resistance of the label-probe-target hybrids to RNase.

Target Amplification

The sensitivity of solution hybridization formats is in the tens or hundreds of thousands of target molecules. One reversible affinity capture system was reported to measure 50,000 targets with a signal to noise ratio of 5 and to detect as few as 15,000 targets (Thompson, et al., 1989). Affinity capture techniques which are reversible ordinarily have detection limits in the hundreds of thousands of target molecules. This presents a serious challenge for using DNA probes in clinical situation, because earlier we established our needed sensitivity at around 1000 target molecules. The only known solution to this dilemma is to amplify assay signal by repetitively copying intended targets or repetitively copying hybridized label-probe. Other theoretical solutions exist which have not worked in practice. The search for a "superlabel" which would give greater sensitivity than ^{32}P is so far a failure. Several fluorescent and luminescent probes have been reported to have greater sensitivity than the ^{32}P, but these isolated reports have never translated into general use. Another theoretical approach is to form probe "networks" such that several label-probes become associated with each other, then with target (Urdea, et al., 1987). Since the number of probes per target is increased, signal increases. In practice, while probe networking probably explains the exquisite and unexpected sensitivity of in situ hybridization (Harper and Saunders, 1981), virtually all other attempts to use probe networking to increase signal and not increase noise have failed.

DNA target amplification (Saiki, et al., 1988) is the repetitive copying of a 200–500 nucleotide region of an intended target by thermostable, DNA dependent DNA polymerase. DNA target amplification has been called polymerase chain reaction (PCR), because the number of copies of target increases exponentially as target copies serve as templates for further copying. The PCR mixture includes targets, primers, deoxynucleoside triphosphates and polymerase. Technically the following steps are involved: cycle 1 (A) heat-denature DNA at 95°C; (B) cool to 55°C to anneal primers for DNA synthesis; (C) heat to 70°C for DNA synthesis; cycle 2, (A) heat-denature DNA at 95°C; (B) cool to 55°C to anneal primers to target; (C) heat to 70°C for DNA synthesis. Theoretically, each "cycle" of denaturation, primer annealing and target copying doubles the number of target sequences. The length of PCR products is determined by the spacing between the two primer sites on the target. The great advantages of PCR are that target sequences are amplified millions or billions of fold starting from as few as one target molecule and what is copied is a sequence which is characteristic of the target, not of a probe. For applying PCR to clinical diagnostics there are still challenges to overcome. PCR is not absolutely specific, in order to be amplified, a target must possess sequences complementary to each of two oligonucleotide primers. If absolutely specific primer-annealing condition can be established, then this requirement would essentially preclude amplification of any target but the intended target. However, primer annealing is conducted at 50°–55°C and DNA polymerization occurs as the sample is being heated to 70°C, so during PCR the specificity-setting conditions are largely indeterminate. In practice, some reports show exclusive amplification of the intended target while others show major misamplification of unintended targets. Conditions are not optimized, so PCR operates at less than theoretical efficiency. This is important because at theoretical efficiency true amplification products are highly selected for. They are the only products which are amplified logarithmically from start to finish. Misamplification products will be amplified at a slower rate during initial cycles. Therefore, in PCR, efficiency is linked to specificity.

Contamination takes on a new meaning in PCR. Contamination of a negative sample by a single molecule of intended target can lead to a false positive result. Worse, this increased problem of contamination is in the context of synthesizing huge quantities of intended target by amplification. Contamination during PCR experiments has proven to be difficult to cope with in a research setting; in a clinical setting, contamination presents a nightmare. One major step to controlling contamination is to aliquot reagents in a single-use size and only use rigorously quality-controlled reagents. A second ingredient is isolating samples from each other; in the long run, sample isolation will be accomplished by automation. However, the best way to limit contamination may be to limit the amplification process to the lowest acceptable levels. This will minimize quantities of contaminants generated and will also minimize the chance that contamination will result in a false positive result. Even so, some presumptive false positive results will need to be confirmed to rule out contamination. This can be done by amplifying from the same specimen a larger region containing the original sequence, by amplifying a different region of the same target or by determining the sequence of the PCR product. RNA target amplification can also be accomplished by PCR, providing the first step in the process is copying of the RNA target into DNA by reverse transcriptase. While there is good evidence that a single DNA molecule in a specimen can be amplified to detectable limits, this sensitivity has not been shown for RNA targets. If RNA, rather than DNA has been amplified, omission of reverse transcriptase will eliminate signal.

A target amplification technique different from PCR and using T7 RNA polymerase promoters has been described (Kwoh, et al., 1989) and is referred to as transcription amplification system (TAS). This system uses primers with T7 promotor sequences attached and requires two rounds of target copying to create a double-stranded target copy adjacent to the promotor. This DNA then serves as a template for transcription by DNA dependent RNA polymerase, creating 50–1000 single-stranded RNA copies of the target sequence. This amplification, too, can be carried through several cycles by copying RNA amplification products about 50–fold. TAS has the advantage over PCR of a larger fold amplification/cycle and the disadvantage of requiring the action of three different enzymes. They differ fundamentally in the fact that PCR products are dsDNA of defined length while TAS products are ssRNA of more or less uniform length.

Probe Amplification

Probe amplification is increasing hybridization probe molecules in number by repetitive copying or by some other means. The first probe amplification to be introduced used probes containing RNA bacteriophage replication sequences and the RNA dependent RNA polymerase, called "replicase" (Chu, et al., 1986). The model bacteriophage for this technique is Qβ and the replicating RNA, MDV, is a 300 nucleotide "midivariant" RNA containing only that portion of the Qβ 5000 nucleotide genome required for replication. Probes contain MDV RNA plus a sequence complementary to target RNA or target DNA. Qβ probes are hybridized with target and probe-target hybrids are purified away from unhybridized probe. Qβ replicase is added and probe replication occurs at a rate of one generation per 20 s. After some fixed period of time the quantity of replicated probe is measured.

Qβ amplification results in a billion-fold probe increase in 20 min, without temperature cycling. The same contamination issues raised for target amplification also apply to probe amplification. In addition, all the unhybridized probe must be removed prior to probe amplification to eliminate false-positive results. Since probe is amplified, there is no direct way to know that target was involved, so confirmation of positive results will be important. A reasonable strategy for analyzing future clinical specimens may be to screen using Qβ amplification and confirm positives with PCR.

Another strategy for probe amplification is ligation amplification. Two oligonucleotides complementary to adjacent target sequences are brought into juxtaposition by hybridization, then ligated enzymatically as described by Landergen, et al. (1988). One oligonucleotide carries an affinity ligand and the other a label, hybridization products can be scored by affinity capture. If the hybridization mixture also contains oligonucleotides complementary to those described above, then amplification can be accomplished as follows: cycle 1 (A) hybridize at 65°C; (B) add ligase and ligate at 37°C; (C) denature at 95°C; cycle 2 (A) hybridize at 65°C; (B) add ligase and ligate at 37°C; (C) denature at 95°C. As in PCR, products formed during ligation amplification serve as substrate for further amplification, so ligation amplification products accumulate logarithmically. Temperature cycling is necessary for ligation amplification. Since all available ligases are thermolabile, the present scenario requires that ligase be added at each cycle.

Many other kinds of amplification are possible and remain to be developed. At present, amplification is the only known way to achieve the sensitivity needed for clinical diagnostics. Since amplification is enzymatic, relatively pure nucleic acids are needed. Amplification from extracts is possible, especially with small samples, but such a strategy may lack the flexibility and reliability needed for routine diagnostics. One approach is to use reversible target capture to achieve purification of targets or probe-target hybrids, then perform amplification on relatively sequence-pure material.

Acknowledgments

This work was supported by grants from the W.W. Smith Foundation and GENE-TRAK Systems. We gratefully acknowledge the collaborative help of GENE-TRAK Systems scientists.

References

Chu, B.C.F., Kramer, F.R. and Orgel, L. E. 1986. Synthesis of an amplifiable reporter RNA bioassay. Nucleic Acids Res. 14: 5591–5603.

Green, M.R., Maniatis, T. and Melton, D.A. 1983. Human β-globin pre-mRNA synthesized *in vitro* is accurately spliced in *Xenopus oocyte* nuclei. Cell 32: 681–694.

Hamaguchi, K. and Geiduschek, E.P. 1962. The effect of electrolytes on the stability of the deoxyribonucleic helix. J. Am. Chem. Soc. 84: 1329–1338.

Harper, M.E. and Saunders, G.F. 1981. Localization of single copy DNA sequences on G-branded human chromosomes by *in situ* hybridization. Chromosoma 83: 431–439.

Kwoh, D.Y., Gavid, G.R., Whitfield, K.H., Chappelle, H.L., DiMichelle, L.J. and Gingeras, T.R. 1989. Transcription-based amplification system and detection of amplified human immunodeficiency virus type 1 with a bead-based sandwich hybridization format. Proc. Nat. Acad. Sci,. USA 86: 1173–1177.

Saiki, R.K., Gelfand, D.H., Stoffel, S., Scharf, S.J., Higuchi, R., Horn, G.T., Mullis, K.B. and Erlich, H.A. 1988. Pimr-directed enzymatic amplification of DNA with a thermostable DNA polymerase. Science 239: 487–491.

Thompson, J. and Gillespie, D. 1987. Molecular hybridization with RNA probes of concentrated solutions of guamide thiocyanate. Anal. Biochem. 163: 281–291.

Thompson, J., Solomon, R., Pellegrino, M., Sakai, K., Lewin, M., Field, M., Castrovinci, M., Sacramone, I. and Gillespie, D. 1989. A noise free molecular hybridization procedure for measuring RNA in cell lysates. Anal Biochem. 181: 371–378.

Urdea, M.S., Running, J.A., Horn, T., Clyne, J., Ku, L. and Warner, B. 1987. A novel method for the rapid detection of specific nucleotide sequences in crude biological samples without blotting or radioactivity: Application to the analysis of hepatitis B virus in human serum. Gene 61: 253–264.

Virtanen, M., Laaksonen, M., Soderland, H., Palva, A., Halonen, P. and Rankis, M. 1983. Novel test for rapid viral diagnosis: Detection of adenovirus in nasopharyngeal mucus aspirates by means of nucleic acid sandwich hybridization. Lancet i: 281–283.

Wallace, R.B., Johnson, M.J., Hirose, T., Miyake, T., Kamashima, E.H. and Itakura, K. 1981. The use of synthetic oligonucleotides as hybridization probes. II. Hybridization of oligonucleotides of mixed sequence to rabbit β-globin DNA. Nucleic Acids Res. 9: 879–894.

White, B.A. and Davidson, N. 1968. Kinetics of renaturation of DNA. J. Mol. Biol. 31: 349–370.

Discussion (DR. GILLESPIE)

Smith: You mentioned some chemistry where you coupled DNAs to supports and then used them to pull things out of solution. Could you elaborate on the type of supports you used and the linkage chemistry?

Gillespie: The support that I eluded to is a magnetic bead with a surface chemistry that has not been published. It has been developed by Advanced Magnetics. It is a conventional plastic chemistry with alkylated carboxyl or amino groups. This work has been a collaboration between people in my lab and people at Genetrak Systems. Genetrak Systems has been responsible primarily for developing the amplification of work that I described to you. The chemistry for linking the ligand to the solid support that we use, since it is an oligonucleotide, is standard chemistry using allyl groups on one end or the other of oligo-dT and linking to the activated carboxyl group of the plastic.

Key Words: Affinity capture, Chaotropic salt, Guanidine isothiocyanate, Molecular hybridization, Nucleation complex, Nucleic acid probe, Oligonucleotide, Polymerase chain reaction, Probe amplification, Riboprobe, Transcription amplification system

Structure-Function Relationships, In Vivo Mutability and Gene Amplification in Human Cholinesterases, Targets for Organophosphorous Poisons

Hermona Soreq, Lewis Neville, Averell Gnatt, Revital Ben-Aziz, Yaron Lapidot-Lifson[1], Gal Ehrlich, Shlomo Seidman, Efrat Lev-Lehman, Rachel Beeri, Dalia Ginzberg and Haim Zakut[1]

Department of Biological Chemistry, The Life Sciences Institute, The Hebrew University of Jerusalem, Israel 91904

[1] *Department of Obstetrics and Gynecology, The Edith Wolfson, Medical Center, Holon, The Sackler Faculty of Medicine, Tel-Aviv University, Tel-Aviv, Israel*

Introduction

The human cholinesterase genes and their protein products have been the focus of intensive research for many years (for comprehensive reviews see Whittaker, 1986; Rakonczay and Brimijoin, 1988; and Soreq and Zakut, 1990) because of the physiological function attributed to these enzymes, which are both capable of hydrolyzing the neurotransmitter acetylcholine. Genetic linkage evidence indicated that two distinct genes, designated ACHE and CHE, encode the two principal forms of cholinesterases, acetylcholinesterase (acetylcholine acetyl hydrolase, AChE, EC 3.1.1.7) and butyrylcholinesterase (acylcholine acylhydrolase, BuChE, EC 3.1.1.8) which differ in their substrate specificities and sensitivities to selective inhibitors. The toxic effects of organophosphorous (OP) poisons, such as common insecticides or nerve gases, are generally attributed to their specific inhibition of cholinesterases, interfering with cholinergic neurotransmission. OP inhibition of cholinesterases occurs through a covalent interaction of the OP compounds with a serine residue in the active esteratic site (Koelle, 1972). However, detailed *structure-function* relationships in this family of enzymes have been hampered by the difficulties in purifying mammalian cholinesterases.

Both AChE and BuChE are potentially involved in the regulation of cholinergic signalling, together with the neurotransmitter acetylcholine and the various acetylcholine receptors. Tightly coordinated expression of the CHE and ACHE genes, encoding BuChE and AChE, respectively, has been observed in multiple biosystems. In addition to tissues where cholinergic innervation is to be expected, like brain and muscle, human cholinesterases are intensively expressed in various developing and tumor cells where no function could be assigned to these enzymes (for recent examples see Malinger, et al., 1989; Zakut, et al., 1990). Furthermore, inhibition of cholin-

D. Kamely et al. (eds), Biotechnology: Bridging Research and Applications, 153–166.

esterase activities appears to affect cellular differentiation in bone marrow cells, where carbamylcholine or physostigmine administration induces pro-megakaryocytopoiesis, AChE production and platelet production (Burstein, et al, 1980, 1985). Other examples include sperm cells, in which OP poisons interfere with motility (Rama Sastry and Sadavongvivad, 1979). To approach the biochemical as well as the biological questions related with these enzymes, a search for the human CHE and ACHE genes and their modes of expression was initiated.

Materials and Methods

The covalent binding of OPs to the esteratic site serine in cholinesterases served for microsequencing of the radioactively labeled active site peptide (Lockridge, 1984). Oligodeoxynucleotide mixtures were thereafter synthesized with the potential to encode this peptide and cDNA libraries from various human tissues were screened with these radioactively labeled oligonucleotide probes (Prody, et al., 1986). The first human cDNA clone which was thus isolated encoded BuChE, as became evident from comparing with peptide sequencing data (Prody, et al., 1987). The cDNA was A,T-rich, as expected from a tissue-specific gene. It included 7 consensus sequences for glycosylation sites and a domain encoding a signal sequence, reflecting the secretory nature of its protein product (Soreq and Gnatt, 1987). The same cDNA was found in brain, muscle, lymphocytes and liver libraries, suggesting that a single mRNA transcript encodes BuChE in all tissues (Dreyfus, et al., 1990).

To isolate the DNA sequence encoding AChE, oligodeoxynucleotides were synthesized which would recognize areas in *Torpedo* AChEcDNA (Schumacher, et al., 1986), but did not appear in human BuChEcDNA. The first clones which were thus found (Soreq and Prody, 1989) served to search for others until a diverse collection of human AChEcDNA clones of various origins and lengths had been assembled. These clones all encoded a protein with >50% homologies to both human BuChE and *Torpedo* AChE. All of the clones terminated within a short domain of approximately 100 nucleotides, downstream from the expected 5′ end of the coding sequence. To reveal the complete coding region, a genomic DNA fragment encoding the missing 5′ domain was isolated. This was combined with the cDNA by enzymatic restriction and ligation to yield an AChE-coding DNA. In striking contrast to BuChEcDNA, the AChEcDNA sequence was found to be extremely rich in G,C residues (Ben-Aziz, et al., 1990; Soreq et al., 1990).

In order to find an explanation for the premature termination of the various AChEcDNA clones, the AChE coding sequence was subjected to secondary structure analysis by the Wisconsin Program. A tightly folded stems and loops structure spanning the common termination domain of the isolated AChEcDNAs ($DG^\circ = -117$ kcal) was revealed. In other systems, for example in the adenovirus genome (Kessler, et al., 1989), structures with free energy of the same order of magnitude were shown to attenuate transcription *in vivo*. It is currently assumed that the attenuator sequence was responsible for the incomplete synthesis of ACHEcDNAs and that it is capable of controlling AChEmRNA synthesis *in vivo*.

Results and Discussion

Comparative Analysis of Cholinesterase Coding Sequences

The tightly G,C-rich folded sequence is a characteristic of AChEcDNA and which distinguishes it from BuChEcDNA. Consequently, the two cDNAs are completely not homologous to each other, which was surprising in view of the > 50% identically aligned amino acid residues in their protein products (Ben-Aziz, et al., 1990; Soreq et al., 1990). In addition, the AChE protein is far less glycosylated than BuChE (3 vs. 7 sites). This further distinction could be related to the different subcellular localizations of the two enzymes, with BuChE being a soluble serum protein whereas AChE is cell surface linked and better protected from proteolytic degradation. It should be emphasized that neither AChE nor BuChE display any sequence homology to serine esterases, which make cholinesterases into a distinct although limited family of proteins.

The highly efficient catalytic activity of cholinesterases led to the development of very sensitive assays for these enzymes, including gel activity staining (Brock and Bader, 1983) based on the precipitation of ferrous-copper thiocholine complexes (Koelle and Friedenwald, 1949). When human serum is thus analyzed, various BuChE activity bands may be observed. The more abundant C_1-C_4 bands were all genetically linked to one locus, designated CHE1 and were found to appear in all humans (Whittaker, 1986). In contrast, the faster migrating C_5 band is only expressed in about 8% of the caucasian population and is linked to a separate locus, designated CHE2 (Whittaker, 1986). It is currently accepted that neither locus is linked to the ACHE locus, controlling AChE production (reviewed by Soreq and Zakut, 1990).

Allelic Variants of Human BuChE: Genetic Linkage and Gene Mapping Information

Detailed biochemical analysis revealed several alleles in the principal CHE1 locus, controlling BuChE production. Most abundant are the "atypical" and the "silent" alleles, producing serum BuChE with altered biochemical properties. The "silent" enzyme cannot hydrolyze any substrate, whereas the "atypical" efficiently hydrolyzes butyrylcholine but is incapable of hydrolyzing its larger analog succinylcholine, which is rather efficiently hydrolysed by the normal BuChE enzyme but not at all by AChE (Whittaker, 1986). The distinction in succinylcholine recognition between AChE and BuChE serves several clinical purposes. Succinylcholine is used via its acetylcholine receptor AChE blocking activity for muscle relaxation at surgery, and is removed from the circulation by BuChE. Therefore, individuals with "atypical" or "silent" CHE1 alleles will suffer prolonged muscle relaxation following surgery, while being perfectly healthy at other times. It has long been known that the variable CHE1 alleles also dictate different sensitivities to dibucaine (Kalow and Genest, 1957). Furthermore, the low level or absence of active BuChE in the serum, a characteristic of the "atypical" and "silent" phenotypes, also implies reduced scavenging activity and therefore higher vulnerability to OP poisons.

Different CHE1 alleles appear in various populations, and at least 8 phenotypes are known which differ in the biochemical properties of serum BuChE (Whittaker, 1986). These phenotypes are all genetically linked to the TF transferrin gene on the long arm of Chromosome No. 3 (Sparkes, et al., 1984). In contrast, the genetic linkage of the CHE2 locus is still a matter of dispute, with a recent genetic linkage to the haptoglobin gene on Chromosome No. 16 observed in the U.S. (Marazita, et al., 1989) and other data indicating linkage to Chromosome No. 2 in a Danish family (Eiberg, et al., 1989). The question therefore arose whether the CHE2 locus displays different linkage properties in different populations.

When BuChEcDNA was [^{35}S]-labeled and hybridized with spread human chromosomes, both Chromosome No. 3 and No. 16 were labeled (Soreq, et al., 1987; Zakut, et al., 1989). However, genomic blots from a number of individuals were indicative of a single gene. An unusual BuChEcDNA was then isolated from a glioblastoma tumor library, which carried two point mutations encoding 2 amino acid substitutions (Asp70 → Gly and Ser425 → Pro) and a 0.5 Kb long 3'-extension **(Gnatt, et al., 1990).** This 3′ fragment of BuChEcDNA appeared to be unique in the genome, which made it an appropriate probe for gene mapping by *in situ* chromosomal hybridization. The 3′-extension fragment bound only to Chromosome 3, demonstrating that one gene at the CHE1 locus suffices to encode for BuChE in all tissues, including the glioblastoma tumors where BuChE was previously shown to be overexpressed (Razon, et al., 1984).

An independent approach to the question of gene mapping involves the use of chromosome blots. These are derived from human-hamster hybrid cell lines, each of which carries only one or few human chromosomes. Hybridization conditions were therefore established under which only the human gene would be detected. When 27 hybrid cell lines were examined using these hybridization conditions, all of the lines carrying human Chromosome No. 3 proved positive, including one line in which the only human chromosome present was No. 3. In contrast, none of the lines carrying Chromosome No. 16 but lacking No. 3 showed a positive band (Gnatt et al., 1990). Therefore, the individual from whom these lines were constructed certainly did not carry a CHE gene on his Chromosome No. 16 and the question whether *some* individuals do – in agreement with the genetic linkage data – remains open.

AChE and BuChE are Encoded by Two Distinct Genes

The CHE gene is rather large, with two very big introns at its 3′-end (Arpagaus, et al., 1990). This raised an intriguing question: could these large introns harbor the AChE coding exons? It should be noted that in contrast with the CHE gene there is no genetic linkage information on the ACHE gene, perhaps because any mutation in it would likely be lethal. To find out whether the two genes reside on the same DNA fragment, cosmid recombination was performed. This approach involves the selection of dual-antibiotic resistant recombinant colonies conferred by cosmid-plasmid clones. Of 4 cosmid clones which were thus isolated, all were apparently derived from the same gene, included the 3′-extension, and did *not* hybridize with AChEcDNA (Gnatt et al., 1991). Therefore, AChE cannot be encoded from the same gene as BuChE, and exact mapping of this gene awaits further experimentation.

Expression of Recombinant Human Cholinesterases in mRNA Microinjected Xenopus Oocytes and Transfected Cells

To approach the issue of structure-function relationships within the cholinesterase gene family, an efficient expression system was required where sufficient amounts of the various cholinesterases could be correctly produced to allow for precise biochemical analysis. Messenger RNA-injected *Xenopus* oocytes served well for this purpose. Without the need for transfection and the establishment of stable lines, one may transcribe synthetic mRNAs *in vitro* from each of the discussed cDNAs (Ben-Aziz and Soreq, 1990) to inject into the oocytes and direct the production of active heterologous enzymes. Biochemical analyses of the recombinant cholinesterases produced in the oocytes proved that BuChEcDNA encodes an enzyme which hydrolyzes butyrylthiocholine better than acetylthiocholine (Soreq, et al., 1989), while the product of AChEcDNA displays substrate specificity for acetylthiocholine (Ben-Aziz, et al., 1990; Soreq et al., 1990) and both interact with various specific inhibitors exactly as expected. Table I summarizes the oocyte expression data. More recently, mammalian cell transfections with the same AChEcDNA sequence revealed similar biochemical properties for the recombinant human AChE (Velan, et al., 1991).

[^{35}S] methionine labeling of oocytes injected each with 1 ng of synthetic BuChEmRNA, followed by gel electrophoresis and autoradiography, revealed that approximately 10% of the nascent proteins produced in the

Table I. Expression of recombinant human cholinesterases in mRNA microinjected *Xenopus* oocytes

Cloned sequence:	*AChEcDNA*	*BuChEcDNA*
Acetylthiocholine hydrolysis	+++	+
Butyrylthiocholine hydrolysis	−	++
BW 284C51 inhibition	+++	−
iso-OMPA inhibition	−	+++
Succinylcholine inhibition	−	+++

Molecularly cloned AChEcDNA and BuChEcDNA were constructed into the pGEM transcription vector and synthetic RNA was transcribed *in vitro* from each of these sequences. *Xenopus* oocytes microinjection was performed as described (Soreq, et al., 1989) and catalytic activities were measured spectrophotometrically for each of the enzymes. Note the specificity of substrates and inhibitors for each of the recombinant cholinesterases as produced in frog oocytes.

oocyte overnight are translated from the injected mRNA. Activity gel staining demonstrated a single stainable band, its intensity being approximately 20% of that observed with 1 μl serum. Indeed, biochemical measurements have shown that homogenates from injected oocytes contain BuChE in levels within the same order of magnitude as that present in human serum. The oocyte-produced recombinant BuChE displays a distinct migration pattern in non-denaturing gel electrophoresis (Soreq and Seidman, 1991). This is possibly due to the oocyte-specific mode of glycosylation, unequal in many cases to the mammalian ones (Soreq, 1985). Altogether, the oocyte-produced recombinant human enzymes may clearly be produced in sufficient amounts for detailed biochemical analyses.

Structure-function Relationships in Recombinant BuChE Variants

Pharmacological studies have shown variable dissociation coefficients for substrates and inhibitors which interact with the naturally occurring cholinesterases and their variants. While AChE only interacts with acetylcholine (ACh), "usual" BuChE would also hydrolyze butyrylcholine (BuCh) and succinylcholine (SucCh). This already implies differences in the active site dimensions, as do the various binding constants of OP inhibitors to AChE and BuChE. It is generally accepted that the catalytic subunit in cholinesterases contains esteratic and anionic sites, both of which are involved in the catalytic reactions and inhibition of these enzymes. Consequently, the common prophylaxis and therapy against OP poisoning involves the use of quaternary oximes such as 2–PAM, assumed to block and protect the anionic site (discussed in Neville, et al., 1990). Comparative examination of the interactions of recombinant cholinesterases with these different ligands was therefore initiated, using the oocyte system for expression.

Several expression vectors were constructed for these experiments, including various BuChEcDNAs with both, one or neither of the mutations from

Table II. Currently discovered variations in the human CHE gene

Number of altered *Residue*	Amino acid *Change*	Nucleotide *Change*
70	Asp → Gly	A → G
114	Tyr → His	A → C
117	Gly → Stop	A inserted
243	Thr → Met	G → T
390	Gly → Val	G → T
425	Ser → Pro	T → C
539	Ala → Thr	G → A
561	Phe → Tyr	T → A

Currently discovered variations in the human CHE gene include published mutations (Arpagaus, et al., 1990; Neville, et al., 1990) and unpublished point mutations discovered in our laboratory. See text for expression studies referring to the Gly70 and Pro425 mutations.

"unusual" (Gnatt et al., 1990) BuChEcDNA clone. Expression in oocytes clearly revealed resistance to both Dibucaine and SuCh in double mutants, and some SucCh resistance in BuChE variants carrying the Gly70 mutation alone (Neville, et al., 1990). This implied disruption of anionic site interactions by the Asp70 → Gly mutation as well as steric relationships between this amino acid residue and the Ser425–containing peptide.

Interestingly, the Gly70 mutation also conferred resistance to iso-OMPA, an OP inhibitor specific for BuChE; and to Bambuterol and Ecothiophate, two common drugs (against asthma and glaucoma, respectively) with CHE inhibitory properties. No difference was recorded, however, with respect to DFP or paraoxon, the toxic metabolite of parathion. This implied that individuals carrying CHE variants may respond differently to several but not all cholinesterase inhibitors. Furthermore, the mutant enzymes also differ in their ability to be reactivated by the commonly used oxime treatment. While 10 mM 2–PAM restores the activity of DFP-inhibited normal enzyme, it is not effective with the Gly70 species (Neville, et al., 1990). Current work proceeds to map specific properties of cholinesterases to precise peptide domains using many more mutants, all of which represent existing and rather common *in vivo* mutations. Table II summarizes the 8 single point nucleotide alterations that are currently known in the human CHE gene and the amino acid substitutions conferred by these mutations. Construction of various recombinations of these mutants may now provide a new tool to examine their synergistic effects using the oocyte system.

A 2–dimensional model for cholinesterases was proposed by P. Taylor and colleagues (1987) based on the *Torpedo* AChE sequence, with good agreement with our data. Our present findings, in the absence of x-ray crystallography, imply that Asp70 and Ser425, localized in the cysteine loops A and C of this model, are near each other in the 3–dimensional structure of the protein but allow for ligands at least 10 Å in diameter to penetrate between them. The dramatic effects exerted by Asp70 mutagenesis in BuChE define its involvement in the active site of cholinesterases in general. In

H.AChE	PLWMGVPHGYEIEFIFGIPL--DPSR
H.BuChE	PEWMGVMHGYEIEFVFGLPL--ERRD
B.AChE	PEWMGVPHGYEIEFIFGLPL--EPSL
T.AChE	PEWMGVIHGYEIEFVFGLPL--VKEL
D.AChE	GEWMGVLHGDEIEYFFGQPL--NNSL
D.Est 6	HRKYGKSPAY--AYVYDNPA--EKGI
Bov.Tg	HSSLELL--TDVLYAFGLPFYPAYEG

Figure 1. The 440 region. Sequence alignment was performed as previously described (Soreq and Prody, 1989). Note the negatively charged residues around the 440 region, probably involved in anionic site interactions.

contrast, the milder effects of Ser425 alterations suggest the involvement of other, nearby negatively charged regions (such as the peptide glu-ile-glu at amino acid residue No. 440) in the anionic site interaction. Figure 1 presents the conserved negatively charged peptides in this domain in various members of the cholinesterase gene family.

CHE Gene Amplification in OP-exposed Individuals

The powerful biochemical dissection which was enabled by the microinjection approach with *in vivo* variants of the CHE gene initiated a search for yet more mutations. It so happens that the abundance of variant CHE alleles in Israel is 2–fold that found in Europe (Szeinberg, et al., 1972), probably due to a founder effect characterizing size-limited populations. In addition, extensive agriculture in Israel is responsible for a relatively high consumption of OP insecticides. The combination of a defective CHE phenotype and prolonged OP exposure was exemplified in the H. family (Prody, et al., 1989), where one brother was brought to intensive care following parathion spraying in a cotton field and his sister suffered from prolonged apnea following SucCh administration at the surgical removal of oocytes for *in vitro* fertilization. The serum BuChE activities in this family were very low, which explained their vulnerability to OP poisoning and SucCh administration. Moreover, 100 copies of the CHE gene were found in peripheral blood DNA from the family member poisoned by parathion as well as in one of his sons. The amplification was most intense in central sequences within the coding region and was to the best of our knowledge unique to this gene.

Gene mapping by *in situ* hybridization revealed labeling 10–fold higher than usual which mapped to Chromosome 3q26–ter in lymphocyte chromosomes from the affected individual. This implied that the amplified DNA was probably inherited, although we could not exclude the possibility that the amplification re-occurred in the son as a result of a parallel insult (i.e., exposure to OP poisons at the critical developmental stage; for further discussion of this point see Soreq and Zakut, 1990).

The findings in this family linked a defective enzyme, exposure to OP agents and gene amplification; to be inherited and to appear in all tissues, this multi-element phenomenon should occur in germ cells – oocytes or, more likely because of their large numbers, sperm. This, in turn, is in agreement with early observations that OP agents decrease sperm motility and therefore fertility (Rama Sastry and Sadavongvivad, 1979).

Cholinesterase Gene Amplification in Tumors

Why or when do genes amplify? The H. family presented the first case of inheritable gene amplification in apparently healthy human individuals. There is, however, ample precedent for gene amplification in tumors. For example, genes encoding target proteins to toxic inhibitors amplify under exposure to such inhibitors (i.e., DHFR in leukemias, (Schimke, 1984)). Many oncogenes, either homologous to growth factors or to their receptors, amplify in rapidly developing cells (i.e., EGFR in gliomas, (Libermann, et al., 1984)). In insects and amphibia, developmentally essential genes amplify when high expression is needed (i.e., chorion genes in *Drosophila*, (Spradling, 1987)). In all these cases, the amplified genes confer a selection advantage to the cells and are subjected to frequent mutability. Consequently, not all of the amplified copies are expressed.

It is not yet clear whether CHE gene amplification in the H. family reflected a response to OP exposure. Even if this was the cause for selection pressure leading to that event, the amplification could be a unique phenomenon caused by the combination of chronic OP exposure and defective CHE phenotype. To examine whether this is the case, various tumor types were screened in search for CHE and/or ACHE gene amplifications. Gene amplification frequently coincides with chromosome breakage in tumors. Therefore, cholinesterase gene amplification was searched for in tumors or disorders where the chromosomal 3q26–ter region breaks often. This occurs in leukemias and platelet disorders (Pintado, et al., 1985). Furthermore, cholinesterase inhibitors such as physostigmine induce platelet production in the mouse (Burstein, et al., 1980), which indicates that the CHE gene is actively involved in megakaryocytopoiesis. DNA hybridization revealed that in leukemias and platelet disorders, *both* the CHE and the ACHE genes amplify. Co-amplification probably occurs in response to a single stimulus, as it was observed in a considerable fraction of the cases examined (Lapidot-Lifson, et al., 1989). There is no oncogene nearby on the long arm of chromosome No. 3 to explain this phenomenon, indicating that the amplification is related to either the functional properties of these two genes or their structural buildup.

Another tissue type where cholinesterase gene amplification was looked for were ovarian tumors, where cytochemical staining demonstrated over-expression (Drews, 1975). Here again, both CHE and ACHE genes were amplified (Zakut, et al., 1990). There was good correlation between the levels of cholinesterase gene amplification in ovarian tumors and the parallel amplification of several oncogenes. However, there was *no* CHE and/or ACHE gene amplification in gall bladder tumors (unpublished observations), demonstrating that this gene amplification phenomenon also depends on the nature of cells where it occurs.

Conclusions

Not much is known about origins of replication in the mammalian genome or what induces them to become active (Stark, 1986). However, it is known that some genes amplify more readily than others. It is further assumed that the amplification occurs spontaneously and in relationship both to expression and to cell division (Tzeng-Horng and Hamlin, 1990) and that it persists there only when conferring a selection advantage (Schimke, 1984). It is therefore possible that the CHE and/or ACHE coding sequences either *contain* such an origin or *induce* the process of DNA replication. It will be intriguing to find out whether disruption of the balance at the level of DNA replication would initiate a series of events that will end up in abundant gene amplifications. Should that be the case, in each tumor where cholinesterase genes are expressed, they would also be vulnerable to amplification. To examine this hypothesis, experiments with transgenic mice are currently being performed.

Acknowledgments

Supported by the U.S. Army Medical Research and Development Command (Contract No. DAMD 17–90–Z-0038, to H.S.).

References

Arpapaus, M., Kott, M., Vatsis, K.P., Bartels, C.F., La Du, B.N. and Lockridge, O. 1990. Structure of the gene for human butyrylcholinesterase: Evidence for a single copy. Biochem. 29: 124–131.

Ben-Aziz, R., Gnatt, A., Prody, C., Lev-Lehman, E., Neville, L., Seidman, S., Ginzberg, D., Soreq, H., Lapidot-Lifson, Y. and Zakut, H. 1990. Differential codon usage and distinct surface probabilities in human acetylcholinesterase and butyrylcholinesterase. In: Proceedings of the Third International Meeting on Cholinesterases. Bacu, F., ed. ACS Books, Washington, D.C. In press.

Ben-Aziz, R. and Soreq, H. 1990. Improving poor *in vitro* transcription from G,C rich genes. Nuc. Acids Res. 18: 3418.

Brock, D.J.H. and Bader, P. 1983. The use of commercial antisera in resolving the cholinesterase bands of human amniotic fluids. Clin. Chem. Acta. 127: 419–422.

Burstein, S.A., Adamson, J.W. and Harker, L.A. 1980. Megakaryocytopoiesis in culture: Modulation by cholinergic mechanisms. J. Cell. Physiol. 54: 201–208.

Burstein, S.A., Boyd, C.N. and Dale, G.L. 1985. Quantitation of megakaryocytopoiesis in liquid culture by enzymatic determination of acetylcholinesterase. J. Cell. Physiol. 122: 159–165.

Drews, E. 1975. Cholinesterase in embryonic development. Prog. Histochem. Cytochem. 7: 1–52.

Dreyfus, P.A., Pincon-Raymond, M., Zakut, H., Seidman, S. and Soreq, H. 1990. Search for the molecular origins of butyrylcholinesterase polymorphism by cDNA screening, deletion mutagenesis and *Xenopus* oocyte co-injections. In: Proceedings of the Third International Meeting on Cholinesterases. Bacu, F., ed., ACS Books, Washington, D.C. In press.

Eiberg, H., Nielsen, L.S., Klausen, J., Dahlen, M., Kristensen, M., Bisgaard, M.L., Moller, N. and Mohr, J. 1989. Linkage between serum cholinesterase 2(CHE2) and γ-crystallin gene cluster (CRYG) assignment to chromosome 2. Clin. Genet. 35: 313–321.

Gnatt, A., Ginzberg, D., Lieman-Hurwitz, J., Zakut, H. and Soreq, H. 1991. Human acetylcholinesterase and butyrylcholinesterase are encoded by two distinct genes. Cell. Mol. Neurobiol. 11: 91–104.

Gnatt, A., Prody, C., Zamir, R. Lieman-Hurwitz, J., Zakut, H. and Soreq, H. 1990b. Expression of alternatively terminated unusual CHEmRNA transcripts mapping to chromosome 3q26–ter in nervous system tumors. Cancer Res., 50: 1983–1987.

Kalow, K. and Genest, K. 1957. A method for the detection of atypical forms of human serum cholinesterase: Determination of dibucaine numbers. Can. J. Biochem. Physiol. 35: 339–346.

Kessler, M., Ben-Asher, E. and Aloni, Y. 1989. Elements modulating the block of transcription elongation at the adenovirus 2 attenuation site. J. Biol. Chem. 264: 9785–9790.

Koelle, G.B. 1972. Anticholinesterase agents. In: The pharmacological basis of therapeutics. Goodman, L.S. and Gilman, A., eds., McMillan, New York. pp. 445–466.

Koelle, G.B and Freidenwald, J.S. 1949. A histochemical method for localizing cholinesterase activity. Proc. Soc. Exp. Biol. Med. 70: 617–622.

Lapidot-Lifson, Y., Prody, C. Ginzberg, D., Meytes, D. Zakut, H. and Soreq, H. 1989. Co-amplification of human acetylcholinesterase and butyrylcholinesterase in blood cells: Correlation with various leukemias and abnormal megakaryocytopoiesis. Proc. Natl. Acad. Sci. USA. 86: 4715–4717.

Libermann, T.A., Razon, N., Bartal, A.D., Yarden, Y., Schlessinger, J. and Soreq, H. 1984. Expression of epidermal growth factor receptors in human brain tumors. Cancer Res. 44: 753–760.

Lockridge, O. 1984. Amino acid composition and sequence of human serum cholinesterase: A progress report. In: Cholinesterase Fundamental and Applied Aspects. Brzin, M., Barnard, E.A. and Sket, D., eds., Walter de Gruyter, New York. pp. 5–12.

Malinger, G., Zakut, H. and Soreq, H. 1989. Cholinoceptive properties of human primordial, preantral, and antral oocytes: *In situ* hybridization and biochemical evidence for expression of cholinesterase genes. J. Mol. Neurosci. 1: 77–84.

Marazita, M.L., Keats, B.J.B., Spence, M.A., Sparkes, R.S., Field, L.L., Sparkes, M.C. and Crist, M. 1989. Mapping studies of the serum cholinesterase-2 locus (CHE2). Hum. Genet. 83: 139–144.

Neville, L.F., Gnatt, A., Padan, R., Seidman, S., and Soreq, H. 1990. Anionic site interactions in human butyrylcholinesterase disrupted by two adjacent single point mutations. J. Biol. Chem. 265: 20735–20739.

Pintado, T., Ferro, M.T., San Roman, C., Mayayo, M. and Larana, J.G. 1985. Clinical correlations of the 3q21;q26 cytogenetic anomaly: A leukemic or myelodysplastic syndrome with preserved or increased platelet production and lack of response to cytotoxic drug therapy. Cancer. 55: 535–541.

Prody, C., Dreyfus, P.A., Zamir, R., Zakut, H. and Soreq, H. 1989. *De novo* amplification within a "silent" human cholinesterase gene in a family subjected to prolonged exposure to organophosphorous insecticides. Proc. Natl. Acad. Sci. USA. 86: 690–694.

Prody, C., Gnatt, A., Zevin-Sonkin, D., Goldberg, O. and Soreq, H. 1987. Isolation and characterization of full length cDNA clones coding for cholinesterases from human fetal tissues. Proc. Natl. Acad. Sci. USA. 84: 3555–3559.

Prody, C., Zevin-Sonkin, D., Gnatt, A., Koch, R., Zisling, R., Goldberg, O. and Soreq, H. 1986. Use of synthetic oligodeoxynucleotide probes for the isolation of a human cholinesterase cDNA clone. J. Neurosci. Res. 16: 25–35.

Rakonczay, Z. and Brimijoin, S. 1988. Biochemistry and pathophysiology of the molecular forms of cholinesterases. In: Subcellular Biochemistry. 12. Harris, J.R., ed., Plenum Press, New York. pp. 335–378.

Rama Sastry, B.V. and Sadavongvivad, C. 1979. Cholinergic systems in non-nervous tissues. Pharm. Rev. 30: 65–132.

Razon, N., Soreq, H., Roth, E. Bartal, A. and Silman, I. 1984. Characterization of activities and forms of cholinesterases in human primary brain tumors. Exp. Neurol. 84: 681–695.

Schimke, R.T. 1984. Gene amplification in cultured animal cells. Cell. 37: 705–713.

Schumacher, M., Camp, S., Maulet, Y., Newton, M., MacPhee-Quigley, K., Taylor, S.S., Freidmann, T. and Taylor, P. 1986. Primary structure of *Torpedo californica* acetylcholinesterase deduced from its cDNA sequence. Nature. 319: 407–409.

Soreq, H. 1985. The biosynthesis of biologically active proteins in mRNA microinjected *Xenopus* oocytes. CRC Crit. Reviews in Biochem. 18: 199–238.

Soreq, H. and Gnatt, A. 1987. Molecular biological search for human genes encoding cholinesterases. Molec. Neurobiol. 1: 47–80.

Soreq, H. and Prody, C. 1989. Sequence similarities between human acetylcholinesterase and related proteins: Putative implications for therapy of anticholinesterase intoxication. In: Computer-assisted modelling of receptor-ligand interactions, theoretical applications to drug design. Golombeck, A. and Rein, R., eds., Alan R. Liss, New York. pp. 347–359.

Soreq, H., Ben-Aziz, R., Prody, C.A., Gnatt, A. Neville, L., Lieman-Hurwitz, J., Lev-Lehman, E., Ginzberg, D., Seidman, S., Lapidot-Lifson, Y. and Zakut, H. 1990. Molecular cloning and construction of the coding region for human acetylcholinesterase reveals a G, C-rich attenuating structure. Proc. Natl. Acad. Sci. USA 87:9688–9692.

Soreq, H. and Seidman, S. 1991. *Xenopus* oocytes microinjection: From genes to proteins. In press in: Methods in Enzymology. Ion channels. Eds, B. Rudy and L. Iverson. Academic Press, NY.

Soreq, H., Seidman, S. Dreyfus, P.A., Zevin-Sonkin, D. and Zakut, H. 1989. Expression and tissue-specific assembly of human butyrylcholinesterase in microinjected *Xenopus* oocytes. J. Biol. Chem. 264: 10608–10613.

Soreq, H. and Zakut, H. 1990. Cholinesterase genes: Multilevelled regulation. Monographs in human genetics. Vol. 13. Sparkes, R.S., ed., Karger, Basel.

Soreq, H., Zamir, R., Zevin-Sonkin, D. and Zakut, H. 1987. Human cholinesterase genes localized by hybridization to chromosomes 3 and 16. Hum. Genet. 77: 325–328.

Sparkes, R.S., Field, L.L., Sparkes, M.C., Crist, M., Spence, M.A., James, K. and Garry, P.J. 1984. Genetic linkage studies of transferrin, pseudocholinesterase, and chromosome 1 loci. Hum. Hered. 34: 96–100.

Spradling, A. 1987. Gene amplification in Dipteran chromosomes. In: Results and problems in cell differentiation: Structure and function of eukaryotic chromosomes. Vol. 14. Hennig, W., ed., Springer-Verlag, Berlin. pp. 199–212.

Stark, G.R. 1986. DNA amplification in drug resistant cells and in tumors. Cancer Surveys. 5: 1–23.

Szeinberg, A., Pipano, S., Assa, M., Medalie, J.H. and Neufeld, H.N. 1972. High frequency of atypical pseudocholinesterase gene among Iraqi and Iranian Jews. Clin. Genet. 3: 123–127.

Taylor, P., Schumacher, M., MacPhee-Quigley, K., Freidmann, T. and Taylor, S. 1987. The structure of acetylcholinesterase: Relationship to its function and cellular disposition. Trends in Neurosci. 10: 93–95.

Tzeng-Horng, L.E.U. and Hamlin, J.L. 1989. High-resolution mapping of replication fork movement through the amplified dihydrofolate reductase domain in CHO cells by gel renaturation analysis. Mol. Cell Biol. 9: 523–531.

Velan, B., Konnan, C., Grosfeld, H., Leitner, M., Gozes, Y., Flashner, Y., Sery, T., Cohen, S., Ben-Aziz, R., Seidman, S., Shafferman, A. and Soreq, H. 1991. Recombinant human acetylcholinesterase is secreted from transiently transfected 293 cells as a soluble globular anzyme. Cell. Molec. Neurobiol. 11: 143–156.

Whittaker, M. 1986. Cholinesterase: Monographs in human genetics. Vol. 11. Beckman, A., ed., Karger, Basel.

Zakut, H., Ehrlich, G., Ayalon, A., Prody, C., Malinger, G., Seidman, S., Kehlenbach, R. and Soreq, H. 1990. Acetylcholinesterase and butyrylcholinesterase genes co-amplify in primary ovarian carcinomas. J. Clin. Invest. In press.

Zakut, H., Zamir, R., Sindel, L. and Soreq, H. 1989. Gene mapping on chorionic villi chromosomes by hybridization *in situ*: Localization of cholinesterase cDNA binding sites to chromosomes 3q21, 3q26–ter and 16q21. Hum Rep. 4: 941–946.

Discussion (DR. SOREQ)

Chakrabarty: Have you looked at people who are not exposed to organophosphate to see whether they ever show amplification?

Soreq: Absolutely. We have looked so far at DNA from at least 100 individuals. We have found two families with this type of amplification. One family had a defective phenotype and prolonged exposure. The other family, not yet published, suffers from idiotypic thrombocytopenia. They have a disturbance in megakaryopoiesis which might be related to the amplification phenomenon. We have just started studying them now. In addition, 25% of the leukemic patients that we looked at had the amplification and about 40% of those with primary ovarian carcinomas. We did not find any amplification in 15 cases of gall bladder carcinoma and about 20 cases of glioblastomas that we looked at. So it is specific, related to the particular type of tumor in which it occurs. Regarding the unexposed individuals, it is very difficult to know whether any Israeli has been completely unexposed to any organophosphorous poison. Samples from the United States and England did not have any amplification. This is the beginning of a study that will have to be epidemiological in nature. Beyond saying that we have found 20 amplifications out of about 100 individuals examined, I have no other answers yet.

Chakrabarty: Assuming that amplification indeed occurred in response to the entry of organophosphorus into the body, do you know what happens after the person has been removed from this exposure?

Soreq: Early in the 1970's people studied chromosomal breakage that occurred in response to occupational exposure or acute poisoning by organophosphates. Twenty-five percent of the exposed individuals suffered from chromosomal breakage in peripheral blood lymphocytes. This was referred to as a transient unimportant phenomenon because it only took six months to disappear but, of course, if you have chromosomal breakage in germ cells that will be transferred to the next generation regardless of its transient nature. So, I do not know how to relate that to this question except that in blood cells it takes six months to remove DNA damage that has occurred in response to such exposure. If it is transferred to the next generation, this is a perpetual change that probably cannot be repaired.

Gillespie: Are any other genes commonly co-amplified with the cholinesterase genes or are they randomly co-amplified?

Soreq: We tried to look at that. First, there is no oncogene in the vicinity of either of the cholinesterase genes as far as we can tell. Second, we have chosen specific oncogenes – both those on the same chromosome 3 such as *raf*, although it is on the other arm, as well as those on another chromosome, such as the *Sis* gene on chromosome 4. In each tumor, we found that the oncogenes that amplified were particular to that type of tumor and the cholinesterase genes appeared in either type of tumor. We suspect that whenever the cholinesterase genes are expressed, they are also vulnerable to amplification, probably because of either structural properties in the genes or something common in the recognition of the cis- or trans-element. But there is no linkage between these genes and a particular oncogene; rather, it relates to the type of tumor.

Eisenstadt: Is there an animal or cell culture model for studying amplification induced by the organophosphates?

Soreq: Yes, there is a *Culex* model where the esterase gene was amplified and conferred complete resistance towards organophosphate poisons. There is another type of *Culex* in which mutagenesis conferred the same resistance. So, I think that we are following in the footsteps of insects in terms of model systems. This is the first example that I know of in *Homo sapiens*.

Key Words: Acetylcholinesterase, Butyrylcholinesterase, DNA transfection, Gene amplification, Mutagenesis, Organophosphorous poisons, Selection advantage, Structure-function relationships, Variant alleles, *Xenopus* oocyte microinjection

Assessment of Genetic Damage in Human Tissue by Immunoassay

Paul T. Strickland[1], **Nathaniel Rothman**[2] **and Miriam C. Poirier**[3]
Departments of Environmental Health Sciences[1] and Epidemiology[2], The Johns Hopkins School of Hygiene and Public Health, 615 North Wolfe Street, Baltimore, MD 21205 USA; Laboratory of Cellular Carcinogenesis and Tumor Promotion[3], National Cancer Institute, NIH, Bethesda, MD 20892 USA

Introduction

Immunological methods have been developed to assess genetic damage in human tissues. This approach has focused primarily on carcinogen-induced DNA modifications. Antibodies have been produced that recognize DNA modified by a variety of chemical or physical agents including polycyclic aromatic hydrocarbons (PAHs), alkylating compounds, aflatoxins, oxidizing agents and UV radiation (Strickland and Boyle, 1984; Poirier, 1984; Santella, et al., 1988). A partial list of these agents is shown in Table 1. Immunoassays based on these antibodies have been used to measure genetic damage in human lung (Perera, et al., 1989), esophagus (Umbenhauer, et al., 1985), skin (Yang, et al., 1987; Bruze, et al., 1989), blood (Harris, et al., 1985; Perera, et al., 1988) and urine (Groopman, et al., 1985; Wild, et al., 1988). Genetic damage in human tissues and body fluids has been documented in cohorts exposed to carcinogens through their occupation, environment, diet, life style, or medical treatment (reviewed in Poirier and Beland, 1987; Bartsch, et al., 1988).

Quantitation of genetic damage by this and other methods should be useful for the assessment of biological dose from occupational, dietary, and environmental exposures to carcinogens. Ultimately, this technology may assist in the estimation of cancer risk and the study of inter-individual differences in response to carcinogen exposure. Two examples of this methodology will be described here: the quantitation of i) UV photoproducts induced in skin by solar simulating radiation and ii) PAH-DNA adducts induced in peripheral blood white blood cells by ingestion of char-broiled food.

D. Kamely et al. (eds), Biotechnology: Bridging Research and Applications, 167–177.

Table 1. Carcinogens used to modify DNA for antibody production

Alkylating compounds	(reviewed in Strickland and Boyle, 1984)
Benzo(a)pyrene diol epoxide	(reviewed in Poirier, 1984)
2-Acetylaminofluorene	(reviewed in Poirier, 1984)
Aflatoxin B1	(reviewed in Strickland and Boyle, 1984)
Cisplatin	(reviewed in Poirier, 1984)
1-Aminopyrene	(Hsieh, et al., 1985)
4-Aminobiphenyl	(Roberts, et al., 1985)
Melphalan	(Tilby, et al., 1987)
UV Radiation	(reviewed in Strickland, 1985)
Psoralen-UVA	(Santella, et al., 1985)
Ionizing Radiation	(reviewed in Strickland and Boyle, 1984)
Osmium tetroxide	(Leadon and Hanawalt, 1983)
Chloroacetaldehyde	(Young and Santella, 1988)
Crotonaldehyde, Acrolein	(Foiles, et al., 1987; Shivapurkar, et al., 1989)

Detection of UV Photoproducts in Human Skin

Experimental studies indicate that the ultraviolet-B (UVB, 290–320 nm) portion of the solar spectrum is primarily responsible for the carcinogenic properties of sunlight (Forbes, et al., 1989). This waveband encompasses the short-wavelength end of the solar spectrum that reaches the earth's surface. The terrestrial spectrum contains both UVB and UVA (320–400 nm) radiation in a ratio of approximately 15:1 to 50:1 (UVA:UVB), depending on a number of factors including atmospheric conditions, latitude, altitude, time of day, time of year, etc. We have recently examined cyclobuta-dithymidine (T< >T) photoproduct induction and repair in human skin exposed *in situ* to environmental levels of sunlight (Bruze, et al., 1989).

A solar-simulating UV apparatus which approximates both the spectrum and intensity of the UV portion of summer midday sunlight at 39°N latitude was used (Strickland and Creasey, 1989). The UV fluence (290–400 nm) rate at the exposure site was 20 J m^{-2} s^{-1}. Untanned gluteal skin of 19 healthy volunteers was exposed to solar UV for 60 minutes (72 KJ m^{-2}) for determination of T< >T photoproducts. This is approximately equivalent to 60 minutes of midday summer sunlight at 39°N latitude.

In 10 individuals, skin biopsies were collected from exposed and unexposed skin immediately (< 5 min) after irradiation for assessment of T< >T induction. In the remaining 9 individuals, biopsies were collected from exposed skin both immediately (< 5 min) and 4 h after irradiation for assessment of T< >T repair. During the 4 h post-irradiation period, skin was covered by clothing, thereby preventing exposure to visible light. Punch biopsies (4 mm diameter) were taken after intradermal injection of 1% xylocaine with epinephrine, and immediately frozen at − 135°C. The epidermis was removed from thawed skin biopsies and DNA was extracted (Bruze, et al., 1989). Antibodies specific for T< >T photoproducts (Strickland and

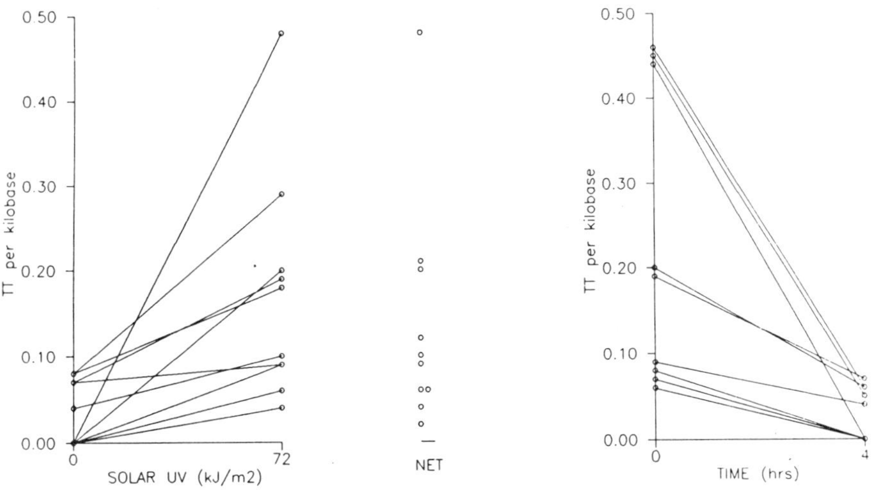

Figure 1. Induction and repair of T< >T photoproducts in human skin by solar-simulating UV radiation. Left Panel: T< >T photoproducts in skin of 10 individuals before and after exposure to 72 KJ m^{-2} solar UV. DNA was extracted from epidermis of 4 mm punch biopsies and T< >T content determined by ELISA. NET yields of T< >T are the difference between exposed (72 KJ m^{-2}) yield and unexposed (O J m^{-2}) yield for each individual. Right Panel: T< >T photoproducts in skin of 9 individuals immediately or 4 h after exposure to 72 KJ m^{-2} solar UV. Yields of T< >T immediately after exposure and not adjusted for background (unexposed) yields of T< >T as in left panel. (Reproduced with permission from Bruze, et al., 1989; Elsevier Press).

Boyle, 1981) were used to assess photoproduct levels by enzyme-linked immunosorbent assay (ELISA) as previously described (Bruze, et al., 1989).

Solar UV radiation (72 KJ m^{-2}) induced detectable amounts of T< >T in all individuals tested (Figure 1). In the first group (left panel), unexposed as well as exposed skin was assayed, and in 5 of 10 individuals some inhibition was detected by ELISA analysis of DNA from unexposed sites. This background level was subtracted from the unadjusted levels of T< >T (in exposed skin) to determine the net level of T< >T photoproduct induced in each case (Figure 1). Most of the samples (7 of 10) contained T< >T levels clustered between 0.02 and 0.12 T< >T per Kb. Three samples contained higher levels (0.20, 0.21, and 0.48 T< >T per Kb).

Repair of T< >T was examined in a second group of individuals (Figure 1, right panel). Unadjusted levels of T< >T ranged from 0.06 to 0.46 T< >T per Kb immediately after irradiation, and declined to between undetectable ($\leq$ 0.01 T< >T per Kb) and 0.07 T< >T per Kb. Percentage loss of T< >T photoproducts at 4 h ranged from 56–98% (80.2 $\pm$ 14.0%; mean $\pm$SD). This rate of T< >T loss is equivalent to a half-time of about 100 min and is comparable to previous measurements in human skin irradiated with UVB (Sutherland, et al., 1980) or broad band UV (D'Ambrosio, et al., 1981). These studies indicate that loss of cyclobuta-dipyrimidines occurs more rapidly in human skin cells *in vivo* than in cultured human keratinocytes or fibroblasts which exhibit much slower repair kinetics. Thus repair rates measured in cultured cells may not accurately reflect absolute repair rates *in vivo*. Similar differences have recently been reported for repair rates of pyrimidine dimers in rat skin cells (Mullaart, et al., 1988).

The amount of T< >T photoproducts measured in skin samples irradiated with 72 KJ m^{-2} of solar UV was compared with the minimal erythemal dose (MED) of the individuals tested (not shown). No consistent relationship between the two UV-induced responses was observed. The small range of MEDs observed and the large increments of dose administered prohibits a definitive statement regarding the association between photoproduct yield and skin sensitivity. Although we did not observe an association between individual MED and net T< >T yield, our value for T< >T photoproducts formed per MED by solar UV of 0.06 ± 0.04 T< >T per Kb per MED (in the 7 clustered samples) is similar to that found by Freeman, et al. (1986) using UVB radiation.

Detection of PAH-DNA Adducts in Blood Cells

Detection of PAH-DNA adducts in peripheral blood leukocytes (PBLs) by immunoassay has been reported in several studies (Harris, et al., 1985; Haugen, et al., 1986; Perera, et al., 1987, 1988; Liou, et al., 1989). These adducts have been attributed to PAH exposure from occupation, smoking and diet.

Carcinogenic PAHs have been identified in many foods including broiled, barbecued, and smoked meat or fish (IARC, 1973; Fazio and Howard, 1983). Diet may constitute a significant proportion of total human exposure to PAHs, depending upon the environmental and occupational sources to which one is exposed (Fazio and Howard, 1983; Lioy, et al., 1988). A recent observational study of PAH-DNA adducts in PBLs from urban fire fighters and matched controls demonstrated an association between adduct levels and several sources of PAH exposure including char-broiled food consumption (Liou, et al., 1989; Strickland, et al., 1990). In order to assess the effect of dietary PAHs on adduct load, we conducted a controlled feeding study in which volunteers consumed char-broiled beef and PAH-DNA adduct levels were monitored in PBLs (Rothman, et al., 1990).

Four healthy caucasian non-smoking male volunteers aged 33–39 adhered to a diet free of broiled meat and smoked foods for one month. During this period, three baseline blood samples (40 ml) were collected at two week intervals. Volunteers then consumed char-broiled ground beef daily for seven days and avoided all other broiled or smoked foods. Each volunteer consumed approximately 280 gm char-broiled beef per day. One individual did not complete the planned schedule and ate char-broiled beef on four days only. The concentration of BP in the seven daily samples was 2.56 ± 0.85 μg kg^{-1} (mean ± SEM) ranging from 0.43–7.0 μ kg^{-1} as determined by thin layer chromatography (Lioy, et al., 1988; Rothman, et al., 1990).

Blood samples were collected immediately before eating char-broiled beef on the first day of feeding (day 1) and on days 2, 5, and 8. During the three weeks following feeding (days 8–31), volunteers again were asked to refrain from eating any broiled meat or smoked foods. During this follow-up period blood samples were obtained on days 12 and 24 (or 31).

Blood samples were immediately centrifuged and DNA was extracted from the PBL fraction by a modification of the method of Miller et al. (1988) involving high salt fractionation followed by chloroform/isoamyl alcohol extraction. The PAH-DNA adduct content was analyzed by ELISA as previously described (Santella, et al., 1988). DNA modified with BP-diolepoxide I (BPDE) at a level of 4.4 fmoles BPDE μg^{-1} DNA was used to generate standard curves. The ELISA used rabbit anti-BPDE-DNA antibody (#33), goat anti-rabbit IgG conjugated to alkaline phosphatase, and the fluorescent enzyme substrate 4–methylumbelliferyl phosphate. Since this antiserum recognizes DNA modified by several PAH-diolepoxides, adduct levels are expressed as BPDE-DNA adduct equivalents in fmole adduct μg^{-1} DNA (Weston, et al., 1988; Perera, et al., 1987).

Baseline levels of PAH-DNA adducts in the month prior to beef consumption were between non-detectable and 0.19 fmole μg^{-1} DNA for all individuals (Figures 2a, 2b). During the week of char-broiled beef consumption, 2 individuals exhibited an increase in PAH-DNA adduct level (Figure 2b). Volunteer A exhibited a 6–fold increase after one day of char-broiled beef consumption and volunteer B exhibited a 3–fold increase after 4 days of char-broiled beef consumption. PAH-DNA adduct levels in volunteers C and D did not increase at any time during the study (Figure 2a). These results indicate that ingestion of PAH is associated with increased PAH-DNA adduct levels in some, but not all, individuals.

The mean daily intake of BP from char-broiled beef during the feeding period was 0.72 μg per person. Previous studies (Lioy, et al., 1988; Vaessen, et al., 1988) have reported values for mean daily BP intake from normal diets of 0.12 or 0.08 μg per person ranging from 0.001–1.15 μg per person. Therefore, the daily BP intake in the experimental study described here was high, but not outside the normal range.

Two individuals exhibited measurable increases in PAH-DNA adduct levels while the two other individuals did not respond. This difference could be due to interindividual differences in constitutive or induced physiological parameters such as digestion, absorption, metabolism, excretion and DNA adduct formation and repair (Harris, 1989). Variation in the inducibility of PAH metabolizing enzymes (Conney, et al., 1977) in hepatocytes and PBL of individuals fed char-broiled beef may explain some of the observed interindividual differences in this study. The rapid disappearance of PAH-DNA

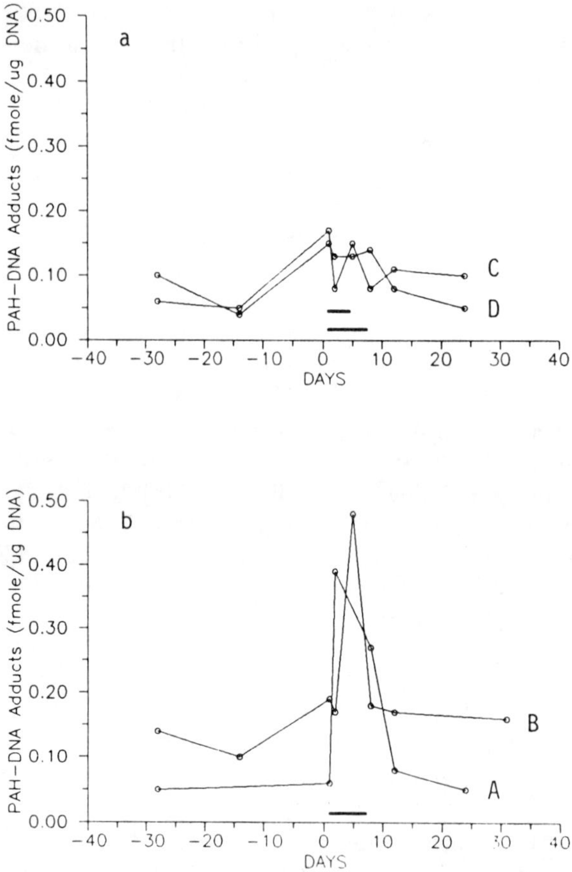

Figure 2. BPDE-DNA antigenicity expressed as PAH-DNA adducts for each individual. Solid bars represent char-broiled beef feeding period. Volunteer D consumed beef for 4 days only. Volunteers A, B, and C consumed beef for 7 days. Blood samples were collected on days -28, -14, and -1 before feeding, days 2 and 5 during feeding and days 8, 12, and 24 (or 31) after feeding of char-broiled beef. The lower limit of sensitivity of the assay was 0.04 fmole adduct μg^{-1} DNA at 15% inhibition. Panel a: no significant increase above background. Panel b: significant increase above background. (Reproduced with permission from Rothman, et al., 1990; IRL Press).

adducts in the two responding individuals is likely due to the relatively short lifetime (< 24 hrs) of the majority of nucleated cells in the peripheral blood. Only 15–30% of PBL are long-lived (weeks to years).

These results indicate that dietary sources of PAH can contribute to PAH-DNA adduct load in PBLs and should be considered when using PAH-DNA adduct level as a marker of exposure to PAH from occupational or environmental sources. Furthermore, the biochemical basis for the observed interindividual differences in PAH-DNA adduct levels should be examined to assess the usefulness of this endpoint as a marker of exposure.

Summary

Recent advances in the detection and measurement of genetic damage, or surrogates such as protein damage, in human tissues and body fluids provide the means to examine human exposure and response to genotoxic agents at the molecular level. Several different immunological and biochemical technologies have been developed (comprehensive reviews include Bartsch, et al., 1988; Garner, 1990; Vanderlaan, 1990).

The ability to measure genetic damage in human tissues allows one to address issues of internal dose. This methodology may be used to study factors governing interindividual variation in levels of genetic damage sustained from similar exposures. For example, significant variation in levels of genetic damage among patients receiving similar doses of chemotherapeutic agents has been reported (Poirier, et al., 1988), however the causes of this variability are unclear. This methodology should also be useful for examining factors that mediate individual susceptibility to cancer. Molecular biomonitoring studies in populations at risk for cancer could examine the association between induction (and repair) of genetic damage and the development of cancer. In summary, the information gained by this methodology should enhance our understanding of the mechanisms by which carcinogens lead to cancer and provide insight into variation in susceptibility to cancer.

Acknowledgments

The authors wish to acknowledge the assistance of the volunteers who participated in these studies. Research supported in part by DHHS grants ES03841, AR38884, ES03819 and the Mellon Foundation. Figure 1 is reprinted with permission from Elsevier Press; Figure 2 is reprinted with permission from IRL Press.

References

Bartsch, H., Hemminki, K. and O'Neill, I.K. 1988. Methods for detecting DNA damaging agents in humans: Applications in cancer epidemiology and prevention. IARC Sci. Publ. No. 89. IARC, Lyon.

Bruze, M., Emmett, E.A., Creasey, J. and Strickland, P,T. 1989. Cyclobuta-dithymidine induction by solar-simulating UV radiation in human skin: II. Individual responses. J. Invest. Dermat. 93: 341–344.

Conney, A.H., Pantuck, E.J., Hsiao, K.C., Kuntzman, R., Alvares, A.P. and Kappas, A.

1977. Regulation of drug metabolism in man by environmental chemicals and diet. Fed. Proc. 36: 1647–1652.

D'Ambrosio, S.M., Slazinski, L., Whetstone, J.W. and Lowney, E. 1981. Excision repair of UV-induced pyrimidine dimers in human skin *in vivo*. J. Invest. Dermatol. 77: 311–313.

Fazio, T. and Howard, J.W. 1983. Polycyclic aromatic hydrocarbons in foods. In: Handbook of Polycyclic Aromatic Hydrocarbons, Vol. 1. Bjorseth, A., ed. Marcel Dekker, New York. pp. 461–505.

Foiles, P.G., Chung, F.L. and Hecht, S.S. 1987. Development of a monoclonal antibody-based immunoassay for cyclic DNA adducts resulting from exposure to crotonaldehyde. Cancer Res. 47: 360–363.

Forbes, P.D., Davies, R.E. and Urbach, F. 1978. Experimental ultraviolet photocarcinogenesis: Wavelength interactions and time-dose relationships. NCI Monograph. 50: 31–38.

Freeman, S.E., Gange, R.W., Matzinger, E.A. and Sutherland, B.M. 1986. Higher pyrimidine dimer yields in skin of normal humans with higher UVB sensitivity. J. Invest. Dermatol. 86: 34–36.

Garner, C., ed. 1990. Biomonitoring and carcinogen risk assessment. Oxford University Press, Oxford.

Groopman, J.D., Donahue, P.R., Zhu, J., Chen, J. and Wogan, G.N. 1985. Aflatoxin metabolism in humans: Detection of metabolites and nucleic acid adducts in urine by affinity chromatography. Proc. Natl. Acad. Sci. USA. 82: 6492–6497.

Harris, C.C., Vahakangas, K., Newman, J.M., Trivers, G.E., Shamsuddin, A., Sinopoli, N., Mann, D.L. and Wright, W.E. 1985. Detection of benzo(a)pyrene diol epoxide-DNA adducts in peripheral blood lymphocytes and antibodies to the adducts in serum from coke oven workers. Proc. Natl. Acad. Sci. USA. 82: 6672–6676.

Harris, C.C. 1989. Interindividual variation among humans in carcinogen metabolism, DNA adduct formation and DNA repair. Carcinogenesis. 10: 1563–1566.

Haugen, A., Becher, G., Benestad, C., Vahakangas, K., Trivers, G.E., Newman, M.J. and Harris, C.C. 1986. Determination of PAHs in the urine, benzo(a)pyrene diol epoxide-DNA adducts in lymphocyte DNA, and antibodies to the adducts in sera from coke oven workers exposed to measured amounts of PAHs in the work atmosphere. Cancer Res. 46: 4178–4183.

Hsieh, P.G., Jeffrey, A.M. and Santella, R.M. 1985. Monoclonal antibodies to 1–aminopyrene-DNA. Carcinogenesis. 6: 1289–1293.

IARC. 1973. IARC monograph on the evaluation of the carcinogenic risk of chemicals to man: Certain polycyclic aromatic hydrocarbons and heterocyclic components. IARC Monogr. 3: 91–136.

Leadon, S.A. and Hanawalt, P.C. 1983. Monoclonal antibody to DNA containing thymine glycol. Mutation Res. 112: 191–200.

Lioy, P.L., Waldman, J.M., Greenberg, A., Harkov, R. and Pietarinen, C. 1988. The total human environmental exposure study (THEES) to benzo(a)pyrene: Comparison of the inhalation and food pathways. Arch. Environ. Health. 43: 304–312.

Liou, S.H., Jacobson-Kram, D., Poirier, M.C., Nguyen, D., Strickland, P.T. and Tockman, M.S. 1989. Biological monitoring of fire fighters: sister chromatid exchange and polycyclic aromatic hydrocarbon-DNA adducts in peripheral blood cells. Cancer Res. 49: 4929–4935.

Miller, S.A., Dykes, D.D. and Polesky, H.F. 1988. A simple salting out procedure for extracting DNA from nucleated cells. Nucleic Acids Res. 16: 1215.

Mullaart, E., Lohman, P.H.M. and Vijg, J. 1988. Differences in pyrimidine dimer removal between rat skin cells in vitro and *in vivo*. J. Invest. Dermatol. 90: 346–349.

Perera, F.P., Santella, R.M., Brenner, D., Poirier, M.C., Munshi A.A., Fischman, H.K. and Ryzin, J.V. 1987. DNA adducts, protein adducts, and sister chromatid exchange in cigarette smokers and nonsmokers. J. Natl. Cancer Inst. 79: 449–456.

Perera, F., Hemminki, K., Young, T.L., Santella, R.M., Brenner, D., Kelly, G. 1988. Detection of polycyclic aromatic hydrocarbon-DNA adducts in white blood cells of foundry workers. Cancer Res. 48: 2288–2291.

Perera, F., Mayer, J., Jaretzki, A., Hearne, S., Brenner, D., Young, T.L., Fischman, H.K.,

Grimes, M., Grantham, S., Tang, M.X., Tsai, W.Y. and Santella, R.M. 1989. Comparison of DNA adducts and sister chromatid exchange in lung cancer cases and controls. Cancer Res. 49: 4446–4451.

Poirier, M.C. 1984. The use of carcinogen-DNA adduct antisera for quantitation and localization of genomic damage in animal models and the human population. Environ. Mutag. 6: 879–887.

Poirier, M.C. and Beland, F.A. 1987. Determination of carcinogen-induced macromolecular adducts in animals and humans. Prog. Exp. Tumor Res. 31: 1–10.

Poirier, M.C., Egorin, M.J., Fichtinger-Schepman, A.M.J., Yuspa, S.H. and Reed, E. 1988. DNA adducts of cisplatin and carboplatin in tissues of cancer patients. In: Methods for detecting DNA damaging agents in humans. Bartsch H., Hemminki, K., O'Neill, I.K., eds. IARC Sci. Publ. No. 89. IARC, Lyon. pp. 313–320.

Roberts, D.W., Benson, R.W., Flammang, T.J. and Kadlubar, F.F. 1985. Development of an avidin-biotin amplified enzyme-linked immunoassay for detection of DNA-adducts of the human bladder carcinogen 4–aminobiphenyl. In: Mechanisms of DNA Damage and Repair: Implications for Carcinogenesis and Risk Assessment. Simic, M., Grossman, L., Upton, A.C., eds. Plenum Press, New York. pp. 479–488.

Rothman, N., Poirier, M.C., Baser, M.E., Hansen, J.A., Gentile, C., Bowman, E.D. and Strickland, P.T. 1990. Formation of polycyclic aromatic hydrocarbon-DNA adducts in peripheral white blood cells during consumption of charcoal-broiled beef. Carcinogenesis. 11:1241–1243.

Santella, R.M., Dharmaraja, N., Gasparro, F.P. and Edelson, R.L. 1985. Monoclonal antibodies to DNA modified by 8–methoxypsoralen and UVA light. Nucleic Acids Res. 13: 2533–2544.

Santella, R.M. 1988. Application of new techniques for the detection of carcinogen adducts in human population monitoring. Mutation Res. 205: 271–282.

Santella, R.M., Weston, A., Perera, F.P., Trivers, G.T., Harris, C.C., Young, T.L., Nguyen, D., Lee, B.M. and Poirier, M.C. 1988. Interlaboratory comparison of antisera and immunoassays for benzo(a)pyrene-diol-epoxide-I-modified DNA. Carcinogenesis. 9: 1265–1269.

Shivapurkar, N.M., Chisholm, S.L., Iype, S.N., Iype, P.T. 1989. Monoclonal antibodies directed against acrolein-DNA adducts. Proc. Am. Assoc. Res. Cancer. 30: 124.

Strickland, P.T. and Boyle, J.M. 1981. Characterization of two monoclonal antibodies specific for dimerised and non-dimerised adjacent thymidines in single stranded DNA. Photochem. Photobiol. 34: 595–601.

Strickland, P.T. and Boyle, J.M. 1984. Immunoassay of carcinogen-modified DNA. Prog. Nucleic Acids Res. Mol. Biol. 31: 1–58.

Strickland, P.T. 1985. Immunoassay of DNA modified by ultraviolet radiation. Environ. Mutag. 7: 599–607.

Strickland, P.T. and Creasey, J.S. 1989. Induction and repair of cyclobuta-dithymidine photoproducts in hamster skin by solar-simulated UV radiation. J. Photochem. Photobiol. B. Biology. 3: 17–24.

Strickland, P.T., Liou, S.H., Poirier, M.C., Nguyen, D., Tockman, M.S. 1990. Contribution of occupation, diet and smoking to PAH-DNA adduct load in peripheral blood cells. In: Biomonitoring and Carcinogen Risk Assessment. Garner, C., ed. Oxford University Press, Oxford. In press.

Sutherland, B.M., Harber, L.C. and Kochevar, I. 1980. Pyrimidine dimer formation and repair in human skin. Cancer Res. 40: 3181–3185.

Tilby, M.J., Styles, J.M. and Dean, C.J. 1987. Immunological detection of DNA damage caused by melphalan using monoclonal antibodies. Cancer Res. 47: 1542–1546.

Umbenhauer, D., Wild, C.P., Montesano, R., Saffhill, R., Boyle J.M., Huh, N., Kirstein, U., Thomale, J., Rajewsky, M.F. and Lu S.H. 1985. O^6-Methyldeoxyquanosine in oesophageal DNA among individuals at high risk of oesophageal cancer. Int. J. Cancer 36: 661–665.

Vaessen, H.A., Jekel, A.A. and Wilburs, A.A. 1988. Dietary intake of polycyclic aromatic hydrocarbons. Toxicol. Environ. Chem. 16: 281–294.

Vanderlaan, M., ed. 1990. Immunoassays for monitoring human exposure to toxic chemicals. Am. Chem. Soc., Washington, DC.
Weston, A., Willey, J.C., Manchester, D.M., Wilson, V.L., Brooks, B.R., Choi, J.S., Poirier, M.C., Trivers, G.E., Newman, M.J., Mann, D.L. and Harris, C.C. 1988. Dosimeters of human exposure to carcinogens: Polycyclic aromatic hydrocarbon-macromolecular adducts. In: Methods for Detecting DNA Damaging Agents in Humans. Bartsch, H., Hemminki, K., O'Neill, I.K., eds. IARC Sci. Publ. No. 89: 181–189.
Wild, C.P., Chapot, B., Scherer, E., Den Engelse, L. and Montesano, R. 1988. Application of antibody methods to the detection of aflatoxin in human body fluids. In: Methods for Detecting DNA Damaging Agents in Humans. Bartsch, H., Hemminki, K., O'Neill, I.K., eds. IARC Sci. Pub. No. 89: 67–74.
Yang, X.Y., DeLeo, V. and Santella, R.M. 1987. Immunological detection and visualization of 8–methoxypsoralen-DNA photoadducts. Cancer Res. 47: 2451–2455.
Young, T.L. and Santella, R.M. 1988. Development of techniques to monitor for exposure to vinyl chloride: monoclonal antibodies to ethenoadenosine and ethenocytidine Carcinogenesis. 9: 589–592.

Discussion (DR. STRICKLAND)

Eisenstadt: Cairns has argued that, because only skin cancer is high in XPs, it is unlikely that environmental exposure to Ames types of mutagens is a major cause of human cancer. It would, therefore, be interesting to assess the level of damage in DNA from the lymphocytes of XP individuals. Has Dr. Grossman done this?

Grossman: Cairns was saying that the repair capacity is uniform in all organs, but tumor formation is limited to the skin. We plan to study 800 Chesapeake watermen by measuring repair directly in their lymphocytes. This will be accompanied by determining pyrimidine dimer adduct levels at the same time.

Strickland: Also, the other implication of your question was "Have we looked for PAH adducts in XP patients?" We have not, but it is an interesting idea.

E. Rosenberg: Thinking about the barbecue dinner that is coming up, has anyone studied the effect of taking certain vitamins or a large dose of vitamin C or carotinine before the barbecue to prevent these adducts?

Strickland: That is another good question. We are planning to do this, but it has not been done. The next experiment in this series is obviously to study more individuals to see whether, in fact, there is the differential response that was suggested. Then there are a whole host of things – vitamins, nutrients, other dietary components – that may influence digestion, absorption, and metabolism which we can look at with this experimental design. One of the problems with working with human subjects is that you are very limited in the type of carcinogens that you can administer; food carcinogens and sunlight are two that are possible.

L. Grossman: To return to the question that Dr. Eisenstadt raised, since the nucleotidic excision repair system is a rather broad spectral pathway, one would expect to see the same level of repair of pyrimidine dimers as for polycyclic aromatic hydrocarbons. We are examining samples of lymphocytes from about a 100 people from the Gambia who are sensitive to aflatoxin. Since aflatoxin adducts are repaired by the same pathway that repairs pyrimidine dimers, we will examine whether the repair of aflatoxin adducts is the same or different than with UV damage. It certainly is easier to irradiate DNA than to synthesize the DNA adduct of aflatoxin and still be socially acceptable.

Lotan: You mentioned aromatic epoxides. It is known that one of the natural defense systems of the organism is to produce more detoxifying enzyme, such as glutathione transferase. Have you noted any parallel between formation of adducts and increase in level of glutathione transferase?

Strickland: That is a good point, but we have not done it yet. We are planning to look at certain metabolic phenotypes that we think are related to PAH metabolism. One of our problems is that we are somewhat limited in what we can measure in humans – which of the P450 isozymes we can measure either systemically or on a molecular level. But we certainly can look at glutathione transferase μ, and we can look at n-acetyl-transferase phenotype.

Lotan: I noted that you mentioned adducts with melphalan. Can you elaborate a little on this?

Strickland: This was a study by Tilby, et al. in London that made monoclonal antibodies to melphalan-modified DNA, published in Cancer Research (1987). The complete reference is in my proceedings paper.

B. Green: You mentioned raising antibodies against various adducts. I was not clear to me if you had antibodies against carcinogens as well. Has anybody investigated whether antibodies against carcinogens or carcinogen DNA adducts could afford any protection?

Strickland: Everything I listed in that table was to modified DNA or modified nucleotides. There has been some work looking at the parent compounds, although not as much. Most of the work that has focused on parent compounds has been on pesticides and herbicides, rather than on the carcinogens. I have not seen any data pro or con that would suggest that antibodies to carcinogens afford protection.

Key Words: Antibodies, DNA adducts, ELISA, Benzo(a)pyrene, Polycyclic aromatic hydrocarbons, UV photoproducts

Field Operable Devices for Immunological, Molecular and Toxicological Diagnosis – A Review on a Unified Approach

Falk Fish
Orgenics Ltd., Yavne, Israel

Introduction

Improving health care and intensive agricultural practices require that *in vitro* diagnosis will be delivered quicker but nevertheless farther away from the experienced laboratory. Developers and manufacturers of *in vitro* diagnostic devices are facing decreasing product lifetime with increasing demands for shorter assay duration and simplified test protocols.

The term "*in vitro* diagnosis" implies detection or visualization of otherwise invisible moieties, such as molecules and genes or their biological combinations, in the form of microorganisms. The visualization involves one or more signal amplification steps. The requirement to perform the diagnostic tests in remote and less equipped locations leads to the limitation of amplification means to more portable, more reliable and less dependent on instrumentation.

Such high demands can nevertheless be satisfied after relatively short development times if all diagnostic reactions are considered equally as receptor-ligand interactions, which can be presented and packaged identically. By logical packaging, by utilization of the basic law of the masses and in the form of traditional technologies, field operable *in vitro* diagnostics can be realized quickly and economically. When keeping to these principles, immunodiagnostics and molecular diagnostics for human and agricultural use can share the same basic reagents and packaging, thereby affording reduced investment without sacrificing performance.

This communication reviews the principles, construction and performance of diagnostic devices incorporating the unified approach.

D. Kamely et al. (eds), Biotechnology: Bridging Research and Applications, 179–204.

Materials and Methods

Antibody Preparation

Antisera against human and animal immunoglobulins were purchased from various suppliers. Monoclonal anti-sulfonated DNA ascites was generated in virus free BALB/c mice (Charles River Biotechnical Services, Margate, UK). Following a rough estimate of their potency (testing serial dilutions against a limited amount of solid phase antigen) an immunoglobulin fraction separated from the sera by precipitation at 40% saturation of neutral ammonium sulfate. In the case of the ascites and for some sera, the immunoglobulin was first separated by caprylic acid treatment (Russ, et al., 1983).

Antibody-enzyme Conjugates

Alkaline phosphatase (Boehringer, Mannheim) was conjugated to antibody preparations by the single-step glutaraldehyde method (Avrameas 1969). Horseradish Peroxidase (HRP) was conjugated to antibodies by the periodate oxidation method (Tijssen, 1985).

Enzyme Immunoassay

Enzyme immunoassays in the microtitration format were carried out as described by Engvall and Perlmann (1971, 1972).

ImmunoComb Enzyme Immunoassay

Enzyme immunoassays in the ImmunoComb format are equivalent to the microtitration format (see above) with the following modifications: (a) specimens and conjugates are suspended in a buffered saline containing 15–25% heat inactivated animal serum and a detergent (usually 0.1% Tween-20); (b) a precipitating Bromo Chloro Indoxyl Phosphate with Nitro Blue Tetrazolium (BCIP-NBT) chromogenic reaction (McGadey, 1970) is employed to visualize bound conjugate. The formulation for a stable mixture of BCIP and NBT was developed by Ritterband (1990).

Cold Labeling of DNA

DNA was labeled and visualized by the reagents included in the Chemiprobe kit (Organics Ltd., Yavne, Israel), which is based on the exposure of DNA to a mixture of bisulfite and O-methylhydroxylamine according to Poverenny et al. (1979).

Molecular Biology Methodology

Extraction, purification and fractionation of nucleic acids were carried out according to protocols contained in Maniatis et al. (1982)

Results and Discussion

The Basic Principles Behind the Comb Unified Packaging Concept for Diagnostics

The Ideal Characteristics of a Diagnostic Device

In devising a universal diagnostic device the needs of the potential customers as well as the capability of the manufacturer should be observed. These can be contradictory: the user may be looking for an ingenious, work-saving and foolproof device; the manufacturer will benefit from simpler manufacturing processes on one hand and some difficult proprietary technology to protect his territory on the other. The manufacturer will be able to respond to market needs in short lead times on new products and will be able to profit from small lot sizes. In addition, the diagnostic device must be compact, transportable and stable in storage. Its performance characteristics should follow the current acceptable industry standards.

The Basic Principles of the Chemistry in the Comb System

All diagnostic assays start with the reaction between a ligand and its complementary receptor. The reaction can be schematically presented in the equilibrium equation:

$$[\text{LIGAND}] + [\text{RECEPTOR}] \leftrightharpoons [\text{LIGAND-RECEPTOR-COMPLEX}]$$

The ligand receptor complex is the epitome of the diagnostic process and its stability and quantity will determine the quality of the device. The law of the masses provides the following guidelines: (a) by increasing the concentration of one or both reactants more complex will be formed and the complex will be more stable; (b) removal of newly formed complex will naturally pull the reaction forward and will result in an increased amount of the complex.

The complex itself is usually undetectable by the senses and requires amplification. The amplification process can be constructed from one or more additional steps, in which the complex of a primary reaction serves as the analyte for the next reaction. Also in this case adherence to the law of the masses will yield a more reliable and stable diagnostic.

Only heterogenous reaction systems (in which a reactant is located on a solid phase) can optimally utilize the above principles, hence their popularity in diagnostics in general and their employment in the Comb system. By attaching the receptor to a solid phase the complex may be separated from the reactants is enabled in a logical and straightforward way. It should be noted here in passing that some ingenious and successful homogenous system diagnostics exist today, most notable is the EMIT (Rubenstein, et al., 1972) system. These, however, impose certain restrictions on the relative sizes of the receptor and analyte and usually require dedicated instrumentation for visualization and analysis of the results. Single step heterogenous systems (e.g., agglutination) feature the rapidity of homogenous systems with a wider compatibility of receptor-analyte pairs, yet suffer from frequent deleterious matrix effects and reduced sensitivity (Grandien, et al., 1989) – since they cannot fully accommodate the law of the masses and require careful balancing of the concentrations of the receptor and analyte.

The Application of the Principles in the Comb System

The Comb is a hardware and chemistry system designed to provide the required specifications of a field-operable device. The Comb utilizes the unique properties of the spot enzyme immunoassay to translate the principles of chemistry into a usable diagnostic format. The receptor is attached to the solid phase as a spot and the signal, the color, is deposited on a 2–dimensional surface. The receptor is concentrated in a small area, thus providing the high concentration required to drive its reaction with the analyte forward. The color spot, which forms at the end of the reaction, is a result of many concerted reactions. The reaction is, therefore, more reliable and can be viewed as infinite number of replicate individual reaction wells. The 2–dimensional characteristic of the spot reaction can be utilized to attach more than one receptor on a defined space so that more than one analyte can be tested for in one reaction mixture. In addition, special control receptor spots are attached to the solid phase. These spots monitor the activity of the reagents or the presence of a specimen (e.g., an anti-IgG receptor will monitor the presence of a serum specimen AND the integrity of the all the reagents in an antibody test).

A further improvement in the chemical characteristics of the spot assay is afforded by the specially formulated polystyrene compound, which is employed as the solid phase surface in the Comb system and is activated by a proprietary process to increase binding of receptor materials to the surface.

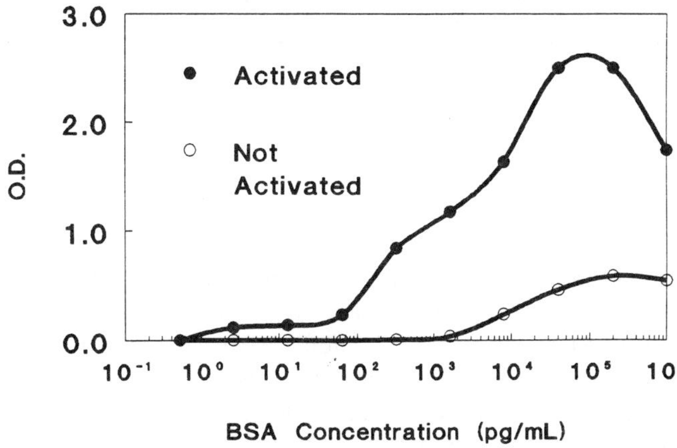

Figure 1. The effect of activation on the protein binding capability of the Comb's plastic surface. Various concentrations of BSA were incubated with activated and non-activated plastic. The relative amount of bound BSA was estimated by incubating each piece of plastic with an alkaline phosphatase conjugated anti-BSA antibodies, followed by washes and incubation with a chromogenic substrate of phosphatase (p-nitro-phenyl phosphate). The optical density was determined in a microtitration photometer (ELISA reader). Full circles: activated plastic; empty circles: non-activated plastic.

Figure 1 depicts the effect of plastic activation on the binding of a model protein (BSA) to the plastic. The activation not only increases the amount of protein bound to the surface but also allows for several orders of magnitude less material to be detectable.

Other components of the Comb system are also delivered in a form which promotes their reaction towards the end-product. The enzyme conjugates are employed at an excess, as determined by preliminary dose-response studies and the components of the chromogenic system are presented in near saturated concentration (Ritterband, 1990). The alkaline phosphatase chromogenic system of the Comb system, namely the coupled reaction between an indoxyl phosphate and a tetrazolium salt, also incorporates an amplification stage (Lodja, 1979).

The Unified Test Card – the Comb. The solid phase of the Comb system which carries the receptor spots, has a card form with 12 projections (teeth, tines) – the Comb. The receptor spots are located on the teeth area of the cards. Thus the Comb constitutes a hybrid between the dispstick and the microtitration formats, incorporating the advantages of the dipstick with those of the microtitration plate.

The dipstick format offers operational simplicity inasmuch as volumetric manipulations are not required: the stick is dipped into the specimen and the reagents as opposed to the transfer of the solutions into reaction wells in the microtitration format. The stick also allows for more than one receptor spot on its surface with the ability to test for more than one analyte in addition to an internal control. Similarly to the microtitration format, the Comb offers a multi-specimen capability with its convenience and error reduction.

Three variations of Combs are available (Figure 2), each has its advantages and disadvantages. The top Comb in Figure 2 ("undivided") must be used as a unit of 12 tests including controls. It has an advantage in a population screening and offers ideal reaction control, because all teeth develop in the same milieu (see below for the discussion on the Developing Plate). The bottom Comb in Figure 2 ("superdivided") is breakable into 12 individual teeth, thus offering complete flexibility in terms of the number of specimens and controls in any test. The biological drawback is the lack of cross-control capability between teeth so that each requires an internal control receptor spot. The middle Comb in the figure is, of course, a hybrid incorporating features of divided and undivided Combs.

The Unified Test Reagent Container – the Developing Plate. The reagents for the Comb system are packed in a multicompartment container, whose dimensions fit those of the Comb. The Developing Plate, as the container was officially named by USDA NVSL staff, has 6 compartments, each corresponding to a reagent and a step in the assay. The Comb is moved from compartment to compartment in order. An aluminum foil seals the plate in a compartment by compartment fashion to prevent cross contamination between them. The foil is pierced by the Comb, serving as a reminder not to use the same compartment again.

In correspondence with the variations in the format of the Comb, three variations of the Developing Plate are presented in Figure 3. In all variations, the first compartment is divided into 12 individual specimen wells. The division of the other compartments complements the corresponding Combs. The top Plate in Figure 3. (undivided) offers the ideal control situation, whereby the controls and specimens can be developed in the same milieu, but the user is forced to consume at least 12 test points in each test run. The bottom (superdivided) offers the opposite characteristics.

The 6 compartments of the Developing Plate can accommodate different assay chemistries, both immunological and molecular. Table 1 summarizes the arrangement and composition of the Developing Plate in various diagnostics. It will demonstrate the universal nature of the Comb format as a platform for all diagnostic needs.

The Comb Assay. Running an assay in the Comb format usually requires only one volumetric step, involving introduction of the liquid specimen (in most cases serum or plasma; in the molecular hybridization assay, a tissue extract) into one of the specimen wells of the first compartment (Compart-

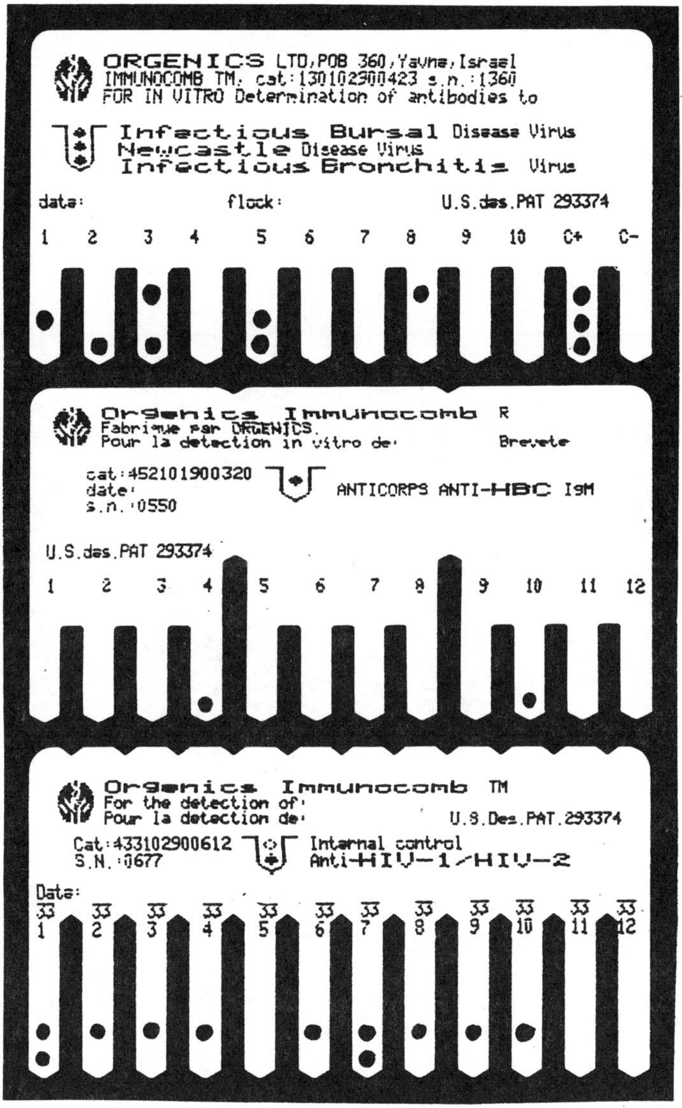

Figure 2. Different Comb formats. The top Format is undivisible. The bottom – is divisible into 12 individual sticks. The middle is a hybrid. See text for discussion·into 12 individual sticks. The middle is a hybrid. See text for discussion on the biological implications of these formats. Note that the printing on the Comb is accomplished by an ink-jet printer. This allows rapid new product implementation and production of small specialized lots.

ment A) of the Developing Plate. In the veterinary diagnostic kits the volumetric step can be replaced by using a prepunched filter paper strip. The skin of the animal is pricked with a lancet to form a drop of venous blood on its skin; the drop is absorbed into one of the punched circles on the

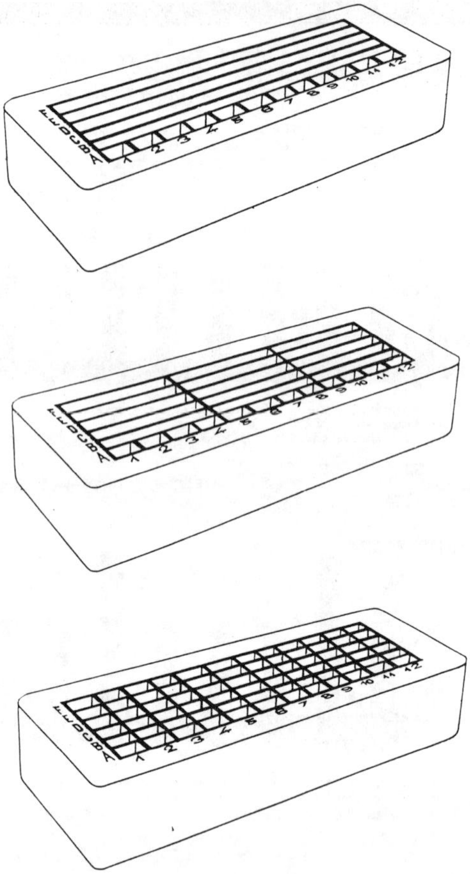

Figure 3. Different Plate formats (corresponding to the Comb formats in Figure 2). See text for discussion on the biological implications of these formats.

absorbent paper strip; the circle is removed from the strip with disposable forceps and inserted into the appropriate specimen compartment.

The further steps do not require any volumetric manipulations: the Comb is inserted into the specimen wells and then moved through the other compartments after specified incubation periods. At the end, color spots develop at the location of the receptors.

Analysis of the Results in the Comb System. The Comb system can accommodate both qualitative and quantitative assays. In a qualitative assay the result is either "+" or "−". Such assay requires the user to visually identify the existence or absence of a spot (e.g., HIV antibody assay) or to compare the intensity of the specimen spot with the intensity of a control or a cutoff. In a quantitative assay, the intensity of the color is related to the quantity or concentration of the analyte.

Table 1. Composition and arrangement of developing plates for various diagnostics

Diagnostic for	Compartment A	B	C	D	E	F	Wash	Other supplied components
Antibody Human	Specimen Diluent	Wash	AlkPhos anti-Ig	Wash	Wash	BCIP-NBT System	Internal	None
IgM Antibody	Stripping Solution	Wash	AlkPhos Anti-IgM	Wash	Wash	BCIP-NBT System	Internal	Stripping Solution
Antibody Poultry	Specimen Diluent	Wash	HRP anti-Ig	Wash	H_2O_2 + Buffer	Stabilizer Solution	External	Chromogen
HBs Antigen	Specimen	Wash	Biotin anti-HBs	AlkPhos Avidin	Wash	BCIP-NBT System	External	
HPV	Hybridization Solution	Wash	Biotin anti-modified DNA	AlkPhos Avidin	Wash	BCIP-NBT System	Internal	Lysis Buffer Modification Hybridization

Abbreviations (*in alphabetical order*):
AlkPhos – alkaline phosphatase conjugate.
BCIP-NBT – Bromo Chloro Indoxyl Phosphate; Nitro Blue Tetrazolium.
HBs – Hepatitis B surface antigen.
HPV – Human Papilloma Virus.
HRP – Horseradish Peroxidase.
Soln – solution.

Intended for both field and laboratory use, the Comb system has 2 result reading options: visual and instrumental.

The visual option involves comparing the intensity of the spot with a 6–color chart, called CombKey. The chart is made of 2 parts: a color wheel with 10 squares of increasing color intensity and an envelope with 6 viewing holes. This construction allows adjustment of the chart's overall density according to the control standard in order to compensate for variations in temperature, timings or operator's errors.

An experiment conducted by Rivetz, et al. (1985) demonstrated that visual reading of the Comb resulted in the same accuracy as an instrumental reading. Thus the human eye is an accurate measuring device and can be relied on when instrumentation is not available or unaffordable.

The CombScan is a dedicated reflectometer for measuring the color density of up to 3 spots on up to 12 teeth of the Comb. The instrument price is low even in small production lots due to its design, relying mainly on standard components. The CombScan operates on 12 Volts DC power and can therefore be used in remote locations by hookup to a car battery. Only 3 photodiodes measure the reflected light from the spots. The movement of the card in front of the photodiode, in order to effect the "scanning" function, is provided by the user's hand similarly to credit card reading devices. The voltage generated by the reflected light is converted to a series of individual readings, each on a scale of 0–255 (i.e., 8–bit conversion). The beginning and end of each tooth is identified by dark and light zones and the readings of the spot are identified out of 30–60 readings of the tooth area as the lowest reflecting center readings. For each spot the absorbance is calculated as 255 – reflectance and from this value the relative absorbance (ratio of specimen absorbance to control absorbance multiplied by 1000) is derived. This function provides adjustment of individual readings according to the control standard in order to compensate for variations in temperature, timings or operator's errors. Instead of including extensive costly mathematical capability, the CombScan translates the relative absorbance to standard units of measurement by comparison to memory stored look-up tables. The look-up tables were derived by prior experiments with the appropriate international standards. Figure 4 presents charts of 4 Comb specificities, summarizing the relations between the relative absorbance function and standard units, each compiled from experiments conducted on various dates, at various temperatures and by various operators. These charts are then used to generate the lookup table.

The CombScore is a means to obtain a profile of the level of an analyte in a population, without the need for a computer with complicated graphics software. "Flock Profiling" is employed by poultry growers to monitor the state of immunity in intensively cultivated flocks in order to get a measure of the success of a vaccination program, on one hand, or to get an early warning in case of infection, on the other (Snyder, et al., 1983). Figure 5 presents the CombScore form and explains its method of usage.

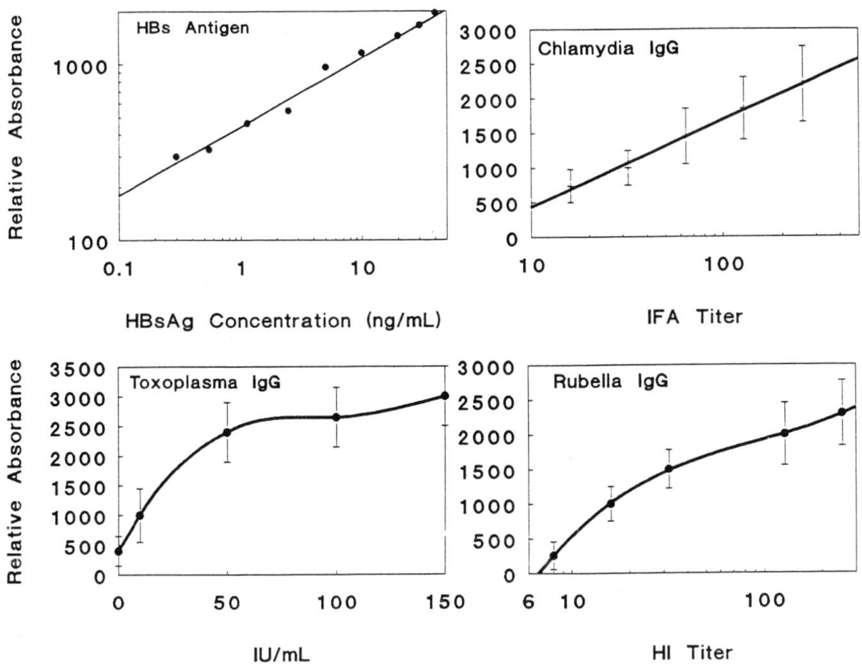

Figure 4. Correlation between CombScan readings and standard units. Each panel corresponds to one specificity, as marked inside the panel. Three serological assays and one antigen assaay are included in this figure.

Employment of the Comb System in Immunodiagnosis

The application of the principles of the Comb to immunodiagnostics is the ImmunoComb system. This part of the review will limit itself to some examples of ImmunoComb devices which appreciably benefit from the characteristics of the Comb system.

Poultry Serology

Due to the intensive cultivation of poultry in densely populated housing, infectious diseases can spread quickly and ruin weeks or months of costly cultivation. The concept of flock profiling was suggested as the means to eliminate the threat of widespread infections and to assess the results of vaccinations by constantly monitoring the antibody level in each flock (Snyder, et al., 1983). Comb kits for 3 common poultry viruses were favourably compared with more laborious agglutination methods (Thayer, et al., 1987; Brown, et al., 1988).

The potential advantage of the Comb system due to its inherent sensitivity and specificity was demonstrated in the case of *Mycoplasma* infections. Poultry flocks in Israel are not usually vaccinated against *Mycoplasma* and infec-

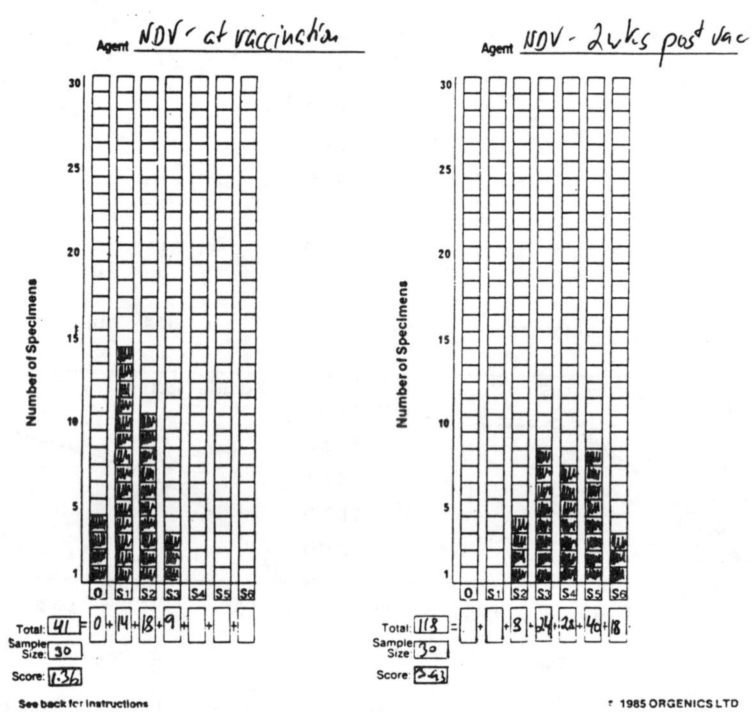

Figure 5. The CombScore form for deriving a profile of antibody level in the population. The individual antibody level measurement is marked in the column corresponding to the particular measurement. A histogram (profile) of the antibody levels in the flock is thus obtained. A population average (the "Score") is obtained by summation of the number of measurements multiplied by their corresponding antibody levels and finally dividing the sum by the sample size.

tion is monitored by appearance of antibodies. In the case of infection the flock has to be eliminated. The current tests for *Mycoplasma* are either the low cost SPA test (rapid slide agglutination), which is sensitive but non-specific and the more specific Hemagglutination Inhibition (HI) test, which is less sensitive and more costly. HI is used as a confirmatory test for the results of the SPA assay.

Table 2 presents the results obtained with the 2 classical tests and the ImmunoComb assay in monitoring natural *Mycoplasma* infection (work done at the regional poultry laboratories of the Israel Ministry of Agriculture, supervised by Dr. Y. Samberg). The table presents the frequency of positive specimens by each of the methods at various days following the start of the experiments. The ImmunoComb detected the presence of full infection more than a month before the other methods could present a confirmed diagnosis.

Chlamydia Serology

Chlamydia trachomatis is an intracellular pathogenic bacterium causing a variety symptoms among which is a frequent sexually transmitted disease.

Table 2. Relative sensitivity of the poultry ImmunoComb kit

DATE	SPA[a]	HI[b]	ImmunoComb[c]	CombScore[d]
Sept 1	7/100	0/7	3/7	2.14
Sept 2	15/100	0/15	15/15	4.20
Sept 7	32/100	0/15	15/15	5.26
Oct 6	42/100	14/20	20/20	4.70

[a] Number of positives/number tested.

[b] Number of positives/number tested. Positives are specimens resulting in HI > = 40.

[c] Number of positives/number tested. Positives are specimens resulting in ImmunoComb reading of S2 or above.

[d] CombScore is the sample average ImmunoComb S value, calculated according to the ImmunoComb instructions manual.

Although a test for antigen is available its sensitivity is variable and may be too low for the detection of deep seated infections. On the contrary, the "old fashioned" serological test of Micro ImmunoFluorescence (MIF) demonstrates good sensitivity (Schachter and Dawson, 1979; Kuo and Tam, 1988). MIF is a slow difficult microscopic test, which is limited to specialised laboratories.

The *Chlamydia* IgG ImmunoComb carries receptor spots containing a cell free extract of a cell culture infected by *Chlamydia* trachomatis and utilizes a single specimen dilution. It was of interest to determine whether such an assay, which is essentially an enzyme immunoassay, will compare well with a microscopic test with whole organisms as the receptor that employs serum dilutions to determine the antibody titer.

Results of such comparative studies are summarized in Table 3. The ImmunoComb ELISA compares well with the other methods for the determination of *Chlamydia* trachomatis antibodies (Mounier, et al., 1989; Scieux, et al., 1989). Another ImmunoComb kit, measuring antibodies to two human infecting *Chlamydia* species (*trachomatis* and *psittaci*) was introduced: this Comb has 2 antigen spots corresponding to the 2 species.

HIV Antibodies

HIV antibody tests constitute one of the most competitive diagnostic markets. Test durations range from a few hours to 10 min for large scale to individual testing respectively.

The receptors for the ImmunoComb HIV antibody card are constructed from synthetic peptides corresponding to the conserved regions of the gp41 and gp120 envelope proteins of the virus. Limiting the receptors to surface proteins reduces the chance for false positive and indeterminates. The potential risk for reduced sensitivity is compensated for by the high concentration of the gp120 peptide in the receptor spot. The results of the clinical studies in Table 4 show that the 35 min Comb test demonstrates maximal sensitivity and specificity. A WHO report also identified the added advantage of the

Table 3. Comparative performance of the ImmunoComb *Chlamydia* kits

ImmunoComb kit	Site of testing	Comparative test method	No. of samples	% sensitivity	% sensitivity	Comments
Chlamydia IgG						
	CHU Dupuytren, Limoges	1. ELISA	295	100	96.7	
		2. IF Spot (BioMerieux)	295	95.04	80.46	
	Hopital St. Louis, Paris (Scieux)	1. Chlamydelisa	163	93.3	97.7	semi-quantitative (−, +, ++, +++ vs. titer)
		2. IF	159	92.6	97.4	
	Hopital Pellegrin, Bordeaux (Bebear)	IF Spot (BioMerieux)	150	98.4	100	
Chlamydia IgA	Hopital St. Louis, Paris (Perol)	IF-INC (antigene Eurobio)	50	100	90.9	Semi-quan. (−, +, ++, +++ vs. titer)

Table 4. Comparative performance of the ImmunoComb HIV-1 kits

ImmunoComb kit	Site of testing (Name of director of the study)	Comparative test method	No. of samples	% sensitivity	% specificity
HIV-1	C.H.U. de Grenoble*	West. Blot Abbott Rec Elavia Past Behring Wellcome	683	99.6	99
	Institut Pasteur de Lyon	West. Blot Elavia Organon	440	100	99.8
	Centre Hospitalier Regional de Toulouse	West. Blot	77	100	100
	Univ. of Wisconsin Center for Health Sci.	West. Blot	74	100	97.6
	C.R.T.S. Strasbourg	West. Blot Abbott 2nd gen	1000	100	97.7
	Pitie Salpetriere	West. Blot	914	99.8	100
	WHO, Geneve	West. Blot	361	99.4	99.0

Comb format for providing testing capability in remote locations (WHO, 1989).

Hepatitis Antigen Test

Tests for Hepatitis B surface antigen (HBsAg) are now becoming more sensitive as they are employed as the single test for determination of actively infected blood. The current acceptable detection level is 0.5 ng ml^{-1} of surface antigen. In order to reach this sensitivity and to reduce overall assay duration two amplification steps were incorporated into the chemistry.

The Comb-bound receptor is an affinity purified polyclonal anti-HBsAg. The second anti-HBsAg antibody is biotinylated and binds an alkaline phosphatase streptavidin conjugate. This combination results in a 2.5 h tes with better than 0.5 ng ml^{-1} sensitivity (see HBsAg panel in Figure 4). The duration of the assay can be further shortened to 1 h with a sensitivity of 1 ng ml^{-1} (92% sensitivity compared to Abbott assays).

The Comb in Genetic Diagnosis – the HybriComb

Genetic diagnosis emloys specific sequences of nucleic acids receptors (probes) and analytes (targets). There are instances where there is an advantage to carry out a genetic diagnosis over immunodiagnosis. This is especially true for viruses which do not induce synthesis of specific antigens, or where the pathogen in question does not possess any unique structural moieties. Yet the methods involving nucleic acid hybridization are so complex as to be performed only in properly equipped laboratories by highly trained personnel.

The first move to popularize DNA technology was the introduction of non-radioactive methods for the labeling and detection of nucleic acids. Cold labeling of nucleic acids was initially introduced by Langer, et al. (1981). It employed enzymatic insertion of biotinylated bases into the nucleic acid probe. The bound probe was visuallized by tagged (enzyme, fluorophor) avidin or anti-biotin antibody.

There were still two problems to solve: (1) biotin tagging is a delicate and costly reaction, requiring careful purification of the nucleic acid; (2) the hybridization protocols themselves (e.g., dot and Southern blots, PCR's) are complicated and require a properly equipped laboratory with a highly experienced staff.

The combination of DNA labeling by chemical sulfonation and the Comb concept resulted in a molecular diagnostic configuration which can solve the above two limitations. The diagnostic is commercially dubbed "HybriComb™" (Nur, et al., 1990), and it employs the principle of Reverse Hybridization.

Reverse Hybridization is possible because of chemical DNA labeling,

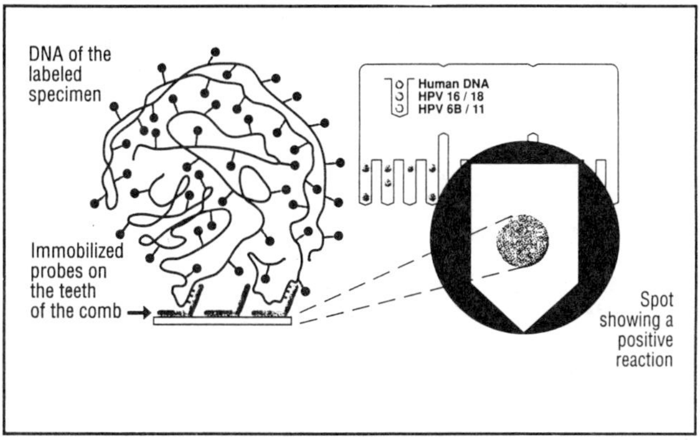

Figure 6. Schematic representation of the Reverse Hybridization principle. Sulfonated specimen DNA hybridizes with the immobilized probe DNA. Amplification is provided by the sulfonated flanking stretches of the specimen DNA which carry extra signal generating sulfonted moieties.

commercially named Chemiprobe (Herzberg, et al., 1987). Labeling is achieved by simply mixing the DNA containing specimen with 2 low cost and stable solutions: sodium bi-sulfite and O-methyl hydroxyl amine (Poverenny, et al., 1979). The DNA is sulfonated at the 5' carbon of some of its cytosine residues and can be visualized with a monoclonal anti-sulfonated DNA monoclonal antibody coupled through a second antibody or an avidin-biotin system coupled to an enzymatic reaction. The labeling method is simple to perform and it does not label any tissue component other than nucleic acids; so that the labeled DNA can be a crude, non-purified preparation.

In the Reverse Hybridization protocol the specimen DNA is labeled by incubating a crude extract of the specimen with the sulfonation reagents. Following removal of unreacted sulfonation reagents, the sulfonated specimen DNA is hybridized with the probe which is immobilized on the solid phase – in this case the Comb. Since more than one spot of receptor DNA can be attached to the surface of the card, any one specimen can be tested for the existence of more than one sequence of interest. Following hybridization and wash, the hybrids are visualized by consecutive incubations with biotinylated anti-sulfonated DNA antibodies, alkaline phosphatase conjugated streptavidin and a chromogenic substrate mixture of BCIP and NBT. The components and their location in the Developing Plate of the HybriComb assay are listed in Table 1.

A built-in amplification capability is realised through the flanking DNA stretches surrounding the analyte sequence. Since they are sulfonated too they will add more signal generating epitopes to the specimen. Figure 6 depicts schematically the principle of Reverse Hybridization.

Figure 7 depicts the dose response relationship between the concentration of the DNA in the specimen and the absorbance of the resulting spot, as

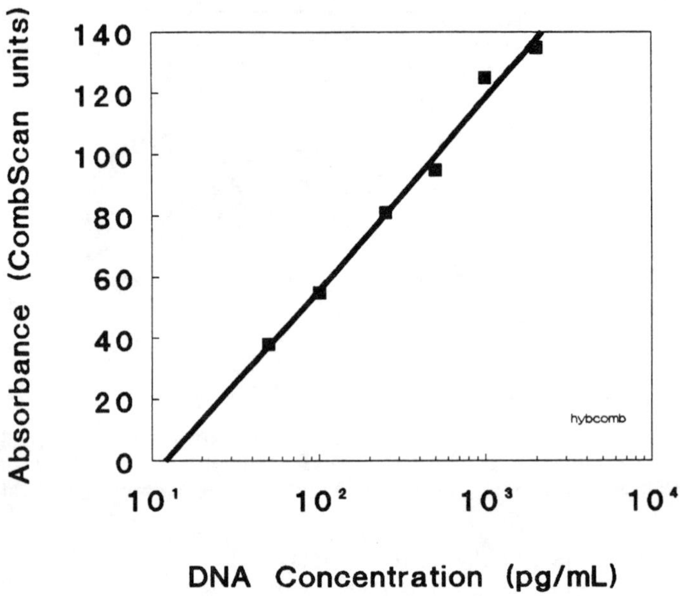

Figure 7. Does respond relationship between DNA concentration in the specimen and the color density of the HybriComb spot. Absorbance was measured by the CombScan reflectometer.

determined for pBR322 DNA model system and a 2 h hybridization step. The lowest level of detection at these conditions is around 10 pg ml^{-1}. The reaction is linear until 100 pg ml^{-1} of DNA.

Human Papilloma Virus (HPV) is gaining interest due to its association with benign and malignant tumors of the cervix, skin and other epithelial tissues (Boshart, et al., 1984). Hence, the first product in the HybriComb product line is a diagnostic for the detection of HPV DNA in human specimens. Each tooth of the HybriComb card carries two spots specific for HPV sequences: the bottom spot is specific for HPV types 6B and 11 (associated with benign tumors); the middle spot is constructed from probes to HPV 16 and 18 (associated with malignant tumors of the cervix). So in one operation diagnosis and typing of HPV is obtained without having to label two probe preparations. An internal control consisting of a probe for human DNA is also attached to each tooth of the HPV HybriComb card. If it does not result in a clear colored spot, it indicates that either there was not sufficient DNA in the specimen or that the reagents failed; in either case, the test is invalid.

Table 5 summarizes the initial study comparing results obtained with the HPV HybriComb to those obtained by dot hybridization, employing ^{32}P labelled probes. Table 6 lists the same results arranged according to the type of detected HPV genome. The study, which was conducted by Prof. Catalan, at Institute Alfred Fournier, Paris, demonstrates an overall 92% agreement between the two methods.

Table 5. Comparison between the HPV HybriComb and ^{32}P dot blot in the detection of HPV in human specimens

		HYBRICOMB +	HYBRICOMB −
RADIOACTIVE DOT BLOT	+	22	3
	−	2	34

$$\text{Sensitivity } \frac{22}{25} \times 100 = 88\%$$

$$\text{Specificity } \frac{35}{37} \times 100 = 94.6\%$$

A study, conducted by Prof. Ritter and Dr. Baldauf at Hopital de Hautepierre, CHRU, Strasburg, compared the colposcopic, cytological and histological observation and the HybriComb determination. Table 7 summarizes the results. Of the 3 HybriComb false positives, one has a past disease history and another was in intimate contact with a condyloma bearing individual. Taking these exceptions into consideration, the HybriComb exhibited 96% sensitivity and 94% specificity as compared to clinical observations. It is thus suggested that molecular diagnosis of HPV associated diseases by the HybriComb can serve as confirmatory diagnostic of pathological observations as well as an epidemiological or screening tool.

Toxicity Detection Devices

This part of the review deviates from the Comb concept and its accommodation of the specific receptor analyte system. Yet, the subject is an important issue for industrial, environmental and medical enterprises, which require sophisticated analytical capabilities away from the fully equipped laboratory.

Detection of general toxicity is not the usual occupation of a diagnostic company. Diagnostic tests and devices for the detection of specific toxicants do exist and these are limited to drugs (including antibiotics) or highly

Table 6. Comparison between the HPV HybriComb and ^{32}P dot blot in the typing of HPV in human specimens

		HYBRICOMB			
		16/18	6B/11	Indeterminate	Negative
	16/18	6	0	0	1
Radioactive	6B/11	1	10	1	2
Dot blot	Indeterminate	3	0	1	0
	Negative	1	1	0	34

92% overall agreement.

Table 7. Confirmation of HybriComb results by pathological observations

Clinical class	Cases	Hybricomb results		
		6B/11	16/18	Neg
Normal	22	0	3[a]	17
Exophytic lesions	2	1[b]	1[c]	0
Exophytic lesions (assoc. with CIN 1)	6	2[d]	4[e]	0
CIN 1	7	–	6	1
CIN 2	7	–	7	0
CIN 3	4	–	4	0

[a] Among 3 patients, one had a partner with exophytic condylomatosis on the penis and a second female patient has been previously treated for CIN 3.
[b] This patient was very weakly positive for HPV 16/18 and showed characteristic lesions associated with HPV on the uterine cervix.
[c] The patient was weakly positive for HPV 6B/11 with the presence of vulvar condylomata acuminata.
[d] Two patients were weakly positive for HPV 16/18 with the presence of flat condylomas on the vagina and on the uterine cervix of one, and solely on the uterine cervix of the second.
[e] One of the patients showed weak positivity for HPV 6B/11 and had condylomata acuminata localized on the vulva.

publicised environmental pollutants and carcinogens. Such tests rely on specific receptors for the toxicants in question; in most cases the receptors are antibodies.

If the nature of the toxicant in the specimen is not known, the diagnostic function is more complicated since the basic element of the diagnostic device, namely the receptor ligand pair, is missing or is unidentifiable. Only a living and metabolically active organism is capable of reacting to a toxic insult. In order to transform this reaction into a diagnostic, one or more amplification steps, each involving its own receptor ligand systems, are called for. The toxicity assay developer is also faced with a reagent stability problem, if the device is intended for field use. The critical ingredient is a living organism and it has to be preserved in a metabolically active state for the shelf life of the device.

We have been concerned with toxicology diagnostics since the inception of the company. The first product in this line was the SOS Chromotest – a colorimetric assay for genotoxic agents (Fish, et al., 1987). The test is based on an engineered *E. coli* strain, ingeniously constructed by Quillardet, et al. (1982) at the Pasteur Institute. The receptor for the genotoxin is naturally the cellular DNA. The DNA damage is detected and the initial signal amplified by the enzymatic cascade of the SOS DNA repair system which leads to the transcription of a lacZ structural gene fused to the sfiA promoter (one of the SOS promoters). The final step of amplification involved the hydrolysis of one or more of the B-galactosidase chromogenic substrates. In order to make the assay usable by minimally equipped, non-biological laboratory, the bacteria are freeze-dried in the logarithmic phase so that they should be able to respond to genotoxicants immediately upon rehydration.

Detection of metabolic toxicants is theoretically more complicated, since there are numerous, known and unknown targets (receptors). In addition, there are toxicants that macroscopically affect only multicellular organisms.

The TOXI-Chromotest (Reinhartz, 1988) was designed to overcome most of these hurdles in a portable, bacteria based assay. The assay protocol provides a maximal number of potential targets (receptor sites) for a variety of toxic mechanisms and ensures sufficient signal amplification by incorporating the following stages:

1. A rough mutant strain of *E. coli* is subjected to freeze-drying, hypertonic and hypotonic stresses. The stress treatment is not lethal but requires synthetic activity to be repaired and to afford continued metabolic activity. The membrane is partially damaged by the stress, allowing entry of particulate toxicants into the bacterial cells. For example, low concentrations of gentian violet in water do not affect intact *E. coli* cells but are detectable by those cells immediately after rehydration from the freeze-dried state (Reinhartz, et al., 1987).
2. The thus stressed *E. coli* is transferred into optimal growth medium, incorporating stress recovery factors. The toxicant containing specimen and an inducer of B-galactosidase (IPTG) are added at this time. A control reaction mix, without the specimen, is set up and run in parallel. In the absence of toxicants the cells will recover from their damages and allow induction and synthesis of the enzyme. If the toxicant in the test specimen will interfere with either the repair of the damage or with any of the induction events the enzyme will not be synthesized.
3. In the last stage of the assay a chromogenic substrate of B-galactosidase is added and the culture conditions are changed to favor extraction of the enzyme from the periplasmic space and to promote enzyme activity. B-galactosidase has a high turnover rate that ensures high sensitivity in detecting even a small number of enzyme producing bacterial cells. A reduction in the amount of color generated in the specimen treated reaction mix indicates presence of toxic activity.

Because a multitude of cellular functions and metabolic pathways are involved in the expression of B-galactosidase in the stressed indicator cells of the TOXI-Chromotest (membrane, signal transduction, signal processing, energy generation, RNA synthesis, protein synthsis and transport) the assay supplies receptor sites for many toxicants, which are usually undetectable by bacterial cells but rather require eukaryotic cells or multicellular organisms to be detected. Table 8 lists some common neurotoxicants and their level of detection (Reinhartz, et al., 1987); these materials are typical toxicants of higher multicellular organisms.

Both Chromotests are presented in the microtitration format and are therefore most convenient in a minimal laboratory environment. A new format of a toxicity assay was recently developed in cooperation with Lumac

Table 8. Minimal inhibitory concentrations of neurotoxicants as determined in the toxi chromotest

Neurotoxicant	Minimal inhibitory concentration[a] in ppm
Silver nitrate	0.75[b]
Gold (colloidal)	0.20
Cobalt chloride	0.45
p-Chloromercuribenzoic acid	0.20
Mercury chloride	0.06
Nickel sulfate	10
Sodium arsenite	0.80
Sodium selenite	5
Zinc chloride	25
Cadmium acetate	15
BHC-Gamma (Lindane)	3
Endosulfan	8[b]
Dieldrin	12[b]
Eldrin	3
Polyoxyethyl alcohol	2.5[b]
Azinophosmexthyl	5
Cypermethyl	3
Deltametrin	2[b]
Methomyl	0.9
2, 4 DiNitroPhenol (DNP)	0.9
Hydroquinone	1
Sodium fluoroacetate	1.5
Sodium azide	2

[a] Minimal inhibition concentration (MIC) is the concentration of the chemical causing 20% toxicity. Toxicity was determined using the TOXI-CHROMOTEST Kit, as described in the Methods section.
[6] These values were obtained when material was tested at pH 7.5; all other values were obtained at pH 5.5.

B.V. (Landgraaf, The Netherlands) to answer all the shortcomings of the existing assay and to provide testing capability away from the laboratory. A gram positive organism was selected for the indicator cell in order to increase the test's sensitivity to beta-lactam antibiotics. The total assay duration was shortened by employing a group of chromogenic enzymes (dehydrogenases), rather than one, to generate the signal and by providing improved conditions for the recovery of the organism from stress and its growth. The color can be measured either photometrically or visually.

Table 9 presents the sensitivity of the new assay format to low concentrations of Penicillin G and demonstrates the effect of the hypotonic shock on the sensitivity of the assay. Table 10 summarizes the sensitivity of the visual test protocol to a variety of antibiotics. Note the somewhat reduced sensitivity to Polymyxin, a typical gram negative toxicant.

Table 9. Effect of hypotonic shock on the sensitivity of bacteria to Penicillin G

Type of reaction medium	Penicillin concentration (mIU/L)	OD at 570 nm	% Inhibition*
	0	0.117	0
Isotonic	4	0.113	3
	7	0.077	34
	0	0.118	0
Hypotonic	4	0.052	56
	7	0.031	74

$$* \text{ \% Inhibition} = \left(1 - \frac{\text{OD + Penicillin G}}{\text{Control OD}}\right) \times 100.$$

Conclusions

A) All diagnostic chemistries involving the basic receptor-ligand interaction and heterogenous complex detection can be packaged and presented in a unified format. This approach can reduce development time and investment.
B) Strict adherence to the basic law of the masses will result in stable and rapid reactions. The reagents involved in such reactions can be stabilized to allow extended storage periods in remote locations.
C) Molecular hybridization reactions can be so simplified as to be conductable in minimal laboratory situations by inexperienced operators.

Acknowledgments

The Comb, Chemiprobe and Chromotest systems are results of the work of our dedicated scientists, including (in alphabetical order) Drs. Rene Elkaim, Max Herzberg, Israel Nur, Avi Reinhartz, Menachem Ritterband, Baruch Rivetz and Fred Samuels. The CombScan reflectometer was designed and

Table 10. Sensitivity of the visual antibiotic assay

Antibiotic	Sensitivity[a]
Penicillin G	3.5 mIU/mL
Ampicillin	2.5 ng/mL
Cloxacillin	2 ng/mL
Chloramphenicol	2.5 μg/mL
Tetracycline	0.2 μg/mL
Oxytetracycline	0.4 μg/mL
Polymyxin	10 μg/mL
Sulfadiazine[b]	20 μg/mL[b]

[a] Lowest concentration causing visually detectable decrease in color.
[b] Test conducted in presence of 10 μg/mL trimethoprim.

built at SRD Medical, with Peter Marin and Dr. Joel Seligson as the major contributors. The author is also indebted to Dr. David Ruben, for compiling the clinical data, to Prof. Alexander Kohn for the critique and suggestions, and to Arlene Bar-Yosef for careful stylistic and typographical review of the manuscript.

References

Avrameas, S. 1969. Coupling of enzymes to proteins with glutaraldehyde. Use of the conjugates for the detection of antigens and antibodies. Immunochem. 6: 43–52.

Boshart, M., Gissman, L., Ikenberg, H., Kleinheinz, A., Scheurlen, W. and zur Hausen, H. 1984. A new type of papillomavirus DNA, its presence in genital cancer biopsies and in cell lines derived from cervical cancer. EMBO J. 3: 1151–1157.

Brown, J., Thayer, S.G. and Fletcher, O.J. 1988. A method to rapidly convert Newcastle Disease Virus and Infectious Bronchitis Virus ImmunoComb scores into conventional Hemagglutination-Inhibition titers. Avian Dis. 32: 517–518.

Engvall, E. and Perlmann, P. 1971. Enzyme-linked immunosorbent assay (ELISA). Quantitative assay of immunoglobulin G. Immunochem. 8: 871–879.

Engvall, E. and Perlmann, P. 1972. Enzyme-linked immunosorbent assay, ELISA. III. Quantitation of specific antibodies by enzyme-labeled anti-immunoglobulin in antigen-coated tubes. J. Immunol. 109: 129–135.

Fish, F., Lampert, I., Halachmi, A., Riesenfeld, G. and Herzberg, M. 1987. The SOS Chromotest kit: a rapid method for the detection of genotoxicity. Toxicity Assess. 2: 135–147.

Grandien, M., Petteson, C.A. and Svensson L. 1989. Latex agglutination tests for the detection of Rota and Adenovirus in diarrheal disease. In: Rapid Methods and Automation in Microbiology and Immunology. Balows, A., Tilton, R.C. and Turano, A., eds. Brixia Academic Press, Brescia. pp.91–99.

Herzberg, M., Reinhartz, A., Ritterband, M., Twizer, S., Smorodinski, N.I. and Fish, F. 1987. New concept in diagnostic procedures based on nonradioactive DNA probes. Chimmicaoggi. 5: 67–71.

Kuo, C. and Tam, M.R. 1988. Immunological diagnosis of *Chlamydia trachomatus* infection. In: Immunological Diagnosis of Sexually Transmitted Diseases. Young, H. and McMillan, A., eds. Marcel Dekker, Inc. New York. pp. 191–211.

Langer, P.R., Waldrop, A.A. and Ward, D.A. 1981. Enzymatic synthesis of biotin labeled polynucleotides: Novel nucleic acid affinity probes. Proc. Natl. Acad. Sci. USA. 78: 6633–6637.

Lojda, Z., Gossrau, R. and Schiebler, T.H. 1979. Enzyme Histochemistry: A Laboratory Manual. Springer-Verlag, Berlin. p. 16.

Maniatis, T., Fritsch. E.F. and Sambrook, J. 1982. Molecular Cloning. A Laboratory Manual. Cold Spring Harbor Laboratory, Cold Spring Harbor, N.Y., 545 pp.

McGadey, J. 1970. A tetrazolium method for non-specific alkaline phosphatase. Histochem. 23: 180–184.

Mounier, M., Benabbou, L., Sangare, A., Gershydamet, G.M. and Denis, F. 1989. Serodiagnostice de *Chlamydia trachomatis*. Evaluation par comparaison de trois methods. Immunocomb, ELISA et MicroImmunoFluorescence. Etude de 295 serums africains. Medecines et Maladies Infectieuses. 19: 384–387.

Nur, I., Elkaim, R. and Herzberg, M. 1990. HybriComb – A novel diagnostic tool for DNA probing. Ann. Biol. Clin. In press.

Poverenny, A.M., Podogorodnichenko, V.K., Bryksina, L.E., Monastyrskaya, G.S. and Sverdlov, E.D. 1979. Immunochemical approaches to DNA structure investigation–I. Immuno-

chemical identification of the product of cytosine modification with bisulfite and O-methylhydroxylamine mixture. Mol. Immunol. 16: 313–316.
Quillardet, P., Huisman, R., D'Ari, R. and Hofnung, M. 1982. SOS Chromotest, direct assay of induction of an SOS function in *Escherichia coli* K12 to measure genotoxicity. Proc. Natl. Acad. Sci. USA. 79: 5971–5975.
Reinhartz, A., Lampert, I., Herzberg, M. and Fish, F. 1987. A new, short term, sensitive bacterial assay kit for the detection of toxicants. Toxicity Assess. 2: 193–206.
Reinhartz, A. 1988. Mirobiological assay kit and method for detecting de novo biomolecular synthesis inhibitors. U.S. Patent 4,774,173.
Reinhartz, A., Lampert, I., Herzberg, M. and Fish, F. 1987. The detection of neurotoxicant activity by a bacterial toxicity assay. In: Model Systems in Neurotoxicology. Alternative Approaches to Animal Testing. Shahar, A. and Goldberg, A.M., eds. Alan R. Liss, Inc., New York. pp. 151–162.
Ritterband, M. 1990. Stable chromogenic substrate mixture of indoxyl phosphate and tetrazolium salt, method of making and using same in biological and diagnostic assays. U.S. Patent Application 07/292,578 (allowed for issuance).
Rivetz, B., Weisman, Y., Ritterband, M., Fish, F. and Herzberg, M. 1985. Evaluation of a novel rapid kit for the visual detection of Newcastle Disease Virus. Avian Dis. 29: 929–942.
Rubenstein, K.E., Schneider, R.S. and Ullman, E.F. 1972. "Homogeneous" enzyme immunoassay. New immunochemical technique. Biochem. Biophys. Res. Comm. 47: 846–851.
Russ, C., Callegaro, I., Lanza, B. and Ferrone, S. 1983. Purification of IgG monoclonal antibody by caprylic acid precipitation. J. Immunol. Meth. 65: 269–271.
Schachter, J. and Dawson, C.R. 1979. Psittacosis-Lymphogranuloma Venerum Agents/TRIC Agents. In: Diagnostic Procedures for Viral, Rickettsial and Chlamydial Infections. Lennette, E.H. and Schmidt, N.J., eds. American Public Health Assoc., Washington, DC. pp. 1021–1059.
Scieux, C., Bianchi, A. and Barateau, V. 1989. Evaluation de deu techniques immunoenzymatiques Immunomb et Chlamydelisa pour la detection des anticorps antichlamydiens de type IgG. Revue Francais des Laboratoires. No. 193.
Snyder, D.B., Marquardt, W.W., Mallinson, E.T. and Russek, E. 1983. Rapid serological profiling by enzyme-linked immunosorbent assay. I. Measurement of antibody activity titer against Newcastle Disease Virus in a single serum dilution. Avian Dis. 27: 161–170.
Thayer, S.G., Nersessian, B., Rivetz, B. and Fletcher, O.J. 1987. Comparison of serological tests for antibodies against Newcastle Disease Virus and Infectious Bronchitis Virus using ImmunoComb solid phase immunosorbent assay, a commercial enzyme-linked immunosorbent assay, and the Hemagglutination-Inhibition assay. Avian Dis. 31: 459–462.
Tijssen, P. 1985. Practice and theory of enzyme immunoassay. Elsevier, Amsterdam, p. 236.
WHO, Global Program on AIDS. 1989. Operational characteristics of commercially available assays to determine antibodies to HIV-1. GPA/BMR/89.4. Geneva. March 1989.

Discussion (DR. FISH)

Smith: DuPont is the competition for this?

Fish: DuPont just went out of the infectious disease diagnostic business and sold it to other companies. There are many competitors, so one always has to try to be the best.

Grossman: Is is possible to put a magnetic tape on the back of the card to record a complete file on the patient?

Fish: You can do anything, but it depends on how much it will cost and how much of that cost the users will bear.

Key Words: Diagnostics, Immunodiagnostics, Molecular hybridization, Toxicology, Antibiotics, Receptor, Ligand, Field operation, *Chlamydia*, Papilloma, Hepatitis, Poultry

Design and Properties of Enzymes Immobilized in Sol-Gel Glass Matrices

Sergei Braun[1], S. Rappoport[1], S. Shtelzer[1], R. Zusman[2], S. Druckmann[2], D. Avnir[2], M. Ottolenghi[2]
[1]*Institute of Life Sciences*
[2]*Institute of Chemistry, The Hebrew University, Jerusalem 91904, Israel*

Introduction

Technical convenience and the urge to lower costs are driving industrial interest, and thus applied research, toward ever-increasing effort to prepare successfully immobilized enzymes. Immobilization allows re-use of enzymes, protects them from harsh external conditions, from microbial contamination, and prolongs their useful lifetime. There are probably as many immobilization methods as there are enzymes. This proliferation of techniques reflects the complexity of the biological material and the variety of its uses. Simple inexpensive general techniques, resulting in stable and active enzyme catalysts are yet in demand (Kennedy and White, 1985a).

An ideal enzyme catalyst should be bound to a mechanically and chemically stable, highly porous hydrophilic carrier. The bond linking the enzyme to the support is required to be stable under the catalyst application conditions to prevent leaching. The strong binding forces also have stabilizing effects on enzyme activity (Martinek and Mozhaev, 1985). The desired immobilization procedure should be simple, mild (non-denaturing) and generally applicable.

Enzymes covalently immobilized on controlled-pore glass beads offer an almost ideal solution to the problems of support and binding force. However, the preparation of catalyst by this immobilization technique is neither simple nor generally applicable. The beads are costly, require tedious chemical derivatization procedures, and lack stability due to the continuous leaching of silica during prolonged usage (Kennedy and White, 1985b).

The most generally applicable immobilization procedure is a simple entrapment of the enzyme in a forming gel of natural or synthetic polymers. The main shortcoming of this technique is the loss of the enzyme by leakage through a nonuniform net of polymer molecules. Rather weak interactions between the enzyme and the matrix result in a relatively nonrestricted diffusional movement of polypeptide chains. This can be of a benefit where

D. Kamely et al. (eds), Biotechnology: Bridging Research and Applications, 205–218.

conformational transitions are required for successful catalysis. Otherwise, this diffusional freedom of motion can negatively affect immobilized enzyme stability.

In this publication, we shall demonstrate (1) that proteins can be trapped within the matrix of a forming sol-gel, (2) that several enzymes can be effectively immobilized in such composite bioactive sol-gel glasses, while retaining high enzymatic activity, (3) that strong binding forces retain the enzyme in the matrix, thus producing a considerable stabilizing effect. Finally, we shall attempt to prove that sol-gel glass enzyme catalyst preparation is a simple, generally applicable, procedure, resulting in an enzyme catalyst whose properties are similar to these of covalently immobilized enzymes.

The application of the sol-gel process (Brinker, et al., 1990) for trapping organic molecules in inorganic glasses is now well-documented and widely demonstrated. It has been recently recognized that a sub-population of trapped molecules can react with external chemical reagents through the pore network of the material. This was used for water detection (Pouxviel, et al., 1989), for the design of photocatalysts for water reduction (Slama-Schwok, et al., 1989), and for the preparation of chemical sensor materials (Zusman, et al., 1990).

Several properties of the sol-gel glasses make them especially attractive as possible enzyme catalyst supports: (a) the ability to entrap large amounts of additives, (b) the thermal and chemical stability of the matrix, (c) simplicity of preparation with no need for covalent modification, (d) easy technological manipulation and production in any desired geometry, including thin films. Recently, Carturan, et al. (1989) demonstrated invertase activity in the aggregates of whole yeast cells trapped in thin sol-gel films, whereas thin films with cell-free invertase preparation were devoid of activity. Braun, et al. (1990) have successfully immobilized a cell-free purified alkaline phosphatase in the bulk of sol-gel.

Materials and Methods

Preparation of Sol-gel Immobilized Enzymes

Enzyme solutions (0.2 ml, 10 mg ml^{-1}) in nonbuffered water, containing various concentrations of NaF were mixed in the cold (4°C) with either methanol or polyethylene glycol (PEG 400). The concentrations of NaF, the volumes of methanol and PEG 400 were as indicated in the legends to tables and figures. Tetra-methoxysilane (TMOS, 1 ml) was then added. The tubes containing the reaction mixture were transferred to a shaking water bath at 8°C. The bath was allowed to reach the room temperature for 2–3 h. In samples containing PEG 400, the polymerization was completed in about 4–5 h. The polymerized sol was allowed to dry for a week at 30°C.

The glass was ground to a size of about 60–100 mesh. In some experiments the immobilized enzyme was packed in a column made from a 2 ml syringe. The column was eluted with 0.5 M NH_4HCO_3 (250 ml), followed by water (250 ml). This cycle was repeated twice. All washing solutions were collected, concentrated by freeze-drying, and assayed for protein content and enzyme activity. Later, when it was found that neither significant enzyme activity nor protein could be detected in the eluates the enzyme containing glasses were washed overnight with running water.

Determination of Enzyme Activities

Water was removed from the washed preparations of bioactive glasses by filtration, and the catalyst was air-dried and weighed. Appropriate assay mixtures were added to the aliquot of air-dried glasses. The test conditions for enzyme activities were as described by Bergmeyer (1983). Enzyme activity was expressed in percentage of the initial soluble enzyme activity. Essential details of assay procedures are stated in the figure and table legends.

Materials

Trypsin (E.C. 3.4.21.4, from bovine pancreas, 11,000 U mg^{-1}) was supplied by RAD Chemicals, Rehovot, Israel. Catalase (E.C. 1.11.1.6 from bovine liver, 30 U mg^{-1}) was from Pharmacia. Peroxidase (E.C. 1.11.1.7, from horseradish, 200 U mg^{-1}), acid phosphatase (E.C. 3.1.3.2, from wheat germ, 0.45 U mg^{-1}) and all of the specialty chemicals were from Sigma.

Results and Discussion

TMOS, Polymerization in Protein-containing Systems

Alkaline phosphatase was recently incorporated into a sol-gel glass (Braun, et al., 1990) applying a system in which NaOH was used to catalyze the polymerization, and methanol was added as a solvent. Actually, as it often is the first attempt, the composition of the polymerization mixture was quite accidental. It was, however, important to precipitate the enzyme before polymerization, in order to protect it from denaturation during the polymerization.

An attempt was made to replace methanol with a more efficient and less denaturing precipitating agent. It was found that PEG 400, a low molecular weight polyethylene glycol, was well suited for this purpose. Higher polymers were not used, since their removal from the sol-gel glass by washing was expected to be difficult. It was found, however, that although PEG 400 mixes with water and TMOS at all proportions, the tri-component systems separate

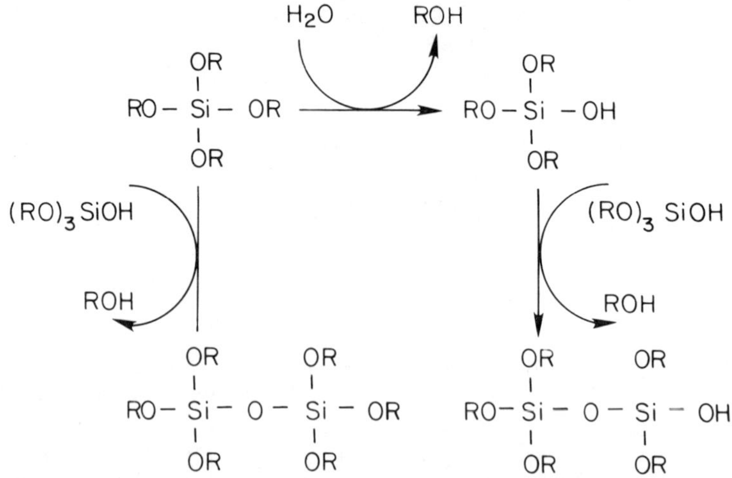

Figure 1. Polycondensation of TMOS.

into two phases at all the concentrations reported here. The precise composition of the two phases was not determined. Nevertheless, judging by the volumes, the upper (organic) phase contained TMOS saturated with PEG and traces of water, while the lower contained mostly PEG and water with little TMOS. Formation of sol-gel from TMOS probably started at the interface, or in the bulk of organic phase, as well, by initial hydrolysis of the TMOS (Figure 1). More hydrophilic products of hydrolysis and polymerization partitioned predominantly into the lower phase, where the gelation occurred. Protein aggregates or other hydrophilic material (e.g., microbial cells) were always found only in the lower phase. Thus, during the formation of glass, the proteins were not in contact with denaturing hydrophobic medium.

After 3–4 h, when solid glass-like material was formed, which totally filled the lower phase, the rate of polymerization slowed down here because of the diffusional barrier for further partitioning of the precursors. Left alone, solid glass devoid of protein formed in the upper phase, as well. We usually removed the upper phase by suction after the solidification took place, thus obtaining homogeneous protein-containing sol-gel.

The polymerization process was considerably speeded up by high concentration of NaF, or in the presence of biological material, such as proteins or microbial cells. Care must be taken, therefore, to minimize the temperature rise caused by large amounts of heat released during the polymerization. The experiments were carried out at low temperatures as shown in the Methods section.

In contrast to the polymerization systems containing PEG, the methanol-based mixtures were homogeneous. Gelation occurred in the whole volume of reaction. During condensation and drying, the gel shrank to about 10%

of its initial volume. The shear generated by this shrinking may rupture protein molecules, as our results indicate.

Entrapment of Enzymes in the Sol-gel Glass

In all our experiments, the enzymes could not be removed from the catalyst particles even following extensive washing with salt solutions at an elevated pH. At a pH above 10, some leaching of both the glass and the enzyme was observed. Otherwise, no protein or enzyme activity could be detected in washing solutions, even after concentration of these solutions by freeze-drying. In order to ascertain the non-leachable character of protein entrapment in sol-gel, glasses containing hemoglobin (3–5 mg ml^{-1} glass) were prepared. The brightly colored aggregates of hemoglobin, occluded in the clear glass, were visible under low magnification. The hemoglobin-containing glass was then immersed in water for as long as one month. No visible spreading of the red hemoglobin was detectable. It seemed that at the concentrations used in these experiments, almost all of the protein was entrapped in the glass.

The scanning electron microscope (SEM) pictures of protein-doped glasses (Figure 2) show spongy uniform glass, dotted with protein aggregates. The degree of aggregation depends upon the protein and upon the conditions of precipitation. At a high concentration of the protein (10–20 mg ml^{-1} glass), the aggregates are no longer discernable in SEM, and the material seems homogeneous with the larger grain, as if aggregates shown in Figure 2 were blended with the silicate.

The same pattern was observed when entrapment of whole *E. coli* cells was attempted. At low cell concentrations, islands of aggregated immobilized cells were visible on the homogeneous glass surface. With the increase of cell concentration, cell aggregates occupied larger areas, until finally, at about 25 mg dry cell weight per ml TMOS, the material seems to consist of large cell aggregates (Figure 3). This phenomenon is probably connected with preferential formation of the glass near the hydrophilic surfaces of protein aggregate or of the cell wall, by partitioning of partially hydrolyzed polymer precursors, as in the polymerization scheme described above.

The probing of the homogeneous glass phase and of the cell aggregate phase by the methods of electron emission spectrography demonstrates that both surfaces are covered by silicate.

Activity of Sol-gel Glass-immobilized Enzymes

In general, in the polymerization mixtures containing methanol, recovery of the enzymes was poor. A better recovery was obtained at higher concentra-

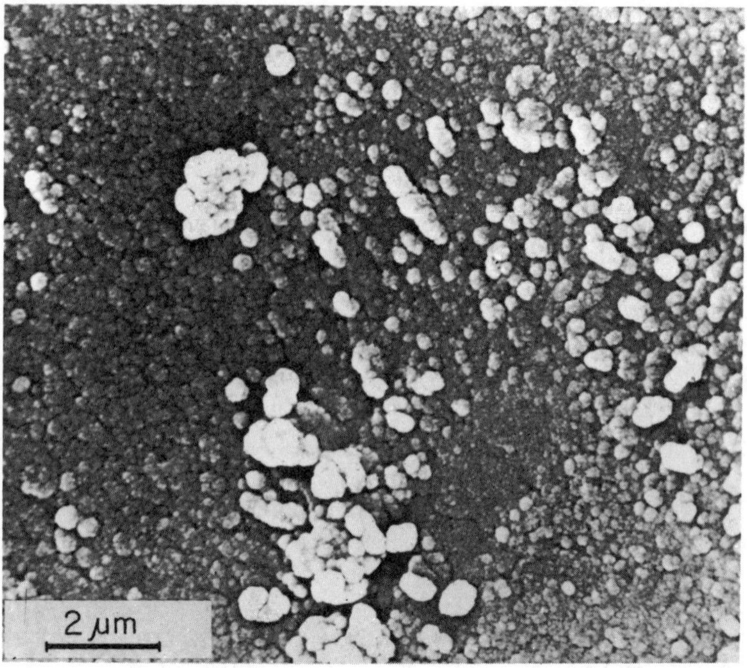

Figure 2. Alkaline phosphatase-doped sol-gel glass, SEM. The sol-gel glass was prepared as described in the Methods. NaF (10 mM) was added to the enzyme solution. Methanol (0.4 ml ml^{-1} TMOS) was used as a solvent. The freshly broken surface of dry unwashed glass was gold-coated and examined in a scanning electron microscope at 20 kV. Protein aggregates (bright dots) of different sizes markedly formed clusters.

tions of NaF. In the gels obtained using PEG 400, enzymes were recovered with higher yields in the mixtures containing larger amounts of PEG.

Higher yields of trypsin activity were observed at lower concentrations of NaF than at high concentrations (Figure 4, left) in the presence of PEG 400. About 40–50% of the initial trypsin activity was recovered when 0.4–0.6 ml PEG 400 was added per ml TMOS. Very similar patterns of activity yields were observed with immobilized catalase, though the yields were lower by an order of magnitude. Sol-gels doped with peroxidase were active, although it was not possible to determine the extent of their activity quantitatively in units of the enzyme, since the dye formed by oxidation of several substrates was adsorbed strongly in the glass. Semi-quantitative comparison of the dye stain shortly after the addition of the assay mixture indicated improved activity yields at higher concentrations of PEG 400. In contrast to trypsin and catalase, sol-gel glasses made at elevated concentrations of NaF were more active. Glasses prepared in methanol-containing mixtures were devoid of peroxidase activity.

High yields (40–60%) of acid phosphatase (Table 1) were obtained in glasses prepared under different conditions. However, no clear trends in

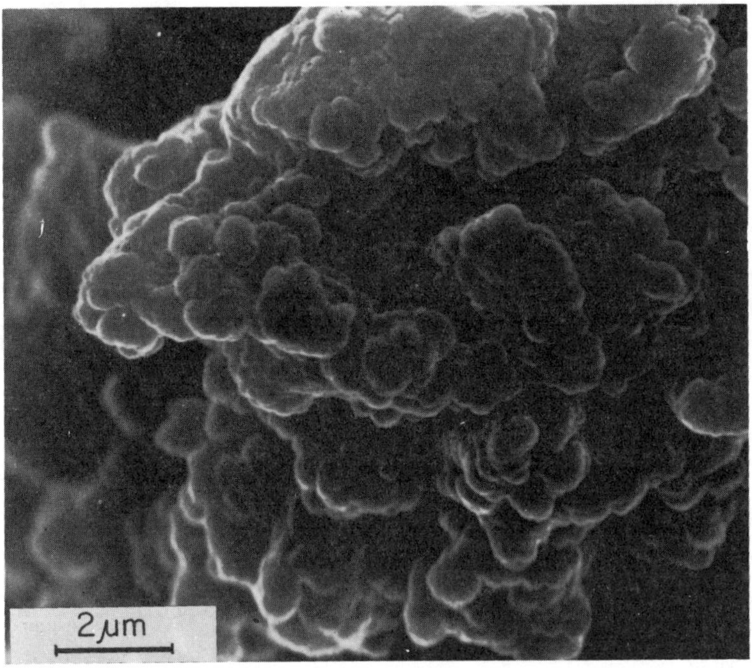

Figure 3. Sol-gel immobilized *E. coli* cells, SEM. The sol-gel glass was prepared as described in the Methods. NaF (10 mM) and PEG 400 (0.4 ml ml^{-1} TMOS) were added to the cells (25 mg dry cell weight per ml TMOS). The freshly broken surface of dry unwashed glass was gold-coated and examined in a scanning electron microscope at 20 kV. Large aggregates of hundreds of cells, as the one shown here, represent well the surface of the very porous pumice-like glass.

yields as a function of the polymerization process parameters were observed. Attempts to immobilize such enzymes as amine oxidase and glycerol oxidase were unsuccessful, either because of failure in the measurement of activity, or due to a failure during the immobilization. We have recently found that quite a few of the sol-gel glasses, prepared either in the presence or in the absence of a protein in the glass, absorb active enzymes from the solution. Thus, it will be possible to differentiate between unsuccessful assay of an immobilized enzyme and unsuccessful entrapment. Properties of bioactive sol-gel glasses obtained by adsorption of enzymes are now being compared to those of sol-gel glass-entrapped enzymes.

The analysis of substrate dose-response curves for sol-gel immobilized alkaline phosphatase (ALP) revealed at least two kinetically different forms of the enzyme (Braun, et al., 1990). The high affinity enzyme component had the Michaelis constant (K_m = 0.08mM), which was also measured for the soluble ALP. About 90% of the enzyme activity, however, had an average K_m of 7mM. The pH maximum of immobilized ALP was about one pH unit higher than for the native enzyme. One has to stress that all activity determinations presented in this work were made at the concentration of

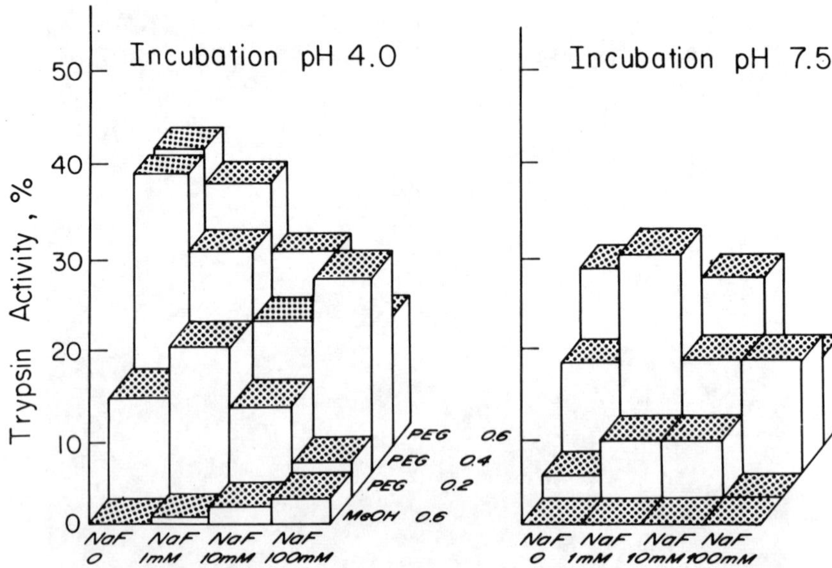

Figure 4. Trypsin-containing sol-gel glasses: Yield of the immobilized enzyme. The sol-gels were prepared as described in the Methods with the additions depicted in the figure. Aliquots of the washed (at pH 5.0) glasses were incubated overnight either at pH 4.0 (left) or at pH 7.5 (right) at 25°C. The assays were performed at the same temperature at pH 8.0 using N^{α}-benzoyl-L-arginine-p-nitroanilide (0.7 mM) as the substrate. The activity was calculated in percentage of soluble enzyme activity used in the preparation of glasses. The enzyme incubated at pH 7.5 shows some autodegradation. An additional incubation at this pH had no effect on enzyme activity.

substrates close to the K_m of these enzymes in solution. The dose-response curves were not investigated. Thus, the existence of kinetically inferior forms would register here as a lower enzyme yield.

Kinetic analysis of the immobilized enzymes is currently underway. However, for trypsin and acid phosphatase, the existence of multiple kinetic variations with the catalytic capacity (V_{max}) of the soluble enzyme will proba-

Table 1. Recovery of acid phosphatase activity in different sol-gel glasses.

Solvent		Catalyst (NaF, mM) 1.0	3.0	10.0
		Acid Phosphatase Yield (%)		
CH_3OH	0.6 ml ml^{-1}	1.9	–	39.2
PEG 400	0.2 ml ml^{-1}	46.6	56.4	21.1
PEG 400	0.4 ml ml^{-1}	39.8	41.0	46.7

The acid phosphatase-containing sol-gel glasses were prepared as described in the Methods with the additions depicted in the Table. The assays were performed on the washed glasses at 25°C at pH 5.6 using p-nitrophenyl phosphate (6 mM) as the substrate. The activity yield was calculated in percents of soluble enzyme activity used in the preparation of glasses.

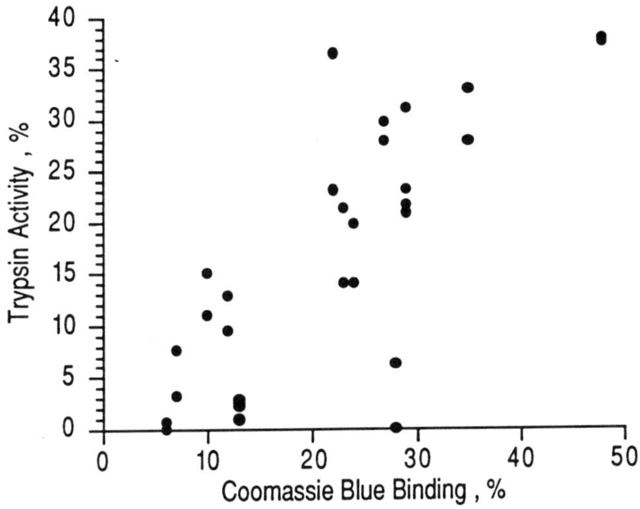

Figure 5. Coomassie blue binding in trypsin-containing glasses: correlation with activity. Trypsin-containing sol-gel glasses were incubated in the staining solution (acetic acid:methanol:water 1:5:4) for 1 h. The glasses were destained in acetic acid (10%) solution in water. The protein bound dye was extracted by methanol containing 10% dimethyl sulfoxide. The concentration of the dye was determined by the optical density of the extract at 510 nm. The binding was calibrated using trypsin precipitated (by 10% trichloroacetic acid) on Whatman 3 MM filters. The filters were subjected to the same staining and destaining procedure as described above. The results of the binding and of the enzyme activity (see Figure 4, left side) were expressed in percentage of soluble enzyme binding and activity used in the preparation of glasses.

bly be disproved. We have compared the ability of the soluble and of the immobilized enzymes to form a complex with the dye Coomassie blue. This dye binds to the positive charges on protein molecules. Coomassie blue binding correlated well (taking into account expected experimental errors) with the activity of the immobilized enzyme (Figure 5). Since we believe that a very large part of the enzyme, added at the onset of polymerization, is present in the glass, the correlation indicates that only a part of enzyme surface is accessible to a small molecule such as Coomassie blue. This part correlates with the accessibility of the enzyme active site to another small molecule, that of the substrate.

Firmness of Interaction Enzyme-matrix and Thermal Stability

It has been shown that sol-gel immobilized ALP is more stable at 70°C than its soluble form (Braun, et al., 1990). This demonstrated that the sol-gel matrix acts as an efficient protective environment. An even more dramatic protective effect was observed in sol-gel immobilized preparations of acid phosphatase, which is much less heat-stable than alkaline phosphatase. The

Table 2. Thermal stability of immobilized acid phosphatase in different sol-gel glasses.

Solvent		Catalyst (NaF, mM) 1.0	3.0	10.0
		Half-life Times at 70°C (min)		
CH_3OH	0.6 ml ml^{-1}	–	–	3.9
PEG 400	0.2 ml ml^{-1}	3.3	3.3	12.0
PEG 400	0.4 ml ml^{-1}	3.2	3.1	3.5

The acid phosphatase containing sol-gel glasses (see Table 1) were incubated in citrate buffer (pH 5.6, 0.1 M) for various periods of time (up to 5 min). The activity of acid phosphatase was determined, as described in the legend to Table 1. The half-life time of the soluble enzyme at the same conditions was below 0.1 min.

acid phosphatase in solution at 70°C lost its activity almost completely in 30 s (half-life time less than 0.1 min). The immobilized acid phosphatase was up to an order of magnitude more stable (Table 2). It is well known that the protective effect of the immobilization matrix against thermal inactivation, or the increase in the melting point of protein polypeptide chain, depends on the amount of binding sites per protein molecule (Martinek and Mozhaev, 1985). Multipoint (3–8 points) attachment of the immobilized enzyme results in a considerable increase of the denaturation temperature. The nonleachable character of the sol-gel immobilization, together with the inability of small molecules (Coomassie blue) to probe the complete surface of the enzyme, indicate a close interaction between large portions of the protein and the matrix. We believe that during sol-gel formation, the matrix and the glass form closely fitted surfaces generating strong binding forces.

The strong character of the binding forces can be demonstrated by diffusional limitations of the immobilized enzyme. Thus, loosely immobilized trypsin can completely degrade itself when left at neutral or slightly basic pH for prolonged periods of time. The sol-gel immobilized trypsin lost no more than 40% of its activity when incubated at 30°C overnight at pH 7.5 (Figure 4, left). Some sol-gels retained their activity completely. This indicates that most of the trypsin molecules are not able to diffuse within the network of the sol-gel.

Conclusions

Preparation and properties of biochemically active sol-gel glasses, obtained by trapping enzymes or whole cells in a polymerizing tetramethoxy-silane are hereby described. Several immobilized enzymes, such as alkaline and acid phosphatases, and trypsin, were recovered in bioactive glasses with 30–60% yield. Catalase, horseradish peroxidase and aspartase yielded either below 30% of initial enzyme activity, or their assay was impossible to measure precisely, due to the adsorption of the substrate or of the product to the sol-gel matrix. The enzymes were bound to the matrix by strong binding forces, and no leaching of the enzyme was observed even at slightly alkaline conditions. No more than 50% of sol-gel immobilized trypsin can be lost to autolysis, indicating that one molecule of the enzyme can degrade no more than one other molecule. Immobilization in sol-gel glass improved stability of enzymes to thermal inactivation. The half-inactivation time for acid phosphatase at 70°C (pH 5.6) increased from about 0.1 min in solution to up to 12 min in sol-gel. Sol-gel entrapment of enzymes seems to be a method generally applicable to the preparation of enzyme catalysts.

Acknowledgments

This work is supported by a grant from the Harry Kay Foundation for applied research of the Hebrew University of Jerusalem. The authors wish to thank Mr. A. Willens for his highly professional advice and help in scanning electron microscopy of the sol-gel glasses.

References

Braun, S., Rappoport, S., Zusman, R., Avnir, D. and Ottolenghi, M. 1990. Biochemically active sol-gel glasses: The trapping of enzymes. Mat. Lett. 10: 1–5.

Bergmeyer, H.U. 1983. Methods of Enzymatic Analysis. Verlag Chemie, Weinheim.

Brinker, C.J. and Scherer, G.W. 1990. Sol-gel Science. Academic Press, San Diego.

Carturan, G., Campostrini, R., Dire, S., Scardi, V. and de Alteriis, E. 1989. Inorganic gels for immobilization of biocatalysts: Inclusion of invertase-active whole cells of yeast (*Saccharomyces cerevisiae*) into thin layers of SiO_2 gels deposited on glass sheets. J. Mol. Cat. 57: L13–L16.

Kennedy, J.F. and White, C.A. 1985a. In: Handbook of Enzyme Biotechnology. Wiseman, A., ed. Ellis Horwood Ltd., Chichester. pp. 147–207.

Kennedy, J.F. and White, C.A. 1985b. In: Handbook of Enzyme Biotechnology. Wiseman, A., ed. Ellis Horwood Ltd., Chichester. pp. 380–420.

Martinek, K. and Mozhaev, V.V. 1985. Immobilization of enzymes: An approach to fundamental studies in biochemistry. Adv. Enzymol. 57: 179–249.

Pouxviel, J.C., Dunn, B. and Zink, J.I. 1989. Fluorescence study of aluminosilicate sols and gels doped with hydroxy trisulfonated pyrene. J. Phys. Chem. 93: 2134–2139.

Slama-Schwok, A., Avnir, D. and Ottolenghi, M. 1989. Photoinduced charge separation across the solid-liquid interface of porous sol-gel glasses: Catalyzed hydrogen generation from water. J. Phys. Chem. 93: 7544–7547.

Zusman, R. Rottman, C., Ottolenghi, M. and Avnir, D. 1990. Doped sol-gel glasses as chemical sensors. J. Non-Cryst. Solids. 122: 107–109.

Discussion (DR. BRAUN)

Fish: What is the maximum size of molecule that can enter into those glass beads? Second, you claim that if you prepare the glass regularly, just in methanol, you get inactivation of the enzyme. However, in one of the last slides did you use a high level of catalyst and get almost the same result?

Braun: The second question first. If you use the large amount of catalyst, you get a lot of hydrolysis and a silicate gel similar to what you would get if you take potassium or sodium silicate and add acid. But it is not the same because of the large volume of glass, probably three times more than with the low amount of catalyst, and it does not shrink that much. The enzymes are mysterious creatures; some like methanol, some do not. What is important is that with the polyethylene glycol system we could successfully immobilize seven or eight enzymes.

There were some enzymes which we could not immobilize. For example, with catalase we got only 10% of the initial activity. We believe that in this system with polyethylene glycol we somehow interfere with the iron molecule which is necessary for enzyme activity. We observe visually a change in the color of the enzyme, so I think that something is happening to the prosthetic group. We could not make more precise measurements.

Sometimes it works with methanol, sometimes it does not. The polyethylene glycol system is always better.

In answer to the first question, the size of the pores is enormous. We never could actually limit it. In the case of cells, for example *E. coli*, the situation may be a little different because I think that the cell wall of *E. coli* catalyzes the process in a different way, and somehow we get smaller pores. So, for example, in order to get active aspartase from the immobilized *E. coli* cells, we must first treat the material with detergent. I think openings in the cell wall of *E. coli* are narrow to begin with. When there are lipids stuck there, you cannot introduce a substrate inside. In any case, there is a problem of diffusion through the framework of the cell wall.

Szybalski: There are two ways to use glass surfaces. One is the way you described for making a covalent attachment. Another way, which Dr. Adam Burkivicz is using in my lab, is that you adsorb DNA quantitatively on the glass fibers, wash it, and then quantitatively elute it under different conditions. That is one of the ways to purify DNA out of cell lysate, in 15 min, with very high efficiency.

Braun: That is correct. Actually, while usual silicates absorb very little protein, the sol-gel glasses adsorb quite a lot of it. You can adsorb trypsin with a very high efficiency on the ready-made sol-gel glass. Our preliminary results indicate that the forces holding trypsin in the adsorption systems are not as strong as when you entrap it within the gel. Therefore, for example, the thermal stability is considerably lower. But we have not completed this study.

Kornguth: I want to compliment you on a very nice demonstration of retention of enzymic activity in the protein immobilized in glass. There are two critical elements, though, in terms of utility. Access is one of these problems, and trypsin can give you a handle on that. You did not mention the substrate that you are using. If it is benzoylarginine ethyl ester, you would expect to have no limitation on access. If you are using a tetra-lysine, you would expect some limitation on access; and if you used hemoglobin that was denatured, you would expect more limitation. What in fact was the substrate and did you use anything larger than BAEE?

Braun: You just stumbled on a very painful spot because we were excited that we can check the diffusion controlled reaction by using different sized substrates. But when we took different lysines, they were completely adsorbed so we could not measure the activity. The story is the same with hemoglobin; it is stuck and we cannot not detect products released in the media. We are now trying to modify the glass, for example, to reverse the charge of the silicate which is charged negatively. Thus, we could reduce the adsorption of polylysine.

Kornguth: There are two interpretations of your observations. Either you have adsorbed the material onto the surface, which is possible with PL, but very unlikely with hemoglobin in a 6 m urea solution in which the trypsin should still be active. Alternatively, the glass pores limited access of the larger substrates to the trypsin. K_m or V_{max} data are required to get at the protein folding and activity question. One would expect that if you had normal activity, there should be constant or similar K_m or V_{max}.

Braun: We did a full K_m analysis only on alkaline phosphatase, not with trypsin. On the basis of partial trypsin activity, either we have 100% of the enzyme active with twice lower K_m or we have the same K_m and 60% of the activity.

B. Green: Can you distinguish between physical entrapment and the possibility of covalent reaction by sugars or hydrophilic residues on the protein? A lot could be interpreted as covalent attachment.

Braun: There was an Italian group who immobilized whole yeast cells in thin layers of sol-gel. They thought that, because of what they saw with electron microscopy, maybe there was a covalent attachment. As an organic chemist, I would think this is unlikely because I would expect that either water would hydrolyze the bond with amino or hydroxy group of the protein or there would be a condensation replacing these groups with Si-O-Si bonds. So, I do not believe in covalent attachment, although this is a possibility, of course.

Gitler: My question is to the same point. Methoxy groups on silanes have been used to attach amino groups to glass, and you can react with amino groups very well. One easy way for you to test whether amino chemistry is involved is to buy acetylated trypsin from Sigma which has all the amino groups blocked by acetyl groups and yet is perfectly enzymatically active. If that is not immobilized essentially the same way, then you would understand what is happening. You are getting diffusion of a very large molecule supposedly through the pores, and yet around the enzyme there must be a matrix which limits some of the accessibility that might be made by the glass reacting with the protein directly.

Braun: Yes, I think that is a very nice experiment, but we have not done it yet. However, it would not exclude, for example, some sort of reaction with the hydroxy group which is also possible. There is another interesting side to this experiment. I believe that part of the attachment may be the ionic compatibility of the silicic acid. During polymerization the charges of arginine and lysine stick together with the charges of the hydroxyls of the silicic acid.

E. Rosenberg: Out of pure ignorance, how do diatoms get immobilized by the glass that they sit in; is that known?

Braun: I am as ignorant as you are. It probably is very similar because there is some sort of directional extrusion of the materials in diatoms. You can disrupt it with any cellular inhibitor, so probably there is some sort of similar system.

L. Smith: Related to covalent attachment, do you have to precipitate your proteins or, if you have them soluble and dispersed, do you still get incorporation?

Braun: The polyethylene glycol 400 which we use is not a very efficient precipitating agent. Depending on the concentration and nature of the protein, pH, etc., in some cases we do not see visible precipitation of the protein, even with hemoglobin which is relatively easily precipitated. I still believe that there is a lot of aggregation because our system has very little free water to solvate the enzyme, and that would move the enzymes together by hydrophobic exclusion. In order to precipitate you have to have aggregates of the size of 1–3 million daltons. If you have aggregates of 0.5 million daltons you probably will not see them, but I believe there is a lot of aggregation in the system. It is not really soluble.

Key Words: Enzyme immobilization, Entrapment, Sol-gel, Xerogel, Glass, Acid phosphatase, Trypsin, Catalase, Peroxidase

Session III
Biological Decontamination

Microbial Degradation of Agent Orange and Mustard Related Compounds

Steven Harvey[1], **Joseph J. DeFrank**[1], **Daphne Kamely**[1], **James J. Valdes**[1] **and Ananda M. Chakrabarty**[2]
[1]*Biotechnology Division, U.S. Army Chemical Research, Development and Engineering Center, Aberdeen Proving Ground, MD 21010; and*
[2]*Department of Microbiology and Immunology, University of Illinois College of Medicine, Chicago, IL 60612.*

Introduction

We live in a highly industrialized society where chemicals play a major role in enhancing industrial and agricultural productivity, in providing physical comforts to our day to day life, as well as in military operations as explosives, ammunitions and agents of chemical warfare. Highly toxic chemicals such as nitrogen and sulfur mustards have been stockpiled for military purposes, and as international treaties leading to their gradual destruction are signed, appropriate technologies will have to be developed for large-scale destruction of such stockpiled chemicals in a cost effective and environmentally safe way. Bioremediation, which uses biological agents mostly in the form of microorganisms or products derived from them, is considered to be highly desirable if appropriate microorganisms can be developed for the rapid degradation of the toxic chemicals. Unfortunately, most toxic chemicals are synthetic and have been introduced in the environment only during the last few decades. Many of these compounds are highly persistent in nature, because natural microorganisms have not had enough time to evolve the appropriate genetic capability to mineralize such compounds. In this article, we briefly consider two types of chemicals that have found military use, *viz.* Agent Orange used as a defoliant in the Vietnam war and sulfur mustard which is a toxic vesicant stockpiled for intended use as a chemical warfare agent.

Biodegradation of Agent Orange

Agent Orange is a mixture of two herbicides, the butyl esters of 2,4–dichlorophenoxyacetic acid (2,4–D) and 2,4,5–trichlorophenoxyacetic acid (2,4,5–

D. Kamely et al. (eds), Biotechnology: Bridging Research and Applications, 221–230.

T). The 2,4–D component is biodegradable, and plasmids encoding the biodegradative enzymes for this compound have been isolated and characterized (Don and Pemberton, 1981). The 2,4,5–T component of Agent Orange is, however, quite recalcitrant and until recently no microorganism was known capable of utilizing it as a sole source of carbon and energy. In order to see if appropriate microorganisms capable of utilizing 2,4,5–T as a sole carbon and energy source could be developed under strong selection in the laboratory, we inoculated microorganisms from various toxic chemical dump sites in a chemostat, provided well characterized bacterial strains harboring known plasmids that specified biodegradation of various aromatic compounds (but not 2,4,5–T), and provided 2,4,5–T as the only major source of carbon in the chemostat. Selection under such conditions over a period of nine to ten months finally yielded a strain of *Pseudomonas cepacia* AC1100 that could utilize 2,4,5–T as its sole carbon and energy source (Kellogg, et al., 1981; Kilbane, et al., 1982). In addition to 2,4,5–T, *P. cepacia* AC1100 is capable of utilizing 2,4,5–trichlorophenol (2,4,5–TCP) as its sole source of carbon and also capable of dechlorinating to varying extent a large number of chlorophenols such as penta, tetra, tri and dichlorophenols (Karns, et al., 1984). It is important to note that even though *P. cepacia* AC1100 can dechlorinate 2,4–D partially, when grown on 2,4,5–T, it is incapable of utilizing 2,4–D as a sole source of carbon, since the enzymes are not induced by 2,4–D, and the products of initial dechlorination appear to accumulate without further oxidation (Haugland, et al., 1990b).

An important question in the directed evolution of the 2,4,5–T pathway in *P. cepacia* AC1100 is the ancestry of the 2,4,5–T degradative (*tft*) genes as well as their regulation. Karns, et al. (1983) demonstrated that the enzymes mediating the uptake (Tup) and the conversion of 2,4,5–T to 2,4,5–TCP (2,4,5–T monooxygenase or TMO) are constitutive, while the enzymes responsible for further conversion of 2,4,5–TCP are inducible by 2,4,5–TCP, but not by 2,4,5–T or pentachlorophenol. Sangodkar, et al. (1988) have shown that 2,4,5–TCP is converted to 5–chloro, 2–hydroxy-hydroquinone (CHQ) instead of a chlorocatechol intermediate. This is reminiscent of the conversion of pentachlorophenol to chlorohydroquinone intermediates by *Flavobacterium* species (Steiert and Crawford, 1986) or by *Rhodococcus chlorophenolicus* (Apajalahti and Salkinoja-Salonen, 1987). Both the *tmo* gene as well as the *chq* genes have been cloned (Sangodkar, et al., 1988), and it is interesting to note that when such gene clusters were used as probes for hybridization with the genomic digests of a large number of pseudomonads, no hybridization was observed (Tomasek, et al., 1989; Sangodkar and Haugland, unpublished observations), suggesting that the *tft* genes *tmo* and *chq* are unique in the cellular background of *P. cepacia* AC1100. The ancestry of these genes is presently unknown.

Tomasek, et al., (1989) demonstrated the presence of three copies of a repeated sequence element near the *chq* genes. This repeat element termed

RS1100 was completely sequenced by Tomasek, et al. (1989) which demonstrated the presence of a 1477–basepair (bp) DNA segment with 38–39–bp terminal inverted repeats immediately flanked by 8–bp direct repeats. The translated sequence of the single large open reading frame of RS1100 showed structural similarity to the phage Mu transposase and other DNA-binding proteins. In this respect, RS1100 shows the characteristics of a typical Insertion Sequence (IS) element. Similar to the *tft* genes, RS1100 did not hybridize to genomic DNA isolated from a variety of *Pseudomonas* strains including *P. cepacia*, again confirming that both the *tft* genes and RS1100 may be of foreign origin recruited in strain AC1100.

The transposability of RS1100 was demonstrated by Haugland, et al. (1990a) who further demonstrated that transposition of RS1100 promotes gene expression at the site of insertion. In addition, RS1100 was shown to mobilize adjacent DNA sequences from the chromosome of *P. cepacia* AC1100 to the plasmid pKT240. Thus, RS1100 has been redesignated as Insertion Sequence 931 (IS931). The fact that the *tmo* gene contains an upstream sequence of IS931, as does the *chq* gene (Tomasek, et al., 1989) and that insertion of IS931 upstream of the promoter-deleted *tmo* gene restores expression of the *tmo* gene appears to suggest that IS931 plays a critical role in the efficient expression of the *tft* genes. Since IS931 can mobilize intervening DNA sequences during transposition (Haugland, et al., 1990a), it is tempting to speculate that IS931 might have also been involved in the recruitment of the *tft* genes from a non-pseudomonal source in *P. cepacia* AC1100; however, no definitive evidence exists at present to show that both the *tft* genes and IS931 have a common ancestry.

Since Agent Orange is a mixture of 2,4–D and 2,4,5–T, and since 2,4–D is known to be degraded by bacteria such as *Alcaligenes eutrophus* JMP134 which harbors the degradative plasmid pJP4 while 2,4,5–T is degraded by *P. cepacia* AC1100, an important question is if a combined culture of JMP134 and AC1100 would allow total degradation of Agent Orange. It is also important to note that the plasmid pJP4 that specifies 2,4–D degradation is a broad-host range plasmid which can be transferred to *P. cepacia* AC1100 allowing it to utilize both 2,4,5–T and 2,4–D. This raises an interesting question if a single culture such as AC1100 harboring the pJP4 plasmid can utilize Agent Orange totally as does the mixture of the two strains JMP134 and AC1100. Haugland, et al., (1990b) demonstrated that the combined cell suspensions of JMP134 and AC1100 could effectively degrade either 2,4–D or 2,4,5–T separately, but could degrade the mixture of 2,4–D and 2,4,5–T poorly because of accumulation of toxic dead-end products of chlorohydroquinones from 2,4–D. In contrast, the pure culture of AC1100 harboring the pJP4 plasmid could effectively degrade the mixture of 2,4–D and 2,4,5–T through simultaneous metabolism of these compounds, suggesting that genetically competent pure cultures may be better suited to degrade mixtures of chemicals than a microbial consortium.

Biodegradation of Thiodiglycol, a Hydrolysis Product of Mustard

Mustard is a toxic vesicant which has been stockpiled in large quantities by several nations. Its hydrolysis proceeds by the following general mechanism (Yang, et al., 1988):

$$\underset{\text{Mustard}}{ClCH_2CH_2SCH_2CH_2Cl} \rightarrow \underset{\text{intermediates}}{\text{Sulfonium ion}}$$

$$\xrightarrow{+H_2O} \underset{\text{Thiodiglycol}}{HOCH_2CH_2SCH_2CH_2OH}$$

Mustard can be synthesized from thiodiglycol with HC1:

$$\underset{\text{Thiodiglycol}}{HOCH_2CH_2SCH_2CH_2HO} + 2\,HCl \xrightarrow{60^\circ} \underset{\text{Mustard}}{ClCH_2CH_2SCH_2CH_2Cl}$$

$$+ 2\,H_2O$$

If possible biological mechanisms of mustard destruction are considered, it is significant that mustard is essentially completely insoluble in aqueous solution. It is also highly toxic to bacterial cells. However, in an agitated aqueous solution, it will hydrolyze to form the thiodiglycol product via the sulfonium ion intermediates. Thiodiglycol is completely soluble in water and has very low toxicity.

Studies by Yang, et al. (1988) have shown that while mustard will quickly react to form the sulfonium ion intermediates, the second step which forms the thiodiglycol is much slower. The first step in the reaction is reversible and the formation of mustard from the sulfonium ions may account for the persistent toxicity of mustard in the environment (Yang, et al., 1988).

Current procedures for the decontamination of mustard typically utilize highly corrosive chemicals. Enzymes capable of catalytically detoxifying mustard might provide an attractive alternative. Two mechanisms can be envisioned by which thiodiglycol-degrading enzymes might be active against mustard. First, because mustard and thiodiglycol are structurally similar, an enzyme might attack the two molecules through the same mechanism. Second, in an aqueous solution, there exists an equilibrium between mustard and thiodiglycol. Degradation of thiodiglycol could therefore drive the reaction toward completion, thus catalyzing the hydrolysis of the mustard.

Isolation of Thiodiglycol-degrading Bacteria

Soil samples were collected from areas suspected of having at one time been contaminated with mustard. The pure culture SH18 was isolated from a chemostat continuous culture to which were added bacteria from many different soil samples. The chemostat was run continuously for nine months with thiodiglycol as the sole source of carbon. SH18 was isolated from the culture

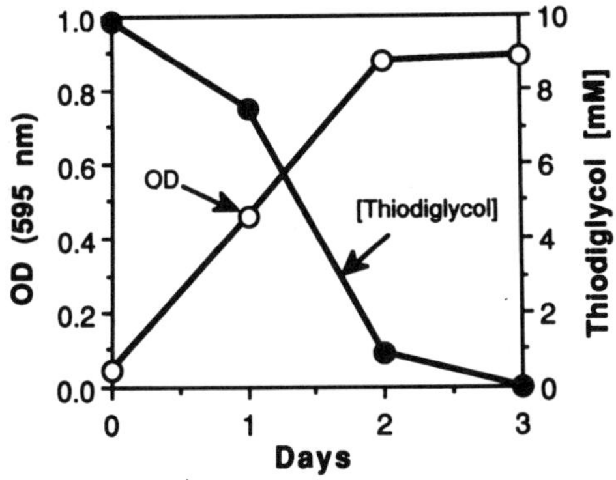

Figure 1. SH18 – Thiodiglycol biodegradation.

medium, streaked three times on L plates for single colonies and reinoculated into 10 mM M9/thiodiglycol medium. SH18 is a Gram-negative, aerobic rod-shaped bacterium. Gas chromatographic (GC) fatty acid analysis (Microbial I.D. Inc.) indicated a 32.2% identity with *Pseudomonas pickettii*.

The pure culture SH41 was isolated as another variant found growing in M9/thiodiglycol medium. It has not been identified but is physiologically and morphologically distinct from SH18. The mixed culture EC7 is from an enrichment culture to which thiodiglycol was provided as the sole source of carbon. It comprises at least three morphologically distinct strains, none of which will grow alone on thiodiglycol.

Figure 1 shows the degradation of thiodiglycol by SH18. The disappearance of thiodiglycol was monitored by GC as the $O.D._{595}$ increased. Depending on its concentration, thiodiglycol produces a peak with retention time between about 3.8 and 4.2 minutes. The measurement on the third day was below the limits of detection. Since 1 mM concentrations can routinely be detected by this method, it was concluded that greater than 90% of the original 10 mM concentration of thiodiglycol had been degraded. The thiodiglycol concentration in the sterile control (not shown) did not decrease at all. Another control containing bacteria and M9 medium (without thiodiglycol) did not increase in turbidity (also not shown). The degradation of thiodiglycol was independently verified by NMR in a separate culture grown in 50 mM thiodiglycol (data not shown).

Growth of SH18 on Hydrolyzed Mustard

An initial attempt was made to determine if SH18 would utilize mustard as its sole carbon source in minimal medium. Mustard in effective concentrations of 10–20 mM (1.6–3.2 mg ml^{-1}) was added to M9 medium and the culture tubes were immediately inoculated with SH18 bacteria which had been grown

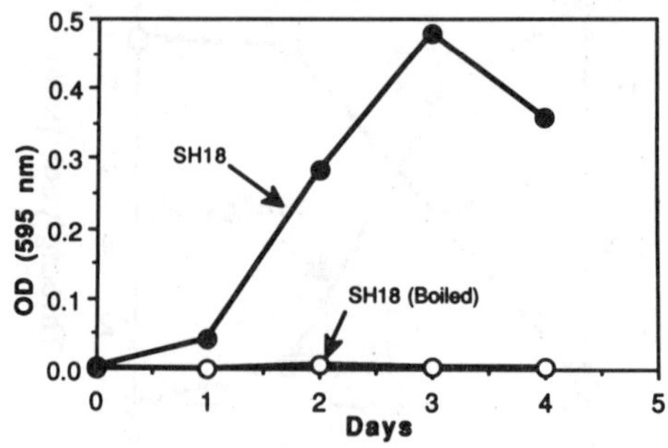

Figure 2. SH18 – Growth on hydrolyzed mustard. (Mustard (20 mM) was added to medium and shaken for 2 days prior to inoculation.)

on thiodiglycol. No growth was seen, presumably due to the toxic effects of mustard. However, when the same amounts of mustard were shaken in the medium for two days prior to inoculation, the cultures became turbid (Figure 2). When the effective concentration of mustard was doubled to 40 mM, no growth was seen even when the mustard was shaken prior to inoculation. NMR analysis showed that in the 20 mM culture at least 73% of the original mustard had hydrolyzed to thiodiglycol at the time the bacteria were added, whereas the 40 mM culture had a maximum of 19% of the original mustard in the form of thiodiglycol. At least 81% of the mustard from this culture was still in the form of sulfonium ions at the time the bacteria were added, indicating that the sulfonium ion intermediates at this concentration do not support growth of SH18.

Growth of Pure Cultures on Possible Metabolic Intermediates

A tentative list of possible intermediates of thiodiglycol metabolism was compiled and SH18 and SH41 were characterized as to which of these compounds they could utilize as sole carbon growth substrates (Table 1). Results show relative differences between SH18 and SH41 with regard to their ability to utilize glycolaldehyde and ethylene glycol.

Conclusions

The development in the chemostat of a strain of *P. cepacia* AC1100 capable of utilizing 2,4,5–T as a sole source of carbon and energy not only signifies the rapid adaptability of microorganisms, but also the important roles played

Table 1. Growth patterns of SH18 and SH41

Compound	SH18	SH41
Dithiodiglycolic acid	−	−
Thiodiglycolic acid	+	++ (20 mM)
Glycolaldehyde	+	−
Oxalic acid	+	+
Isethionic acid	++	++
Glycolic acid	++	++ (20 mM)
Glyoxylic acid	++	++
Ethylene glycol	++	−
Thioglycolic acid	−	−

− = No growth; + = Marginal growth; ++ = Good growth.
Growth of SH18 and SH41 on possible intermediates on thiodiglycol metabolism as sole sources of carbon. Concentrations of 10, 20 and 40 mM were attempted for all compounds. Positive results indicate growth on 10 mM substrate except where other minimum concentrations are indicated.

by strong selection pressure to enhance the evolutionary processes for the genes encoding degradation of synthetic compounds. It is interesting to note that the evolved 2,4,5–T degradative genes are unstable, and undergo frequent and extensive deletion because of the presence of a large number of copies of insertion sequence elements such as IS931. That the IS931 genes, as well as the 2,4,5–T degradative genes, do not show any significant hybridization to the genomic digests of a large number of pseudomonads, including laboratory strains of *P. cepacia*, implies strongly that such genes might have been recruited by *P. cepacia* AC1100 from a non-pseudomonal source and the expression of such genes in a foreign host is aided by promoters carried by the IS931 type of elements. To what extent IS931 types of elements have been involved in the recruitment of the 2,4,5–T genes by *P. cepacia* AC1100 is presently unknown.

The directed evolution of 2,4,5–T genes in the chemostat raises an interesting question regarding the generality of this phenomenon. Is it possible to isolate other microorganisms capable of degrading other toxic chemicals, and will the degradative genes show a similar pattern of evolution as the 2,4,5–T degradative genes? As a first step, the biodegradation of thiodiglycol has been demonstrated with two different pure microbial cultures and one mixed culture which utilize it as their sole source of carbon. One of the pure cultures (SH18) has also been demonstrated to be capable of growth on mustard which has first been shaken in the aqueous medium for two days. Based on the results of NMR analysis of the organic components of the medium, it would appear that SH18 is only capable of growth when a significant proportion of the original mustard was hydrolyzed to thiodiglycol. High concentrations of sulfonium ions appear to inhibit growth. Further studies will be necessary to localize the thiodiglycol degradative genes and study their regulation and various characteristics.

Acknowledgments

We are very grateful to Linda L. Szafraniec and William T. Beaudry who did the NMR analysis, to Jude Height for his work with the mustard and Johnnie Albizo for his help with the gas chromatography. The Mustard work was supported under an Intergovernmental Personnel Act (IPA) with the U.S. Army Chemical Research, Development and Engineering Research Center and the University of Illinois. The Agent Orange work was supported by a Public Health Service grant ES04050 from the National Institute of Environmental Health Sciences to AMC.

References

Apajalahti, J.H.A. and Salkinoja-Salonen, M. 1987. Complete dechlorination of tetrachlorohydroquinone by cell extracts of pentachlorophenol-induced *Rhodococcus chlorophenolicus*. J. Bacteriol. 169: 5125–5130.

Don, R.H. and Pemberton, J.M. 1981. Properties of six pesticide degradation plasmids isolated from *Alcaligenes paradoxus* and *Alcaligenes eutrophus*. J. Bacteriol. 145: 681–686.

Haugland, R.A., Sangodkar, U.M.X. and Chakrabarty, A. M. 1990a. Repeated sequences including RS1100 from *Pseudomonas cepacia* AC1100 function as IS elements. Mol. Gen. Genet. 220: 222–228.

Haugland, R.A., Schlemm, D.J., Lyons, R.P., Sferra, P.R. and Chakrabarty, A.M. 1990b. Degradation of the chlorinated phenoxyacetate herbicides 2,4–dichlorophenoxyacetic acid and 2,4,5–trichlorophenoxyacetic acid by pure and mixed bacterial cultures. Appl. Environ. Microbiol. 56: 1357–1362.

Karns, J.S., Duttagupta, S. and Chakrabarty, A.M. 1983. Regulation of 2,4,5–trichlorophenoxyacetic acid and chlorophenol metabolism in *Pseudomonas cepacia* AC1100. Appl. Environ. Microbiol. 46: 1182–1186.

Karns, J.S., Kilbane, J.J., Chatterjee, D.K. and Chakrabarty, A. M. 1984. Microbial biodegradation of 2,4,5–trichlorophenoxyacetic acid and chlorophenols. In: Genetic Control of Environmental Pollutants. G.S. Omenn and A. Hollaender, eds., Plenum Press, New York, pp. 3–21.

Kellogg, S.T., Chatterjee, D.K. and Chakrabarty, A.M. 1981. Plasmid-assisted molecular breeding: New technique for enhanced biodegradation of persistent toxic chemicals. Science 214: 1133–1135.

Kilbane, J.J., Chatterjee, D.K., Karns, J.S., Kellogg, S.T. and Chakrabarty, A. M. 1982. Biodegradation of 2,4,5–trichlorophenoxyacetic acid by a pure culture of *Pseudomonas cepacia*. Appl. Environ. Microbiol. 44: 72–78.

Sangodkar, U.M.X., Chapman, P.J. and Chakrabarty, A.M. 1988. Cloning, physical mapping and expression of chromosomal genes specifying degradation of the herbicide 2,4,5–T by *Pseudomonas cepacia* AC1100. Gene 71: 267–277.

Steiert, J.G. and Crawford, R.L. 1986. Catabolism of pentachlorophenol by a *Flavobacterium* sp. Biochem. Biophys. Res. Communs. 141: 825–830.

Tomasek, P.H., Frantz, B., Sangodkar, U.M.X., Haugland, R.A. and Chakrabarty, A.M. 1989. Characterization and nucleotide sequence determination of a repeat element isolated from a 2,4,5–T degrading strain of *Pseudomonas cepacia*. Gene 76: 227–238.

Yang, Y.C., Szafraniec, L.L., Beaudry, W.T. and Ward, J.R. 1988. Kinetics and mechanism of the hydrolysis of 2–chloroethyl sulfides. J. Org. Chem. 53: 3293–3297.

Discussion (DR. CHAKRABARTY)

Szybalski: There were a few very serious oil spills recently in Alaska and Texas. Why is there not more use of microorganisms for clean-up? Is it because of strong opposition by pseudoenvironmental activists or because there are no suitable microorganisms engineered as yet?

Chakrabarty: When I used to work with General Electric we developed a microorganism which worked quite well in the lab but was never field-tested because of this debate about release of genetically engineered microorganisms and also because of the concern on the part of General Electric that the company might be sued. A company with substantial resources tries to avoid any kind of controversial applications of technology which are not the major thrust of its business. It is interesting to note, however, that recently the state of Texas allowed use of "natural organisms" in an oil spill, but I do not know exactly what sorts of microorganisms were used. If you ask me, natural microorganisms are not designed to do the job efficiently. Mixtures of microorganisms seldom work very efficiently, and usually you cannot know what organisms are in there because you cannot culture all of them in the lab. So you cannot determine the toxicity of the microorganisms that you are using. There are all kinds of problems with regard to application of natural microorganisms, and that is why you have to be a little careful about bioremediation. Many companies are getting into the act in the name of bioremediation and advocating natural microorganisms that may or may not work. I am a little concerned that if we do not use the right microorganisms and if we therefore do not get the expected results, there will be a great deal of loss of faith in bioremediation.

Kamely: There have been many arguments in the past that genetically engineered organisms, such as mustard or oil degrading microbes, are weaker than their counterparts. As a result, a genetically engineered organism may be less effective. At present in the United States it is difficult to genetically engineer an organism and release it into the environment, so naturally selected organisms are often preferred. If you had the choice, do you think that a more efficient way would be to genetically engineer an organism for the purpose of biodegrading chemical and biological materials?

Chakrabarty: In our work on the surfactant we switched from using living microorganisms to microbial product, because surfactant is a chemical which falls under a different category. This question has bothered me for a long time. What is a genetically engineered microorganism? If you develop an organism in a chemostat under strong selection pressure, do you call it a genetically engineered microorganism? It is a natural organism; it simply has evolved certain genes in response to a very difficult growth environment that you have selected. Microorganisms shuffle genes all the time in response to their needs, so why do we think that a chemostat-selected organism is "genetically engineered"? I think the most important thing to do is to define a genetically engineered microorganism.

Ellis: In terms of some of these microorganisms that are developed, is there selectivity with regard to the halogen? If you substitute fluoride, bromide or iodide will it also work?

Chakrabarty: It depends on the size of the halogen atom. Fluorine resembles the hydrogen atom in terms of its size, so it is a lot easier for the microorganisms to remove a fluorine than a chlorine. It is easier to remove a chlorine than a bromine, and iodine is very difficult.

Grossman: Can these microorganisms be used secondarily as a cheap source of protein supplement for feed?

Chakrabarty: Sure, the initial reason for developing genetically engineered microorganisms was not to use them in an oil spill, but to make protein from petroleum. In the late 1960's, it was supposed to be the "in" thing to do because oil was cheap and bacteria are 80% protein. So if you grow them on crude oil, you get 80% protein-containing organisms instead of oil. That changed in the early 1970's when the price of oil increased.

Er-El: Could you say something about the rate of degradation? We have seen many natural emulsifiers and many ideas about degradation of chemicals, but they are too slow for very large stockpiles.

Chakrabarty: I cannot give you those numbers right now, but in the lab if I use 2,4,5–T at the concentration of $1\,mg\,ml^{-1}$, the strain will remove more than 90% over a period of 48–72 h.

Juris: I would like to add something to what you mentioned about the transfer of the chemical agents stocks from Europe and Japan. It is my understanding that those stocks are going to be transferred to Johnston Island, Hawaii.

Key Words: Genetic organization, Toxic chemicals, Directed evolution, Gene expression

Surface Active and Drag-Reducing Bacterial Polymers

Eugene Rosenberg
Department of Microbiology, George S. Wise Faculty of Life, Sciences, Tel Aviv University, Ramat Aviv, Israel 6997

Introduction

Bacteria are specialists. In any particular ecological niche one group, or a limited number of groups, are dominant. These bacteria are adapted to survive in that niche during nutrient starvation and to grow more rapidly than their competitors during the short time that nutrients become available. The fast growth rate of bacteria is due, in part, to an efficient uptake of nutrients and release of waste products. Rapid influx of nutrients and eflux of waste materials require a high surface to volume ratio. The price that the bacterium pays for a higher surface to volume ratio is that it is highly exposed to the local environment. Thus, all of the components which are on the cell surface or extracellular must be able to function under the conditions of the local environment. For example, if the bacterium is adapted to grow at pH 5, then the cell surface and extracellular enzymes, flagella, fimbriae, binding proteins, capsules, etc., must function effectively at pH 5. On the other hand, the inside of the cell is buffered. This is the reason that the unity of biochemistry is expressed inside the cell, whereas diversity of the microbial world is generally to be found on the outside of cells.

The high surface to volume ratio of bacteria and the diversity of their exocellular components make tham a rich source of specialized polymers, some of which have potential technological applications.

The studies that will be reviewed here involve certain surface active and drag-reducing bacterial exocellular polymers that have demonstrated potential applications. The amphipathic surface active polymers produced by *Acinetobacter calcoaceticus* (emulsans) stabilize oil-in-water emulsions and can be used for treating oil pollution, cleaning oil tanks, oil recovery, lowering the viscosity of heavy oil and in the preparation of cosmetics, foods and paints. Another surface active polymer, biodispersan, can lower the energy costs in grinding limestone and titanium dioxide and dispersing them in water. Bacterial extracellular and capsule polysaccharides can reduce the friction of water flowing in the turbulent state.

Microbial surfactants have been reviewed by Zajic and Panchal (1976), Zajic and Seffens (1984), Cooper and Zajic (1980), Zajic and Mahomedy

D. Kamely et al. (eds), Biotechnology: Bridging Research and Applications, 231–248.

(1984), Singer and Finnerty (1989), Rosenberg (1986) and Rosenberg and Kaplan (1986). Emulsan is an industrial product and has been reviewed by Shabtai, et al., (1985). Drag reduction by biopolymers was reviewed by Hoyt (1985).

Materials and Methods

Bacterial Strains and Growth Conditions

A. calcoaceticus RAG-1 (ATCC 31012) was originally isolated from the Mediterranean coast after enrichment on crude oil (Reisfeld, et al., 1972). *A. calcoaceticus* BD4, isolated by Taylor and Juni (1961), has been described previously (Kaplan and Rosenberg, 1982). *A. calcoaceticus* A2 was isolated by enrichment culture (Rosenberg, et al., 1988). *A. calcoaceticus* MR-481 is a spontaneous mutant of RAG-1 which is deficient in its ability to adhere to hydrocarbons (Rosenberg and Rosenberg, 1981). *Vibrio* sp. NS-31 was isolated from the side of a Mediterranean barracuda (Sar and Rosenberg, 1989).

Unless stated otherwise, *A. calcoaceticus* strains were grown in minimal medium containing (per liter of water): $K_2HPO_4 \cdot 3H_2O$, 22.2 g; KH_2PO_4, 7.26 g; $MgSO_4 \cdot 7H_2O$, 0.2 g; $(NH_4)_2SO_4$, 4 g; ethanol, 20 g. The bacteria were cultivated in shake flasks at 30°C and growth was monitored by absorbance at 540 nm. PUM buffer contained (per liter of water): $K_2HPO_4 \cdot 3H_2O$, 22.2 g; KH_2PO_4, 7.26 g; $MgSO_4 \cdot 7H_2O$, 0.2 g; urea, 1.8 g. PUM-hexadecane medium consisted of 4.2 ml PUM buffer overlaid with 0.5 ml hexadecane. *Vibrio* sp. NS-31 was grown in HS-E medium containing (per liter of water): $K_2HPO_4 \cdot 3H_2O$, 15 g; KH_2PO_4, 2 g; $(NH_4)_2SO_4$, 4 g; $MgSO_4 \cdot 7H_2O$, 0.2 g; NaCl, 30 g; ethanol, 15 ml; final pH 7.2. The ethanol was added in 2 steps: 0.75% before inoculating and 0.75% when the culture reached 500 K.U.

Materials

RAG-1 emulsan was purified from the cell free supernatant fluid of an ethanol grown culture by ammonium sulfate precipitation and cetyltrimethyl ammonium bromide fractionation (Zosim, et al., 1982). Protein-free BD4 emulsan (Kaplan, et al., 1987) and PS-6 (Sar and Rosenberg, 1989) were prepared by hot phenol treatment.

Preparation of Biodispersan from *A. calcoaceticus* A2 was described previously (Rosenberg, et al., 1988). Agha-Jari crude oil and gas oil was obtained from the Ashkelon-Eilat Pipeline Co., Israel. The limstone rock used in this study was a white chalk of high purity ($>$ 99% $CaCO_3$) quarried in the south of Israel.

Analytical Procedures

Hydrocarbon-in-water emusification was assayed by adding 0.1–0.5 ml of emulsifier to a mixture of 0.1 ml of hexadecane/2–methylnaphthalene (1:1, vol/vol) in a 100 ml flask containing TM buffer (20 mM Tris-HCl buffer, pH 7.0 and 10 mM $MgSO_4$) to a final volume of 7.5 ml. The assay mixture was incubated at 30°C with reciprocal shaking (160 strokes/min) for 1 hr. Turbidities were determined in a Klett-Summerson photometer (green filter). One unit of emulsifying activity per mililiter is that concentration which yields 100 K.U. Cell-associated emulsan was measured by the ELISA test as previously described (Goldman, et al., 1982).

Drag reduction measurements were performed at 25°C in a turbulent flow rheometer built according to a similar system described by Rosen and Cornford (1971). The fluid velocity in the 1.71 mm diameter tube was 808 cm/sec, corresponding to a pipe Reynolds number of 17,300. The pressure drop for sea water of HS- E medium was 0.40 atm. Drag reduction is expressed as percent decrease in the pressure drop of the sample compared to that of the suspending medium.

Limestone grinding experiments were performed with 10% suspensions in a 1.9 l Norton porcelain jar mill containing cylindrical Burundum grinding pellets. The jars were placed on a Norton model 753 RM jar mill which was operated at a roller speed of 290 rpm. Particle size reduction was measured as described previously (Rosenberg, et al., 1989).

Results and Discussion

RAG-1 Emulsan

Acinetobacter calcoaceticus RAG-1 produces a unique polysaccharide/lipid with an average molecular weight of 10^3 Kd (Zuckerberg, et al., 1979). The linear polysaccharide is composed of the repeating trisaccharide N-acetyl D-galactosamine, N-acetyl L-galactosamine uronic acid and an unidentified diaminosugar. Fatty acids, C-12 to C-18, comprising 15% of the emulsan dry weight, are attached to the polysaccharide backbone via O-ester and N-acyl linkages (Belsky, et al., 1979). The amphipathic nature of the polymer is due to hydroxyl and carboxyl (1.5 μmol mg^{-1}, pK′ 3.05) groups in the polysaccharide (hydrophilic) and long chain fatty acid side chains (hydrophobic).

During the exponential phase of growth of RAG-1 the emulsan is present on the cell surface in the form of a minicapsule (Figure 1, Table 1). Using an immunocytochemical test, all of the emulsan was cell-associated during early growth (0–10h). Addition of chloramphenicol was bacteriostatic and caused the release of the emulsan into the extracellular fluid. During the end

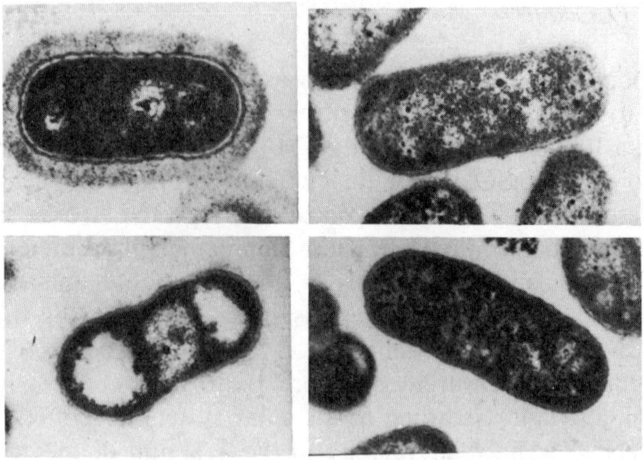

Figure 1. Transmission electron microscopy of *A. calcoaceticus* RAG- 1 and emulsan negative mutants. (Upper left) Exponential phase RAG-1. (Lower left) Stationary phase RAG-1. (Upper and lower right) Exponential phase emulsan negative mutants. (Source: Adapted with permission from Pines, 1983).

of the growth phase and stationary phase, emulsan continued to be synthesized and excreted, reaching close to 40% of the cell dry weight.

RAG-1 emulsan enhanced both the formation and stability of hydrocarbon-in-water emulsions. The most remarkable features of emulsan as an emulsifier are (1) it is effective at low concentrations (0.001 to 0.01%), (2) it shows substrate specificity (Table 2) and (3) once emulsions are formed they do not break by coalescence of oil droplets, but rather by creaming of the emulsified oil (Figure 2). Emulsan did not emulsify pure aliphatics (e.g.,

Table 1. Cell-bound and extracellular RAG-1 emulsan[a]

Fermentation[b] (h)	Turbidity (A_{540} nm)	Emulsan (% cell dry wt)	
		Cell-bound[d]	Extracellular[e]
0	0.2	12	Trace
6	1.8	13	Trace
12	4.2	12	2
24	7.4	9	7
48	7.9	4	20
72	8.0	2	38
12 + CAP[c]	4.3	0.5	13

[a] Data adapted from Rubinovitz et al., 1982, and Goldman et al., 1982.
[b] An overnight culture of *A. calcoaceticus* RAG-1 was diluted 1:20 into fresh ethanol minimal medium and incubated iwth shaking at 20°C.
[c] Chloramphenicol (50 μg ml^{-1}) was added to the culture and allowed to incubate for 7 h.
[d] Cell bound emulsan was measured by the microplate ELISA procedure (Goldman et al., 1982).
[e] Extracellular emulsan was estimated by measuring hexosamine content of the dialyzed extracellular culture fluid.

Table 2. Emulsification of hydrocarbons by RAG-1[a] and BD4[b] emulsan

Substrate	Addition	Emulsan turbidity (K.U.)	
		RAG-1	BD4
Hexadecane	None	20	25
Methylnaphthalene	None	0	5
Hexadecane	Methylnaphthalene	284	320
Crude oil	None	780	30
Kerosine	None	42	85
Kerosine	Hexadecane	68	–
Kerosine	Methylnaphthalene	950	500

[a] Data taken from Rosenberg, et al., 1979. Emulsan concentration was 0.05 mg ml^{-1}.
[b] Data taken from Kaplan and Rosenberg, 1982. Emulsan concentration was 0.05 mg ml^{-1}.

hexadecane), aromatics (e.g., methylnaphthalene), or cyclic (e.g., dicyclohexane) hydrocarbons. However, all mixtures which contained an aliphatic and an aromatic (or cyclic) hydrocarbon were emulsified effectively. Petroleum fractions which contain an appropriate mixture of aliphatics and ring compounds, such as Agha Jari crude oil or gas oil, were emulsified, whereas fractions such as kerosine and gasoline, which contain mainly aliphatics, were emulsified only if a ring compound was added. In addition to the specificity for an aliphatic and aromatic hydrocarbon, both the RAG-1 and BD4 emulsan required the presence of a divalent cation.

Two suggestions have been proposed for the natural role of RAG-1 emulsan: emulsification and desorption. Many hydrocarbon- degrading microor-

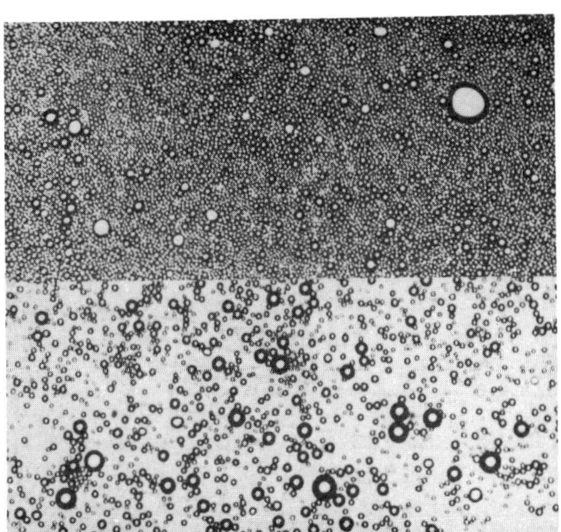

Figure 2. Light microscopy of a hexadecane-in-water emulsion stabilized by 0.001% emulsan. The lower field shows the "emulsanosol" (cream) that was formed after several days standing (Magnification × 390).

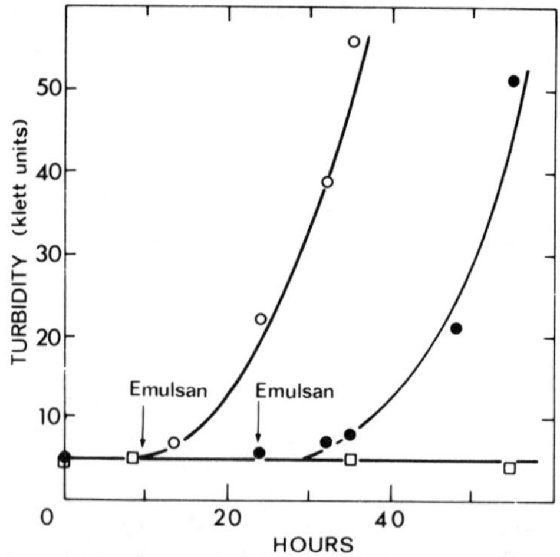

Figure 3. Induction of growth of MR-481 on hexadecane by emulsan addition. Klett tubes containing 4.2 ml of PUM-hexadecane media were inoculated with MR-481 to an initial turbidity of 5 Klett units and incubated at 30°C. Emulsan was added to a final concentration of 50 fg ml^{-1} at 10 (○) or 24 hr (●). Tubes containing no emulsan served as controls (□).

ganisms produce extracellular emulsifying agents, the inference being that emulsification plays a role in growth on water-insoluble substrates. The evidence for this proposal is indirect. First, theoretical and experimental studies on the production of single cell protein from hydrocarbons in fermentors demonstrated that growth rate can be limited by the interfacial surface area between water and oil (Erickson and Nakahara, 1975). When the surface area becomes limiting, biomass increases arithmetically rather than exponentially. However, fermentor studies, with a view towards achieving high growth rates and high cell yields on pure substrates, utilize strong agitation, forced aeration and optimized temperature, pH and nutrient addition. These conditions are probably not relevant to cell-oil interactions as they occur in nature. Second, there is a good correlation between emulsifier production and growth on hydrocarbons (Itoh and Inouye, 1982; Hisatsuka, et al., 1972). In the case of *A. calcoaceticus* sp. HO1–N extracellular membrane vesicles partition exogenously supplied hydrocarbon in the form of a microemulsion (Kappeli and Finnerty, 1980). Third, and most significantly, mutants which do not produce emulsifer grew poorly on hydrocarbons (Itoh and Suzuki, 1972). In the case of an *A. calcoaceticus* RAG-1 emulsan deficient mutant MR-481, the addition of the purified emulsan stimulated growth on hexadecane (Figure 3).

Desorption is an important part of the natural growth cycle of microorganisms. After microbial attachment to and growth on a surface for a period of time, the conditions on the surface often become unfavorable for further development. For example, toxins may accumulate and crowded conditions may prevent transport of necessary nutrients. It would then be advantageous for the microorganism to desorb and find a new habitat. In fact, mechanisms for detachment seem to be essential for all attached microorganisms in order to facilitate dispersion and colonization of new surfaces (Shilo, 1989). One of the natural roles of an emulsifier may be in regulating desorption of the producing strain from hydrophobic surfaces. Consider *A. calcoaceticus* RAG-1 growing on the surface of crude oil in sea water. During the growth phase, the bacteria accumulate emulsan in the form of a minicapsule on the cell surface. This bacterium utilizes only relatively long chain n-alkanes for growth. After these compounds are utilized, RAG-1 becomes starved, although it is still attached to the oil droplet, enriched in aromatic and cyclic paraffins. Starvation of RAG-1 causes release of the minicapsule of emulsan. It is postulated that this released emulsan forms a polymeric film on the n-alkane-depleted oil droplet, thereby desorbing the starved cell. In effect, the "emulsifier" frees the cell to find fresh substrate. At the same time, the depleted oil droplet has been "marked" as used, by now having a hydrophilic outer surface to which the bacterium cannot attach.

Emulsan stabilizes emulsions by binding tightly to the hydrocarbon/water interface, forming a strong polymeric film of greater than 20 Å thickness (Zosim, et al., 1982). The structure of these emulsan-stabilized hydrocarbon droplets explains many of the useful properties of emulsan; the anionic heteropolysaccharide binds considerable amounts of water and, together with the strong negative surface charge, prevents coalescence of the droplets and adhesion of the droplets to most surfaces. The fact that the outer surface of the droplets becomes hydrophilic explains both the cleaning action of emulsan (Gutnick and Rosenberg, 1977) and the ability of the polymer to greatly reduce the viscosity of heavy oils (Basta, 1984).

Structure-function relationships of emulsan were studied using a specific emulsan depolymerase (Shoham and Rosenberg, 1983). The enzyme splits the heteropolysaccharide backbone by an eliminase reaction, without hydrolyzing O-ester or N-acyl linkages. When the molecular weight of emulsan was reduced, to 330,000, 125,000 and 62,000, the emulsifying activity decreased to 75%,40% and 26%, respectively, of its initial values. Thus, high molecular weight emulsan is required for maximum activity.

If an emulsan stabilized hydrocarbon in water emulsion is allowed to stand for several days, or centrifuged briefly, the oil droplets "coated" with emulsan and water rise, forming an "emulsanosol" (Figure 2). Not surprisingly, emulsanosols have properties characteristic of neither the pure hydrocarbon nor aqueous solutions of emulsan. For example, an emulsan-specific bacteriophage, ap3, binds to emulsanosols, but not to emulsan free in solution (Pines and Gutnick, 1984). Emulsanosols bind and thereby concentrate certain water-soluble cations, such as the dye Rhodamine B (Zosim, et al., 1982) and uranyl ion (Zosim, et al., 1983). Up to 90% of the bound uranium (800 mg g^{-1} emulsan) can be desorbed from emulsanosol at pH 2 with practically no release of the biopolymer.

Low concentrations of emulsan (0.05–0.2 mg ml^{-1}) inhibited the adhesion

Table 3. Reconstitution of BD4 emulsan[a]

Fraction	Emulsifying activity (U/mg)
BD4 emulsan	220
Pure BD4 polysaccharide	0
Purified BD4 protein	5
Polysaccharide + protein[b]	210

[a] Data taken from Kaplan et al., 1987.
[b] Ratio of polysaccharide to protein was 3.5.

of *A. calcoaceticus* RAG-1, *Streptococcus pyogenes*, and other bacteria to hydrophobic surfaces, such as hydrocarbons and epithelial cells (Rosenberg, et al., 1983). The anti-adherence properties of emulsan are due to the formation of surface films, which render hydrophobic surfaces hydrophilic and thus prevent attachment of bacteria through hydrophobic interactions. More generally, polyanions have been shown to inhibit the adherence of *Vibrio parahaemolyticus* to intestinal cells (Carruthers and Anderson, 1979).

BD4 Emulsan

A high percentage of *A. calcoaceticus* strains produce extracellular polymeric emulsifiers (Sar and Rosenberg, 1983). The extracellular emulsifier produced by *A. calcoaceticus* BD4 consists of a protein/polysaccharide complex. The polysaccharide is composed of the repeating anionic heptasaccharide:

→3)-α-L-Rha-(1→3)-α-D-Man-(1→3)-α-L-Rha-(1→3)-α-L-Rha-(1→3)-β-D-Glc-(1→
2
↑
1
α-L-Rha-(1→4)-β-D-glcUA

The protein component of the emulsifier complex was purified seven-fold on a Bio-Gel A-1.4 m column from the extracellular fluid of a polysaccharide deficient mutant. As summarized in Table 3, neither the pure polysaccharide nor the purified protein were active as emusifiers. However, when they were both present, a complete reconstitution of emulsifying activity was obtained. Thus, the BD4 emulsan apparently derives its amphipathic properties from the association of an anionic hydrophilic polysaccharide with proteins.

On a weight basis, the BD4 emulsan complex is as active as the RAG-1 emulsan. Although the RAG-1 and BD4 emulsans differ chemically, they share several properties: Both originate from the cell capsule, require divalent cations and show substrate specificity (Table 2). Moreover, in capsule form, both the BD4 and RAG-1 emulsan polysaccharides prevent cell adhesion to hydrocarbons (Rosenberg, et al., 1983b). How then do the RAG-1 and BD413 polysaccharide emulsans assist in emulsification? When the emulsans are released from the capsular structures, they contain noncovalently linked complexed proteins. In addition, RAG-1 emulsan contains covalently linked fatty acids. The protein component plays a crucial role in the emulsifying activity of the BD4 emulsan. In the case of the RAG-1 emulsan, the protein is not absolutely required for emulsifying activity, presumably because hydrophobic sites, covalently linked fatty acids, contribute significantly to its activity. Characteristic of both emulsans is the fact that pure polysaccharide backbones, lacking hydrophobic fatty acids esters and complexed proteins, showed little or no emulsifying activity. Therefore, the ability of the emulsans to adsorb at hydrocarbon-water interfaces and to form stable interfacial films is due to amphipathic properties derived from a unique combination of a hydrophilic polysaccharide backbone with covalently or noncovalently hydrophobic components.

Biodispersan

In addition to emulsifying agents, certain strains of *Acinetobacter* produce extracellular polymers that disperse inorganic solids, such as limestone or titanium dioxide (Rosenberg, et al., 1988, 1988b). These materials, referred to as biodispersans, are hydrophilic and are unable to emulsify oil in water. Biodispersan is produced during the stationary phase growth of *A. calcoaceticus* A2, and yields as high as 12 g l^{-1} have been obtained in large fermentors.

The purified biodispersan of strain A2 had an average molecular weight of 51,400. Titration of the polymer gave two inflection points: $pK_1 = 3.1$ (1.15 fEq/mg) and $pK_2 = 8.0$ (0.4 fEq/mg). Strong acid hydrolysis of the purified polymer yielded four different monosaccharides: Glucosamine, galactosamine uronic acid, a 6–methyl aminohexose and an unidentified amino sugar. ^{13}C-nuclear magnetic resonance spectrophotometry indicated that the polymer consisted of a repeating tetrasaccharide, partially N-acetylated.

The dispersing power of biodispersan is illustrated in Figure 4. In the absence of the dispersing agent, milled limestone formed flocs in water of 5–40 fm in diameter, which settled rapidly. When biodispersan was added, at a weight ratio of 1:2000 of limestone, the flocs were dispersed into particles of 1–4 fm diameter, with a concomitant large decrease in the rate of settling. Biodispersan was also effective in dispersing titanium dioxide and certain clays. The dispersion of a 10% slurry of limestone in water (pH 9.1) was inhibited by lowering the pH (50% inhibition at pH 8.0 and complete inhibition at pH 6.7). Although biodispersan induced dispersion were not affected by monovalent anions and cations, a 50% inhibition was obtained with 2 mM phosphate or 8 mM magnesium ions.

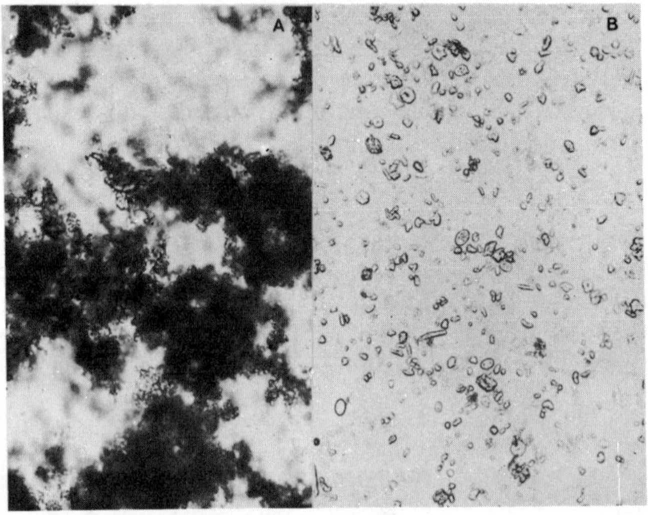

Figure 4. Phase micrograph of a 10% suspension of limestone powder in the absence (A) and presence of 50 fg ml^{-1} Biodispersan (B). Magnification × 750.

Use of Biodispersan in Grinding Limestone and Making Paper

Calcium carbonate (limestone) constitutes a large fraction of the weight of paper produced by the modern, neutral/alkaline process (Hagermeyer, 1984). The quality of the paper is controlled by the size and shape of the limestone particles. The major cost in preparing fine and ultra-fine limestone powder is grinding. When used as a filler, it must be ground so that it contains a uniform fineness of approximately 70% smaller than 2 μ. Commercially available chemical dispersants and grinding aids are generally not used in this process because they decrease the quality of the paper by interfering with the floculation reaction used for retention of limestone. In addition, the presence of chemical dispersants creates a problem in recycling the water, as well as a potential ecological waste hazard.

In collaboration with the American-Israel Paper Mills Ltd, Hadera, Israel, we tested the use of Biodispersan in manufacturing paper. During the course of this investigation it was discovered that Biodispersan is not only a surfactant, but also is a *surfractant*, aiding in the fracturing of limestone during the grinding process (Figure 5). The grinding time was reduced with increasing concentrations of Biodispersan. For example, time required for 50% size reduction was decreased from 6 hr in the control to 4 hr and 3 hr in the presence of 0.05% and 0.1% Biodispersan, respectively. Grinding for 9 hr in the presence of Biodispersan resulted in a large decrease in mean diameter, standard deviation and percentage of limestone paricles under 2 fm (Table 4). Commercially available dispersants, such as Dispex N-40 and Cataflot P-

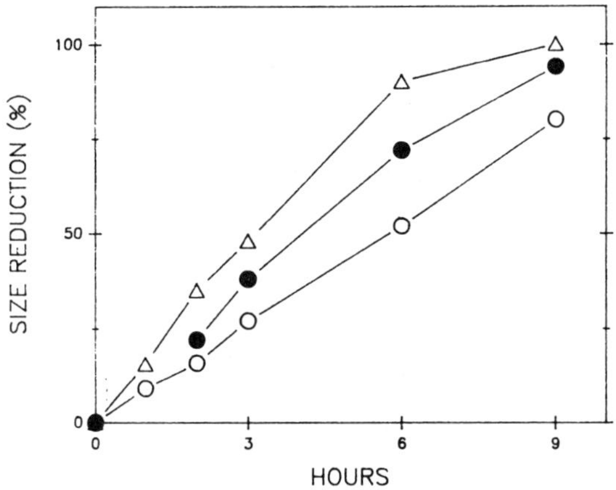

Figure 5. Biodispersan as a grinding aid. Limestone (200 g) was mixed with 200 ml water, containing 200 mg Biodispersan (△), 100 mg Biodispersan (●) or no Biodispersan (○). After different periods of grinding, samples were removed and the extent of particle size reduction was estimated. The ordinate is the percent size reduction relative to that achieved with 0.1% Biodispersan at 9 h.

40, had no effect on the grinding of limestone under the conditions used in this experiment.

For use in making paper, limestone was ground for 8 hr in a jar mill with Biodispersan (1:1000), Dispex N-40 (1:500), or without any dispersant (control). The average particle size at the end of the grinding was 3.1 μ for the control and limestone ground with Dispex N-40. In contrast, the average particle size of Biodispersan-ground limestone was 1.7 μ. In addition to the smaller particle size of the Biodispersan-ground limestone, the size distribution of the particle was more homogeneous, as indicated by its lower standard deviation. Using the above samples of ground limestone as filler, several small batches of paper hand sheets were produced by the standard procedures of the Hadera Paper Mill. Typical analyses of paper made with three different limestone powders are shown in Table 5. Biodispersan improved retention of the limestone and produced a paper with a higher opacity and brightness than the control. The addition of Dispex N-40 interfered with retention of

Table 4. Particle size distribution of limestone milled with increasing concentrations of Biodispersan[a]

Biodispersan (mg ml^{-1})	Mean diameter ± S.D. (μm)	Particles < 2 μm (%)
0	2.7 ± 2.7	62
0.05	2.3 ± 2.2	68
0.1	2.2 ± 2.1	69
0.2	2.1 ± 2.1	71

[a] Adapted from Rosenberg et al., 1989. Limestone (10% water) was ground for 9 h in a jar mill. Particle size was carried out electronically, in a Laser Galai-cis-1 computerized inspection system.

Table 5. Biodispersan in making paper[a]

Parameter	Control	Biodispersan[b]	Dispex[c]
Particle size (average)	3.1 μm	1.7 μm	3.1 μm
$CaCO_3$ retained in paper	70%	79%	55%
Opacity	86%	88%	84%
Brightness	86%	88%	85%

[a] Taken from Rosenberg et al., 1989.
[b] Biodispersan was added at a ratio of 1 part per 1000 parts of limestone at the time of grinding.
[c] Dispex N-40 was used at 2 parts per 1000 parts limestone.

the limestone and decreased the quality of the paper, as indicated by all the parameters that were studied.

The mechanism by which Biodispersan acts as a grinding aid is unknown. It does not appear to be due exclusively to its dispersing activity because other excellent dispersants, such as Dispex and Cataflot, do not act as grinding aids. It may be that Biodispersan acts as a surfractant by binding to the limestone, diffusing into microfissions and thereby lowering surface energies. In this regard, the effect of pH is interesting. During the grinding, the pH is approx. 9.0; at this pH, Biodispersan contains ionized carboxyl groups and unionized amino groups. When used as a pigment filler for making paper, the ground limestone is diluted into water, pH 7.0. At this pH, the amino groups are largely ionized, and the localized positive charges may be involved in assisting the floculation of the limestone particles onto the negatively charged cellulose fibers. This may help to explain the high retention of Biodispersan-ground limestone in making paper.

Drag-reducing Polymers from Fish Skin Bacteria

Virtually all fish are covered with a mucous secretion. The mucus of some (but not all) fish have been shown to reduce frictional drag in turbulent pipe flow (Rosen and Cornford, 1971). This effect is attributed to the "Toms phenomenon" (Toms, 1948), in which dilute solutions of extended polymers (synthetic and biological) flowing in tubes yield friction-drag coefficients less than values predicted for Newtonian fluids (Kennis, 1968; Hoyt, 1985; Sar and Rosenberg, 1987). Direct measurements of velocity profiles of the water flow over the surface of rainbow trout indicated that the mucus reduced significantly the overall drag experiences by the fish (Daniel, 1981).

We have been actively investigating the question: Are the drag-reducing properties of fish mucus due to the fish secretions or to bacteria which inhabit this ecological niche? To answer this question, bacteria were isolated from the skin of freshly captured, healthy (barracuda-like) *Sphyraena sphyraena* and tuna from the Mediterranean Sea and from *Fistularia* sp. (pipe fish) from the Red Sea. Two interesting classes of bacteria were obtained in pure culture. The first group (mainly from the tuna) produced lawns on agar

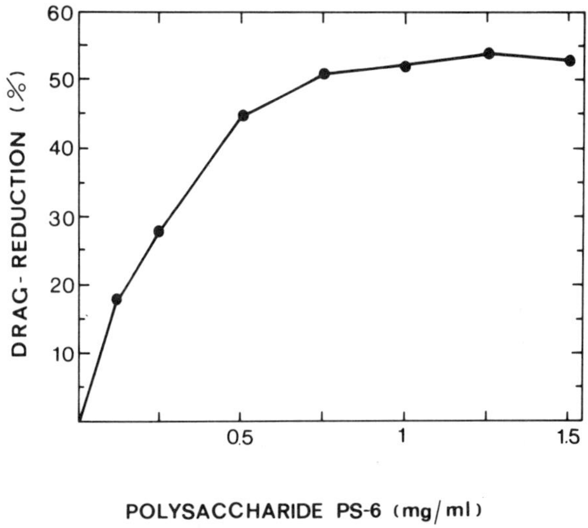

Figure 6. Drag reduction by the purified mucopolysaccharide from a *Vibrio* sp. Drag reduction was determined in a turbulent flow rheometer.

medium which were as hydrophobic as polystyrene (Sar and Rosenberg, 1987). The high colonial hydrophobicity may be advantageous for both the bacteria (by helping it adhere to the skin) and the fish (by further smoothing the skin surface). It is known that fricational drag reduction can be achieved by minimizing the surface area (Webb, 1978).

The second group of fish skin bacteria produced drag-reducing polymers (Sar and Rosenberg, 1989). The synthesis and properties of one of these polymers, produced by *Vibrio* sp. that was isolated from *S. sphyraena*, has been studied. Acid-base titration of the deproteinized polymer gave inflection points at $pK_1 = 3.3$ (2.4 meq g^{-1}) and $pK_2 = 9.7$ (0.8 meq g^{-1}). ^{13}C-NMR spectroscopy resolved 25 peaks, including three methyl groups, three carboxyl groups and four signals in the anomeric region (99–103 ppm), indicating the presence of four different monosaccharides. Acid hydrolysis yielded an amino acid, pyruvate, hexosamine, uronic acid, glucose and galactose. The purified polymer had an intrinsic viscosity of 1280 $cm^3\ g^{-1}$ and reduced frictional drag of water in a turbulent flow (Reynolds number of 17,000) rheometer by 55% (Figure 6).

It is interesting to note that certain intact bacteria (no extracellular polymer) are extremely effective in reducing frictional drag. *A. calcoaceticus* BD4 at a density of 3.6×10^9 cells ml^{-1} reduced drag by 55% (Figure 7). This drag reduction was due to the tightly bound capsule (0.4 mg ml^{-1}). Capsule-deficient mutants of BD4 failed to reduce drag. The cell-bound polysaccharide had a three-fold higher drag reducing activity than the purified polymer which was free in solution.

Recently, we have isolated a number of capsule-containing bacteria from

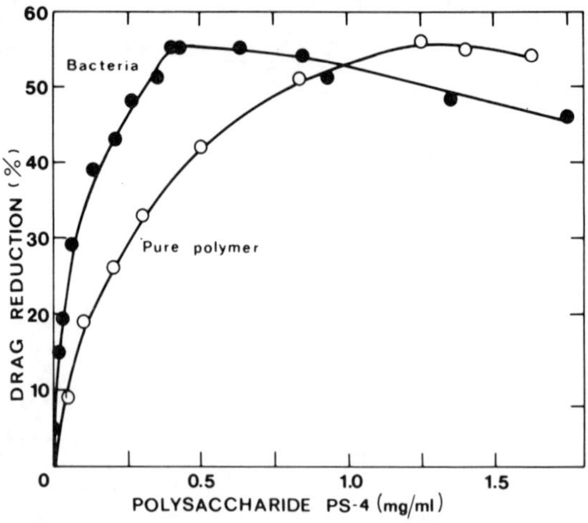

Figure 7. Drag reduction by the capsulated bacterium *A. calcoaceticus* BD4 (●) and the purified capsule polysaccharides PS-4 (○). Drag reduction was determined in a turbulent flow rheometer.

the pipe fish which are able to grow on fish mucus and demonstrate extremely high drag reduction at low concentrations (Bernadsky and Rosenberg, unpublished). Although the circumstantial evidence points to these drag-reducing bacteria being responsible for the reduction in frictional drag of fish mucus, the definitive experiments have not yet been performed.

Conclusions

Bacteria produce a wide variety of surface active and drag- reducing exopolymers. Table 6 summarizes the polymers that were presented in this article. *A. calcoaceticus* RAG-1 and BD4 produced capsular polysaccharides of M.W.

Table 6. Surface active and drag reducing bacterial polymers

Bacterial polymer	Polymer	Properties	Potential applications
A. calcoaceticus RAG-1 emulsan	Polysaccharide: lipid	Stabilizes oil-in-water emulsions; Binds cations	Oil cleaning Viscosity reduction UO_2^{++} concentration General emulsifier
A. calcoaceticus BD4 emulsan	Polysaccharide proetin complex	As above	As above Drag-reduction
A. calcoaceticus A2 biodispersan	Polysaccharide	Dispersant Surfactant	Dispersant grinding aid
Vibrio sp. polymer	Mucopolysaccharide	Drag-reduction Delay turbulence	Energy saving Faster speeds

around 10^6 that are released into the medium during unbalanced growth. The hydrophobic portion of the amphipathic RAG-1 emulsan is fatty acid esters, whereas a specific hydrophobic protein appears to play this role in the BD4 emulsan. The enormous diversity of microbial exopolymers make them an interesting group of biodegradable materials for potential industrial applications. One of the major challenges of this field is to find those particular microbial polymers which have the right combination of properties for specific industrial or medical applications. Emulsans stabilize oil-in-water emulsions and can be used to remove oil from surfaces, reduce the viscosity of heavy oils, concentrate uranyl ions from water, and as general emulsifiers for the food, paint, detergent and cosmetic industries. *A. calcoaceticus* A2 produces a polysaccharide of M.W. 52,400 that binds to limestone, preventing floculation and reduces the energy required to produce finely ground limestone. Bacteria isolated from the skin or mucus of fast swimming fish produced exocellular polymers that are effective in reducing the friction of turbulent water. One of these polymers has been purified and shown to be a high molecular weight mucopolysaccharide. Drag-reducing polymers are potentially useful in reducing the costs of pumping water and increasing the speed of large objects moving through water.

Acknowledgments

Much of the work presented in this paper was carried out jointly with D.L. Gutnick, E.Z. Ron, Z. Zosim, C. Rubinowitz, M. Rosenberg, N. Kaplan, O. Pines, A. Gottlieb, R. Legmann, and N. Sar. The emulsifier work was supported in part by a grant from Petroleum Fermentation, Curacao, Netherlands Antilles. The Biodispersan studies were supported by grants from Mineral Biotech Israel and Israel Chemicals Ltd.

References

Basta, N. 1984. Biopolymers challenge petrochemicals. High Technol. Feb: 67–70.

Belsky, I., Gutnick, D.L. and Rosenberg, E. 1979. Emulsifier of *Arthrobacter* RAG-1: Determination of emulsifier-bound fatty acids. FEBS Lett. 101: 175–178.

Carruthers, M.M. and Anderson, B. 1979. Inhibition by polyanions of adherence of Kangawa-positive *Vibrio parahaemolyticus*: A physiochemical effect. J. Infect. Dis. 140: 119–122.

Cooper, D.G. and Zajic, J.E. 1980. Surface active compounds from microorganisms. Adv. Appl. Microbiol. 26: 229–259.

Daniel, T.L. 1981. Fish mucus: *In situ* measurement of polymer drag reduction. Biol. Bull. 160: 376–382.

Erickson, L.E. and Nakahara, T. 1975. Growth in cultures with two liquid phases: Hydrocarbon uptake and transport. Process. Biochem. 10: 9–13.

Goldman, S., Shabtai, Y., Rubinovitz, C., Rosenberg, E. and Gutnick, D.L. 1982. Emulsan in *Acinetobacter aclcoaceticus* RAG-1: distribution of cell-free and cell-associated cross-reacting material. Appl. Environ. Microbiol. 44: 165–170.

Gutnick, D.L. and Rosenberg, E. 1977. Oil tankers and pollution: A microbiological approach. Ann. Rev. Microbiol. 31: 379–396.

Hagemeyer, R.W. 1984. In: Pigments for Paper. Hagemeyer, R.W., ed., Tappi Press, Atlanta. p. 53.

Hisatsuka, K., Nakahara, T. and Yamada, K. 1972. Protein-like activator for n-alkane oxidation by *Pseudomonas aeruginosa* S_7B_1. Agric. Biol. Chem. 36: 1361–1369.

Hoyt, J.W. 1985. Drag reduction in polysaccharide solutions. Trends Biotechnol. 3: 17–21.

Itoh, S. and Suzuki, T. 1972. Effect of rhamnolipids on growth of *Pseudomonas aeruginosa* mutant deficient in n-paraffin utilizing ability. Agric. Biol. Chem. 36: 2233–2235.

Ito, S. and Inoue, S. 1982. Sophorolipids from *Torulopsis bombicola*: Possible relation to alkane uptake. Appl. Environ. Microbiol. 43: 1278–1283.

Kaplan, N. and Rosenberg, E. 1982. Exopolysaccharide distribution of and bioemulsifier production by *Acinetobacter calcoaceticus* BD4 and BD413. Appl. Environ. Microbiol. 44: 1335–1341.

Kaplan, N., Zosim, Z. and Rosenberg, E. 1987. Reconstitution of emuslsifying activity of *Acinetobacter calcoaceticus* BD4 emulsan by using pure polysaccharide and protein. Appl. Environ. Microbiol. 53: 440–446.

Kappeli, O. and Finnerty, W.R. 1980. Characteristics of hexadecane partition by the growth medium of *Acinetobacter* sp. Biotechnol. Bioeng. 22: 495–503.

Kennis, R.P. 1980. Drag reduction by bacterial metabolites. Nature. 217: 940–942.

Pines, O. 1983. Functional association of emulsan with the bacterial cell surface. Ph.D. Thesis, Tel Aviv University.

Pines, O. and Gutnick, D.L. 1984. Specific binding of a bacteriophage at a hydrocarbon-water interface. J. Bacteriol. 157: 179–183.

Reisfeld, A., Rosenberg, E. and Gutnick, D. 1972. Microbial degradation of crude oil: Factors affecting the dispersion in sea water by mixed and pure cultures. Appl. Environ. Microbiol. 24: 363–368.

Rosen, M.W. and Cornford, N.E. 1971. Fluid friction of fish slimes. Nature. 234: 49–51.

Rosenberg, E., Perry, A., Gibson, D.T. and Gutnick, D.L. 1979. Emulsifier of *Arthrobacter* RAG-1: Specificity of hydrocarbon substrate. Appl. Environ. Microbiol. 37: 409–413.

Rosenberg, M. and Rosenberg, R. 1981. Role of adherence in growth of *Acinetobacter calcoaceticus* RAG-1 on hexadecane. J. Bacteriol. 148: 51–57.

Rosenberg, E., Gottlieb, A. and Rosenberg, M. 1983. Inhibition of bacterial adherence of hydrocarbons and epithelial cells by emulsan. Infect. & Immun. 39: 1024–1028.

Rosenberg, E., Kaplan, N., Pines, O., Rosenberg, M. and Gutnick, D. 1983b. Capsular polysaccharides interfere with adherence of *Acinetobacter calcoaceticus* to hydrocarbon. FEMS Lett. 17: 157–161.

Rosenberg, E. 1986. Microbial surfactants. CRC Crit. Rev. Microbiol. 3: 109–132.

Rosenberg, E. and Kaplan, N. 1986. Surface-active properties of *Acinetobacter* exopolysaccharides. In: Bacterial Outer Membranes as Model Systems. Inouye, M., ed., John Wiley & Sons, New York. pp. 311–342.

Rosenberg, E., Rubinovitz, C., Gottlieb, A., Rosenhak, S. and Ron, E.Z. 1988. Production of biodispersan by *Acinetobacter calcoaceticus* A2. Appl. Environ. Microbiol. 54: 317–322.

Rosenberg, E., Rubinovitz, C., Legmann, R. and Ron, E.Z. 1988b. Purification and chemical properties of *Acinetobacter calcoaceticus* A2 Biodispersan. Appl. Environ. Microbiol. 54: 323–326.

Rosenberg, E., Schwartz, Z., Tennenbaum, A., Rubinovitz, C., Legmann, R. and Ron, E.Z. 1989. A microbial polymer that changes the surface properties of limestone: Effect of Biodispersan in grinding limestone and making paper. J. Dispersion Sci. & Technol. 10: 241–250.

Rubinovitz, C., Gutnick, D.L. and Rosenberg, E. 1982. Emulsan production by *Acinetobacter calcoaceticus* RAG-92 in the presence of chloramphenicol. J. Bacteriol. 152: 126–132.

Sar, N. and Rosenberg, E. 1983. Emulsifier production by *Acinetobacter calcoaceticus* strains. Curr. Microbiol. 9: 309–314.

Sar, N. and Rosenberg, E. 1987. Fish skin bacteria: colonial and cellular hydrophobicity. Microb. Ecol. 13: 193–202.

Sar, N. and Rosenberg, E. 1987. Drag reduction by *Acinetobacter calcoaceticus* BD4. Appl. Environ. Microbiol. 53: 2269–2270.
Sar, N. and Rosenberg, E. 1989. Fish skin bacteria: production of friction-reducing polymers. Microb. Ecol. 17: 27–38.
Shabtai, J., Pines, O. and Gutnick, D. 1985. Emulsan: A case study of microbial capsules as industrial products. Developments in Indust. Microbiol. 26: 291–307.
Shilo, M. 1989. The unique characteristics of benthic cyanobacteria. In: Microbial Mats. Cohen, Y. and E. Rosenberg, eds. ASM, Washington, D.C. pp. 207–213.
Shoham, Y. and Rosenberg, E. 1983. Enzymatic depolymerization of emulsan. J. Bacteriol. 156: 161–167.
Singer, M.E. and Finnerty, W.R. 1984. Microbial metabolism of straight-chain and branched alkanes. In: Petroleum Microbiology. Atlas, R.M., ed. Macmillan, New York. pp. 1–59.
Taylor, W.H. and Juni, E. 1961. Pathways for biosynthesis of a bacterial capsular polysaccharide: Characterization of the organism and polysaccharide. J. Bacteriol. 81: 688–693.
Toms, B.A. 1948. Some observations on the flow of linear polymer solutions through straight tubes at large Reynolds numbers. Proc. Intern. Rheol. Cong., Holland. 2: 135–141.
Webb, W.P. 1978. Hydrodynamics: Nonscombroid fish. In: Fish Physiology VII. Hoari, W.S. and Randall, D.J., eds. Academic Press, New York. pp. 189–237.
Zajic, J.E. and Panchal, C.J. 1976. Bioemulsifiers. CRC Cit. Rev. Microbiol. 5: 39–66.
Zajic, J.E. and Seffens, W. 1984. Biosurfactants. CRC Crit. Rev. Microbiol. 1: 87–101.
Zajic, J.E. and Mahomedy, A.Y. 1984. Biosurfactants: Intermediates in the biosynthesis of amphipathic molecules in microbes. In: Petroleum Microbiology. Atlas, R.M., ed., Macmillan, New York. pp. 221–298.
Zosim, Z., Gutnick, D. and Rosenberg, E. 1982. Properties of hydrocarbon-in-water emulsions stabilized by *Acinetobacter* RAG-1 emulsan. Biotechnol. Bioeng. 24: 281–292.
Zosim, Z., Gutnick, D.L. and Rosenberg, E. 1983. Uranium binding by emulsan and emulsanosols. Biotechnol. Bioeng. 25: 1725–1735.
Zuckerberg, A., Diver, A., Peeri, Z., Gutnick, D.L. and Rosenberg, E. 1979. Emulsifier of *Arthrobacter* RAG-1: Chemical and physical properties. Appl. Environ. Microbiol. 37: 414–420.

Discussion (DR. E. ROSENBERG)

Kornguth: It is questionable whether these compounds have evolved to increase the acceleration of the fish in the water; from your data they would have to be shed at the mg ml^{-1} level. That strikes me as relatively high for what might be produced in a fish swimming in water.

E. Rosenberg: That is a good question. The ones that we are now studying at 2–10 $\mu g\ ml^{-1}$ will reduce friction by as much as 30–40%. Second, almost all of these fish are hunters; the tuna which swims fast its entire life does not have these kinds of bacteria. The ones that seem to have it are the hunters which swim without turbulence for a very short time as they sprint to get their prey. It probably pays only for burst swimmers to use this kind of technique; for those that are continuously swimming fast, I do not think it would pay. I do not know if they are releasing it. I am not so sure it has to be in the water; I am not so sure that on the surface layer it could not decrease turbulence. It could be a wall effect to a large extent. I have talked to some experts, and this is not clear. It is an unknown area, but it is interesting that these bacteria are only found in and on the slime of fast-swimming fish and sprinters. So I think there is a correlation that is interesting.

Ward: Is anything known at all about the biosynthetic pathways of these polysaccharides?

E. Rosenberg: Some of the bacterial pathways are just now being worked on by Kaplan and Gutnick. They have genetic mutations but really nothing is known at the biochemical level. There is a lot known on the physiological level of regulation of the polysaccharide. People are thinking about polysaccharide engineering and so forth in this regard.

Ellis: I was intrigued by the limestone experiments that you have done. Can you give me any better feeling for mechanism? Can there be a local change in pH associated with this, for example, helping to dissolve the solid?

E. Rosenberg: The only thing I can think of that could work as a surfractant to lower energy is what is called the Riebender effect. A Russian physical chemist had indicated that if you have flaws in crystals, polymers that are negatively charged can migrate into these little cracks, producing a double repulsive effect that would lower the energy and more easily break the material. That is the only thing that I know about. What is interesting and unusual is that the biodispersan polymer contains some free amino groups. Therefore when you grind, you are grinding at pH 9. At pH 9 you have a free amino group, but when you make the paper you dilute the material into water at pH 7. The amino now becomes an NH^{3+} which is why we have better flocculation on the paper.

Braun: It was claimed by, for example Atkinson and David (Adv. Biochem. Eng. (1976) 4: 41–124), that the mechanism of bacterial adhesion to hulls of ships is based on high hydrophobicity. Would you comment on this slowing down of ships versus speeding up of tuna by hydrophobic bacteria?

E. Rosenberg: It is an interesting question. A large portion of bacterial adhesion to surfaces is due to hydrophobicity, but certainly not all of it. Probably the initial layer of bacteria is hydrophobic, but in a film you have a lot of glycocalyx and polysaccharide which makes it hydrophilic. So the surface is hydrophilic, while the surface of these tuna bacteria are hydrophobic and have a smoothing effect.

Eisenstadt: Do you know anything about the ability of those polysaccharides to withstand fouling or degradation in the marine environment?

E. Rosenberg: No, I do not.

Key Words: Surfactant, Emulsifier, Dispersant, *Acinetobacter*, Emulsan, Biodispersan, Drag-reduction, Hydrocarbon, Fish bacteria

Towards Catalytic Antibodies for the Degradation of Toxic Agents

Bernard S. Green[1] **and Michael Glikson**[2]
[1]*The Hebrew University, Faculty of Medicine, School of Pharmacy, Department of Pharmaceutical Chemistry, P.O. Box 12065, Jerusalem 91120, Israel*
[2]*Department of Chemical Immunology, The Weizmann Institute of Science, Rehovot 76100, Israel, and Israel Defense Forces Medical Corps*

Introduction

A large variety and number of harmful and toxic agents are released into the environment and represent major health threats (Salkinoja-Salonen, et al., 1989). The destruction of unwanted stores of toxic substances in ways which are environmentally-safe is a problem facing all industrial nations. In addition, ingestion of toxic substances requires therapeutic approaches to detoxify them. The accidental exposure to insecticides or the intentional poisoning by organophosphorus "nerve gases" represent, respectively, common occurrences and an ever-present threat. There are many ways of dealing with these widely different kinds of problems and no single approach will be the panacea for all detoxification problems. The purpose of this paper is to call attention to a radically new and rapidly emerging technology, catalytic antibodies, as a possible approach to detoxification. We review the principles involved in eliciting catalytic antibodies and discuss the application of these ideas to organophosphorus detoxification in more detail.

Several considerations may make the antibody approach attractive relative to alternatives. For example, one of the explanations put forward for the problems in the biodegradation of persistent organic chemicals in soil is the possible low threshold value for biodegradation by the enzymes present in microorganisms. (Salkinoja-Salonen, et al., 1989). The Km values for these enzymes may be insufficient to readily bind the substrate. If true, the tight binding which is the hallmark of antibodies may make catalytic antibodies especially attractive for situations where the substance to be degraded is present in very low concentrations. In such cases, high catalytic efficiency (large k_{cat} values) may not be so important. One might also find advantages for catalytic antibodies in situations where it is undesirable to introduce

D. Kamely et al. (eds), Biotechnology: Bridging Research and Applications, 249–264.

reproducing bacterial systems. The release of new, biotechnologically engineered microorganisms is sometimes problematic and the use of a catalytic antibody may have advantages in such situations. In addition, no source of energy is required for the antibody.

Catalytic Antibodies

Although the idea of catalyzing a chemical reaction using selectively raised antibodies had been considered for many years (Jencks, 1969; Raso and Stollar, 1975; Kohen, et al., 1980), it is only very recently that this concept has been unequivocally established. Initially, stoichiometric reactions within antibody combining sites were demonstrated (Kohen, et al., 1980; Tramontano, et al., 1986), but in the last few years an impressive number of reaction types have been found to be catalyzed by monoclonal antibodies. (For reviews, see Schultz, 1988; Green, 1989; Green and Tawfik, 1989; Schultz, et al., 1990; Shokat and Schultz, 1990; Lerner and Benkovic, 1990).

Monoclonal Antibodies

The single, most significant advance which has propelled catalytic antibody research forward has been the development of hybridoma technology. This involves the fusion of antibody-secreting cells from an immunized animal with myeloma tumor cells to produce large numbers of stable, immortal cell lines (hybridomas) each of which secretes a molecularly homogeneous antibody. The many advantages of this approach over polyclonal mixtures of antibodies from the serum of an immunized animal, include the availability of large amounts of antibody having defined specificity, the reproducibility of conditions (in one's own, or in another laboratory), the reliability of results, and the possibility of readily characterizing, defining, and purifying the antibody. Furthermore, any physical or chemical study of the antibody molecules by methods such as cleavage to smaller binding fragments (Fab), elucidation of the amino acid sequence of the antibody combining site (the complementarity determining regions, CDRs), X-ray crystallography to determine the three-dimensional structure, and site-directed mutagenesis to modify selected amino acids all demand the use of monoclonal antibodies (MABs). Understanding of the mechanism or specificity of an antibody-catalyzed reaction is also facilitated by monoclonal antibodies. The most significant advantage from the point of view of catalysis has been that one can produce hundreds or thousands of individual, defined monoclonal antibodies and, if one can devise a suitable assay for catalysis, test them all for the desired catalytic reactions. The advantages of large numbers–the immune system is capable of producing hundreds of millions of different antibodies–is especially striking when compared to the total number of known enzymes (estimated at about 1,500 (Shokat and Schultz, 1990)) and provides confidence that many, if not all, catalytic properties as well as binding properties may be found in an individual antibody molecule.

Approaches to Catalytic Antibodies

Several different approaches may be taken to elicit catalytic antibodies and examples of most of these directions have been explored. The majority have led to at least some degree of success; the use of transition state analogs has resulted in remarkable advances.

Transition State Analogs as Haptens

The most straight-forward method thus far for obtaining catalytic antibodies is based on the Pauling postulate (Pauling, 1948), derived from transition state theory, which ascribes enzyme catalysis to the complementarity of the enzyme to the transition state, not the substrate, of the catalyzed reaction. Indeed, the degree of catalysis, k_{cat}/k_{uncat}, (the ratio of the rate of the catalyzed reaction to that of the uncatalyzed reaction) can be shown to be proportional to the affinity constants of the enzyme for the transition state (K_t) and the substrate (K_s), respectively (Fersht, 1985; Kraut, 1988): $k_{cat}/k_{uncat} = K_t / K_s$.

Because of their fleeting existence at an energy peak along the reaction pathway, transition states cannot be isolated. However, stable substances which mimic the size, shape, and charge distribution of a given transition state have been synthesized and, as expected, bind tightly to enzymes which catalyze the reaction proceeding via that transition state; such substances are termed transition state analogs. If enzyme binding to the transition state reduces the activation energy for the chemical reaction involving that transition state, it may be expected that an antibody which binds strongly to a given transition state analog will catalyze the reaction or reactions involving the transition state which is mimicked by that analog (Jencks, 1969).

The phosphonate and phosphonamidate functional groups are established, excellent transition state analog inhibitors for ester and amide hydrolyzing enzymes. Monoclonal antibodies raised against haptens incorporating these phosphorus-containing elements were found to act as selective catalysts for substrates having structures analogous to those of the transition state analogs (Tramontano, et al., 1988; Janda, et al., 1988; Shokat, et al., 1989; Pollack and Schultz, 1987; Tawfik, et al., 1990). These reactions display moderate to high rate accelerations, are substrate-specific, and are specifically inhibited by the immunizing hapten.

For example, MABs raised against the hapten **1** catalyzed hydrolysis of the corresponding ester **2**; the tetrahedral, charged transition state **3** is closely

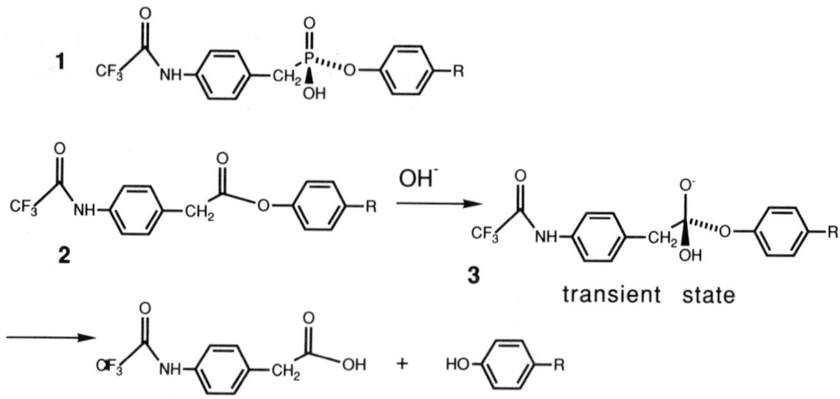

Scheme 1.

mimicked by hapten **1**, the phosphonic acid being ionized at physiological pH (Tramontano, et al., 1988).

The rate enhancement over background hydrolysis for the reaction of **2** was ca. 6×10^6. In this and in several other MAB-catalyzed reactions, the antibody is clearly doing more than simply stabilizing the transition state; acid-base catalysis is undoubtedly also involved.

Haptens which are expected to elicit antibodies capable of transition state stabilization have also been synthesized and used successfully in cyclization reactions (Napper, et al., 1987), a Claisen rearrangement (Jackson, et al., 1988; Hilvert and Nared, 1988; Hilvert, et al., 1989a), and Diels-Alder reactions (Hilvert, et al., 1989b; Braisted and Schultz, 1990). The latter illustrate the possibility of effectively using catalytic monoclonal antibodies (catMABs) for reactions which are probably not catalyzed by enzymes.

Substrate Analogs Incorporating Functional Groups Complementary to Those Desired as Haptens

The idea here is to decide which functional groups within the CDR would be effective in catalyzing a desired reaction and then to include in the hapten appropriately placed complementary functional groups. This haptenic functionality will tend to select for the desired complementary functional groups in the antibody. For example, placement of hydrogen donors, apolar groups, carboxylic acids, etc., in the CDRs of at least some of the produced MABs should be accomplished by incorporating hydrogen acceptors, apolar groups, amine functions, etc., respectively, in the haptens used to elicit the antibodies. This approach was used for the strategic placement of a carboxylate in a MAB which subsequently abstracted a proton from a structurally related substrate and led to elimination of hydrogen halide and formation of a double bond (Shokat, et al., 1989). Thus, hapten **4** generated MABs

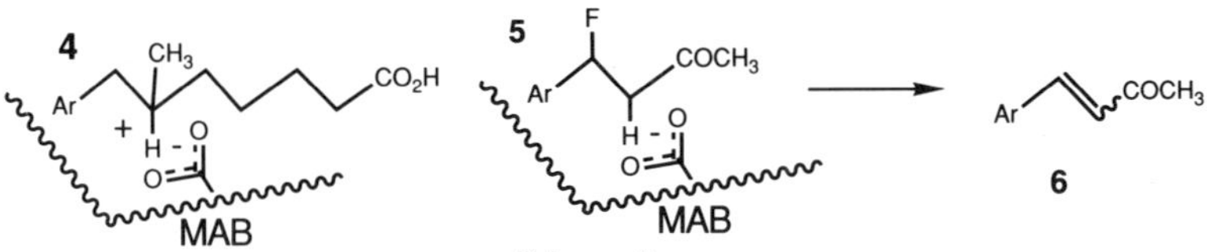

Scheme 2.

which catalyzed the β-elimination of **5** to yield **6** (stereochemistry not established).

The interesting generation of ester-hydrolyzing catMABs was also achieved using the cationic hapten **7** which combines a simple secondary alcohol to mimic the tetrahedral transition state geometry and the cationic group which presumably generated a complementary carboxylate group on the MAB. High rate accelerations were observed (ca. 10^6) for hydrolysis of the ester **8** at nearly neutral pH (Janda, et al., 1990).

Haptens Eliciting Combining Sites which Recognize Two Different Molecules or Two Different Regions of a Molecule and Act as a Template to Favor Reaction

Since tight, selective binding is one of the hallmarks of an antibody, it would seem possible to use antibodies raised against a molecule $A \times B$ for the binding of the separate entities, A and B, in order to carry out the reaction: $A + B \rightarrow A - B$. One of the first approaches towards catalytic antibodies included this idea (Raso and Stollar, 1988). A photochemical dimerization reaction has been achieved using this approach (Balan, et al., 1988). Anti-fluorescein (**9**) antibodies might be expected to have combining sites which can bind two molecules, a xanthenoid **10** and negatively charged species such as sulfite **11** (Janjic and Tramontano, 1989) and mediate a reaction between them.

The recently reported Diels-Alder reactions (Hilvert, et al., 1989b; Braisted and Schultz, 1990) may also be viewed as examples of $A + B \rightarrow AB$ reactions proceeding within an antibody template.

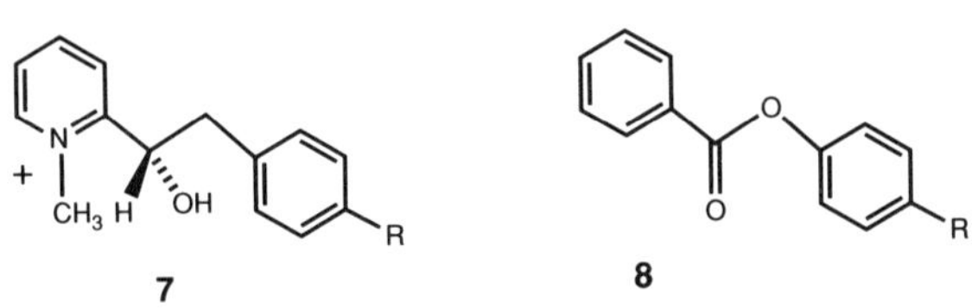

Scheme 3.

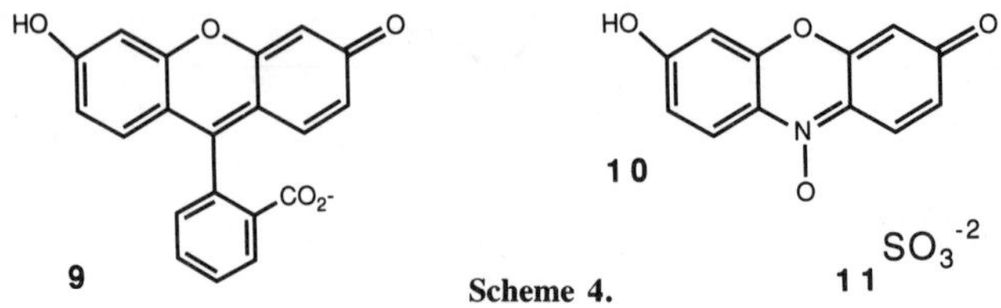

9 10 11

Scheme 4.

Covalent Linking of a Catalytic Function At or Near the CDR

The combination of antibody specificity and inorganic or other chemical catalysts has been achieved in semisynthetic catalytic antibodies, where the antibody is chemically derived at or near its combining site (Pollack and Schultz, 1989). Unlike site-directed mutagenesis, this approach does not require detailed information of the antibody structure or amino acid sequence.

Haptens that Elicit CDRs which Bind Substrate Plus Cofactor

Since many enzymes use cofactors to catalyze reactions, the incorporation of cofactors into catalytic antibodies may be expected to expand the scope and efficiency of the catalyzed reactions. Pyridoxal cofactor was attempted many years ago (Raso and Stollar, 1975). The use of a stable Co(III) peptide complex to elicit antibodies gave some of the most promising metal ion cofactor chemistry: When peptides related to the hapten were exposed to the MABs in the presence of a metal complex, specific peptide cleavage was observed (Janda, et al., 1988).

Seek Naturally Occurring Antibodies

A sensitive and reliable assay is critical in order to detect such antibodies. Paul and coworkers have used radiolabelled VIP (the 28 amino acid vasoactive intestinal peptide) to discover that serum from patients with autoimmune disease may produce catalytic antibodies with extraordinarily high catalytic activity (Paul, et al., 1989). It would be very interesting to seek further examples of such catalytic antibodies in humans and learn whether they have any function. It will be especially intriguing to elucidate the mechanism of this peptidase reaction.

Site-Directed Mutagenesis to Introduce a Catalytically Active Amino Acid into an Antibody Combining Site

By selectively changing an appropriate amino acid in an antibody combining site, conventional binding antibodies may be converted into catalytic antibodies. This requires at least some knowledge of the CDR residues positioned at the substrate bond to be reacted. By substituting His for Tyr^{34} in the variable region of the light chain, in the well-studied dinitrophenol-binding MAB, MOPC315, and forming an Fv fragment, appreciable catalytic activity was introduced (Baldwin and Schultz, 1989).

Anti-Idiotypic Anti-Enzyme Antibodies

This idea has probably occurred to many: in the same way that antibodies raised against anti-receptor antibodies may mimic the receptor, antibodies raised against antibodies which are specific for the active site of an enzyme may display a measure of enzyme activity. The concept is fascinating but undoubtedly difficult in practice.

Select Catalytic Antibodies from Large Numbers of Randomly Produced Antibodies

One of the major achievements in antibody production has been the "combinatorial library approach" in which libraries of the gene segment which code for the heavy- and light-chains of Fab fragments are randomly combined so as to express, in principle, all possible CDRs present in the genetic repertoire of the mouse (Huse, et al., 1989). These are expressed in bacteria and can thus be generated rapidly, in large amounts, and further modified (e.g., random or site-directed mutagenesis). The major problem lies in the development of general screens for selecting the bacterial clones expressing the desired antibody activity but this and related methods provide a powerful approach to antibody catalysts.

"Short" Transition State Analog Approach

The use of monoclonal antibodies as catalysts requires an efficient screening procedure in order to select the hybridomas which secrete catMABs out of the much larger number of clones secreting antibodies which only bind the hapten used for immunization. A very general approach for obtaining antibodies with desired properties involves the design of molecules which differ from the immunizing haptens in structure such that one can select for a desired property or reject an unwanted property by testing all of the

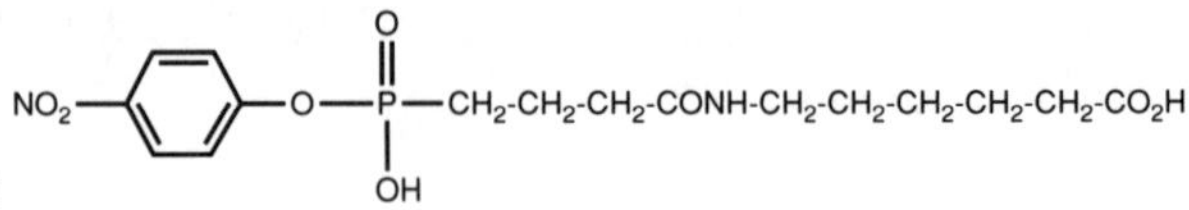

12

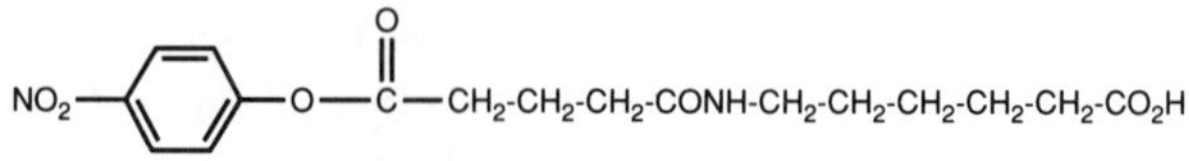

13

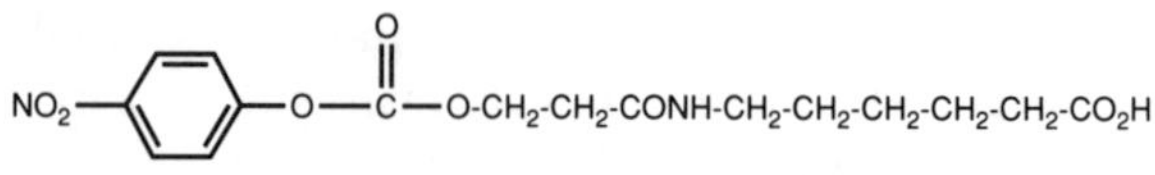

14

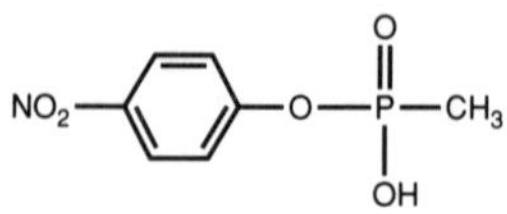

15

Scheme 5.

antibodies for binding to these substances. The assay of MAB binding is rapidly and easily performed using ELISA by measuring the competitive inhibition of binding by the test substances and the hapten. A simple method used in our laboratory for selecting catMABs, based on this approach, has been the use of "short transition state analogs" (Tawfik, et al., 1990). Such substances are truncated analogs of the immunizing hapten (the transition state analog, TSA, for the reaction to be catalyzed by the antibody) which contain all the unique elements of the transition state. Such molecules should bind strongly to the MAB and their structures should minimize the elements which are common to substrate and TS (by the earlier definition of catalysis: strong binding to the TS; weak binding to the substrate) and elements common to TS and products, ensuring that the catMABs display turnover; minimization of product inhibition.

For example, MABs raised against the hapten **12** were expected to catalyze the hydrolysis of the analogous ester **13** and carbonate **14**. By selecting those MABs which bind most tightly to the short TSA **15** (from all the large number of MABs which bind tightly to the hapten **12**), a number of effective catMABs were identified directly, all of which exhibited significant rate enhancement ($k_{cat}/k_{uncat} \sim 10^4$) for hydrolysis of ester **13** as well as substrate specificity and turnover.

Problems in Eliciting Catalytic Antibodies

The above suffices to demonstrate the broad application of antibodies as potential catalysts for essentially any chemical reaction. What must be done in order to achieve this in practice? The three main problems are: 1) hapten design and synthesis; 2) screening of MABs; and 3) catalytic efficiency.

Design and Synthesis of Haptens

Relatively few types of haptenic structures have been reported as effective producers of catalytic antibodies. This is undoubtedly due to the newness of this field and the relatively few laboratories thus far involved in catalytic antibody research. The transition state structures of some reactions, such as ester or amide hydrolysis, are very well established, while the structures of rate-determining transition states for other reactions are poorly understood. The synthesis of stable molecules which accurately mimic the inherently non-isolable transition state structures represents a formidable problem and it may be anticipated that considerable synthetic organic ingenuity will be devoted to the successful design and synthesis of sophisticated transition state analogs for effective catalytic antibody production.

Screening for Catalytic Activity

The short TSA approach (Tawfik, et al., 1990) represents one way of reducing the number of antibody-secreting clones that must be examined, but this relies on the highly questionable premise that the TS analog closely mimics the true TS. The ideal approach would be to screen the large number of hybridomas (or the much larger number of bacteria containing the combinatorial repertoire of possible CDRs) directly for catalytic activity. This will require novel approaches (Pollack, et al, 1989) and must be able to detect catalytic activity with high sensitivity and selectivity.

Catalytic Antibody Efficiency

Although several examples of high catalytic activity have been reported (Tramontano, et al., 1988; Janda, et al., 1990), the reactions involve relatively labile substrates. The majority of antibody-catalyzed reactions thus far reported have rate-enhancement values of 10^2–10^4, which is quite weak in comparison to enzymes. Again, this field is still very young, and the development of more effective haptens, as well as the combination of more than one element or approach used thus far (e.g., transition state stabilization; cofactor; site directed mutagenesis; etc.), may soon provide for more effective catMABs for more challenging reactions, such as peptide bond hydrolysis.

Toxic Agent Decontamination

The use of catMABs may be considered for a variety of harmful substances. Some of the challenging applications include the detoxification of anti-cholin-

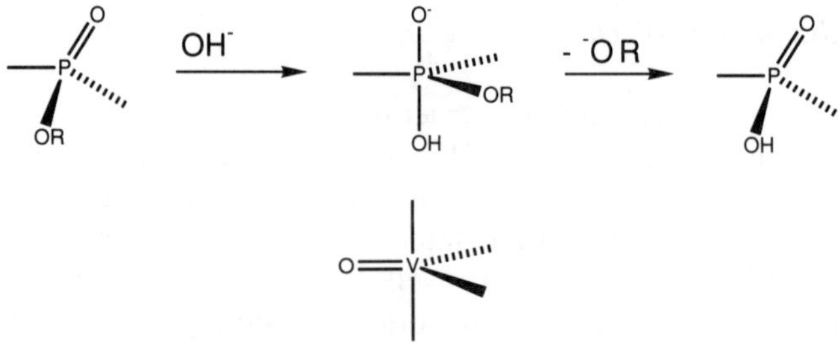

Scheme 6.

esterase organophosphate poisons, such as soman, or of peptide or protein toxins, such as ricin. In order to raise antibodies for such applications, suitable haptens have to be devised. We have considered vanadium compounds as transition state analogs to mimic the trigonal bipyramide transition state involved in phosphate or phosphonate hydrolysis. Although the synthesis of appropriate chemically and configurationally stable trigonal bipyramide vanadium transition state analogs has not yet been successful, we have synthesized a number of ground state analogs of soman and raised MABs against these, some of which display anti-soman binding activity (Glikson, et.al., to be published). The synthesis of haptens effective in eliciting peptide-hydrolyzing catMABs is one of the major challenges facing the catalytic antibody field and many directions toward this goal are being investigated today.

A more detailed discussion of the factors involved in the use of catMABs might benefit from the discussion of a specific example of a problem, such as the toxic organophosphate compounds, especially those used as chemical warfare agents.

Potential Systemic Use of Catalytic Antibodies for Organophosphate Intoxication

Organophosphate intoxication involves the irreversible inhibition of the enzyme acetylcholinesterase and results in a fatal outcome, unless treatment is started very shortly after the intoxication (Hirshberg and Lerman, 1984). The nerve agent soman is especially troublesome, due to rapid "aging" of the inhibited enzyme, which makes it resistant to any known dephosphorylating medication (Dunn and Sidell, 1989). These difficulties in medical treatment, as well as the expected difficulty of providing adequate treatment in cases of mass casualties, make the suggestion of immunizing persons at high risk of exposure to organophosphates very attractive. Antibodies may also serve as adjuvants to post-exposure pharmacological antidotal treatment.

The three potential uses of antibodies in toxicology include: 1) active, 2) passive pre-exposure immunization and 3) post-exposure immunotoxicotherapy. This last method is commonly used in other intoxications, including Digitalis overdose and snake envenomation (Scherrmann, et al., 1989).

Passive Immunization

In research performed with small animals, various non-toxic analogs of organophosphates were used, linked to protein carriers, to raise protective antibodies (Lenz, et al., 1984; Brimfield, et al., 1985; Buenafe and Rittenberg, 1987). Polyclonal as well as monoclonal antibodies were produced but the results have thus far been disappointing with regard to protection of animals passively immunized with these antibodies (Dunn and Sidell, 1989; Lenz, et al., 1984). The two main problems were a) low immunogenicity of the analogs, resulting in low affinity antibodies (Lenz, et al., 1984; Brimfield, et al., 1985); and b) when calculated stoichiometrically, a large amount of antibody is required to neutralize a lethal dose of the toxic agent. For example, in order to bind 1 mg of soman, which is approximately the lethal human dose, more than 400 mg of specific antibody would be required, assuming 100% binding.

Another potential problem involves the reversible formation of organophosphate-antibody complexes which may retard the natural elimination of the toxin by renal excretion or by other esterases in the body. This may therefore create a circulating depot of the poison which cannot be excreted by the kidneys and may be a source for later recurrence of the intoxication.

In principle, all of these problems may be solved by using catalytic antibodies. Increasing the efficacy of organophosphate neutralization several hundredfold by antibody-mediated catalysis would compensate for the low affinity and enable one to use a much lower antibody dose. It may also prevent depot formation due to rapid degradation of the toxic compound after binding to the antibody.

In spite of these theoretical advantages of the catalytic approach, some general problems regarding the use of antibodies in humans remains. The most important is the use of heterologous antibody which induces various acute and subacute hypersensitivity reactions which are more severe on repeated administration (Dillman, et al., 1986). Although heterologous antibodies are in use for various malignant diseases (Dillman, 1989) and in cases of severe intoxications (Scherrmann, et al., 1989), its widespread use in large populations may impose a serious problem, and almost certainly will limit its use to a single administration. This cardinal problem may be solved by various modifications of the antibody, such as using the smaller Fab fragments (Scherrmann, et al., 1989), human monoclonal (Gaffar, et al., 1986), or chimeric antibodies (Morrison, 1985).

Active Immunization

Inducing active immunization against organophosphates in humans is more complicated due to limitations on the use of adjuvants and protein carriers. Moreover, the high titer of antibodies necessary to bind poison is even more difficult to achieve by active immunization. Eliciting catalytic antibody activity by active immunization in an experimental animal has not yet been reported. However, after finding an especially effective transition state analog for inducing organophosphate-hydrolyzing catalytic antibodies, its ability to induce active immunization should also be assessed. Once this has been established and studied in an animal system, its administration to humans may be planned, using antigen presentation systems compatible with human use.

Immunotoxicotherapy

This is achieved by administering antibodies as soon as possible after intoxication in order to bind (or to catalyze destruction of) the toxin and neutralize it. This is the principle underlying current treatment of massive Digitalis overdose as well as various envenomations (Scherrmann, et al., 1989; Smith, et al., 1976). This concept may be problematic when applied to organophosphates due to the very short time interval between intoxication and irreversible binding to cholinesterase (Dunn and Sidell, 1989). Therefore, antibodies would have to be administered very soon, within minutes, after exposure. Nevertheless, antibodies may serve as adjuvant therapy following pharmacologic antidotes, and catalytic antibodies may have all the same advantages over binding antibodies as detailed under "passive therapy" above.

In conclusion, many drawbacks of the immunological approach to organophosphate poisoning may be solved if effective catalytic antibodies to these compounds can be produced. Once such antibodies are available, various additional improvements can also be attempted, such as directing the antibodies to the more toxic stereoisomer.

Conclusions

Elimination of unwanted or harmful materials is the basic job for which antibodies were created and it is therefore quite appropriate that we should be considering catalytic antibodies for improved methods of detoxification. Some advantages might be the mild nature and the high specificity of such reactions in comparison to conventional decontamination agents. As a novel approach for immunotherapy, catalytic antibodies may provide significant advantages over conventional immunotherapy in that lower doses and higher reaction rates may be possible.

We anticipate that future developments in hapten design and synthesis and other areas of catalytic antibody research will lead to effective catalysts for detoxification of specific toxins as well as more general, class-specific degradation catalysts. For example, the recent reports of MABs which bind metalloporphyrins (Schwabacher, et al., 1989; Keinan, et al., 1990) may lead to catalytic antibodies which mimic the important P450–type enzymic detoxification reactions.

Acknowledgments

We thank Prof. Yehuda Danon, Surgeon General of the Israel Defense Forces, for his warm support and interest.

References

Balan, A., Doctor, B.P., Green, B.S., Torten, M., Ziffer, H. 1988. Antibody combining sites as templates for promoting selective organic chemical reactions. J.C.S. Chem. Comm. pp. 106–108.

Baldwin, E., Schultz, P.G. 1989. Generation of a catalytic antibody by site-directed mutagenesis. Science 245: 1104–1107.

Braisted, A.C., Schultz, P.G. 1990. An antibody-catalyzed biomolecular Diels-Alder Reaction. J. Am. Chem. Soc. 112: 7430–7431.

Brimfield, A.A., Hunter, K.W., Lenz, D.E., et al. 1985. Structural and stereochemical specificity of mouse monoclonal antibodies to soman. Mol. Pharmacol. 28: 32–39.

Buenafe, A.C. and Rittenberg, M.B. 1987. Combining site specificity of monoclonal antibodies to the organophosphate hapten soman. Mol. Immunol. 24: 401–407.

Cochran, A.G., Sugasawara, R., Schultz, P.G. 1988. Photosensitized cleavage of a thymine dimer by an antibody. 110: 7888–7890.

Dillman, R.O. 1989. Monoclonal antibodies for treating cancer. Ann. Int. Med. 111: 592–603.

Dillman, R.O., Beauregard, J.C., Halpern, S.E., Clutter, M. 1986. Toxicities and side effects associated with intravenous infusions of murine monoclonal antibodies. J. Biol. Response Mod. 5: 73–84.

Dunn, M.A. and Sidell, F.R. 1989. Progress in medical defense against nerve agents. J. Amer. Med. Assoc. 262: 649–463.

Gaffar, S.A., Royston, I., Glassy, M.C. 1986. Strategies for the design and use of tumor reactive human monoclonal antibodies. BioEssays 4: 119–123.

Green, B.S. 1989. Monoclonal antibodies as catalysts and templates for organic chemical reactions. In: Advances in Biotechnological Processes. Mizrachi, A., ed. Vol. 7. pp. 359–393.

Green, B.S. and Tawfik, T.S. 1989. Catalytic monoclonal antibodies; tailor-made, enzyme-like catalysts for chemical reactions. Trends in Biotech. 7: 304–310.

Hirshberg, A. and Lerman, Y. 1984. Clinical problems in organophosphate insecticide poisoning: The use of a computerized information system. Fund. Appl. Toxicol. 4: S209–S214.

Hilvert, D. and Nared, K.D. 1988. Stereospecific Claisen rearrangement catalyzed by an antibody. J. Am. Chem. Soc. 110: 5593–5594.

Hilvert, D., Carpenter, S.H., Nared, K.D., Auditor, M-T.M. 1989a. Catalysis of concerted reactions by antibodies: The Claisen rearrangement. Proc. Natl. Acad. Sci. USA 85: 4953–4955.

Hilvert, D., Hill, K.W., Nared, K.D., Auditor, M.-T. M. 1989b. Antibody catalysis of a Diels-Alder reaction. J. Am. Chem. Soc. 111: 9261–9262.
Huse, W.D., Sastry, L., Iverson, S.A., Kang, A.S., Alting-Mees, M., Burton, D.R., Benkovic, S.J., Lerner, R.A. 1989. Generation of a large combinatorial library of the immunoglobulin repertoire in phage lamda. Science 246: 1275–1281.
Jackson, D.Y., Jacobs, J.W., Sugasawara, R., Reich, S.H., Bartlett, P.A., Schultz, P.G. 1988. An antibody catalyzed Claisen rearrangement. J. Am. Chem. Soc. 110: 4841–4842.
Jacobs, J., Schultz, P.G., Sugasawara, R., Powell, M. 1987. Catalytic antibodies. J. Am. Chem. Soc. 109: 2174–2176.
Janda, K.D., Benkovic, S.J., Lerner, R.A. 1989. Catalytic antibodies with lipase activity and R or S substrate selectivity. Science 244: 437–440.
Janda, K.D., Schloeder, D., Benkovic, S.J., Lerner, R.A. 1988. Induction of an antibody that catalyzes the hydrolysis of an amide bond. Science 241: 1188–1191.
Janda, K.D., Weinhouse, M.I., Schloeder, D.M., Lerner, R.A., Benkovic, S.J. 1990. Bait-and-switch strategy for obtaining catalytic antibodies with acyl-transfer capabilities. J. Am. Chem. Soc. 112: 1274–1275.
Janjic, N., Tramontano, A. 1989. Antibody-catalyzed redox reaction. J. Am. Chem. Soc. 111: 9109–9110.
Jencks, W.P. 1969. Catalysis in Chemistry and Enzymology. McGraw-Hill, New York. p. 288.
Keinan, E., Sinha, S.C., Sinha-Bagchi, A., Benory, E., Ghozi, M.C., Eshhar, Z., Green, B.S. 1990. Towards antibody-mediated metallo-porphyrin chemistry. Pure Appl. Chem. 62: 2013–2019.
Kohen, F., Kim, J.B., Lindner, H.R., Eshhar, Z., Green, B.S. 1980. Monoclonal immunoglobin G augments hydrolysis of an ester of the homologous hapten: An esterase-like activity of the antibody containing site? FEBS Lett. 111: 427–431.
Lenz, D.E., Brimfield, A.A., Hunter, K.W., et al. 1984. Studies using monoclonal antibodies against soman. Fund. Appl. Toxicol. 4: S156–S164.
Lerner, R.A., Benkovic, S.J. 1990. Observations in the interface between immunology and chemistry. Chemtracts-Organic Chemistry 3: 1–36.
Morrison, S. 1985. Transfectomas provide novel chimeric antibodies. Science 229: 1202–1207.
Napper, A.D., Benkovic, S.J., Tramontano, A., Lerner, R.A. 1987. A stereospecific cyclization catalyzed by an antibody. Science 237: 1041–1043.
Paul, S., Volle, D.J., Beach, C.M., Johnson, D.R., Powell, M.J., Massey, R.J. 1989. Catalytic hydrolysis of vasoactive intestinal peptide by human autoantibody. Science 244: 1158–1162.
Pauling, L. 1948. Chemical achievement and hope for the future. Am. Scientist 36: 51–58.
Pollack, S.J., Hsuin, P., Schultz, P.G. 1989. Stereospecific hydrolysis of alkyl esters by antibody. J. Am. Chem. Soc. 111: 5961–5962.
Pollack, S.J. and Schultz, P.G. 1989. A semisynthetic catalytic antibody. J. Am. Chem. Soc. 111: 1929–1931.
Raso, V. and Stollar, B.D. 1975. The antibody-enzyme analogy. Biochem. 14: 584–591.
Salkinoja-Salonen, M., Briglia, P.M.M., Valo, R., Maggblom, and McBain, A. 1989. Cleanup of old industrial sites. In: Advances in Biotechnology IV. Kamely, D., Chakrabarty, A., and Omenn, G.S., eds. Gulf Publishing. Houston. pp. 347–367.
Scherrmann, J. M., Terrien, N., Urtizberea, M., et al. 1989. Immunotoxicotherapy: Present status and future trends. Clin. Toxicol. 27: 1–35.
Schultz, P.G. 1988. The interplay between chemistry and biology in the design of enzymatic catalysts. Science 240: 426–433.
Schultz, P.G., Lerner, R.A., Benkovic, S.J. 1990. Catalytic antibodies. Chem. Eng. News, pp. 26–40.
Schwabacher, A.W., Weinhouse, M.E., Auditor, M-T.M., Lerner, R.A. 1989. Metalloselective antiporphyrin monoclonal antibodies. J. Am. Chem. Soc. 111: 2944–2946.
Shokat, K.M., Leumann, C.J., Sugasawara, R., Schultz, P.G. 1989. A new strategy for the generation of catalytic antibodies. Nature 338: 269–271.
Shokat, K.M., Schultz, P.G. 1990. Catalytic antibodies. Ann. Rev. Immunol. 8: 335–363.

Smith, T.W., Haber, E., Yeatman, L., Butler, V.P. 1976. Reversal of advanced digoxin intoxication with Fab fragments of digoxin specific antibodies. New Engl. J. Med. 294: 797–800.

Tawfik, D.S., Zemel, Z.Z., Arad-Yellin, R., Green, B.S., Eshhar, Z. 1990. Simple method of selecting catalytic monoclonal antibodies that exhibit turnover and specificity. Biochem. 29: 9916–9921.

Tramontano, A., Ammann, A.A., Lerner, R.A. 1988. Antibody catalysis approaching the activity of enzymes. J. Am. Chem. Soc. 110: 2282–2286.

Tramontano, A., Janda, K.D., and Lerner, R.A. 1986. Proc. Natl. Acad. Sci. USA 83: 6736–6740.

Discussion (DR. B. GREEN)

Er-El: I appreciate the importance of this type of research for the understanding of protein and enzyme catalysis. I fail to see practical applications to catalysis of tonnage quantities or destruction of tonnage quantities of bonds. We now have the technology for producing monoclonal antibodies on the order of 1–10 kg. With the development of *E. coli* antibodies we may be able to increase that a bit. Still, on a molecular weight basis, an antibody is sometimes several-fold larger than an esterase. Could you elaborate on that? Also, it would be extremely interesting and practical to use nature itself, as you briefly mentioned, by injecting human beings with transient organophosphorus to make their systems immune to the effect of organophosphates.

B. Green: It is always exciting to talk about the future. The possibility of immunizing with a transition state analogue and inducing antibodies that have catalytic capability is especially intriguing. One can imagine eventual novel active immunization whereby enzyme deficiency diseases might be treated with such antibodies, or peptide-hydrolyzing antibodies might break down AIDS or other viral coat proteins. I feel certain that this is one direction. To come back to your first point, the wide scope of potential antibody applications makes me feel that antibodies might have application even in large scale operations. One reason for skepticism is that enzymes themselves are not used for very many industrial applications, although as we heard yesterday there is increasing interest in that direction. But recall that only a few years ago, neither catalytic antibodies nor genetically engineered antibodies existed, so what today seems unlikely might be reality very soon, including applications in large scale production. I certainly anticipate applications where the enzyme-catalyzed reaction does not exist, such as the selective modification of a single functional group in a molecule containing many similar functionalities. I agree that in the immediate future it looks as if catalytic antibodies will not be used for very large scale applications.

Ellis: You touched briefly on 5–coordinate intermediates, which reminded me that a lot of those species are stereochemically non-rigid; axial and equatorial positions switch. Is that a problem in mimicking these systems?

B. Green: In the case of the vanadium compounds, we have had problems because of their complex chemistry – different oxidation states and both the trigonal bipyramidal and square pyramidal structures. The pseudo-rotation about the phosphorus is believed to be critical for assuring that the incoming and outgoing groups are at the apical positions, and the antibody catalyzed reaction must presumably also be able to mediate this.

Salemme: I am perhaps more optimistic than you are about being able to use antibodies. Single chain antibodies have a molecular weight of about 20,000, which is about the same size as industrial enzymes that are produced in multi-ton quantities every year. I do not think molecular weight will ultimately be a limitation. On the other hand, transition state stabilization contributes only part of the activation energy for typical enzyme catalyzed reactions. Consequently, it should be pointed out that to bring catalytic antibodies up to the kind of rates that characterize conventional enzymes, they will need to be extensively engineered. This will require a crystal structure determination of a complex between a substrate analogue and a catalytic antibody, to add to the antibody the additional features of general acid-based catalysis, proximity and so forth, which can add the additional orders of magnitude of catalytic acceleration that characterize most enzymes, in addition to transition state stabilization.

B. Green: There are several different approaches that one might use to get catalytic antibodies; your point is very well taken. But note that Lerner has reported catalytic antibodies having rate enhancements of several million using simple transition state stabilization, so one can get quite significant rate enhancements. For many practical applications such as decontamination, detoxification, or chemical modification reactions, tremendous rate enhancements may not be so critical.

Fish: I am disappointed that you did not test all those low affinity antibodies. I would think that the best enzyme might be one which would reject the transition state rather than have high affinity for it. Maybe one should try to select antibodies which have extremely high affinity for the substrate and less affinity for the transition state.

B. Green: Experimentally and theoretically, as I emphasized, if you bind the substrate too tightly you increase the energy of activation and do not get enhanced reaction. You do have to get strong binding to the transition state. In cases where the transition state analogue is a poor mimic of the true transition state, you might be right about checking the low affinity antibodies.

Key Words: Biocatalyst, Detoxification, Enzyme-like catalysis, Monoclonal antibodies, Transition state, Organophosphate

Bacterial Heavy Metal Resistance Systems and Possibility of Bioremediation

Simon Silver
Department of Microbiology and Immunology, University of Illinois College of Medicine, Box 6998, Chicago, IL 60680 U.S.A.

Introduction

Bioremediation of heavy metal wastes by either bioaccumulation or by enzymatic detoxification has not as yet been put to practical use. The potential for the former has been demonstrated by a couple of biotechnology start-up companies (Brierley, et al., 1986, 1989; Darnall, 1989; Darnall, et al., 1989). For enzymatic detoxification, mercury-contaminated sludge has effectively been decontaminated at a laboratory scale (Hansen, et al., 1984; Hansen, personal communication). Here in a Symposium on Applied Biotechnology (in particular as related to problems faced by the military), I will describe the potential for these two quite different methods for bioremediation of toxic mineral wastes: firstly bioaccumulation, for which there is an applied microbiology literature but less basic science and theoretical understanding, and then the redox conversion of toxic heavy metals, for which bioremediation might be accomplished by oxidation or reduction from a more toxic form to a less toxic (or readily removed from the environment) form. These three bioconversions are reduction of Hg(II) and Hg(I) ions to metallic Hg(0), the oxidation of more toxic arsenite [As(III)] to arsenate[As(V)]; and the reduction of more toxic chromate [Cr(VI)] to less toxic and more readily removed chromium ions [Cr(III)]. Laboratory understanding of the redox processes is better developed than more applied or field development. The molecular biology of bacterial resistances to mercury, arsenic and chromate has recently been developed in considerable detail, and that information will be summarized. For the mercury system, the resistance mechanism is the detoxification mechanism. For currently known examples of arsenic and chromate resistance, microbial plasmid-determined heavy metal resistances do not involve bio-transformations, but rather cell membrane transport mechanisms that assure lower intracellular concentrations for the toxicant than is found in the external environment. (Related reviews of this area will appear in the published proceedings of two conferences, the International Conference on Biotechnology and Environmental Science, Bangkok, Thai-

D. Kamely et al. (eds), Biotechnology: Bridging Research and Applications, 265–287.

land, and the Eighth International Biodeterioration Symposium, Windsor, Canada, both held also during the summer of 1990).

Bioaccumulation

The use of microbes to absorb toxic heavy metals for bioremediation and to accumulate minerals of economic value (e.g., gold) has been developing over the last decade but has not yet become a predictable usable technology. Thus, further applied research in this area is very much called for. One early and promising report was by Strandberg, et al. (1981), who found that both yeast and bacterial cells "bioaccumulated" uranium from dilute aqueous solutions to levels up to 15% of the dry cell weight for the bacteria or 30% for the yeast. The uptake was rapid and highly specific. The heavy metal accumulated extracellularly in yeast (Strandberg, et al., 1981), probably associated with cell wall carbohydrates as studied by Beveridge (1989) and Macaskie and Dean (1989; see below). However, the accumulation of uranium salts by *Pseudomonas aeruginosa* cells was intracellular. Subsequently Strandberg and Arnold (1988) measured the accumulation of ^{237}Neptunium, a uranium-238 product, by microbial cells; and in this case bacterial strains absorbed more neptunium (up to 1.5% cell dry weight) than did yeast cells. It is clear from these efforts that different microbes might be better adapted for particular toxic metals of interest (i.e., there was specificity for both the metal ion and the microbe involved.)

Beveridge (1989 and references therein) and Gadd and White (1989 and references therein) have studied the accumulation of heavy metal compounds by prokaryotic and eukaryotic microbes, respectively. Beveridge (1989) followed mainly by electron microscopy the deposition of a range of metal cations on the anionic cell wall layers of Gram positive and Gram negative eubacteria and also on archaebacteria. Whether this surface deposition shows the specificity or high capacity needed to function for bioremediation needs to be tested. The studies of Gadd and White (1989) on bioaccumulation of toxic metal cations and radionuclides by fungal and yeast cells showed high capacity (up to 50% of cell mass) for such materials as thorium. Intracellular energy-dependent uptake was sometimes seen, as well as cell-surface deposition.

Macaskie and Dean (1989) and Macaskie, et al. (1987) have been developing microbial systems for detoxification of liquid wastes. Using columns packed with immobilized cell mass, they demonstrated bioaccumulation of a wide range of heavy metals, some uranium-related and others such as cadmium, lead and silver. Highly efficient accumulation was demonstrated in several cases. Pregrown *Citrobacter* cells were incorporated in polyacrylamide gels (which were shredded and packed in columns) or grown as a biofilm on a solid (glass bead) support. The glass bead adsorbed cells functioned optimally. With *Citrobacter* accumulating cadmium or lead, a cell-bound acid phosphatase released inorganic phosphate, resulting in cadmium or lead phosphate precipitates on the cell surface. A comparable uranium phosphate precipitate was also formed. The differences in immobilizing methods and the dependence upon cellular metabolism, even for extracellular precipitation, demonstrated the meticulousness with which the immobilized cell systems must be tuned to the environmental problem at hand. The need to handle large volumes, sometimes up to 10^7 liters per day, also guides the development of these bioabsorbant microbial systems (Macaskie and Dean, 1989).

Darnall (Darnall, 1989; Darnall, et al., 1989; Greene and Darnall, 1989; Greene, et al., 1986) and the Brierley's (Brierley, et al., 1986; 1989; Hutchins, et al., 1986) formed companies to develop and market pelletized microbial cell mass as specific and effective bio-absorbants. After encouraging early developments, the Brierley company "Advanced Mineral Technologies" went out of business, apparently for economic rather than technological reasons. Their products were granulated, non-living cellular biomass from mixed microbial species, which were used in either fixed-bed canisters or fluid-bed reactor systems. Silver, cadmium, copper, zinc and lead were accumulated from dilute solutions, reducing the residual levels by more than 99% (Brierley, et al., 1986; Hutchins, et al., 1986). Gold was also accumulated by these cell-mass pellets at up to 39% (g/g) from dilute gold cyanide solutions. The metals could be stripped from the granules, which could then be reused. The use of dried non-live cell mass affords advantages when stripping the pellets with harsh (generally acid) solutions and also eliminates concerns about metal toxicity, which clouds the alternative use of live cells. Granules made with different microbial mixtures showed somewhat different specificities and capacities. Overall microbial biomass seemed more effective than algal biomass (Brierley, et al., 1986) and than conventionally manufactured anion exchange resins. In one trial run on industrial waste water, more than 99% of lead (in the range of several mg l^{-1}) was still being effectively removed after 39 days.

The system developed by Darnall and colleagues at Bio-recovery Systems, Inc. differs from that of Advanced Mineral Technologies primarily in that algal cell mass is being used (Greene, et al., 1986; Greene and Darnall, 1989). It is not clear whether this difference will be of continuing importance. Since there are differences even within a group of microbes in capacity and in specificity from mineral to mineral, both Advanced Mineral Technologies and Bio-recovery Systems have developed a range of cell-mass formulations. The Bio-recovery Systems granulated cell material has been prepared both from eukaryotic algae and from prokaryotic cyanobacteria. No consistent patterns of differences were reported. The cell-mass biosorbants were developed initially for harvesting gold from dilute gold cyanide solutions. The algal cells immobilized in a silica polymer matrix and called "AlgaSORB" were useful for detoxifying industrial waste solutions and polluted ground water containing cadmium, copper, lead, and mercury.

Microbial polythiol polypeptides (called metallothioneins and phytochelatins) are potential bioaccumulation products that are smaller and better defined than whole cells. Olafson, et al. (1979, 1980, 1984, 1988) reported preliminary characterization of metallothionein-like polythiols from prokaryotes, especially cyanobacteria. These efforts have not been followed up adequately. Olafson, et al. (1979) found a cadmium-inducible protein of apparent molecular weight of 8100 in a *Synechococcus* isolate. The polypeptide contained 11/80 residues as cysteine (Olafson, et al., 1980). Only a single Cd^{2+} cation was bound per polypeptide, much less than animal metallothioneins or the comparable polypeptide reported by Higham, et al. (1984). Subsequent reports revised the amino acid composition (Olafson, et al., 1980) and showed that the polypeptide was made during cell growth on the highest tolerated levels of Cd^{2+}. The amino acid sequence was not homologous with metallothioneins from yeast on animals (Olafson, et al., 1988).

Higham, et al. (1984) provided the strongest evidence for a bacterial metallothionein polypeptide. Higham, et al. (1984) isolated a series of 3 related cation-binding polypeptides of apparent molecular weights from 3,600 to 6,900 from a cadmium-resistant isolate of *Pseudomonas putida*. Metal-ion analysis showed 4 to 7 Cd^{2+}, Zn^{2+}, or Cu^{2+} atoms per polypeptide and a combination of NMR and other physical-chemical analysis led to a model tetra-thiol bound Cd (the amino acid analysis showed 10–22% cysteines in the polypeptides). Unfortunately, after this very strong initial report, subsequent papers on this topic did not add substantially (Higham, et al., 1985, 1986a,b). Apparently (P.D. Sadler, personal communication), the ability of the bacterial strain to produce the metallothionein-like polypeptides has been lost. When the *P. putida* strain was functioning, it bioaccumulated Cd^{2+} (Higham, et al., 1985). Resistance and bioaccumulation ability were rapidly lost during growth in the absence of cadmium. Additional efflux mechanism(s) appeared to function during lag and exponential growth with bioaccumulation occurring during stationary phase. During the lag phase before growth of Cd^{2+}-adapted cells on high cadmium, extensive blebbing from the cell surface occurred and the cells contained numerous internal electron-dense aggregates (Higham, et al., 1986a) which were identified as containing cadmium-polyphosphate (Higham, et al., 1986b).

An earlier report of metallothionein-like cadmium-binding components produced adaptively by *E. coli* (Mitra, et al., 1975) were followed through with preliminary characterization of the components (Khazaeli and Mitra, 1981), but unfortunately it is not clear whether these materials are metallothionein-like or phytochelatin-like.

Periodically, reports appear of the production of excess hydrogen sulfide and associated extracellular precipitation of sulfide complexes as the basis for bacterial resistance and detoxification of such cations as Hg^{2+}, Cu^{2+} and Cd^{2+}. Copper sulfide precipitation was reported as the mechanism for plasmid-determined copper resistance in *Mycobacterium scrofulaceum* (Erardi, et al., 1987), and the same plasmid that governs sulfide production also determined mercury resistance by volatilization (i.e., reduction) (Meissner and Falkinham, 1984). Whether precipitation and volatilization of mercury are occurring simultaneously in the same bacterial cell was unclear. Aiking, et al. (1985) showed detoxification of mercury, cadmium and lead as sulfide or phosphate complexes (depending on the relative availability of S or P) in a *Klebsiella aerogenes* strain. Production of H_2S by *Desulfovibrio* may be responsible for immobilization of mercury in anaerobic estuarine environments (Belliveau and Trevors, 1989).

Mercury Reduction and Volatilization

While there has been considerable progress on the molecular genetics and protein engineering of the mercurial detoxification systems (below), efforts in using these systems for bio-remediation of toxic mercurial wastes have been slow to start. Additional work needs to be done here. With mercurial-polluted soil, the addition of mercury volatilizing *P. aeruginosa* cells (with or without added nutrients) had little effect on mercury elimination, presumably because the soil-bound mercury is well fixed on particles and inaccessible to the bacteria (Silver, unpublished data). Addition of soluble thiol compounds may potentiate bioremediation of mercury-polluted soil.

With well-stirred aqueous systems, the opposite occurs; there is the essentially complete and continuous elimination of inorganic mercury waste by the bacterial system (Hansen, et al., 1984). In more recent efforts with bioreactors, Hansen (personal communication) has been able to achieve long term (up to 2 weeks) continuous removal of mercury from 75 mg l^{-1} (0.37 mM) down to 120 $\mu g\ l^{-1}$ (0.6 μM; more than 99% removal). The reduced mercury could be recovered either as a sludge containing essentially 50% Hg^0 by weight or was volatilized from the system, depending upon aeration. Volatilized Hg^0 could be recovered on a activated charcoal column, if desired.

By far the best understood heavy metal resistance system is that for mercuric ions and organomercurial compounds. The mechanism of resistance in bacteria is enzymatic detoxification. More toxic Hg^{2+} is converted to less toxic Hg^0, which is volatilized away from the contaminated site (and the bacteria). Organomercurials such as phenylmercury and methylmercury are first cleaved enzymatically to inorganic Hg^{2+} plus an organic compound (benzene and methane respectively for the two examples). The enzymes responsible for these conversions are mercuric reductase and organomercurial lyase (Figure 1).

Eight mercury resistance systems have been cloned and their DNA sequences determined. In each case, the system consists of a number of genes (approximately 6), whose products carry out a carefully regulated process of mercury uptake by the cell, delivery of the toxic Hg^{2+} ions to the intracellular enzyme mercuric reductase, with the subsequent volatilization of the product, gaseous metallic Hg^0.

I. Hg^{2+} Reductase
8 sequenced: 478, 545 to 564 or 632 amino acids
Dimer
Contains 2 FADs
Requires NADPH and thiol

II. ORGANOMERCURIAL LYASE
4 sequenced: 212 to 218 amino acids
Monomer
Requires thiol

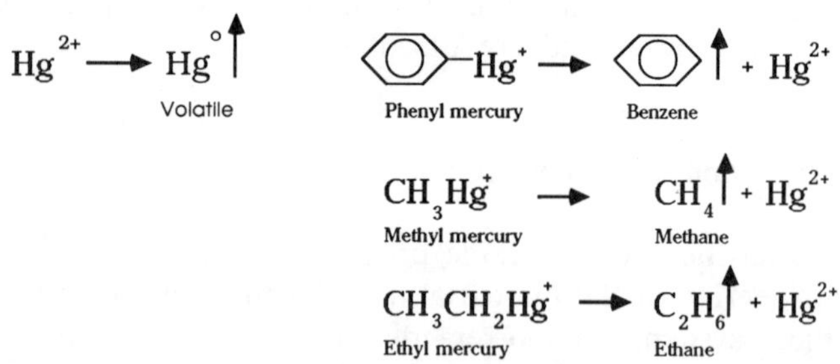

Figure 1. The enzymes for mercury and organomercurial degradation.

Figure 2 shows the organization of this genetic system in the Gram negative bacteria. While that from the Gram positive bacteria is somewhat different, the overall process remains the same. The *mer* operon begins with a divergently transcribed regulatory gene *merR* (Foster, et al., 1979; Foster and Brown, 1985), the product of which is a 144 amino acid polypeptide, which functions in the cell as a dimer (O'Halloran, et al., 1989; O'Halloran, 1989) and which binds to the next region, the operator/promoter segment (itself less than 75 base pairs long). In the absence of Hg^{2+} ions, the merR protein represses mRNA synthesis of the next (structural) genes. MerR also auto-regulates its own synthesis from a leftwardly oriented promoter in the

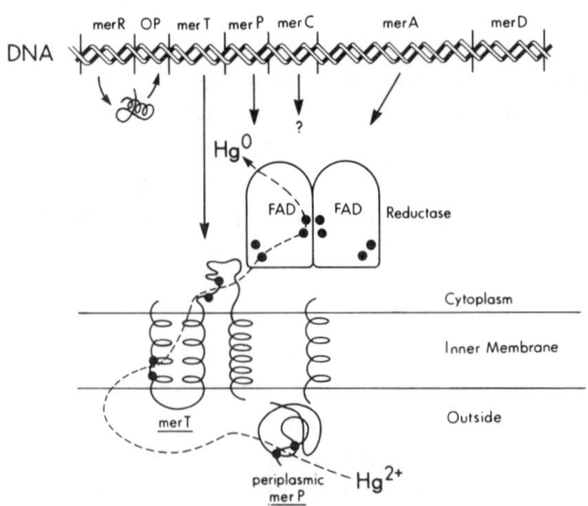

Figure 2. The mercury/organomercurial resistance system of Gram negative bacteria.

same region (Lund and Brown, 1989). When Hg^{2+} ions are added, the MerR protein functions in a positive regulatory fashion to stimulate mRNA synthesis more than 100x fold, "turning on" the mercuric resistance system. The *Bacillus* MerR dimer has been shown to form a tri-cysteine complex with a single Hg^{2+} ion bound by Cys_{79} on one subunit and Cys_{114} and Cys_{123} on the other subunit (Helmann, et al., 1990). With MerR responding to organomercurials as well as to inorganic Hg^{2+}, a deletion of the C-terminal 10 amino acids eliminated the in vivo response to organomercurials but not to Hg^{2+} (Nucifora, et al., 1989a). Hg^{2+} binding is exceeding tight and specific for Hg^{2+} in vivo, although Cd^{2+}, Zn^{2+} and other divalent cations work less well in vitro (Ralston and O'Halloran, 1990; Helmann and Walsh, 1990). The MerR protein binds to the same operator site on the DNA in the presence of or in the absence of mercuric ions, but when mercuric ions are present, MerR associates with RNA polymerase, initiating MRNA synthesis from the promoter site (O'Halloran, et al., 1989; O'Halloran, 1989). The operator and promoter sites for the *mer* operon overlap and contain an usual long 19 base pair distance between canonical −35 and −10 transcriptional initiation sites (Lund and Brown, 1989). Elongated spacer regions (20 or 21 base pairs) prevented induction and shorter spacers (17 or 18 base pairs) were constitutive for *mer* operon transcription (Parkhill and Brown, 1990). These results were interpreted as consistent with a change in DNA conformation induced by Hg^{2+} binding to MerR, resulting in optimum interaction between RNA polymerase and the transcriptional initiation signals (O'Halloran, 1989; Frantz and O'Halloran, 1990; Parkhill and Brown, 1990).

Then follows a series of 4 to 6 genes (5 in Tn21; Figure 2), whose products carry out the transport and detoxification of mercury. In the first three *mer* operons of Gram negative bacteria whose DNA sequences have been published, the first gene product is MerT, an integral membrane protein, which functions as a Hg^{2+} transport system (Foster, et al., 1979; Nakahara, et al., 1979), bringing mercury from outside of the cell to inside where the mercuric reductase enzyme resides. A small 72 amino acid (after the 19 amino acid "leader sequence" has been cut) polypeptide MerP in the periplasmic space functions (as had been proposed) both as a scavenging mechanism to bind mercury outside the cell (and thus protect the cellular surface) and as a delivery vehicle to bring the MerP-bound mercury to the MerT transport system. The mercury moves from (a) free external in solution, to (b) bound to MerP outside the cell, to (c) transported across the cell membrane on MerT, to (d) an intracellular site. Since intracellular Hg^{2+} would be toxic, it is proposed that the mercury bound to MerT is directly transferred to the mercuric reductase enzyme, which may be loosely associated with the cellular membrane. The N-terminal portion of mercuric reductase shares sequence homologies to the MerP periplasmic polypeptide (and to the beginning of the *cadA* Cd^{2+} ATPase), thus leading to the suggestion that it functions intracellularly as part of the mercury "bucket brigade" (Figure 2).

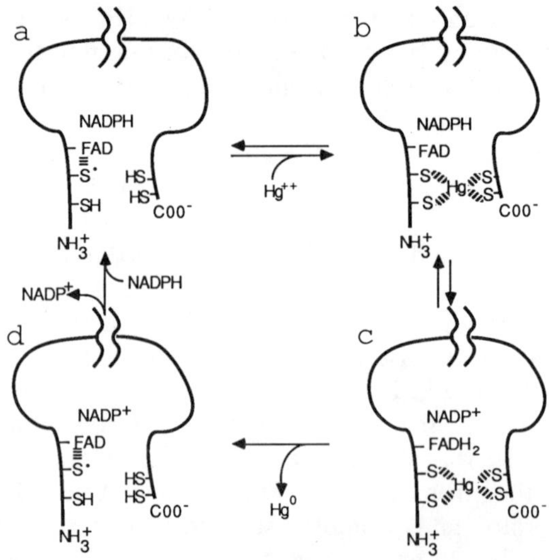

Figure 3. Reaction mechanism of mercuric reductase (adapted from Miller, et al., 1989, and Distefano, et al., 1990).

The Hg^{2+} ion is then transferred to a highly conserved cysteine pair at the C-terminal and of the mercuric reductase enzyme. Those cysteines are required for enzymatic activity (Moore and Walsh, 1989). It is thought that the Hg^{2+} may be tetragonally coordinated (an unusual structure for mercury) between the carboxyl-terminal cysteine pair and the cysteine pair located at the active site of the enzyme (Figure 3). It was recently shown that the tetragonal mercury bridge is shared between the subunits of the dimer, since mutant heterodimers (missing one pair of cysteines–Cys_{135} Cys_{140}– on one subunit and the other pair of cysteines–Cys_{558} Cys_{559}– on the second subunit) are functional both in vivo and in vitro (Distefano, et al., 1990), whereas mutant homodimers do not work.

The active site cysteine pair (Cys_{135} and Cys_{140}) occurs in a stretch of about 15 amino acids, which is highly conserved in all 8 currently-known versions of mercuric reductase and in the closely related oxido-reductase enzymes glutathione reductase, trypanothione reductase and lipoamide dehydrogenase. The carboxyl-terminal cysteine pair is also found in all 8 sequences, but not in glutathione reductase or lipoamide dehydrogenase. The carboxyl end of glutathione reductase is, however, involved in substrate binding and is found close to the active site in the structure derived from X-ray crystallographic analysis (Thieme, et al., 1981; Karplus and Schulz, 1987).

Hg^{2+} is reduced (Figure 1; Figure 3) by electrons which have been transferred from NADPH to the enzyme-bound FAD and then to the Hg^{2+} ion. The Hg^0 is released from the enzyme, and diffuses rapidly from the cell (apparently without a protein carrier). In a rapidly stirred system, Hg^0 volatilizes into the atmosphere. In the absence of aeration (such as in a still ecosystem), the metallic Hg^0 remains bound to cellular lipids, hydrocarbons or sulfur-containing materials. In an unstirred anaerobic laboratory culture, the mercury appears as a gray sludge (Hansen, personal communication), associated with the cells.

The reaction mechanism for mercuric reductase has been studied in detail. The stage of an activated EH_2NADPH form of the enzyme has a charge transfer complex between Cys_{140} and the bound FAD. The carboxyl-terminal cysteine pair Cys_{558} Cys_{559} (contributing to the active site from the other subunit; Distefano, et al., 1990) is fully reduced. Hg^{2+} is added to the enzyme (either from soluble thiols, such as mercaptoethanol in cell-free reactions, or perhaps from the N-terminal cysteine pair Cys_{10} Cys_{13} in vivo; Brown, 1985; Silver and Misra, 1988). Both in vivo and in vitro, the N-terminal cysteines can be removed by mutation or with trypsin, without affecting mercury resistance or cell-free enzyme properties (Moore and Walsh, 1989). Yet, the N-terminal cysteine pair is conserved in 7 of the 8 sequenced versions of mercuric reductase (Silver and Misra, 1988), and it is difficult to understand the repeated occurrence of this sequence if it lacks a function. With the homology between the N-terminal region (including the Cys_{10} Cys_{13} pair) and the MerP periplasmic mercury-binding protein, the role of the N-terminus in passing Hg^{2+} from the membrane MerT protein to the C-terminal cysteine pair was hypothesized (Brown, 1985; Silver and Misra, 1988).

Hg^{2+} is tetragonally-coordinated between the Cys_{135} Cys_{140} on one subunit and the Cys_{558} Cys_{559} pair on the other subunit. Next, the bound NADPH is oxidized, reducing the FAD near the active site. Finally, the electrons are transferred from the $FADH_2$ to mercury; the reduced Hg^0 is released from the enzyme spontaneously, forming the enzyme with oxidized FAD and $NADP^+$. Exchange of reduced NADPH for oxidized $NADP^+$ completes the cycle. In addition to the reaction leading to reduction of mercury ions, a process of activating the enzyme from a form lacking the charge transfer complex to one with the complex and a series of partial reactions in the absence of Hg^{2+} have been characterized (Sahlman, et al., 1984, 1986; Distefano, et al., 1989; Miller, et al., 1989). Overall, mercuric reductase is now a well understood enzyme, both from spectroscopy studies of the reaction mechanism and with a range of mutants in crucial residues. With the recent success in obtaining x-ray diffraction quality crystals of mercuric reductase (Moore, et al., 1989), one can hope for direct structural information, as is already available for the related enzyme glutathione reductase (Thieme, et al., 1981; Karplus and Schulz, 1987). From the structure, some aspects of this model will be directly verified.

After the *merA* gene determining the enzyme mercuric reductase, the *merB* gene determining the enzyme organomercurial lyase has been identified and sequenced in plasmid pDU1358 (Gram negatives) and in the 3 Gram positive sources of the mercurial resistance operon. The amino acid sequence alignment of these polypeptides (Figure 4) show a family of enzymes 212 to 218 amino acids in length. One version of this enzyme has been purified and shown to occur as a monomer (Begley, et al., 1986a). The amino acid sequences are 41 to 73% identical, when compared 2 at a time. Highly conserved sequences are found throughout the polypeptide, but mostly in the central half, especially between positions 75 and 170 (Figure 4). Begley, et al. (1986b) and Walts and Walsh (1988) worked on the reaction mechanism of organomercurial lyase, especially varying the specific substrate. The results summarized in Figure 5 suggest that a crucial cysteine (either Cys_{96}, Cys_{117} or Cys_{159} in the *E. coli* sequence) initially binds the organomercurial substrate. A concerted S_E2 mechanism of carbon-mercury bond breakage and carbon-hydrogen bond formation is proposed (Begley, et al., 1986b; Walts and Walsh, 1988; Walsh, et al., 1988), with the hydrogen donated by an additional residue driving off the "R" group (benzene from phenylmercury or methane from methylmercury) as the mercury associates with a second conserved cysteine residue. Then the thiol compound required for in vitro enzyme activity (usually cysteine in vitro, but either glutathione or the enzyme mercuric reductase itself in vivo) releases the Hg^{2+} product from the lyase.

The last gene in the four sequenced Gram negative versions of the *mer* operon is *merD*, which was initially defined by insertion mutations following *merA* that led to slightly reduced levels of mercury resistance. More recently, it has been proposed that *merD* is involved in regulation of the *mer* operon, partially because of sequence homologies between MerD and MerR amino acid sequences (Brown, et al., 1986) and partially from studies of regulation with gene fusions (Nucifora, et al., 1989b).

Arsenite Oxidation and Resistance

I know of no microbial system for reduction of As(III) or As(V) to As(0). If such a system exists, analogous to the microbial mechanism for reduction of sulfate to elemental sulfur, then it might function as a useful system for bioremediation. There is a much information on the molecular biology and biochemistry of plasmid determined As(III) and As(V) resistances and on the oxidation of more toxic arsenite [As(III)] to less toxic arsenate [As(V)]. Unfortunately, plasmid-determined resistance does not involve redox chemistry (see below) and there are no recent studies of the arsenite oxidation system.

The plasmid determined resistance system for arsenate and arsenite (which also confers resistance to antimonate) has been studied in detail both with *E. coli* and *S. aureus*. Earlier results were summarized by Silver and Misra (1988). Resistance to arsenite did not involve oxidation to arsenate (Silver, et al., 1981). There was no evidence of extracellular detoxification or chel-

```
                *           *              *    **   ***
S. aureus     MK  NISEFSAQLDQTFDQGEAV  SMEWLFRPLLKM  33
Bacillus      MKTEIQEIVTRLDQQSNKGEGGESMKWLFRPLLQM   35
Streptomyces  MDSQAQQLATRLTTAFNGGGAASSRPWLWRPLLQL  35
E. coli       MK     LAPYILERLTSVNRTNGTAD     LLVPLLRE  30

              **  *   *      *           *    *
S. aureus     LAEGDPVPVEDIAAETGKPVEEVKQVLQTLPSVEL   68
Bacillus      LAGGESVTIEDMATTTGKPVEEVKKVLQSLPSVEI   70
Streptomyces  LAQGRPVTVEQIAQATDRTPDQVREALAANPDTEY  70
E. coli       LAKGRPVSRTTLAGILDWPAERVAAVLEQATSTEY   65

              *  *   *  ***   *      *    **  ******
S. aureus     DEQGRVVGYGLTLFPTPHRFEVDGKQLYAWCALDT  103
Bacillus      DEQGRVVGLGLTLIPTPHHFTVDGKQLYAWCALDT  105
Streptomyces  DERGRITGSGLTQNPTPHHFEVDGQQLYTWCALDT  105
E. coli       DKDGNIIGYGLTLRETSYVFEIDDRRLYAWCALDT  100

              * ***  **     * **  * *    * *  *
S. aureus     LMFPALIGRTVHIASPCHGT GKSVRLTVEPDRVVS  138
Bacillus      LIFPALIGRSVNIESPCHSTGEPIRLNVEPDHIVS   140
Streptomyces  LIFPAILGRPAHVTSPCHATGTPVRLTVEPDQVTS   140
E. coli       LIFPALIGRTARVSSHCAATGAPVSLTVSPSEIQA   135

              ***     ** *  *           **   ****
S. aureus     VEPSTAVVSIVTPDEMASVRSAFCNDVHFFSSPSA   173
Bacillus      VEPSTAVVSIVTPDDMSSIRTAFCNEVHFFSSPNA   175
Streptomyces  VEPATAVVSIVTPDAPASIRTAFCNQVHFFATPDA   175
E. coli       VEPAGMAVSLVLPQEAADVRQSFCCHVHFFASVPT   170

                        *           *    *   **
S. aureus     AQDWLNQHP    ESSVLPVEDAFELGRHLGARYEE  205

Bacillus      AEDWLDQHP    GGKVLSVEDAFELGRLMGTRYEE  207

Streptomyces  GKGYLEEHP    VATVLPVADAYQLGRPLTEALLT  207

E. coli       AEDWASKHQGLEGLAIVSVHEAFGLGQEFNRHLLQ  205

                      **
S. aureus     SGPTNGSCCNI  216
Bacillus      SRPANGSCCHI  281
Streptomyces  GDTPPG CC    215
E. coli       TMSSRTP      212
```

Figure 4. Organomercurial lyase sequence alignment (references for the 4 sequences are Laddaga, et al., 1987; Y. Wang, et al., 1989; J. Altenbuchner, et al., personal communication; and Griffith, et al., 1987, respectively).

ation of arsenite or arsenate. Resistant cells accumulated less arsenate than sensitive cells, and that alone accounted for the resistance (Silver, et al., 1981). The reduced accumulation was not due to a change in properties of arsenate uptake (which occurs via the cellular phosphate transport systems),

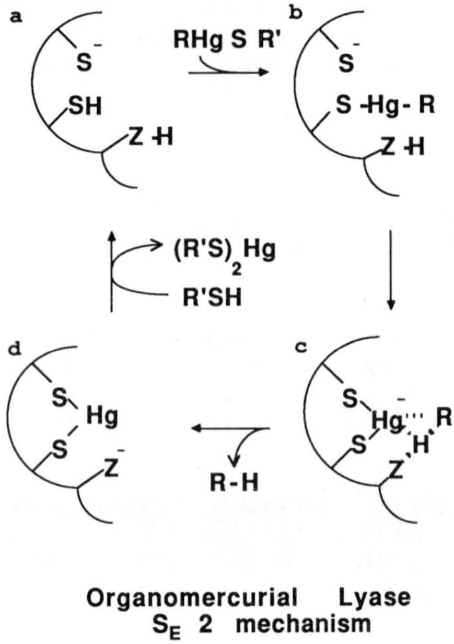

Figure 5. Reaction mechanism of organomercurial lyase (adapted from Begley, et al., 1986b, and Walsh, et al., 1988).

but rather accumulated arsenate is rapidly effluxed from the cells by an inducibly-synthesized plasmid-governed system (Silver and Keach, 1982; Mobley and Rosen, 1982).

In the dozen years since we reviewed the literature on microbial arsenite oxidation (Summers and Silver, 1978), there have been no additional studies on the biochemistry or genetics of this process. Such studies are essential, if the process is to be used. Arsenite oxidizing bacteria were recognized in 1918 with arsenite-tolerant bacteria isolated from a cattle dip solution, where arsenite was used as an anti-cattle tick agent. Similar bacteria were found in Australia during the 1940s. The bacteria were mainly Pseudomonads. Osborne and Ehrlich (1976) isolated an arsenite oxidizing *Alcaligenes* strain from soil. The synthesis of the arsenite oxidizing activity was induced by growth on arsenite. Whole cell oxidation showed saturation kinetics with a K_m of 1.5 mM AsO_2^-. Crude cell-free preparations showed arsenite oxidizing activity associated with both particulate and soluble fractions. It was suggested that arsenite oxidation was membrane associated via the cytochrome oxidase (Osborne and Ehrlich, 1976). In similar independent studies, Phillips and Taylor (1976) isolated an arsenite-oxidizing *Alcaligenes* strain from arsenite-containing sewage. These efforts need to be initiated again with purification of the enzyme activity and isolation of the gene(s) involved.

The molecular genetics of the arsenate/arsenite/antimony resistance determinant of plasmid R773 (from Gram negative bacteria) has progressed re-

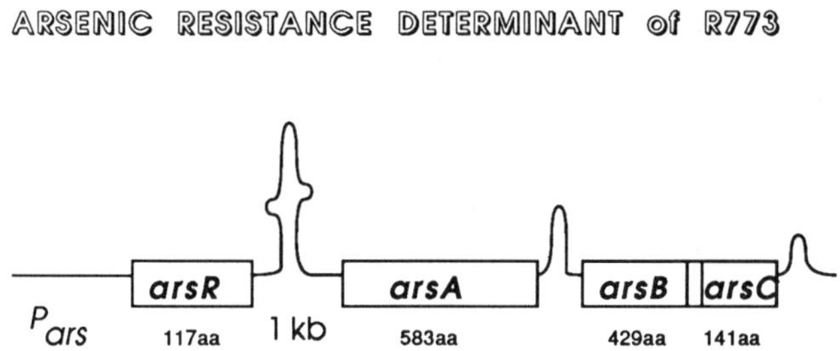

Figure 6. Arsenic resistance determinant of *E. coli* plasmid R773 (adapted from Chen, et al., 1986 and Owolabi and Rosen 1990).

cently with purification of the protein components, cloning and sequencing of the genes and mutational analysis of gene function and component interactions. This work was recently summarized by Tisa and Rosen (1990). The three genes that together determine the arsenic resistance efflux ATPase were cloned and sequenced by Chen, et al. (1986). The additional upstream regulatory gene *arsR* was identified by San Francisco, et al. (1990). This system is diagrammed in Figure 6. Transcription into mRNA starts at the left at a promoter identified by Owolabi and Rosen (1990) and continues for 4.4 kb. The first gene, *arsR*, determines the 117 amino acid regulatory protein, which is 30% identical in sequence to the comparable gene product from a Gram positive Staphylococcal plasmid (Rosenstein and Gotz, 1990) (Figure 7). After the *arsR* gene is a 1 kb region of DNA lacking open reading frames (San Francisco, et al., 1990), but with a potential regulatory hairpin loop. Then follow the three genes for the components of the ATPase, *arsA* (for the ATPase subunit itself), *arsB* (for an integral membrane protein presumably containing the anion transmembrane channel) and *arsC* (for a small soluble polypeptide needed for arsenate–but not for arsenite–resistance) (Chen, et al., 1986; Tisa and Rosen, 1990). There are inverted repeat sequences as candidates for regulatory control between the *arsA* and *arsB* genes and after the *arsC* gene as well (Figure 6). The ArsA ATPase subunit has been purified and characterized physically (Hsu and Rosen, 1989). It

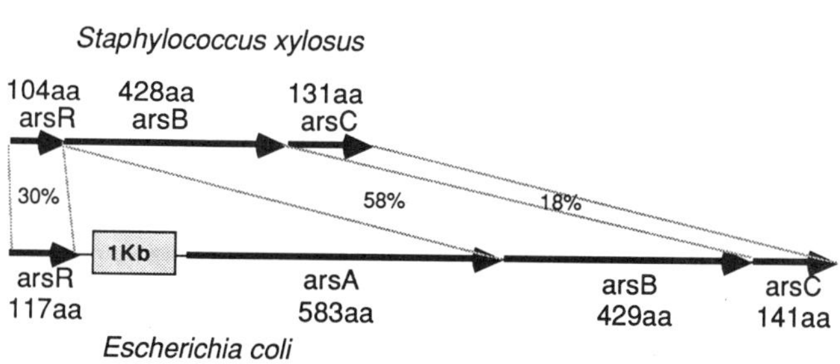

Figure 7. Comparison of plasmid arsenic resistance determinants from *Staphylococcus* and *E. coli* (from Silver and Laddaga, 1990, with permission).

is an arsenite (and antimonate)-stimulated ATPase (Rosen, et al., 1988). Conformational changes upon addition of ATP or oxyanions was indicated by protection from proteolysis (Hsu and Rosen, 1989). At the sequence level, the ArsA protein appears to be a fusion dimer, with the N-terminal half closely homologous with the C-terminal half (Chen, et al., 1986) and both halves containing potential ATP-binding motifs. Mutational analysis showed that the N-terminal motif $G_{15}KGGVGKTS_{23}$ is required for efflux activity and resistance (Karkaria and Rosen, 1990). It is not known whether the second ATP-binding motif is essential and whether both play enzymatic roles, or one region is enzymatic and the other regulatory. The next gene product, ArsB, is a membrane protein (San Francisco, et al., 1989) that is made in smaller amounts than ArsA (Tisa and Rosen, 1990), apparently because of mRNA processing leading to a shorter mRNA missing the *arsB* segment (Owolabi and Rosen, 1990). The ArsB protein anchors the otherwise soluble ArsA ATPase subunit to the cell membrane (Tisa and Rosen, 1989). Whereas the complex of ArsA and ArsB suffice to confer resistance to (and presumedly efflux of) arsenite and antimonate, the third gene *arsC* is needed for arsenate resistance (and efflux) (Chen, et al., 1986).

Chromate Reduction and Resistance

For chromate (as for arsenite), there is both microbial detoxification (reduction of chromate to Cr^{3+}) and plasmid-determined chromate resistance, but the two processes are unrelated. Chromate resistance results from lowered accumulation by the cells (Ohtake, et al., 1987). It is not yet clear whether this results from a direct block on uptake (via the chromosomally-determined sulfate transport system) or accelerated efflux following uptake.

Two plasmid-determined chromate resistance systems were recently cloned and sequenced in our laboratory (Cervantes, et al., 1990; A. Nies, et al., 1989, 1990). Since the DNA sequences were sufficiently different that a relationship could not be recognized by Southern DNA/DNA hybridization (Cervantes, et al., 1990) (only 55% identical bases, when the sequences were available), it was of interest that the gene organization and predicted protein products were closely related (Figure 8). The *Alcaligenes eutrophus* system (as cloned) consists of two genes *chrA* and *chrB* that are needed for inducible chromate resistance plus a third partial open reading frame (that is apparently not needed, but which is complete in the second, *Pseudomonas aeruginosa* chromate resistance system). The long 401 or 416 amino acid *chrA* protein is found in the cell membrane (Cervantes, et al., 1990; A. Nies, et al., 1990). It may be the only structural component of the membrane transport resistance determinant. Other than hydrophobicity, the amino acids sequences give no suggestion for functional regions, and in particular lack discernible ATP-binding motifs. The upstream *chrB* gene of the *Alcaligenes* system is likely to be the regulatory gene, since it is present in the cloned

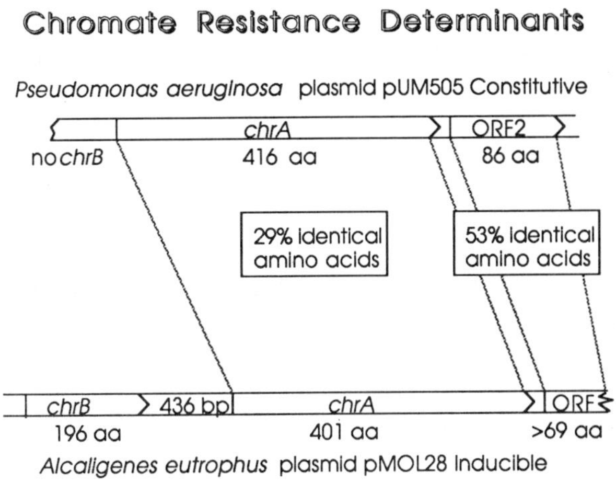

Figure 8. Comparison of plasmid chromate resistance determinants from *Pseudomonas* and *Alcaligenes* (from Cervantes and Silver, 1990, with permission).

Alcaligenes system (inducible) (A. Nies, et al., 1990) but absent in the cloned Pseudomonas system (constitutive) (Cervantes, et al., 1990). A *chrA-lacZ* gene fusion has been shown to respond specifically to chromate induction, but not to other tested oxyanions (D. Nies, personal communication). Whether gene regulation occurs upstream of the *chrB* gene and/or within the 436 base pairs between *chrB* and *chrA* needs to be determined. The shorter 86 amino acids open reading frame in the *Pseudomonas* clone and sequence is apparently not required for chromate resistance under some conditions, because it was disrupted during cloning of the *Alcaligenes* system. Yet, 53% of the first 69 amino acids in the predicted gene products are identical (Figure 8). Clearly work has just started on this new system and there is much to learn.

The *P. fluorescens* strain that provided the first chromate-resistance plasmid also has the capacity to reduce chromate to Cr^{3+} (Bopp and Ehrlich, 1988), but the two phenomena are not related. A strain cured of the plasmid retains full chromate reductase activity but has lost chromate resistance (Bopp and Ehrlich, 1988; Ishibashi, et al., 1990). The potential for use of microbes to remove toxic chromate from water was patented (Bopp, 1984).

Recent progress toward development of a useful chromate reductase activity has been achieved by Ohtake and coworkers with an anaerobic membrane respiratory chain system from a newly-isolated *Enterobacter cloacae*. This chromate-resistant *E. cloacae* strain was isolated from activated sludge (P.C. Wang, et al., 1989). Although it was resistant to chromate under both aerobic and anaerobic conditions, chromate reduction occurred only anaerobically (P.C. Wang, et al., 1989). The resistance and ability to reduce chromate was lost on anaerobic growth on nitrate, suggesting that the nature of the anaerobic respiratory pathway was crucial for chromate reduction.

The reduced chromate produced insoluble chromium hydroxide. Chromate reductase activity was inhibited by oxygen (P.C. Wang, et al., 1989; Komri, et al., 1989, 1990a). Initial levels of 1–2 mM CrO_4^{2-} were rapidly detoxified, although higher concentrations inhibited cellular activity. A variety of carbon sources functioned as electron donors for chromate reduction and the activity was inhibited by other oxyanions such as molybdate, vanadate and tellurate (Komori, et al., 1989). Cyanide and membrane uncouplers inhibited chromate reductase activity. Subcellular right-side-out membrane vesicles from *E. cloacae* contained the cellular chromate reductase activity (P.C. Wang, et al., 1990) and ascorbate-phenazine methosulfate functioned as electron donor with membrane preparations (P.C. Wang, et al., 1990).

Attempts to use the *E. cloacae* strain to detoxify chromate-containing effluents are being made. Up to 90% of 4 mM CrO_4^{2-} was removed and precipitated within dialysis bags containing intact bacteria (Komori, et al., 1990a, b), but the process took several days. For reduction of waste chromate, a small supplement of broth was needed to provide energy (Ohtake, et al., 1990). It was also sometimes necessary to dilute the chromate-containing waste prior to exposure to the bacteria, presumably to lower the levels of other toxic materials present. These efforts are only preliminary, and further work using bacteria within dialysis bags or cells immobilized on columns may lead to the development of useful chromate detoxification technology.

Ishibashi, et al. (1990) provided the newest (and still preliminary) description of a bacterial soluble chromate reductase activity that functions aerobically. The K_m for this system was about 40 μM CrO_4^{2-}, and the activity was highly specific for chromate. Sulfate and nitrate were without effect, although mercury and silver cations noncompetitively inhibited chromate reductase activity. This system may be identical to that reported by Horitsu, et al. (1987) from a different Pseudomonad isolate. There have been several additional reports of chromate reduction as a possible means of bioremediation (e.g. Romanenco and Kkoren'kov, 1977; Lebedeva and Lyalikova, 1979; Horitsu and Kato, 1980; Shimada and Matsushima, 1983; Kvasnikov, et al., 1985; Gvozdyak, et al., 1986), but these have not been developed further.

Summary

This relatively relaxed "walk through" the Periodic Table of bacterial resistance mechanisms to toxic heavy metals and their unfulfilled potential (sometimes) for bioremediation leads to several overall conclusions. Firstly, for essentially all toxic heavy metals that occur in the environment as cations or anions, bacterial cells have evolved highly specific resistance mechanisms in order to survive in a frequently polluted world. The genes governing these resistance mechanisms frequently occur on plasmids and transposons, facilitating molecular genetic studies and movement from cell to cell. Secondly, there are only a few overall mechanisms of resistance: (a) efflux pumping the toxic material out from the cell; (b) bioaccumulation in a physiologically inaccessible form (either intracellular or extracellular), and (c) redox chemistry converting a more toxic ion species into one less toxic. Redox chemistry, although more rare in bacterial cells, perhaps affords the most promising systems for bioremediation of toxic wastes. Reduction and volatilization of mercury from inorganic and organomercurials is the best understood of these resistance mechanisms. It is ready for scale up research and practical applications to aqueous mercury sources. Chromate reduction and arsenite oxidation are other such potential systems, since these convert more toxic to less toxic forms. The biochemistry and molecular biology behind these processes are not understood. Bioaccumulation also may be put to practical use, but this has not been accomplished to date.

Efflux pumping appears the least promising for practical uses, but even here there is the potential for building membranes with highly specific permeabilities to toxic salts of interest. Until we understand the fundamental biochemical and genetic processes that bacterial cells use to achieve resistances to toxic metal ions, then it is premature to rule out the possibility of these processes being useful in cleaning up environmentally polluted materials.

Acknowledgments

The work in our laboratory has been supported by grants from the National Science Foundation and National Institutes of Health. Many people have contributed to 20 years progress. In more recent years, these include N.L. Brown, C. Hansen, D. and A. Nies, T.K. Misra, L. Chu, C. Cervantes, G. Nucifora, H. Ohtake, B.P. Rosen, and R.P. Novick.

References

Aiking, H., Govers, H. and van't Riet, J. 1985. Detoxification of mercury, cadmium, and lead in *Klebsiella aerogenes* NCTC 418 growing in continuous culture. Appl. Environ. Microbiol. 50: 1262–1267.

Begley, T.P., Walts, A.E. and Walsh, C.T. 1986a. Bacterial organomercurial lyase: Overproduction, isolation and characterization. Biochem. 25: 7186–7192.

Begley, T.P., Walts, A.E. and Walsh, C.T. 1986b. Mechanistic studies of a protonolytic organomercurial cleaving enzyme: Bacterial organomercurial lyase. Biochem. 25: 7192–7200.

Belliveau, B.H. and Trevors, J.T. 1989. Mercury resistance and detoxification in bacteria. Appl. Organometallic Chem. 3: 283–294.

Beveridge, T.J. 1989. Interactions of metal ions with components of bacterial cell walls and their biomineralization. In: Metal-Microbe Interactions. Poole, R.K. and Gadd, G.M., eds. IRL Press, Oxford. pp. 65–83.

Bopp, L.H. 1984. Microbial removal of chromate from contaminated waste water. U.S. Patent # 4,468,461, issued August 28,1984.

Bopp, L.H., Chakrabarty, A.M. and Ehrlich, H.L. 1983. Chromate resistance plasmid in *Pseudomonas fluorescens*. J. Bacteriol. 155: 1105–1109.

Bopp, L.H. and Ehrlich, H.L. 1988. Chromate resistance and reduction in *Pseudomonas fluorescens* strain LB300. Arch. Microbiol. 150: 426–431.

Brierley, C.L., Brierley, J.A. and Davidson, M.S. 1989. Applied microbial processes for metals recovery and removal from waste water. In: Metal Ions and Bacteria. Beveridge, T.J. and Doyle, R.J., eds. John Wiley & Sons, N.Y. pp. 359–382.

Brierley, J.A., Brierley, C.L. and Goyak, G.M. 1986. AMT- BIOCLAIM: A new waste water treatment and metal recovery technology. In: Fundamental and Applied Biohydrometallurgy. Lawrence, R.W., Branion, R.M.R. and Ebner, H.G., eds. Elsevier, Amsterdam. pp. 291–304.

Brown, N.L. 1985. Bacterial resistance to mercury: Reductio ad absurdum. Trends Biochem. Sci. 10: 400–403.

Brown, N.L., Misra, T.K., Winnie, J.N., Schmidt, A., Seiff, M. and Silver, S. 1986. The nucleotide sequence of the mercuric resistance operons of plasmid R100 and transposon Tn501: Further evidence for *mer* genes which enhance the activity of the mercuric ion detoxification system. Molec. Gen. Genet. 202: 143–151.

Cervantes, C., Ohtake, H., Chu, L., Misra, T.K. and Silver, S. 1990. Cloning, nucleotide sequence, and expression of the chromate resistance determinant of *Pseudomonas aeruginosa* plasmid pUM505. J. Bacteriol. 172: 287–291.

Cervantes, C. and Silver, S. 1990. Inorganic cation and anion transport systems of *Pseudomonas*. In: *Pseudomonas*: Biotransformations, Pathogenesis and Evolving Biotechnology. Silver, S., Chakrabarty, A.M., Iglewski, B. and Kaplan, S. eds. American Society for Microbiology, Washington, D.C. pp. 359–372.

Chen, C.-M. Misra, T.K., Silver, S. and Rosen, B.P. 1986. Nucleotide sequence of the structural genes for an anion pump. The plasmid-encoded arsenical resistance operon. J. Biol. Chem. 261: 15030–15038.

Darnall, D.W. 1989. Removal and recovery of heavy metal ions from waste waters using a new bioabsorbant: AlgaSORB. In: Innovative Hazardous Waste Treatment Technology. Freeman, H., ed. Technomic Publishing Company, Lancaster, PA. In press.

Darnall, D.W., Gabel, A.M. and Gardea-Torresday, J. 1989. AlgaSORB: A new biotechnology for removing and recovering heavy metal ions from ground water and industrial waste water. In: Hazardous Waste Treatment: Biosystems for Pollution Control, Proceedings of the 1989 A & WMA/EPA International Symposium. EPA, Cincinnati, Ohio. pp. 113–124.

Distefano, M.D., Au, K.G. and Walsh, C.T. 1989. Mutagenesis of the redox-active disulfide in mercuric ion reductase: Catalysis by mutant enzymes restricted to flavin redox chemistry. Biochem. 28: 1168–1183.

Distefano, M.D., Moore, M.J. and Walsh, C.T. 1990. Active site of mercuric reductase resides at the subunit interface and requires Cys135 and Cys140 from one subunit and Cys558 and Cys559 from the adjacent subunit: Evidence from in vivo and in vitro heterodimer formation. Biochem. 29: 2703–2713.

Erardi, F.X., Failla, M.L. and Falkinham III, J.O. 1987. Plasmid- encoded copper resistance and precipitation by *Mycobacterium scrofulaceum*. Appl. Environ. Microbiol. 53: 1951–1954.

Foster, T.J. and Brown, N.L. 1985. Identification of the *merR* gene of R100 by using *mer-lac* gene and operon fusions. J. Bacteriol. 163: 1153–1157.

Foster, T.J., Nakahara, H., Weiss, A.A. and Silver, S. 1979. Transposon A-generated mutations in the mercuric resistance genes of plasmid R100–1. J. Bacteriol. 140: 167–181.

Frantz, B. and O'Halloran, T.V. 1990. DNA distortion accompanies transcriptional activation by the metal-responsive gene-regulatory protein MerR. Biochem. 29: 4747–4751.

Gadd, G.M. and White, C. 1989. Heavy metal and radionuclide accumulation and toxicity in fungi and yeasts. In: Metal-Microbe Interactions. Poole, R.K. and Gadd, G.M., eds. IRL Press, Oxford. pp. 19–38.

Greene, B. and Darnall, D.W. 1989. Microbial oxygenic photoautotrophs (cyanobacteria and algae) for metal ion binding. In: Microbial Mineral Recovery. Ehrlich, H. ed. McGraw Hill, N.Y.

Greene, B., Hosea, M., McPherson, R., Henzl, M., Alexander, M.D. and Darnall, D.W., 1986. The interaction of gold(I) and gold(III) with algal biomass. Envir. Sci. Technol. 20: 627–632.

Griffin, H., Foster, T.J., Silver, S. and Misra, T.K. 1987. Cloning and DNA sequence analysis of the mercuric and organomercurial resistance determinants of plasmid pDU1358. Proc. Natl. Acad. Sci. 84: 3112–3116.

Gvozdyak, P.I., Mogilevich, N.F., Ryl'skii, A.F. and Grishchenko, N.I. 1986. Reduction of hexavalent chromium by collection strains of bacteria. Mikrobiologiya 55: 962–965.

Hansen, C.L., Zwolinski, G., Martin, D. and Williams, J.W. 1984. Bacterial removal of mercury from sewage. Biotech. Bioengin. 26: 1330–1333.

Helmann, J.D., Ballard, B.T. and Walsh, C.T. 1990. The MerR metalloregulatory protein binds mercuric ion as a tricoordinate, metal-bridged dimer. Science 247: 946–948.

Helmann, J.D. and Walsh, C.T. 1990. Metal dependent transcriptional activation: Binding of metal ions by the *Bacillus* species RC607 MerR protein. Biochem., In press.

Higham, D.P., Sadler, P.J. and Scawen, M.D. 1984. Cadmium resistant *Pseudomonas putida* synthesizes novel cadmium proteins. Science 225: 1043–1046.

Higham, D.P., Sadler, P.J. and Scawen, M.D. 1985. Cadmium resistance in *Pseudomonas putida*: Growth and uptake of cadmium. J. Gen. Microbiol. 131: 2539–2544.

Higham, D.P., Sadler, P.J. and Scawen, M.D. 1986a. Effect of cadmium on the morphology, membrane integrity and permeability of in *Pseudomonas putida*. J. Gen. Microbiol. 132: 1475–1482.

Higham, D.P., Sadler, P.J. and Scawen, M.D.. 1986b. Cadmium binding proteins in *Pseudomonas putida*: Pseudothioneins. Environ. Health Perspect. 65: 5–11.

Horitsu, H. and Kato, H. 1980. Comparisons of characteristics of cadmium-tolerant and bacterium *Pseudomonas aeruginosa* G-1 and its cadmium-sensitive mutant strain. Agric. Biol. Chem. 44: 777–782.

Horitsu, H., Futo, S., Miyazawa, Y., Ogai, S. and Kawai, K. 1987. Enzymatic reduction of hexavalent chromium by hexavalent chromium tolerant *Pseudomonas ambigua* G-1. Agric. Biol. Chem. 51: 2417–2420.

Hsu, C.M. and Rosen, B.P. 1989. Characterization of the catalytic subunit of an anion pump. J. Biol. Chem. 264: 17349–17354.

Hutchins, S.R., Davidson, M.S., Brierley, J.A. and Brierley, C.L. 1986. Microorganisms in reclamation of metals. Ann. Rev. Microbiol. 40: 311–336.

Ishibashi, Y., Cervantes, C. and Silver, S. 1990. Chromium reduction by *Pseudomonas putida*. Appl. Environ. Microbiol. 56: 2268–2270.

Karkaria, C.E. and Rosen, B.P. 1990. Mutagenesis of a nucleotide binding site of an anion-translocating ATPase. J. Biol. Chem. 265: 7832–7836.

Karplus, A. and Schulz, G.E. 1987. Refined structure of glutathione reductase at 1.54 Å resolution. J. Mol. Biol. 195: 701–729.

Khazaeli, M.B. and R.S. Mitra. 1981. Cadmium-binding component in *Escherichia coli* during accommodation to low levels of this ion. Appl. Environ. Microbiol. 41: 46–50.

Komori, K., Wang, P-C., Toda, K. and Ohtake, H. 1989. Factors affecting chromate reduction in *Enterobacter cloacae* strain HO1. Appl. Microbiol. Biotechnol. 31: 567–570.

Komori, K., Rivas, A., Toda, K. and Ohtake, H. 1990a. Biological removal of toxic chromium using an *Enterobacter cloacae* strain that reduces chromate under anaerobic conditions. Biotechnol. Bioengin. 35: 951–954.

Komori, K., Toda, K. and Ohtake, H. 1990b. Effects of oxygen stress on chromate reduction in *Enterobacter cloacae*. J. Ferment. Bioeng. 69: 67–69.

Komori, K., Rivas, A., Toda, K. and Ohtake, H. 1990c. A method for removal of toxic chromium using dialysis-sac cultures of a chromate-reducing strain of *Enterobacter cloacae*. Appl. Microbiol. Biotechnol. 33: 117–119.

Kvasnikov, E.I., Stepanyuk, V.V., Klyushnikova, T.M., Serpokrylov, N.S., Simonova, G.A., Kasatkina, T.P. and Pachenko, L.P. 1985. A new chromium-reducing, gram-variable bacterium with mixed type flagellation. Mikrobiologiya 54: 83–88.

Laddaga, R., Chu, L., Misra, T.K. and Silver, S. 1987. Nucleotide sequence and expression of the mercurial resistance operon from *Staphylococcus aureus* plasmid pI258. Proc. Natl. Acad. Sci. 84: 5106–5110.

Lebedeva, E.V. and Lyalikova, N.N. 1979. Reduction of crocoite by *Pseudomonas chromatophila* sp. nov. Mikrobiologiya 48: 517–522.

Lund, P.A. and Brown, N.L. 1989. Regulation of transcription from the *mer* and *merR* promoters of the transposon Tn501. J. Mol. Biol. 205: 343–353.

Macaskie, L.E. and Dean, A.C.R. 1989. Microbial metabolism, desolubilization, and deposition of heavy metals: Metal uptake by immobilized cells and application to the detoxification of liquid wastes. Advances in Biotechnological Processes 12: 159–201.

Macaskie, L.E., Dean, A.C.R., Cheetham, A.K., Jakeman, R.J.B. and Skarnulis, A.J. 1987. Cadmium accumulated by a *Citrobacter* sp.: The chemical nature of the accumulated metal precipitate and its location in the bacterial cell. J. Gen. Microbiol. 133: 539–544.

Meissner, P.S. and Falkinham III, J.O. 1984. Plasmid-encoded mercuric reductase in *Mycobacterium scrofulaceum*. Appl. Environ. Microbiol. 157: 669–672.

Miller, S.M., Moore, M.J., Massey, V., Williams, C.H. Jr., Distefano, M.D., Ballou, D.P. and Walsh, C.T. 1989. Evidence for participation of Cys558 and Cys559 at the active site of mercuric reductase. Biochemistry 28: 1194–1205.

Mitra, R.S., Gray, R.H., Chin, B. and Bernstein, I.A. 1975. Molecular mechanisms of accommodation in *Escherichia coli* to toxic levels of cadmium. J. Bacteriol. 121: 1180–1188.

Mobley, H.L.T. and Rosen, B.P. 1982. Energetics of plasmid- mediated arsenate resistance in *Escherichia coli*. Proc. Natl. Acad. Sci. 79: 6119–6122.

Moore, M.J., and Walsh, C.T. 1989. Mutagenesis of the N- and C- terminal cysteine pairs of Tn501 mercuric ion reductase: consequences for bacterial detoxification of mercurials. Biochemistry 28: 1183–1194.

Moore, M.J., Distefano, M.D., Walsh, C.T., Schliering, N. and Pai, E.F. 1989. Purification, crystallization, and preliminary x-ray diffraction studies of the flavoprotein mercuric ion reductase from *Bacillus* sp. strain RC607. J. Biol. Chem. 264: 14386–14388.

Nakahara, H., Silver, S., Miki, T. and Rownd, R.H. 1979. Hypersensitivity to Hg^{2+} and hyperbinding activity associated with cloned fragments of the mercurial resistance operon of plasmid NR1. J. Bacteriol. 140: 161–166.

Nies, A., Nies, D.H. and S. Silver. 1989. Cloning and expression of plasmid genes encoding resistances to chromate and cobalt in *Alcaligenes eutrophus*. J. Bacteriol. 171: 5065–5070.

Nies, A., Nies, D.H. and S. Silver. 1990. Nucleotide sequence and expression of a plasmid-encoded chromate resistance determinant from *Alcaligenes eutrophus*. J. Biol. Chem. 265: 5648–5653.

Nucifora, G., Chu, L., Silver, S. and Misra, T.K. 1989a. Mercury operon regulation by the *merR* gene of the organomercurial resistance system of plasmid pDU1358. J. Bacteriol. 171: 4241–4247.

Nucifora, G., Silver, S. and Misra, T.K. 1989b. Down regulation of *mer* operon function by the product of the *merD* gene of plasmid pDU1358. Molec. Gen. Genet. 220: 69–72.

O'Halloran, T.V. 1989. Metalloregulatory proteins: metal responsive molecular switches governing gene expression. In: Metal Ions in Biological Systems. Sigel, H., ed. Marcel Dekker, New York. Vol. 25, pp. 105–145.

O'Halloran, T.V., Frantz, B., Shin, M.K., Ralston, D.M. and Wright, J.G. 1989. The *merR* heavy metal receptor mediates positive activation in a topologically novel transcription complex. Cell 56: 119–129.

Ohtake, H., Cervantes, C. and Silver, S. 1987. Decreased chromate uptake in *Pseudomonas fluorescens* carrying a chromate resistance plasmid. J. Bacteriol. 169: 3853–3856.

Ohtake, H., Komori, K., Cervantes, C. and Toda, K. 1989. Chromate resistance in a chromate-reducing strain of *Enterobacter cloacae*. FEMS Microbiol. Let. 67: 85–88.

Ohtake, H., Fujii, E. and Toda, K. 1990. Reduction of toxic chromate in an industrial effluent by use of a chromate-reducing strain of *Enterobacter cloacae*. Environ. Technol. Lett. 11: 663–668.

Olafson, R.W. 1984. Prokaryotic metallothionein. Internat. J. Peptide Protein Res. 24: 303–308.

Olafson, R.W., Abel, K. and Sim, R.G. 1979. Prokaryotic metallothionein: Preliminary characterization of a blue-green alga heavy metal-binding protein. Biochem. Biophys. Res. Commun. 89: 36–43.

Olafson, R.W., Loya, S. and Sim, R.G. 1980. Physiological parameters of prokaryotic metallothionein induction. Biochem. Biophys. Res. Commun. 95: 1495–1503.

Olafson, R.W., McCubbin, W.D. and C.M. Kay. 1988. Primary- and secondary-structural analysis of a unique prokaryotic metallothionein from a *Synechococcus* sp. cyanobacterium. Biochem. J. 251: 691–699.

Osborne, F.H. and Ehrlich, H.L. 1976. Oxidation of arsenite by a soil isolate of *Alcaligenes*. J. Appl. Bacteriol. 41: 295–305.

Owolabi, J.B. and Rosen, B.P. 1990. Differential mRNA stability controls relative gene expression within the plasmid-encoded arsenical resistance operon. J. Bacteriol. 172: 2367–2371.

Parkhill, J. and Brown, N.L. 1990. Site-specific insertion and deletion mutants in the *mer* promoter-operator region of Tn501: the nineteen base-pair spacer is essential for normal induction of the promoter by *MerR*. Nucleic Acids Res. 18: 5157–5162.

Phillips, S.E. and Taylor, M.L. 1976. Oxidation of arsenite to arsenate by *Alcaligenes faecalis*. Appl. Environ. Microbiol. 32: 392–399.

Ralston, R.M. and O'Halloran, T.V. 1990. Ultrasensitivity and heavy-metal selectivity of the allosterically modulated *MerR* transcription complex. Proc. Natl. Acad. Sci. 87: 3846–3850.

Romanenco, V.I. and Kkoren'kov, V.N. 1977. A pure culture of bacteria utilizing chromates and bichromates as hydrogen acceptors in growth under anaerobic conditions. Mikrobiologiya 46: 414–417.

Rosen, B.P., Weigel, U., Karkaria, C. and Gangola, P. 1988. Molecular characterization of an anion pump. The *arsA* gene product is an arsenite (antimonate)-stimulated ATPase. J. Biol. Chem. 263: 3067–3070.

Rosenstein, R. and Götz, F. 1990. Nucleotide sequence and expression of arsenic resistance genes of *Staphylococcus xylosus*. Unpublished manuscript in preparation.

Sahlman, L., Lameir, A.-M., Lindskog, S. and Dunford, H.B. 1984. The reaction between NADPH and mercuric reductase from *Pseudomonas aeruginosa*. J. Biol. Chem. 259: 12403–12408.

Sahlman, L., Lamieir, A.-M. and Lindskog, S. 1986. Rapid-scan stopped-flow studies of the Ph dependence of the reaction between mercuric reductase and NADPH. Eur. J. Biochem. 156: 479–488.

San Francisco, M.J.D., Tisa, L.S. and Rosen, B.P. 1989. Identification of the membrane component of the anion pump encoded by the arsenical resistance operon of plasmid R773. Molec. Microbiol. 3: 15–21.

San Francisco, M.J.D., Hope, C.L., Owolabi, J.B., Tisa, L.S. and Rosen, B.P. 1990. Identification of the metalloregulatory element of the plasmid-encoded arsenical resistance operon. Nucleic Acids Res. 18: 619–624.

Shimada, K. and Matsushima, K. 1983. Isolation of potassium chromate-resistant bacterium and reduction of hexavalent chromium by the bacterium. Bull. Faculty Agriculture Mie Univ. 67: 101–106.

Silver, S., Budd, K., Leahy, K.M., Shaw, W.V., Hammond, D., Novick, R.P., Willsky, G.R., Malamy, M.H. and Rosenberg, H. 1981. Inducible plasmid-determined resistance to arsenate, arsenite and antimony(III) in *Escherichia coli* and *Staphylococcus aureus*. J. Bacteriol. 146: 983–996.

Silver, S. and Keach, D. 1982. Energy-dependent arsenate efflux: the mechanism of plasmid-mediated resistance. Proc. Natl. Acad. Sci. 79: 6114–6118.

Silver, S. and Laddaga, R.A. 1990. Molecular genetics of heavy metal resistances in *Staphylococcus* plasmids. In: Molecular Biology of the Staphylococci. Novick, R.P., ed. VCH Publishers, New York. pp. 531–549.

Silver, S. and Misra, T.K. 1988. Plasmid-mediated heavy metal resistances. Ann. Rev. Microbiol. 42: 717–743.

Strandberg, G.W., Shumate II, S.E. and Parrott Jr., J.R. 1981. Microbial cells as biosorbents

for heavy metals: Accumulation of uranium by *Saccharomyces cerevisiae* and *Pseudomonas aeruginosa*. Appl. Environ. Microbiol. 41: 237–245.
Strandberg, G.W. and Arnold Jr., W.D. 1988. Microbial accumulation of neptunium. J. Indus. Microbiol. 3: 329–331.
Summers, A.O. and Silver, S. 1978. Microbial transformations of metals. Ann. Rev. Microbiol. 32: 637–672.
Thieme, R., Pai, E.F., Schirmer, R.H. and Schulz, G.E. 1981. Three-dimensional structure of glutathione reductase at the 2 Å resolution. J. Mol. Biol. 152: 763–782.
Tisa, L.S. and Rosen, B.P. 1989. Molecular characterization of an anion pump: The *arsB* protein is the membrane anchor for the ArsA protein. J. Biol. Chem. 265: 190–194.
Tisa, L.S. and Rosen, B.P. 1990. Transport systems encoded by bacterial plasmids. J. Bioenerg. Biomembr. 22: 493–507.
Walsh, C.T., Distefano, M.D., Moore, M.J., Shewchuk, L.M. and Verdine, G.L. 1988. Molecular basis of bacterial resistance to organomercurial and inorganic mercuric salts. FASEB J. 2: 124–130.
Walts, A.E. and Walsh, C.T. 1988. Bacterial organomercurial lyase: Novel enzymatic protonolysis of organostannanes. J. Amer. Chem. Soc. 110: 1950–1953.
Wang, P.C., Mori, T., Komori, K., Sasatsu, M., Toda, K. and Ohtake, H. 1989. Isolation and characterization of an *Enterobacter cloacae* strain that reduces hexavalent chromium under anaerobic conditions. Appl. Environ. Microbiol., 55: 1665–1669.
Wang, P.C., Mori, T., Toda, K. and Ohtake, H. 1990. Membrane- associated chromate reductase activity from *Enterobacter cloacae*. J. Bacteriol. 172: 1670–1672.
Wang, Y., Moore, M., Levinson, H.S., Silver, S., Walsh, C. and Mahler, I. 1989. Nucleotide sequence of a chromosomal mercury resistance determinant from a *Bacillus* sp. with broad-spectrum mercury-resistance. J. Bacteriol. 171: 83–92.

Discussion (DR. SILVER)

Braun: Thank you for a very interesting presentation. In order for a system removing mercury to work efficiently and to have a fast exchange rate between the thiols, it has to be built either on very high mercury binding site concentration or on the organized system with short diffusion distances. Which is the case?

Silver: Both the bacteria work with the MerP extra-cellular binding protein and the membrane transport proteins. They can pull the mercury away from other thiolated compounds by higher affinities and routing pathways, but they do not need the small soluble thiols to work. If there are other thiols around, the whole bacteria still take up the mercury. The reductase enzyme really needs soluble thiols or the membrane proteins to feed it. Therefore what you say is indeed our experience.

Szybalski: Hg^{2+} is quite toxic; Hg^0 is really not a very good way of detoxifying because it goes into air and we breathe it. How about Hg^{1+} which is not toxic?

Silver: Hg^0 is non-toxic as Hg^0. The problem is that when we breathe in Hg^0 it passes across the lung into the blood and is oxidated immediately by catalase in the red blood cells to Hg^{2+} which is toxic. If you swallow metallic mercury, it is apparently harmless. Fakirs in India used to do this and jump up and down to feel it because at 14 g ml^{-1} metallic mercury, as it feels nice in the tummy. Hg^0 is very non-toxic, essentially inert. In fact, the bacterial system also reduces Hg^{1+} to Hg^0. The enzyme that works on Hg^{2+} (that we usually use) also reduces Hg^{1+} very quickly, and Ag^+ and Au^{3+} very slowly (overnight). There is an additional problem with Hg^{1+} in that it dismutates abiotically,

[$2\ Hg^{1+} \rightarrow Hg^{2+} + Hg^0$], confusing the enzymatic reduction. One additional point can be made to answer your question. Mercury reduction occurs anaerobically as well as aerobically. All you need is cells with NADPH. Anaerobically, without shaking we see a silvery sludge of Hg^0 associated with the bacteria. This Hg^0 can be extracted by organic solvents. Anaerobic reduction works just as well as aerobic to volatilize it. Cell-associated Hg^0 is non-toxic to the bacteria.

Key Words: Arsenite oxidation, Bioaccumulation, Bioremediation, Chromate reduction, Enzymatic detoxification, Heavy metal resistance, Mercury reduction, Metallothioneins, Polythiol polypeptides, Redox conversion, Volatilization

The Cholinergic Binding Site: Structure and Application

Bella Ohana[1]**, Rachel Mosckovitz**[1] **and Jonathan M. Gershoni**[2]
[1]*Department of Biophysics Weizmann Institute of Science, Rehovot, Israel*
[2]*National Cancer Institute, National Institutes of Health, Bethesda, MD*

Introduction

The nicotinic acetylcholine receptor (AChR) is the target of numerous ligands. The agonist, acetylcholine, binds to the α-subunit of the receptor eliciting the opening of an ion channel which leads to membrane depolarization and eventual muscle contraction. The antagonists are impressively chemically diverse and often block the effect of acetylcholine through competitive inhibition. The peptide α-bungarotoxin, the alkaloid d-tubocurarine and even rabies virus, are examples of such competitive ligands. Thus the question arises, what mechanisms enable these different molecules to bind specifically to the same domain of the receptor? What are the rules that govern this process of biorecognition?

Through studies of the cholinergic binding site we hope to gain insight as to how ligands bind to their receptors and, as will be discussed, apply this knowledge in the design of receptor-based therapeutics.

Methodology

The AChR is especially abundant in the electric organ of the ray *Torpedo californica* and has been affinity purified and extensively studied biochemically, as well as cloned and sequenced (for reviews see Conti-Tronconi and Raftery, 1982; Popot and Changeux, 1984; Hucho, 1986; McCarthy, et al., 1986; Maelicke, 1988). Through these studies it was illustrated that the antagonist binding site is situated in the α-subunit of the receptor (Karlin, 1983). Toxin-overlays of protein-blots of the receptor has enabled the mapping of the binding site to the region of α180–200 (Gershoni, 1987a). The corresponding cDNA sequences of this area have been cloned into *E. coli* expression vectors, e.g., pATH2 (Dieckmann and Tzagoloff, 1985), which has allowed the systematic analysis of recombinant binding sites (Gershoni,

D. Kamely et al. (eds), Biotechnology: Bridging Research and Applications, 289–297.

```
                187                               200
                 |                                 |
R1737            W  V  Y  Y  T  C  C  P  D  T  P  Y  L  D
R2037         H  W  V  Y  Y  T  C  C  P  D  T  P  Y  L  D
R2337      K  H  W  V  Y  Y  T  C  C  P  D  T  P  Y  L  D
R4137   W  K  H  W  V  Y  Y  T  C  C  P  D  T  P  Y  L  D
R4134   W  K  H  W  V  Y  Y  T  C  C  P  D  T  P  Y  L
R4131   W  K  H  W  V  Y  Y  T  C  C  P  D  T  P  Y
R4128   W  K  H  W  V  Y  Y  T  C  C  P  D  T  P
R4125   W  K  H  W  V  Y  Y  T  C  C  P  D  T
R4122   W  K  H  W  V  Y  Y  T  C  C  P  D
        |                                |
       184                              195
```

Figure 1. A series of cholinergic binding sites were prepared as fusion proteins in *E. coli* by cloning the appropriate oligonucleotide pairs into the Sma1 site of the pATH2 expression vector. The sites were designated according ot the length of the oligonucleotides used. Thus for example R4137 was prepared from a 41 base oligonucleotide paired with a 37 base oligonucleotide.

1987; Aronheim, et al., 1988). In principle, recombinant binding sites have been produced by generating oligonucleotide pairs (sense and anti-sense joined by a short 8–9 base stretch of homology) that were filled-in and cloned into the Sma1 site of the vector. This strategy is particularly useful in that one can generate novel constructs by mixing the pairs of oligonucleotides thus creating chimeric sites (see below). The recombinant proteins were then semi-purified and tested for toxin binding through Scatchard analysis. The bound versus free ligand was separated by a modified filtration assay and the background binding was determined by including a 1000–fold excess of unlabeled toxin into the reaction mixtures (Aronheim, et al., 1988; Ohana and Gershoni, 1990).

Results

Two series of constructs were prepared; the first represented the systematic removal of NH_2 or COOH terminal residues from the sequence α184–200, while the second was a survey of the binding sites derived from the receptors of *Torpedo*, *Xenopus*, chick, mouse, calf, human and *Drosophila* (residues α183–204) [Figures 1 and 2; Ohana and Gershoni, 1990].

The recombinant binding sites were found to bind ^{125}I-labeled α-bungarotoxin at various levels. The results of the Scatchard analyses pertaining to the sequences described in Figure 1 are given in Figure 3 (each construct was prepared and tested 4–6 times).

As is shown, the effect of removing the terminal Asp-200 is only marginal, whereas the removal of the first Trp-184 reduces the binding of toxin from 0.7×10^{-7}M to 2.3×10^{-7} M. Through such an analysis one can predict that a minimal binding site, starting from His-186 to Tyr-198, should have an affinity for toxin within the realm of 5×10^{-7} M. Indeed the construct R2031,

SOURCE	NH2 HALF										COOH HALF												
TORPEDO	G	W	K	H	W	V	Y	Y	T	C	C	P	D	T	P	Y	L	D	I	T	Y	H	
CHICK									A														
XENOPUS														K									
MOUSE							F		S				T										
CALF							F		A				S										
HUMAN					S		T		S														
DROSOPHILA	A	V	R	N	E	K	F		S			-	E	E						V	F	N	L

Figure 2. The binding sites derived form the nictotinic acetylcholine receptor of different organisms were prepared as recombinant fusion proteins. The sequence of the *Torpedo* receptor is given and only the residues that differ from it are indicated. The dashed line separates the NH_2 and COOH halves of the sites.

which corresponds to the sequence H186–Y198, binds toxin with an affinity of 4.7×10^{-7} M (data not shown).

It is apparent that the *Torpedo* sequence continues to be the most efficient for toxin binding (Figures 3 and 4). This might reflect the fact that over the years of evolution, fish have been less threatened by terrestrial venomous snakes than other vertebrates. From these analyses it is also clear that both the NH_2 and the COOH halves of the site are important in toxin binding. Particularly notable is the relatively weak binding of the human site. Intact human AChR binds a 1000–fold more efficiently than the isolated site (Ohana

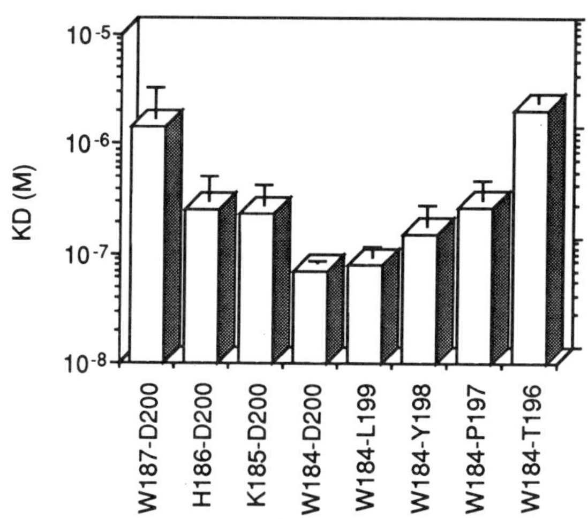

Figure 3. The binding of α-bungarotoxin to recombinant binding sites.

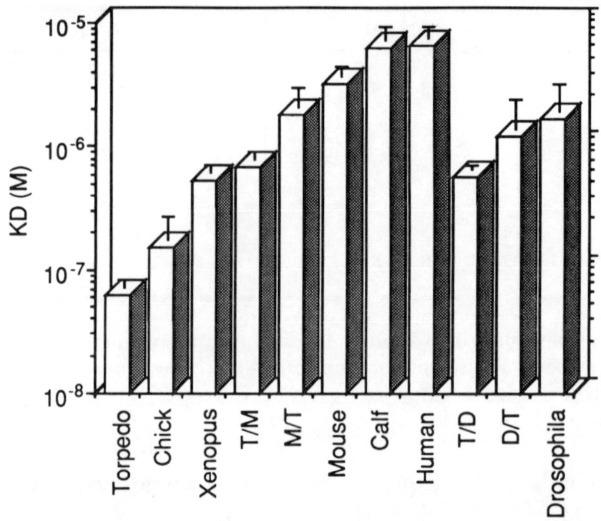

Figure 4. The binding of α-bungarotoxin to recombinant cholinergic binding sites derived from different organisms and their chimeras. The sites have been arranged in the order of increasing number of changes with respet to the *Torpedo* sequence.

and Gershoni, 1990). This may reflect the involvement of W187 and the apparent requirement for a hydrophobic patch following the positive subsite provided by K185 or H186.

To evaluate the specific contribution of some of the residues, chimeric sites were also constructed. This was achieved by pairing the oligonucleotide corresponding to the NH_2 half of the site of one organism, with that corresponding to the COOH half of another.

In this manner chimeras such as T/M or its reciprocal M/T (*Torpedo*/mouse and mouse/*Torpedo* respectively) were tested for toxin binding. In doing so, it was possible to isolate the effect of exchanging the D195 for serine or threonine and also illustrate the flexibility of the site as is seen in the *Torpedo*/*Drosophila* chimeras (T/D and D/T). These experiments, together with the previous analysis in Figures 1 and 3, emphasize the significance of the area of α180–200 for ligand binding. In our recent study (Fraenkel, et al., 1990) we show via selective T1 NMR relaxation measurements that this site not only specifically binds antagonsits but binds the agonist acetylcholine as well. Thus the region under study is a major component of the cholinergic binding site.

Discussion

The analysis of recombinant cholinergic binding sites has provided some insight as to the general requirements for ligand binding. It appears that a

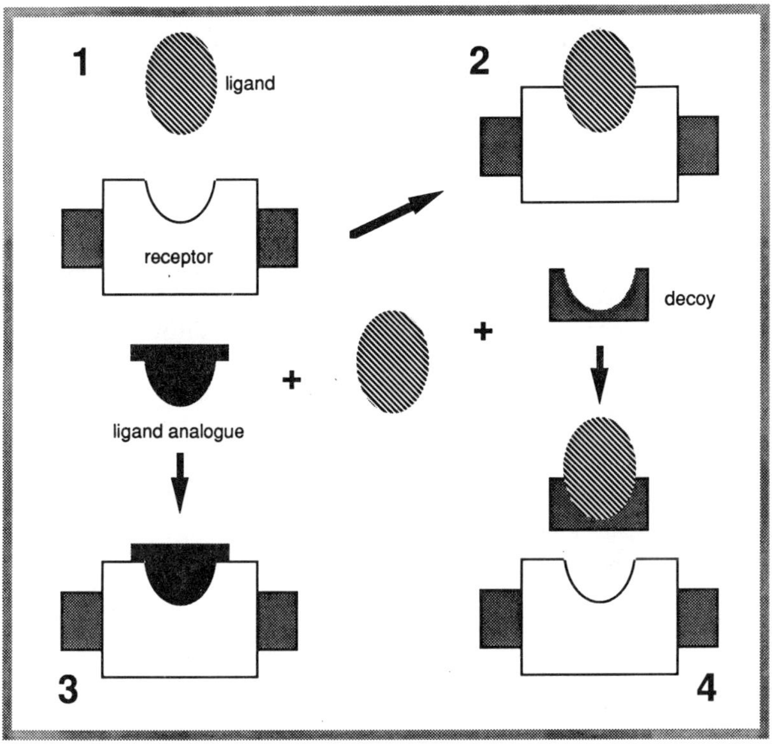

Figure 5. The ligand binds to its receptor to generate the pathogenic complex in (2). This can be prevented with a ligand analogue (3), however this would interfere with receptor function. Such interference does not occur when a decoy is employed (4).

positive sub-site, satisfied by K185 and H186, needs to be followed by a hydrophobic patch which precedes a nucleophilic domain provided by D195 and D200. The latter does not, however, necessarily need to be a negatively charged group as is illustrated by the fact that the T/M chimera continues to bind toxin, albeit with lesser affinity. This general topography of the ligand binding site would complement nicely the carbonyl group of acetylcholine, its cluster of hydrophobic moieties and the quaternary amine of the choline itself (Behling, et al., 1988). In order to further test the binding site, a new series of single point mutations of selected residues are being made and evaluated. In this manner the specific involvement of the tryptophan residues is being questioned (see also Radding, et al., 1988) as well as the issue of the postulated disulfide bridging between Cys192 and Cys193 (Kao and Karlin, 1986; see also Mosckovitz and Gershoni, 1988).

While we are still trying to understand the mechanism of ligand binding, we can simultaneously test the utility of the site we have already isolated and produced in mass. The application that we have been testing for mimic receptors has been in the area of drug design. As was mentioned previously, rabies virus binds to the AChR. A mimic cholinergic receptor could act as a molecular decoy and intercept a pathogen such as rabies or a neurotoxin before these ever reached their target, namely the neuro-muscular junction. The rationale for "molecular decoyance" is illustrated in Figure 5. Thus far, we have tested this hypothesis in mice and found that animals treated with recombinant cholinergic decoys become protected against α-neurotoxins such α-cobratoxin and d-tubocurarine (Gershoni and Aronheim, 1988). The questions that arise in these experiments deal with our ability to create more effective decoys. How does one increase the affinity of the decoy for its target? How do we increase the *in vivo* half life of the decoy and what is the best avenue for delivery? It is our belief that, through the systematic analyses that we have been conducting on the cholinergic binding site, we will be able to not only learn the principles responsible for biorecognition but also how to use these principles towards designing better and more effective decoys.

References

Aronheim, A., Eshel, Y., Mosckovitz, R. and Gershoni, J. M. 1988. Characterization of the binding of α-bungarotoxin to bacterially expressed cholinergic binding sites. J. Biol. Chem. 263: 9933–9937.

Behling, R. W., Yamane, Y., Navon, G. and Jelinski, L. W. 1988. Conformation of acetylcholine bound to the nicotinic acetylcholine receptor. Proc. Natl. Acad. Sci. U.S.A. 85: 6721–6725.

Conti-Tronconi, B. M. and Raftery, M. A. 1982. The nicotinic cholinergic receptor: Correlation of molecular structure with functional properties. Ann. Rev. Biochem. 51: 491–530.

Dieckmann, C. L. and Tzagoloff, A. 1985. Assembly of the mitochondrial membrane system. J. Biol. Chem. 260: 1513–1520.

Fraenkel, Y., Navon, G., Aronheim, A. and Gershoni J. M. 1990. Direct measurement of agonist binding to genetically engineered peptides of the acetylcholine receptor by selective T1 NMR relaxation. Biochemistry 29: 2617–2622.

Gershoni, J. M. 1987a. The application of ligand overlay of protein blots to the study of the nicotinic acetylcholine receptor. Electrophoresis 8: 428–431.

Gershoni, J. M. 1987b. Expression of the α-bungarotoxin binding site of the nicotinic acetylcholine receptor by *Escherichia coli* transformants. Proc. Natl. Acad. Sci. U.S.A. 84: 4318–4321.

Gershoni J. M. and Aronheim, A. 1988. Molecular decoys: Ligand-binding recombinant proteins protect mice from curarimimetic neurotoxins. Proc. Natl. Acad. Sci. U.S.A. 85: 4087–4089.

Hucho, F. 1986. The nicotinic acetylcholine receptor and its ion channel. Eur. J. Biochem. 158: 211–226.

Kao, P. N. and Karlin, A. 1986. Acetylcholine receptor binding site contains a disulfide cross-link between adjacent half-cystinyl residues. J. Biol. Chem. 261: 13735–13743.

Karlin, A. 1983. The anatomy of a receptor. Neurosci. Commen. 1: 111–123.

Maelicke A. 1988. Structure and fuction of the nicotinic acetylcholine receptor. In: Handbook of Experimental Pharmacology. 86: 267–313.

McCarthy, M. P., Earnest, J. P., Young, E. F., Choe, S. and Stroud, R. M. 1986. The molecular neurobiology of the acetylcholine receptor. Ann. Rev. Neurosci. 9: 383–413.

Moskovitz, R. and Gershoni, J. M. 1988. Three possible disulfides in the acetylcholine receptor α-subunit. J. Biol. Chem. 263: 1017–1022.

Ohana B. and Gershoni, J. M. Comparison of the toxin binding site of the nicotinic acetylcholine receptor from *Drosophila* to human. Biochemistry. 29: 6409–6415.

Popot, J.-L. and Changeux, J.-P. 1984. Nicotinic receptor of acetylcholine: Structure of an oligomeric intergral membrane protein. Physiol. Rev. 6: 1162–1239.

Radding, W., Corfield, P. W. R., Levinson, L. S., Hashim, G. A. and Low, B. W. 1988. α-Toxin binding to acetylcholine receptor α179–191 peptides: intrinsic fluorescence studies. FEBS Lett. 231: 212–216.

Discussion (DR. GERSHONI)

Er-El: You have shown several toxins which greatly vary in size. Have you tried to look at minimizing the receptor for the peptide toxins, rather than minimizing the ligand?

Gershoni: The whole definition of what is a receptor and what is a ligand is interesting and has changed to some degree over the years. Initially, I was taking a complex of 2,500 amino acids (the intact receptor), and looking at the binding of a small 74–residue protein. But now, by chopping down our receptor, we are left with a 13–amino acid peptide which binds the 74 amino acid-long ligand. Who then is binding whom? Of course, we can also cut down the size of the toxins per se; the central "loop-2" appears to be important. There have been recombinant loops and synthetic peptides which represent this loop which continue to bind the nicotinic receptor. Tom Lentz, a colleague at Yale, has shown, in fact, that he can make synthetic peptides of "loop-2" which bind to synthetic peptides of the receptor, but unfortunately the affinities are very low. I think, for the purpose of trying to understand the molecular interactions, one can exploit the huge diversity of alpha-neurotoxins derived from snakes – easily, 20–30 toxins are available and about 100 different sequences are known. These are divided into short and long toxins, each slightly different in their structure. It actually is a very convenient system for looking at a diversity of toxins binding to a single substrate.

M. Rosenberg: You mentioned that the 20–mer has three orders of magnitude less affinity than the full receptor, therefore you are at 0.1% of the actual binding of this system. Thus, what concentrations did you need to be able to get any kind of protection or blocking?

Gershoni: Around equimolar with the toxin, or at least within one order of magnitude, we get a pharmacological effect. I would like to comment on what is 0.1%. Obviously, the free energy drop is only a few kilocalories because the difference is related to the log of Kd. So under the circumstances, we should not be misled by the drop from 10^{-10} to 10^{-7} or 10^{-8}. The issue is not so much the affinity, but rather, can we intercept the pathogen in an effective way? In the case of the neuromuscular junction, there is compartmentalization, so if there is sufficient amount of decoy in the blood stream which can intercept the toxin before it actually reaches the diaphragm (the ultimate target) and then clear it, we would be in a pretty good shape. The problem is going to be much more severe in intervening with a viral infection. There, taking the example of CD4 and gp120, we are confronted with yet a more complicated problem. Today, CD4 is interacting with gp120 with an affinity of 10^{-9}, and the concentration used in most neutralization experiments is about 1 μg of CD4 to protect 10^6 cells (about one plate). What is the concentration of T4 positive cells in the lymph node? There, there is a marked increase; in order to protect the lymph node from infection we are going to need a higher affinity or a higher concentration of CD4. Neither has been developed thus far. So, the point is very well taken as to what is going to be the balance of an effective decoy and the pathogen and its ultimate target – these three components are going to dictate the competition and in what way the balance is going to work, in our favor or not.

Amitai: Could you elaborate on the kinetics of the binding of the toxins to the synthetic peptide ligands since we know the alpha-bungarotoxin is a very fast-acting ligand, with a bimolecular rate constant of 10^6–10^7 M^{-1} min^{-1}? Did you look for delayed death in your in vivo experiments?

Gershoni: I will answer your second question first. Our overall rule of thumb was if the animal was alive two or more weeks after the end of the experiment, it "survived." In the case of curare, I never saw a delayed death beyond 1–2 hr. Once the curare was cleared, that was the end of it. For cobratoxin, 24 hr was taken as the end point. Most of these experiments were over in the first eight hours; periodically you would see another death or so. But from 24 hr to two weeks, we saw no effect.

As far as the kinetics are concerned, that is an extremely well-taken and interesting issue. One has to assume that the complex between the ligand and its interface has some sort of thermodynamic advantage. The bound state must be energetically advantageous over the dissociated state. The question is in essence, therefore, an issue of kinetics. We have not as yet asked the question, comparing the 17–mer vs. the alpha subunit vs. the intact receptor. Do we have accessory phenomena that could accelerate the kinetics, as the ligand interacts with the receptors? On the other hand, we could possibly see the reverse. It may be that the binding site on the intact receptor is more buried, and the minute you excise it, making it more accessible, the kinetics may improve while the affinity may drop. Those are very fundamental questions which are being addressed today.

Lotan: The combination of biochemistry and genetic engineering here achieved is really impressive. I also have a few questions. First, what is the difference between the concept of molecular decoy and the concept of competitive binding of which we have been aware for quite some time? Second, the compartmentalized structure of your system is certainly a problem, but there is a beautiful engineering tool, namely in compartmental analysis which handles exactly such systems. Plug in all the characteristics, and you can predict the distribution of toxin and decoys in each compartment. We have done this for enzymic detoxification, and it works wonders.

Gershoni: My father taught engineering at the Technion, so I have often been told that I should be thinking in terms of engineering as well. Dr. Fong, from Smith-Kline, has done an enormous amount of very elaborate pharmacokinetics looking at the different possible compartments, the clearance, and where CD4 is cleared, and so on. This is a very difficult issue, especially due to the complexity of the human being as apposed to the relative simplicity of the machine model. In principle, there may be a lot of similarities and much to be learned. Unfortunately, I am not an expert there; maybe I should listen more to my father.

With regard to your first question, the novelty of a decoy, there is really very little which is new under the sun. Obviously, we are looking at a binding site, competition, and everything we have all learned about classical enzyme competition. The difference is that as a basic goal, we define the native receptor, identify that aspect which is involved in the ligand binding, in order to learn directly from nature's "blue print" to generate something novel which will intercept pathogens in a beneficial manner. That entity, which we define as a decoy, is derived from the native receptor and must have the appropriate pharmacological characteristics to be useful. It is a semantic issue, and a patent issue as well. In the long run we are trying to define an area where we can make something which is novel.

Key Words: Toxin, Acetylcholine receptor, Drug design, Decoy, Anti-viral agents, Ligand binding site, Biorecognition

Session IV

Material Biotechnology

Outdoor Cultivation of Microalgae in a Closed System for the Production of Valuable Biochemicals

Shoshana (Malis) Arad and Ephraim Cohen
The Institutes for Applied Research, Ben-Gurion University of the Negev, P.O. Box 1025, Beer-Sheva 84110, Israel

Introduction

General

A rapidly emerging area of biotechnology is the cultivation of microalgae, i.e., the use the photosynthetic machinery for the production of valuable chemicals. Attempts have been made to apply the mass culturing of microalgae for the production of biochemicals, proteins, and fuels (Soeder and Binsack, 1978; Shelef and Soeder, 1980; Barclay and McIntosh, 1986; Richmond, 1986). However, only a few commercial successes have been reported, the most important being the production of β-carotene from the halotolerant alga *Dunaliella* (Ben Amotz and Avron, 1989). One of the major drawbacks limiting the mass cultivation of microalgae on a commercial scale is the problem of outdoor cultivation of these organisms on a large scale.

In recent years our laboratory at the Institutes of Applied Research of Ben-Gurion University has been involved in developing processes for the production of valuable biochemicals from various unicellular red algae.

Red Microalgae

The red microalgae include about eight genera, the best known of which are *Porphyridium*, *Rhodella*, and *Rhodosorus*. Morphologically, they are the simplest of all the red algae, and their reproduction is asexual. In most species the algal cells are brown as a result of their chlorophyll and phycoerythrin contents (Gantt, et al., 1968). Many strains grow in fresh water, while others grow in seawater. A great deal of work has been devoted to the genus *Porphyridium*. The ultrastructure of *Porphyridium* cells (Gantt, et al., 1968; Ramus, 1972) and the conditions for growth have been well documented (Brown and Richardson, 1968; Hoogenhault, 1963; Sommerfeld and Nichols, 1970).

D. Kamely et al. (eds), Biotechnology: Bridging Research and Applications, 301–316.

The cells of the red microalgae are encapsulated within a sulfated polysaccharide in the form of a gel. The capsules are thinnest during the logarithmic phase of growth and thickest during the stationary phase (Ramus, 1972). When the rate of solubilization of polysaccharide from the cell surface was measured as a function of its production, it was found that during the stationary growth phase the rate of production was higher than the rate of solubilization in the external medium (Ramus, 1973). This process facilitates the creation of a capsule, and as a consequence cells reach their maximum size at this stage. It seems that the thickness of the capsule is influenced by three factors: the rate of polysaccharide production, the degree of polysaccharide solubility, and the cell surface area.

During growth of *Porphyridium* in a liquid medium, the viscosity of the medium increases due to the release of polysaccharide from the cell surface. Sulfate from the growth medium is utilized by *Porphyridium* in the synthesis of polysaccharide inside the cell. It has been shown by electron microscopy that production and transfer of the polymer to the cell surface is mediated by the golgi apparatus (Ramus and Groves, 1974; Ramus and Robins, 1975; Ramus, 1972, 1973, 1977, 1980). Although it is believed that the capsule is homogenous, it seems possible that the inner layer differs from the outer layer, as it does in the red macroalgae.

The influence of environmental conditions on polysaccharide production in *Porphyridium* has been previously studied. The effect of nitrate on polysaccharide production was studied by a number of workers (Thepenier, et al., 1985; Adda, et al., 1986; Arad, et al., 1988; Arad, 1988). Kost, et al. (1984) and Wanner and Kost (1984) have investigated the effect of nitrate and sulfate starvation on the cell ultrastructure. It was shown (Lerental-Brown and Arad, unpubl.) that sulfate starvation inhibits growth and leads to increased production of polysaccharide content per cell.

Biochemicals from Red Microalgae

Dyes

The red algae are characterized by a group of accessory pigments known as phycobiliproteins, which include phycoerythrin (red-violet) and phycocyanin (blue) (Gantt, 1980). The high fluorescence of these compounds makes them attractive for diagnostics. Recently, phycobiliproteins have been applied in diagnostics in multiple color flow cytofluorimetric analysis, in which conjugates of phycobiliproteins and molecules with specific biological activity bind specifically to their target molecules making them highly fluorescent. The method is very sensitive because the specific absorbance of phycoerythrin is very high, even at very low concentrations. Glazer and Stryer (1984) developed a series of conjugates, designated phycofluor probes, that are composed of fluorescent phycobiliproteins (mainly phycoerythrins) covalently bound to biologically active molecules. These fluorescent markers include phycoerythrin-immunoglobulins, phycoerythrin-protein A, phycoerythrin-avidin, and phycoerythrin-biotin. Ryan, et al. (1984) succeeded in increasing the sensitivity of this method to enable the detection of leukemia at early stages by means of antibodies conjugated to phycoerythrin. They were able to detect 5×10^6 positive cells as compared with 308×10^6 by former methods. The use of phycoerythrin together with a second fluorescent marker enables multiple cytotoxic examination. The phyocerythrin conjugates have a number of other advantages: their fluorescence is not affected by the majority of natural molecules, they are soluble in water, and they can be stored for long periods (Ol, et al., 1982). Today, phycoerythrin which is commercially available as the pure compound or as conjugates, fetches high prices (up to $89.50 per mg), and its diagnostic uses are widening constantly.

In the food industry the tendency to use natural, rather than synthetic, dyes has encouraged the search for alternative dyes (Taylor, 1984). The advantages of the phycobiliproteins over other dyes currently in use are their intense color, their high solubility in water, and their stability to changes in pH. To date, phycocyanin, which is extracted from the blue-green algae, is the only algal dye used in the food industry. There are, in fact, some Japanese patents describing the use of the blue color of phycocyanin in various food products (Patent no. 79–95770). Since many red and blue pigments are thought to be carcinogens, phycoerythrin and phycocyanin might have promising potential as food dyes.

Sulfated Polysaccharides

Despite interest in the functional roles of the polysaccharides of the red microalgae, knowledge concerning their chemical composition and structure is limited, particularly because their chemical structure is more complex than that of other marine polysaccharides such as agar or carrageenan.

The constituent sugars have been identified from total hydrolysates of the polysaccharides produced by different species of unicellular red algae. For all the species studied so far, the bulk of the polymer has been found to be composed of xylose, glucose, and galactose in various ratios (Arad, 1988; Deason, et al., 1983; Dubinsky, et al., 1988; Heaney-Kieras and Chapman, 1976; Heaney-Kieras, et al., 1976, 1977; Percival and Foyle, 1979). Additional minor sugar constituents (i.e., rhamnose and methylated sugars) have also been identified, as have the polysaccharides contain glucuronic acid and half sulfate ester groups, which confer acidic properties on the polymers (Arad, 1988; Deason, et al., 1983; Dubinsky, et al., 1988; Heaney-Kieras and Chapman, 1976; Heaney-Kieras, et al., 1976, 1977; Percival and Foyle, 1979). Another feature common to all the biopolymers is the presence of a protein attached to the polysaccharides (Heaney-Kieras and Chapman, 1976; Heaney-Kieras, et al., 1976, 1977; Percival and Foyle, 1979). We have found that the polysaccharides of all the red algae investigated by us–*Porphyridium* sp., *P. cruentum*, *P. aerugineum* and *R. reticulata*–have a basic disaccharide building block in common. This disaccharide is an aldobiouronic acid with the formula 3–*O*-(α-D-glucopyranosyluronic acid)-L-galactopyranose. We have recently conducted a comprehensive ^{1}H- and ^{13}C-NMR study of this disaccharide, which was isolated from the extracellular polysaccharide of *R. reticulata* by mild acid-catalyzed hydrolysis (Glaser, 1988; Jaseja, et al., 1989). The structure was elucidated by 1D- and 2D-NMR spectroscopy, and the disaccharide was shown to exist in aqueous medium as an ~ 1:2 mixture of the α- and β-anomers, respectively (Glaser, 1988; Jaseja, et al., 1989). Monosugar sulfates were also isolated and characterized by IR and NMR spectroscopy (Lupescu, et al., 1990). At present a trisaccharide is under investigation.

In addition to the complexity of defining the repeating unit in these polysaccharides, it is also thought that the mucilages probably consist of more than one polysaccharide. Electrophoresis of the isotopically labelled (^{14}C and ^{35}S) polysaccharide from *P. aerugineum* showed the presence of two fractions with differing amounts of sulfate. Other attempts to isolate different entities by analytical centrifugation (Heaney-Kieras and Chapman, 1976; Heaney-Kieras, et al., 1976, 1977) or anion-exchange chromatography (Percival and Foyle, 1979; Medcalf, et al., 1975; Evans, et al., 1974) have failed. We have recently shown that the polysaccharides of *R. reticulata* and *Porphyridium* sp. are composed of two fractions that differ in their charge and chemical composition (Dubinsky, et al., 1990). In some species attempts have been made to determine the molecular weight of the polymers, but the values have to be considered as rough estimates (Arad, 1988; Heaney-Kieras and Chapman, 1976; Heaney-Kieras, et al., 1976, 1977; Percival and Foyle, 1979). Lately, we have found that the molecular weight of *Porphyridium* sp. polysaccharide to be $5–7 \times 10^{6}$ daltons.

Although the polysaccharides of *Porphyridium* sp., *P. aerugineum* and *R. reticulata* differ in composition, their rheological characteristics were found to be similar (Geresh and Arad, 1990). Aqueous solutions of the biopolymers were stable over a wide range of pH values (2.0–10.0) and temperatures (20–90°C) and were compatible with monovalent cations. Mixtures of the algal polysaccharides with locust beam gum exhibited synergism and syneresis. When the gel strength was compared with that of agar gel at the same concentration, the polysaccharide gels were found to be weaker.

The polysaccharide of *P. aerugineum* has been found to be superior to Kelzan (xanthan gum) and has already found application in tertiary oil recovery (Savins, 1978). In the food industry, this polysaccharide has shown potential as a thickener (Israel Chemicals Ltd., pers. comm.).

Unsaturated Fatty Acids

Relatively high quantities of essential fatty acids have been demonstrated in *Porphyridium* (Ahern, et al., 1983; Nichols and Appleby, 1969). The most important of these is arachidonic acid (5,8,11,14–eicosatetraenoic acid), a highly unsaturated fatty acid that is only rarely found in plants and in other food sources. In addition to being an essential fatty acid in the human diet, arachidonic acid is the natural precursor of a large family of structurally related C-20 compounds that include the prostaglandins, thromboxanes, leukotrienes, and prostacyclins, all of which are potent biological regulators.

Arachidonic acid and its C-20 derivatives are usually obtained from animal tissue. The use of this family of compounds is becoming increasingly important in clinical research and pharmacy, since arachidonic acid derivatives are synthesized and released sparingly by many tissues and are not concentrated in any organs to a significant degree. To overcome this shortage new sources must be found. In heterotrophic microorganisms, fatty acids have been reported to account for no more than 9% of the dry weight of the organisms examined, and the C-20 fatty acids constitute less than 2.5% of total fatty acids. In *Porphyridium*, arachidonic acid alone constitutes as much as 36% of the total fatty acids (Nichols and Appelby, 1969; Ahern, et al., 1983). Thus, *Porphyridium* has been suggested as a new source of arachidonic acid (Ahern, 1984).

The C-20 polyunsaturated fatty acids are traditionally consumed in the human diet in the form of fish oil. However, the fish do not synthesize it themselves but acquire it from the algae they eat (Klausner, 1986). By extracting these polyunsaturated acids directly from algae, the less desirable components of fish oil, such as related fatty acids and cholesterol, as well as the fish odor, can be avoided.

A species of red microalga was found by us to contain a highly unsaturated fatty acid, eicosapentaenoic acid (EPA), amounting to about 4% of their dry matter and about 40% of their lipids. In feeding experiments performed by us on mice and rats, animals that were fed these dried algae had reduced blood cholesterol and triglyceride levels. In sugar-loading experiments we found that the levels of glucose and insulin in animals fed with the algae were much lower than those in control animals. We are now developing the dried algal product as a health food or food supplement and are negotiating with health care authorities to perform feeding experiments on human volunteers.

Outdoor Cultivation

Contamination

One of the major problems in outdoor cultivation is contamination by bacteria, fungi, and other algae as well as predation by zooplankton (Canter and Lund, 1968).

Contaminant organisms can hinder growth and production of the desirable algae, impair product quality, and cause collapse of the culture. In addition to the technological approach (i.e., sleeves) discussed later, we are also attacking the problem of maintenance of clean cultures by applying two biological concepts: a) selection for resistance of a particular alga to herbicides to which its competitors are sensitive; and b) determining selective growth conditions that are favorable to the alga but not to the predator.

In the first biological approach, we selected a *Porphyridium* mutant resistant to the herbicide sulfometuron methyl (SMM), the active ingredient of the commercial preparation Oust (Du Pont) (van Moppes, et al., 1989). Such a culture may be treated with the herbicide that will selectively kill the competitors but not the mutant alga.

Physiological, biochemical, and genetic studies have shown that the primary mode of action of SMM in bacteria and higher plants is inhibition of the enzyme acetohydroxy acid synthase (AHAS), which catalyzes the first common step in the biosynthesis of the branched-chain amino acids, valine, leucine and isoleucine. SMM reduced growth and maximum cell concentration of the wild-type *Porphyridium* sp., its maximum effect being exerted at 55 M. Addition of valine or isoleucine to the growth medium did not cause relief in the inhibition probably because of the inability of the autotrophically grown algae to take up the amino acids.

According to the frequency of mutation (3×10^7), it seems likely that the mutant is a result of a spontaneous single gene mutation. The mutant was found similar to its parent strain of *Porphyridium* sp. with respect to growth and production of cell wall polysaccharides and appearance of colonies and cell morphology. *Porphyridium* sp. AHAS was found to be the target of SMM. We have thus obtained a biochemically defined mutation, which may be used for maintaining monoalgal culture and may also serve as genetic marker for *Porphyridium*. This is the first mutation known in unicellular red algae. The existence of genetic markers is essential for molecular biology studies.

The SMM-resistant mutant alone is not sufficient for solving the problem of contamination outdoors, since the competitors might also develop resistance. Therefore, we intend to introduce into the mutant an additional mutation for resistance to another herbicide. A double mutant will have better chances outdoors. This subject is under investigation in our laboratory.

In the second biological approach we are seeking selective environmental conditions that are favorable to the desired alga but not to its predators or competitors (Ucko, et al., 1989). The main contaminant in our outdoor cultures of *Porphyridium* sp. is the dinoflagellate *Gymnodinium*. This predator ingests algal cells and can cause collapse of the cultures. Up to 15 cells of *Porphyridium* sp. are ingested by a single *Gymnodinium* cell, its volume being increased accordingly. In order to find differential requirements for the predator and its prey, we compared the effects of salinity, temperature and pH on *Gymnodinium* and *Porphyridium* sp. It seems that *Porphyridium* sp. is the more resistant of the two organisms. It can tolerate and grow under a wider range and more extreme conditions of pH, temperature, and salinity. We assume that the polysaccharide envelope around the *Porphyridium* sp.

cells enables the cells to survive under extreme conditions. This greater tolerance is already being applied to confer an advantage on the alga over its predator.

We have found that *Gymnodinium* contains enzymatic activity that can degrade the sulfated polysaccharide of *Porphyridium* sp. This activity is specific to *Porphyridium* sp. polysaccharide and does not affect the polysaccharides of the other red microalgae tested by us, *P. aerugineum* and *R. reticulata*. This finding is in accordance with the inability of *Gymnodinium* cells to ingest these red algal cells. This enzymatic activity is already being applied by us for the production of protoplasts from *Porphyridium* sp. with the aim of manipulating this alga genetically and for the determination of the composition and structure of the complex cell wall polysaccharide of *Porphyridium* sp.

Cultivation in Closed Systems–Sleeves

Microalgae are usually cultivated outdoors in open ponds. Low productivity and contamination seem to be the factors limiting this technology. It appears that productivity is affected by: limited light availability to the cells in the deeper layers of the ponds (Goldman, 1979); above optimal day temperatures in the summer and suboptimal temperatures in the winter (Richmond 1986); photoinhibition in the summer caused by very high light intensities in the upper layer of the pond (Powles, 1984); increased respiration and hence a loss of biomass caused by high night temperatures in the summer (Grobellaar and Soeder, 1985); fluctuating salinity caused by intensive evaporation (Richmond, 1986); and inhibition of growth as a result of elevated oxygen concentrations (Powles, 1984).

As a result of the various problems, there has been a tendency to change from open to closed systems, e.g., different types of horizontally-arranged glass or polyethylene tubes (Chaumont, et al., 1988; Dicorato, 1980; Torzillo, et al., 1986) or a covered convector (Anderson and Eakin, 1985). A closed system consisting of polyethylene sleeves, as proposed by Trotta (1980) for rotifers, was adapted by us for outdoor cultivation of red microalgae. We have found higher growth and production in algal cultures grown in the sleeves as compared with those grown in the ponds (Cohen and Arad, 1989).

Description of the Cultivation Systems

Single Sleeves

Polyethylene sleeves, each consisting of a transparent polyethylene tube (150 cm long, 0.2 mm thick, and 32 cm in diameter), were sealed at one end

Figure 1. A unit of polyethylene sleeves.

and hung from the other end according to Cohen and Arad (1989). The sleeves were filled with 25 l of medium. The cultures were mixed by means of an air stream containing 3–4% CO_2 pumped into the sleeves at 4–5 l min^{-1} via glass tubing. The pH of the cultures ranged between 7–8.

Unit of Connected Sleeves

This unit consisted of a long transparent polyethylene sleeve, 0.2 mm thick and 10 cm in diameter. The sleeve was looped around iron tubes connected to a suitable structure so as to form a row of vertical columns 2 m high. Each pair of columns was separated from the next at the point of suspension on the iron frame. Each column was filled with 25 l of algal culture via a plastic tube and mixed by means of an air stream containing 3–4% CO_2. The unit was composed of 20 columns, with a total volume of 500 l (Figure 1).

Open Ponds

For purposes of comparison, plastic containers of capacity 50 l, each having a surface area of 0.22 m^2, were used (Cohen and Arad, 1989). The containers were filled with 25 l of culture (12.5 cm deep). Mixing was achieved by means of paddle wheels, and CO_2 was supplied at a flow rate of 30 ml min^{-1} via porous tubing laid at the bottom of the containers.

Table 1. Temperature (°C) and light intensity (μeinsteins $m^{-2}s^{-1}$) in the sleeves and ponds during the daytime throughout the experiment.

Day	0800		1200		1700	
	Sleeves	Ponds	Sleeves	Ponds	Sleeves	Ponds
A. Temperature						
0	18	10	29	23	33	28
4	11	9	23	18	23	18
12	17	12	33	26	30	27
B. Light Intensity						
0	420	145	300	250	250	90
4	70	25	25	30	40	30
12	15	8	7	10	10	4

The experiment was carried out during March.

Growth and Polysaccharide Production in Porphyridium *sp. Grown Outdoors in Sleeves and Ponds*

We have conducted a series of experiments in which we compared growth and production of the unicellular red algae *Porphyridium*. The experiments were carried out during March. Temperatures measured in the morning, at noon, and in the afternoon were 5–8°C higher in the sleeves than in the ponds. The light intensities in the sleeves and ponds were similar at noon, but higher in the sleeves in the morning and afternoon (Table 1).

Cell number and biomass produced by *Porphyridium* sp. growing in batch culture were compared in sleeves and ponds (Table 2). *Porphyridium* sp. grown in the ponds remained viable for only 13 days. After this time the culture became contaminated by dinoflagellates and cell number started to decrease. In this experiment, no dinoflagellate contamination was detected in the cultures grown in the polyethylene sleeves. Maximum cell number was higher in the sleeves than in the ponds, i.e., about 130% higher. In terms of biomass the difference between the sleeves and the ponds was even higher, i.e., 300% after 10 days of growth.

The pattern of changes in polysaccharides – soluble, bound, and total – is shown in Figure 2 (a, b and c, respectively). Most of the polysaccharide is concentrated in the bound fraction which comprises 60–70% of the total polysaccharide. The higher concentration of the bound polysaccharide also contributed to the increase in the biomass. Since the ponds were contamin-

Table 2. Growth and biomass production in *Porphyridium* sp. grown in ponds as compared with sleeves.

	Maximal cell no. ($\times 10^6 ml^{-1}$)	Biomass ($g l^{-1}$)
Ponds	12	0.3
Sleeves	28	1.2

Cell number and biomass were determined on day 10 after innoculation. The experiment was carried out during March.

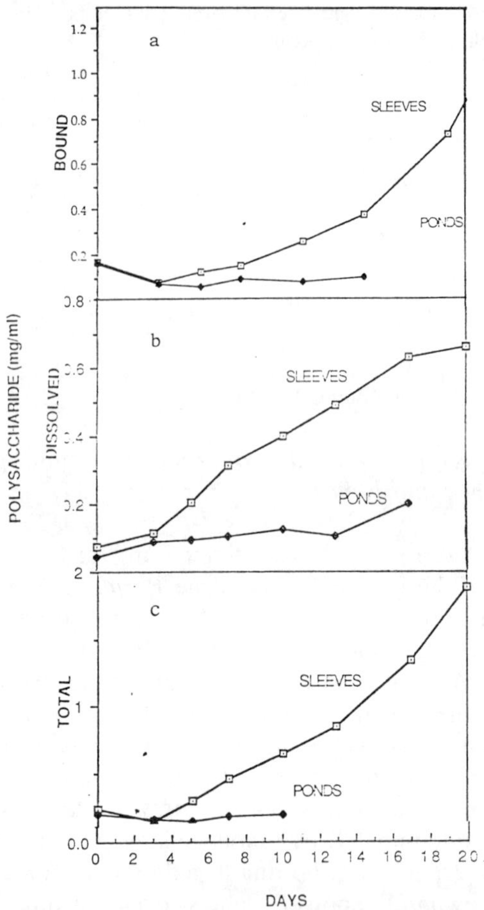

Figure 2. Polysaccharide production in *Porphyridium* sp. a. Bound; b. Dissolved; c. Total.

ated on day 14, cell number and biomass could not be measured; however, polysaccharides were measured up to day 17 (Figure 2). It is interesting to note that the increase in dissolved polysaccharides started as early as the day 3, whereas the increase in the bound polysaccharide was detected only on day 10.

Large-scale Cultivation

The encouraging results obtained with the sleeves led us to proceed with the development of a system for large-scale cultivation. The large-scale unit consisted of a number of connected sleeves, as shown in Figure 1. To investigate the effect of scaling up, *R. reticulata* was grown in a 2.5 m-long sleeve

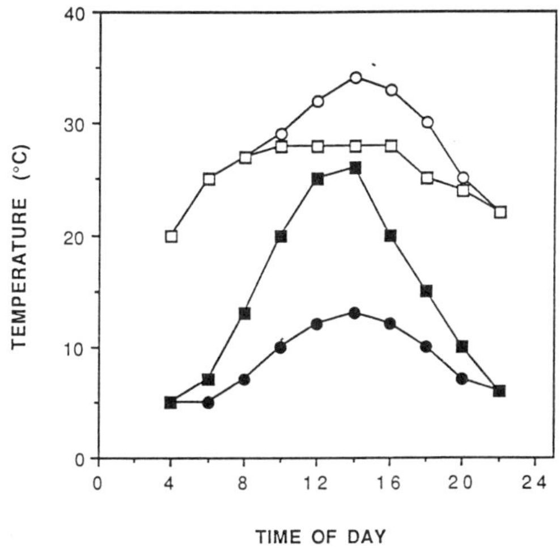

Figure 3. Daily temperature measured in the sleeve unit and minipond duirng algal growth. · = pond water, ■ = pond summer, □ = sleeves summer. The experiment was conducted at 'en Yahav during the summer. The temperature in the sleeves was prevented form rising above 28°C by spraying the sleevers with water from an automatic device.

unit composed of 20 sleeves (each with a diameter of 12 cm), containing a total culture volume of 500 l. The algae were cultured in a semi-continuous mode, i.e., when a culture density of 5–6 × 10^6 cells ml^{-1} was reached, 75% of the total volume was harvested. In this experiment, six harvests at intervals of 4–5 days yielded 79 g day^{-1} of biomass and 67 g day^{-1} of polysaccharide. A comparison was not made with the results for open ponds, since we were unable to culture *R. reticulata* outdoors in open ponds.

Summer and winter temperatures recorded at 'En Yahav in the sleeve units and mini-ponds are presented in Figure 3. In the winter, the low morning temperatures in the sleeves increased rapidly to reach an optimal value of 25°C at noon. The maximum winter temperature in the mini-ponds reached only 10°C. In the summer, the temperature in the ponds rose to 33°C, but that in the sleeves was prevented from rising above 28°C by spraying the outer surface with water.

The productivities predicted for *Porphyridium* sp. cultured continuously in sleeve units and ponds were calculated and are presented in Table 3. For a surface area of 100 m^2 the volume of culture in a sleeve unit with 1.5 m intervals between the rows was 33% higher than that in the ponds, whereas when 2 m intervals were allowed between the rows, the culture volume calculated for the sleeve unit was the same as that for the ponds. Biomass production in sleeve units with inter-row intervals of 1.5 and 2 m can be predicted to be 210% and 310%, respectively, higher than that in the ponds. Similar results were obtained for calculations of polysaccharide production.

Table 3. Predicted productivity of *Porphyridium* sp. calculated for 100 m^2 surface area of sleeves and ponds.

Surface area of 100 m^2 provided by:	Volume (m^3)	Production in continuous growth[b]	
		Biomass ($g\,m^{-2}\,day^{-1}$)	Polysaccharide ($g\,m^{-2}\,day^{-1}$)
Sleeve unit with intervals of 1.5 m between rows	20[a]	35.4	14.8
Sleeve unit with intervals of 2 m between rows	15[a]	26.5	11.1
Pond	15	11.4	6.0

[a] The volume was calculated for sleeves containing 60 1 (20 cm in diameter and 2 m high). Each row of sleeves (10 m long) contained 50 sleeves.
[b] Values were calculated from results obtained in continuous cultivation of *Porphyridium* sp. in sleeves.

Discussion

It seems that algal cultivation in sleeves is superior to that in mini-ponds. This was shown for the red microalgae *Porphyridium* sp. and *P. aerugineum* (Cohen and Arad, 1989) and for *R. reticulata* as well as for other unicells, e.g., *Nanochloropsis* in experiments performed in Beer-Sheva and 'En Yahav (results not shown here). The main factors contributing to the advantage of the sleeves over the ponds are the suitable temperatures and improved light availability. The differences in temperature between the sleeves and the ponds are already evident early in the morning. Thus, in the winter effective photosynthesis occurs over about 9 h of the day in the sleeves, as compared with only 3–4 h in the ponds. During the summer, cooling of the sleeves by spraying with water can prevent collapse of the cultures, which is a possibility in the desert where high temperatures prevail. The surface/volume ratio, which affects light availability, is better in the sleeves than in the ponds. By changing the sleeve diameter light intensity throughout the algal culture can be varied. By decreasing the sleeve diameter, productivity (biomass and polysaccharide) in *Porphyridium* sp. was increased (results not shown in this paper). Although the productivities increased as the sleeve diameter decreased, as was found by Pirt, et al. (1983), we did not experiment with diameters of less than 10 cm, since such small diameters do not seem feasible from practical point of view for large-scale operation. Other factors also contribute to the superior productivity obtained in the sleeves: 1) better turbulence improves light/dark cycles and reduces photo-oxidation, 2) evaporation is prevented by virtue of the system being closed, 3) contamination is reduced, and 4) polyethylene sleeves are inexpensive and can be used continuously for number of runs (their replacement a few times a year is simple).

The sleeve unit has certain advantages over both single sleeves and ponds. Since its installation and operation are simple, operating costs, which are high for single sleeves, are reduced. The relative volume of culture in the sleeve unit is higher than or similar to that in the ponds and is dependent on the distance between the rows. Thus, a pond of surface area of 100 m^2 contains 15 m^3 of culture volume, whereas a sleeve unit of the same surface area contains 15 or 20 m^3 for inter-row distances of 2 or 1.5 m, respectively.

Calculation of predicted productivity based on results obtained in the Negev for various algal species indicates that productivities 24 g m^{-2} day^{-1} can be achieved. Although the sleeve system appears to have advantages over ponds, we do recognize the problems raised in such a comparison, i.e., the productivity of ponds, usually described in g m^{-2} day^{-1}, cannot simply be transferred to the sleeve system. We make use of these productivity units since they are units commonly quoted in the literature (and are used by us only for purposes of comparison). Productivity per volume or per surface area can also be used. In addition, the comparison of the different factors affecting productivity in sleeves and ponds, especially light availability, are not always simple, raising the problems of measurement values, e.g., light per area, or per cell, or per day. All these and other problems have not yet been completely resolved. However, the sleeve unit represents in principle a totally new approach to algal cultivation, i.e., vertical (sleeves) vs. horizontal (ponds). [The horizontal concept also includes the tubular system suggested previously (Chaumont, et al., 1988; Torzillo et. al., 1986).

At this stage, we do not feel that the system has been completely perfected. The development and optimization of the sleeve unit towards large-scale production is now under way.

References

Adda, M., Merchuk, J.G. and Arad (Malis), S. 1986. Effect of nitrate on growth and production of cell wall polysaccharides by the unicellular red alga *Porphyridium*. Biomass 10: 131–140.

Ahern, T.J. 1984. Plant-derived catalysts and precursors for use in prostaglandin synthesis. JAOCS 61: 1754–1758.

Ahern, T.J., Katoh, S. and Sada, E. 1983. Arachidonic acid production by the red alga *Porphyridium cruentum*. Biotechnol. Bioeng. 25: 1057–1070.

Anderson, D.B. and Eakin, D.E. 1985. A process for the production of polysaccharides from microalgae. Biotechnol. Bioeng. Symp. 5: 533–547.

Arad (Malis), S. 1988. Production of sulfated polysaccharides from red unicellular algae. In: Algal Biotechnology. Stadler, T., Mollion, J., Verdus, M-C., Karamanos, Y., Morvan, H. and Christiaen, D., eds. Elsevier Applied Science, London.

Arad (Malis), S., Friedman (Dahan), O., Rotem, A. 1988. Effect of nitrogen on polysaccharide production in *Porphyridium* sp. Appl. Environ. Microbiol. 54: 2411–2414.

Baird, J.K., Sandford, P.A. and Cottrell, I.W. 1983. Industrial applications of some new microbiol polysaccharides. Bio/Technology 778–83.

Barclay, W.R. and McIntosh, R.P. 1986. Algal Biomass technologies. An Interdisciplinary Perspective. J. Cramer Beiheffe zur Nova Hedwigia 83.

Ben Amotz, A. and Avron, M. 1989. The biotechnology of mass culturing *Dunaliella* for products of commercial interest. In: Algal and Cyanobacterial Biotechnology, Cresswell, R.C., Rees, T.A.V. and Shah, N., eds. Honeyman Scientific and Technical.

Brown, T.E. and Richardson, F.L. 1968. The effect of growth environment on the physiology of algae: Light intensity. J. Phycol. 4: 38–12.

Canter, H.M. and Lund, J.W.G. 1968. The importance of protozoa in controlling the abundance of planktonic algae in lakes. Proc. Limn. Soc., Lond. 179: 203–207.

Chaumont, D., Thepenier, C., Gudin, C. and Junjas, C. 1988. Scaling up a tubular photoreactor for continuous culture of *Porphyridium cruentum* from laboratory to pilot plant (1981–1987). In: Algal Biotechnology. Stadler, T., Mollion, J., Verdus, M-C., Karamanos, Y., Morvan, H. and Christiaen, D., eds. Elsevier Applied Science, London, p. 199–208.

Cohen, E. and Arad (Malis), S. 1989. A closed system for outdoor cultivation of *Porphyridium*. Biomass 18: 59–67.

Deason, T.R., Butler, G.C. and Rhyne, C. 1983. *Rhodella reticulata* sp. Nov., A new coccoid rhodophytan alga (Porphyridiales). J. Phycol. 19: 104–111.

Dicorato, A. 1980. Coltivazione sperimentals in sistema tubolare in CNR ed. Prospective della coltura di Spirulina in Italia. Florenz, Italy, p. 261.

Dubinsky, O., Lerental (Brown), Y., Christiaen, D., Glaser, R., Barak, Z. and (Malis) Arad, S. 1988. Production and characterization of polysaccharides in the unicellular red alga *Rhodella reticulata*. In: Algal Biotechnology. Stadler, T., Mollion, J., Verdus, M-C., Karamanos, Y., Morvan, H. and Christiaen, D., eds. Elsevier Applied Science, London. pp. 451–461.

Dubinsky, O., Simon, B., Geresh, S., Barak, Z. and Arad (Malis), S. 1990. Composition of the cell-wall polysaccharide of the unicellular red alga *Rhodella reticulata* at two phases of growth. The 5th International Conference of the Society of Applied Algology. Recent Advances in Algal Biotechnology. 1990, Tiberias.

Gantt, E. 1981. Phycobilisomes. Ann. Rev. Plant. Physiol. 32: 327–347.

Gantt, E., Edwards, M.R. and Conti, S.F. 1968. Ultrastructure of *Porphyridium aerugineum*, a blue-green colored rhodophytan. J. Phycol. 4: 65–71.

Geresh S. and Arad (Malis), S. 1990. The extracellular polysaccharide of the red microalgae: Chemistry and rheology. Biomass. In press.

Glaser, R. 1988. Some illustrative examples of methodology for configurational and conformational assignment of disaccharide stereochemistry. In: Polysaccharides from Microalgae, a New Industry. Ramus, R. and Jones, M-C., eds. Duke University Marine Laboratory, Beaufort, North Carolina, pp. 98–105.

Glazer, A.N. and Stryer, L. 1984. Phycofluor probes. Trends Biochem. Sci. 9: 423–427.

Goldman, J.C. 1979. Outdoor algal mass cultures. I. Applications. Water Res. 13: 1–19.

Grobellaar, J.U. and Soeder, C.J. 1985. Respiration losses in planktonic green algae cultivated in raceway ponds. J. Plankt. Res. 7: 497–506.

Heaney Kieras, J. and Chapman, D.J. 1976. Structural studies on the extracellular polysaccharide of the red alga *Porphyridium cruentum*. Carbohydr. Res. 52: 169–177.

Heaney Kieras, J., Kieras, F.J. and Bowen, D.V. 1976. 2–O-Methyl -D-glucuronic acid, a new hexuronic acid of biological origin. Biochem. J. 155: 181–185.

Heaney Kieras, J., Roden, L. and Chapman, D.J. 1977. The covalent linkage of protein to carbohydrate in the extracellular protein-polysaccharide from the red alga *Porphyridium cruentum*. Biochem J. 165: 1–9.

Hoogenhaut, H. 1963. Naturwissenschaften 50: 456.

Jaseja, M., Perlin, A.S., Dubinsky, O., Christiaen, D., Arad (Malis), S. and Glaser, R. 1989. NMR structure determination of 3–O-(α-D-glucopyranosyluronic acid) L-galactopyranose, an aldobiouronic acid isolated from the unicellular red alga *Rhodella reticulata*. Carbohyd. Res. 186: 313–319.

Jones, R.F., Speer, H.L. and Kury, W. 1963. Studies on the growth of the red alga *Porphyridium cruentum*. Physiol. Plant., 16: 636–43.

Klausner, A. 1986. Algaculture: Food for thought. Nature 4: 947–953.

Köst, H.P., Senser, M. and Wanner, G. 1984. Effect of nitrate and sulphate starvation on *Porphyridium cruentum* cells. Z. Pflanzenphysiol Bd. 1135: 231–249.

Lupescu, N., Geresh, S. Arad, (Malis), S., Bernstein, M. and Glazer, R. Isolation and structure determination of some sulfated monosugars obtained from the extracellular polysaccharide of *Porphyridium* sp. unicellular red alga. Carbohyd. Res. In press.

Nichols B.W. and Appelby R.S. 1969. The distribution of arachidonic acid in algae. Phytochemistry 8: 1907–1915. Ol, V.T., Glazer, A.N. and Stryer, L. 1982. Fluorescent phycobiliprotein conjugates for analyses of cells and molecules. J. Cell Biol. 93: 981–986.

Percival, E. and Foyle, R.A.J. 1979. The extracellular polysaccharide of *Porphyridium cruentum* and *Porphyridium aerugineum*. Carbohyd. Res. 72: 165–176.

Percival, E. and McDowell, R.H. 1967. Chemistry and Enzymology of Marine Algal Polysaccharides. Academic Press, London.

Pirt, S.J., Lee Y.K., Walach, M.R., Pirt, M.W., Bayuzi, H.H.M. and Bazin, M.F. 1983. A tubular bioreactor for photosynthetic production of biomass from CO_2 design and performance. J. Chem. Tech. Biotechnol. 33B: 35–58.

Powles, B.S. 1984. Photoinhibition of photosynthesis induced by visible light. Ann. Rev. Plant Physiol. 35: 15–44.

Ramus, J. 1972. The production of extracellular polysaccharide by the unicellular red algae *Porphyridium aerugineum*. J. Phycol. 8: 97–111.

Ramus, J. 1973. Cell surface polysaccharides of the red alga *Porphyridium*. In: Biogenesis of Plant Cell Wall Polysaccharides. Loewus, F., ed. Academic Press, pp. 333–359.

Ramus, J. 1980. Algae biopolymer production. U.S. Patent No. 4,236,349.

Ramus, J. 1986. Rhodophytes unicells: Biopolymer physiology and production. In: Algal Biomass Technologies. Barclay, W.R. and McIntosh, R.P., eds. J. Cramer, Berlin.

Ramus, J. and Groves, S.T. 1974. Precursor-product relationship during sulfate incorporation into *Porphyridium* capsular polysaccharide. Plant Physiol. 53: 434–439.

Ramus, J. and Robins, D.M. 1975. The correlation of Golgi activity and polysaccharide secretion in *Porphyridium*. J. Phycol. 11: 70–74.

Rees, D. 1972. Shapely polysaccharides. Biochem. J. 126: 257–73.

Richmond A. 1986. Handbook of Microalgal Mass Culture, CRC Press.

Richmond, A.E. 1986. Microalgaculture. CRC Critical Reviews in Biotechnol. 4: 369–438.

Ryan, D.H., Mitchell, S.J., Hennessy, L.A., Bauer, K.D., Horan, P.K. and Cohen, H.J. 1984. Improved detection of rare CALLA-positive cells in peripheral blood using multiparameter flow cytometry. J. Immunol. Methods 74: 115–128.

Savins, J.G. 1978. Oil recovery process employing thickened aqueous driving fluid. Mobil Oil Corporation, N.Y. U.S. Pat 4,079,544.

Schlosser, U.G. 1982. List of strains, Sammlung von algenkulturen. Pflanzenphysiologisches. Institut der Universitat Gottingen (SAG) Ber Deutsch. Bot. Ges. 95: 272.

Shelef, G. and Soeder, C.J. 1980. Algal biomas: Production and use. Elsevier North Holland Biomedical Press.

Soeder C.J. and Binsack, R. 1978. Microalgae for food and feed, a status analysis. E. Schweizerbart'sche Verlags, Stuttgart. Ergebnisse der Limnologie 11.

Sommerfeld, M.R. and Nichols, H.W. 1970. Comparative studies in the genus *Porphyridium aerugineum*. J. Phycol. 6: 67–78.

Taylor, A.J. 1984. Natural colours in food. Dev. Food Colours 2: 150–206. Thepenier, C., Gudin, C. and Thomas, D. 1985. Immobilization of *Porphyridium cruentum* in polyurethane foams for the production of polysaccharide. Biomass 7: 225–240.

Torzillo, G., Pushparaj, B., Bocci, F., Balloni, W., Materassi, R. and Florenzano, G. 1986. Production of *Spirulina* biomass in closed photobioreactors. Biomass 11: 61–74.

Trotta P.A., 1980. Simple and inexpensive system for continuous monoxenic mass culture of marine microalgae. Aquaculture 22: 383–387.

Ucko, M. Cohen, E. Gordin, H. and Arad, (Malis) S. 1989. Relationship between the unicellular red algae *Porphyridium* sp. and its predator the dinoflagellate *Gymnodinium* sp. Appl. Environ. Microbiol. 55: 2990–2994.

van Moppes, D., Barak, Z., Chipman, D.M., Gollop, N. and Arad (Malis), S. 1988. An herbicide (sulfometuron methyl) resistant mutant in *Porphyridium* (Rhodophyta). J. Phycol. 25: 108–112.

Wanner, G. and Köst, H.P. 1984. Membrane storage of the red alga *Porphyridium cruentum* during nitrate and sulphate starvation. Z. Pflanzenphysiol. Bl. 1135: 251–262.

Discussion (DR. ARAD)

Er-El: Scaling up into the sleeves is almost linear because you have a very long way for oxygen and other things to travel. I can see the economical, logical impetus to work in open ponds, but if you go to much more sophisticated instrumentation that requires more manpower, maybe it is feasible to transfer the genes to yeast or something else and produce them in a regular fermentation facility where you can really control the large scale operations with minimum manpower. You can probably scale the sleeves into 300–400 l, but a few of the chemicals you are producing are quite inexpensive. This means that you may have to go to thousands of liters, which could be very difficult in sleeves.

Arad: That is a complicated question with a complicated answer. First of all, we are talking about very high priced products. For example, the price of beta-carotene today is $2,000 per kg. I agree for polysaccharides it might be problematic, but phycoerythrin sells for $80 per mg. For low-priced products, you might be right. The productivity we receive today is so low; it is very difficult to increase it. Our theoretical calculations, taking into consideration all the losses, show that we can get 24 g m^{-2} per day. The productivity today is 5 g m^{-2} per day at best. To answer the polyethylene sleeves, the engineers say that it is very primitive but also very cheap. Our economic calculations show that the construction is less costly than ponds. This is based on that what we have now, but it is not very developed yet, so I am sure it will get even better. That is the way to go. Transferring the genes to microorganisms can be done, but the genetics of the algae is still so undeveloped that it will take many years.

Key words: Microalgae, Natural dyes, Polyethylene sleeves, Polysaccharides, *Porphyridium*, *Rhodella*, Rhodophyta, Unsaturated fatty acids

Chemical Sensing Applications of Semiconductor Photoluminescence

Elizabeth R. M. Luebker[1], Larry K. Leung[1], Catherine J. Murphy[1], George C. Lisensky[2], and Arthur B. Ellis[1]*

[1]*Department of Chemistry, University of Wisconsin-Madison, Madison, Wisconsin 53706*
[2]*Department of Chemistry, Beloit College, Beloit, Wisconsin 53511*
*Author to whom correspondence should be addressed

Introduction and Theory

Semiconductor-derived interfaces have applications in chemical sensing because the bulk photoluminescence (PL) intensity of the semiconductor can be perturbed by chemical adsorption at the surface. Using changes in PL intensity, we have created semiconductor-based systems for non-destructive, *in situ*, "on-line" chemical sensing (Meyer, et al., 1987; Lisensky, et al., 1988). This paper presents practical applications of PL-based chemical sensors. We have investigated the use of semiconductor PL in a headspace analysis, and in the detection of compounds dissolved in hydrocarbon solution and in aqueous solution. We have constructed two portable systems using semiconductor sensors, where the excitation source, semiconductor, and detection electronics can be transported and used as an on-line system at remote locations. These sensors can be used to detect various classes of compounds in both the liquid and gas phases.

A variety of semiconductors with PL in the visible or near-IR region form adducts with compounds present either in gas or liquid phase ambients. We have demonstrated that the PL intensity from single-crystal n-CdS and n-CdSe is reversibly perturbed by surface adduct formation with gaseous ambients (Meyer, et al., 1988, 1989a, 1989b; Leung, et al., 1990). These PL effects are sensitive to the acid/base properties of the gaseous ambient: Lewis bases such as amines and butenes enhance the PL intensity, and Lewis acids decrease the PL intensity. Also, we have demonstrated that relative to a hydrocarbon reference solvent, solutions of Lewis acidic lanthanide β-diketonate complexes in the same solvent decrease the PL intensity of n-CdSe (Murphy and Ellis, 1990b), and Lewis bases such as phosphines (Murphy and Ellis, 1990a) and amines (Lisensky, et al., 1990) increase the PL.

The electronic structure of semiconductors can be described using band

D. Kamely et al. (eds), Biotechnology: Bridging Research and Applications, 317–331.

theory (Smith, 1978). There are two relevant bands, the valence band and the conduction band, separated by a moderate band gap. The valence band contains primarily occupied electronic energy levels and the conduction band contains mainly unoccupied electronic energy levels. When the semiconductor is illuminated with light having energy greater than or equal to the energy of the band gap, an electron can be promoted from the valence band to the conduction band, leaving a hole (the absence of an electron) in the valence band. The promoted electron can recombine with the hole radiatively or non-radiatively. Radiative recombination produces light of roughly band gap energy; this is the band edge PL.

The PL intensity can be perturbed by electronic states, lying energetically within the band gap, that are present at the semiconductor surface. These "surface states" can trap electrons or holes at the surface. For the n-type semiconductors used in our studies, trapping of conduction band electrons in lower energy surface states creates an electric field in the near-surface region of the solid. Because semiconductors have a relatively low density of charge carriers, this field can extend hundreds to thousands of Angstroms into the solid, defining the depletion region. Moreover, the non-radiative, surface electron-hole pair recombination rate, characterized by a parameter called the surface recombination velocity, can also be affected by surface states.

The increase or decrease in PL intensity observed in the presence of an adduct-forming compound that affects the energetic distribution and electronic population of surface states arises either from changes in surface recombination velocity, changes in the electric field thickness within the semiconductor, or both. If there is no change in surface recombination velocity between ambients, or if the surface recombination velocity is very large in both ambients, the PL variations can be interpreted as changes in the semiconductor's electric field thickness using a dead-layer model (Burk, et al., 1986). This model assumes that the electric field rapidly separates electron-hole pairs formed within the depletion region so that radiative recombination is improbable; consequently, the depletion region is "dead" to luminescence. The quantitative form of the dead-layer model is $PL_1 / PL_2 = \exp(-\alpha' \delta D)$, where PL_1 and PL_2 are the PL intensities in the presence of the adsorbing ambient and the reference, respectively; α' is the absorptivity of the solid, corrected for self-absorption; and δD is the variation in dead-layer thickness resulting from the replacement of the reference ambient with an ambient containing the adduct-forming compound.

In the CdSe systems investigated to date, varying the interrogating wavelength produces similar values of δD for a given adduct-forming compound, indicating that the PL intensity variations observed with adduct formation can be explained as changes in the electric field thickness. In the presence of bases that donate electron density to the semiconductor, the dead-layer thickness shrinks and the PL intensity increases; conversely, in the presence of acids that withdraw electron density, the dead-layer thickness expands

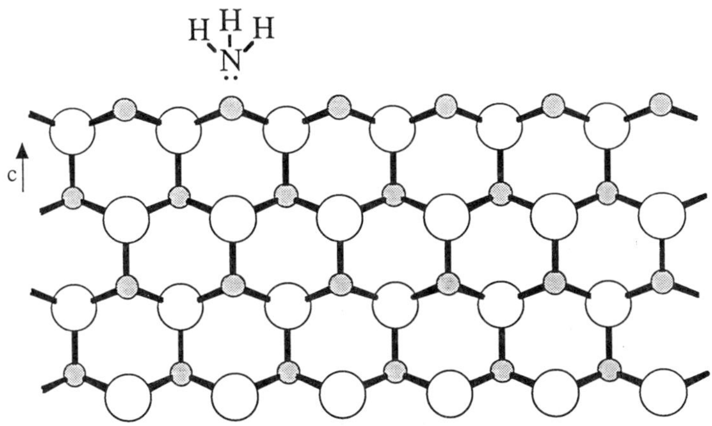

Figure 1. Possible binding mode of ammonia to the idealized (0001) face of CdSe. The smaller circles represent Cd^{2+} sites; the larger circles represent Se^{2-} sites. The (0001) face is perpendicular to the plane of the paper (the crystallographic *c* axis lies vertically in the plane of the paper).

and the PL intensity decreases. Time-resolved experiments have shown little change in the excited-state decay time as a function of ambient for these systems, implying little change in surface recombination velocity, and supporting the electric field origin of the PL changes (Leung, et al., 1990).

In our investigations, we have seen the PL intensity increase (for bases) or decrease (for acids) with concentration and eventually reach a saturation point. If we assume that there are a limited number of surface binding sites available for adduct formation (Figure 1), we can interpret the saturation in PL intensity as saturation of these sites on the semiconductor.

The Langmuir adsorption isotherm model is used to extract equilibrium constants for adduct formation at the semiconductor surface (for example, see Lisensky, et al., 1990). The quantitative expression of the Langmuir model is

$$\theta = K\,C_A\,(1 + K\,C_A)^{-1} \text{ or } \theta^{-1} = 1 + (K\,C_A)^{-1}.$$

The fractional surface coverage, θ, is related to changes in the dead-layer thickness, proportional to ln (PL_1/PL_2); θ is defined as $\ln(PL_x/PL_i)/\ln(PL_s/PL_i)$, where PL_x/PL_i is the ratio of the PL intensity at each concentration relative to the reference ambient, and PL_s/PL_i is the ratio of the PL intensity at saturation to that in the reference state (Figure 2). Plotting θ as a function of the concentration or partial pressure of the compound of interest yields the Langmuir adsorption isotherm. A linear double reciprocal plot of θ^{-1} vs. $C_A{}^{-1}$ allows extraction of adduct-formation equilibrium constants from the reciprocal of the slope for each compound studied.

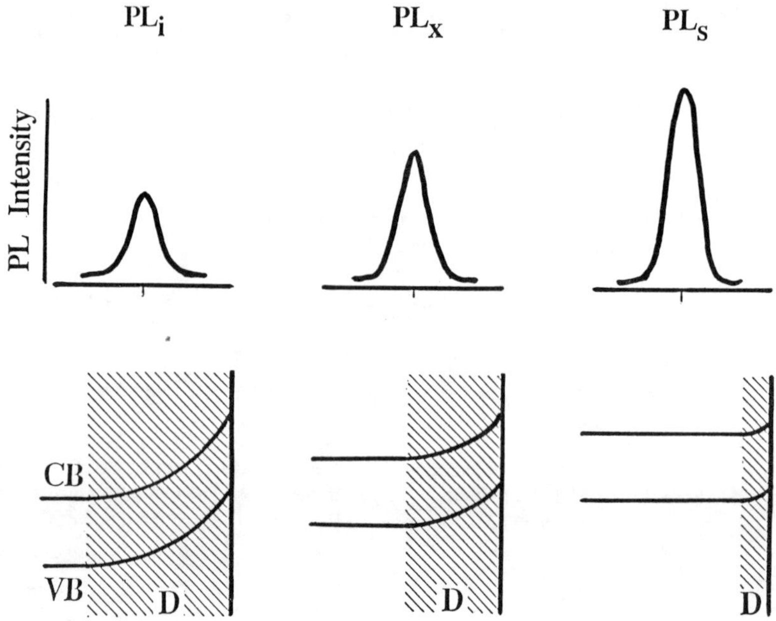

Figure 2. Relationship between PL intensity and the semiconductor's electric field (represented by the bent bands) in the presence of an adduct-forming compound. PL_i is the PL intensity in the presence of the reference ambient, PL_x is the intensity in the presence of an electron-donating compound (Lewis base), and PL_s is the maximum PL intensity observed in the presence of the Lewis base. In the band diagrams, energy is plotted vertically vs. distance from the semiconductor surface (represented by the heavy vertical line). The top of the valence band, VB, and bottom of the conduction band, CB, are indicated by the heavy horizontal lines. The shaded region represents the dead-layer thickness, D.

Materials and Methods

Materials and Sample Preparation

Methanol, bromine (Fisher, 99.9%), H_2SO_4 (Fisher), and 30% H_2O_2 (Baker) were used as received for etching. Aniline (Mallinckrodt) was vacuum-distilled prior to use. Ammonium carbamate (Aldrich) was used as received. Ammonium sulfide (20% by weight in H_2O, Aldrich) was used as received for passivation (*vide infra*). Single-crystal, vapor-grown *c*-plates (10 × 10 × 1 mm) of n-CdSe and n-CdS:Te (Te-doped CdS; Te concentration of ~100 ppm) with resistivity of ~2 Ω-cm were obtained from Cleveland Crystals, Inc., Cleveland, OH. Samples, scored and broken to ~4 × 4 × 1 mm, were etched in a dilute (~1:60 v/v) Br_2/MeOH solution until the shiny 0001 Cd-rich face was observed and then ultrasonicated in MeOH. Low-doped n-GaAs was grown by vapor-phase epitaxy on Si and provided by Dr. A. A. Burk, Jr., of Westinghouse. Samples were scored and broken to ~4 × 4 mm and etched with 1:8:500 (by volume) H_2SO_4:30%H_2O_2:H_2O, followed by a H_2O rinse until the GaAs PL was observed. The GaAs was passivated by placing 1 or 2 drops of aqueous $(NH_4)_2S$ on the surface and drying in a stream of air or nitrogen, repeating the procedure until a surface with high PL intensity was obtained. Semiconductor samples were edge-suspended with DUCO cement or black epoxy on a glass rod, and left in air or under nitrogen until use. For studies in hydrocarbon solution, the semiconductor was stored in the reference solution.

Apparatus and Optical Measurements

Irradiation sources in the laboratory setting included a Coherent Innova 90-5 Ar^+ laser (457.9 and 514.5 nm), a Coherent K-3000 Kr^+ laser (520.9 nm), and Melles Griot HeNe lasers (632.8 nm and 543.5 nm). Laser plasma lines were eliminated by passing the beam through Oriel interference filters (centered at the laser line, fwhm of 10 nm). Except for the HeNe lasers, laser beams were expanded by passage through a 10× beam expander and masked to illuminate only the semiconductor surface.

Headspace Analysis

The glass cell used for the ammonium carbamate headspace analysis is shown in Figure 3.

Organic Solution Detection

The glass cell used for the CdS:Te organic solution experiments has been described previously (Murphy and Ellis, 1990b). The portable apparatus for remote sensing using CdS:Te is diagrammed in Figure 4. It consists of a xenon light source with a CdS photocell and voltage divider, or Si photodiode and operational amplifier, connected to an X-Y recorder used in the time-base mode.

Aqueous Solution Detection

The glass cell used was the same as in the organic solution experiments. The portable apparatus for aqueous sensing with GaAs, shown in Figure 5, uses a HeNe laser as the excitation source and an optical fiber coupled to an RCA Photon Counting Module SPCM-100 for detection.

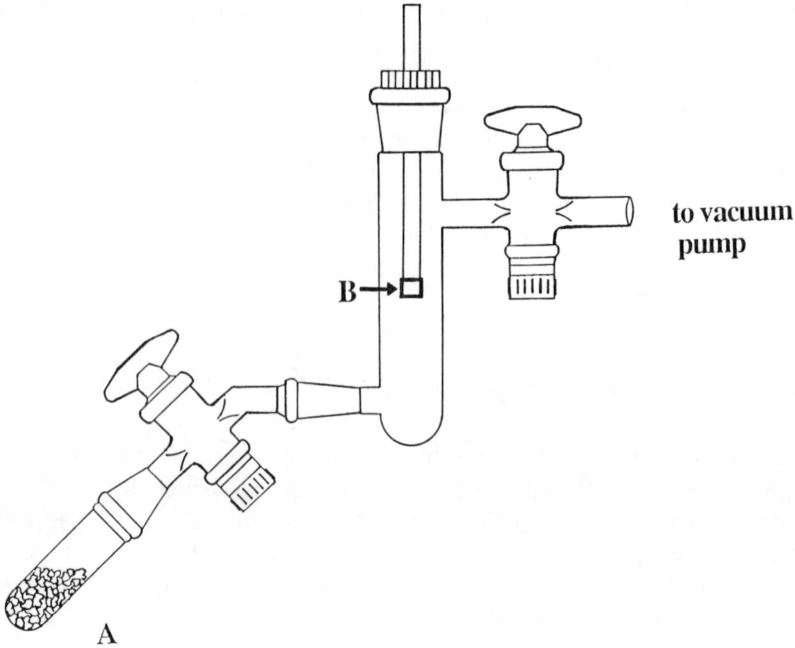

Figure 3. Glass cell used in ammonium carbamate headspace experiment. Compartment A contains the ammonium carbamate. The semiconductor B is mounted on a glass rod using DUCO cement or black epoxy.

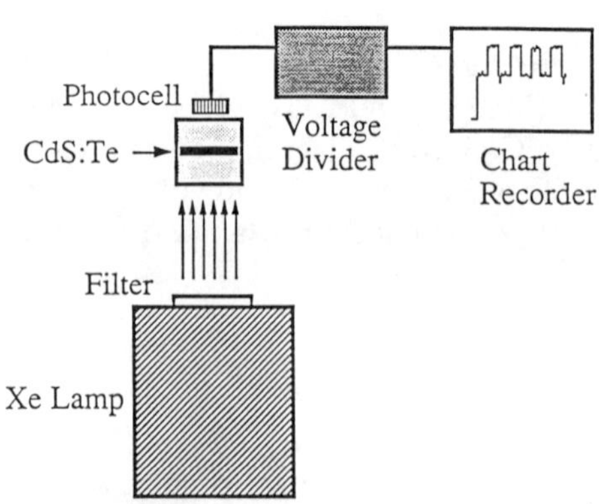

Figure 4. Portable apparatus for remote organic solution sensing using CdS:Te. A Si photodiode and operational amplifier can be substituted for the CdS photocell and voltage divider with similar results.

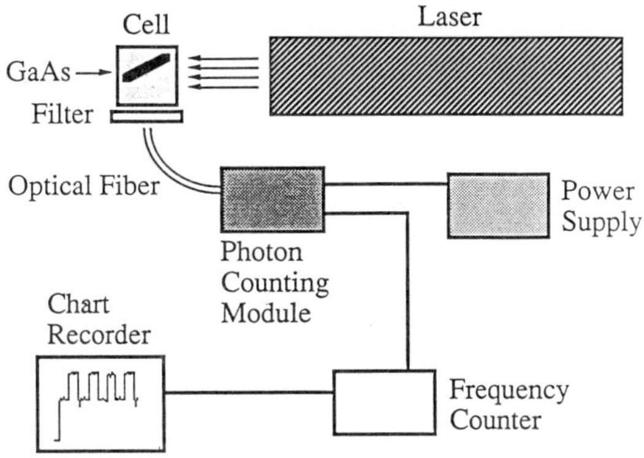

Figure 5. Portable apparatus for remote aqueous solution sensing using $(NH_4)_2S$-passivated GaAs. Portable red (632.8 nm) or green (543.5 nm) HeNe lasers have been used successfully in this arrangement.

Results and Discussion

Headspace Analysis

We have demonstrated that adduct-induced PL perturbations enable semiconductors to serve as gas chromatographic detectors (Lisensky, et al., 1988). These results prompted us to investigate the use of emissive semiconductors in headspace experiments. In the headspace experiment, which employs the apparatus of Figure 3, the semiconductor is positioned in a closed, evacuated cell and selectively exposed to volatile compounds produced in a separate chamber. Because the CdSe-ammonia adduct has been characterized in gaseous adsorption experiments (Meyer, et al., 1988), we used CdSe in the headspace analysis as a selective sensor for NH_3 produced by ammonium carbamate decomposition.

Ammonium carbamate decomposes into ammonia and carbon dioxide according to the reaction:

$$NH_4CO_2NH_{2(s)} \Leftrightarrow 2\ NH_{3(g)} + CO_{2(g)}.$$

We detect the ammonia produced in this decomposition reaction using the PL intensity increase resulting from reversible CdSe-ammonia adduct formation. The CdSe PL intensity in control experiments increased with the partial pressure of ammonia and saturated by about 10% NH_3 (Figure 6). In contrast, the presence of carbon dioxide did not affect the PL intensity.

In the first part of the headspace experiment, the glass sample bulb (Figure 3), containing a few grams of ammonium carbamate, is closed while equilibrium is established between the solid and its gaseous decomposition products. As can be seen in Figure 6, opening the bulb to the evacuated CdSe chamber immediately increases the PL intensity to the level observed with 10% NH_3. The fast response is typical; the kinetics of adduct formation

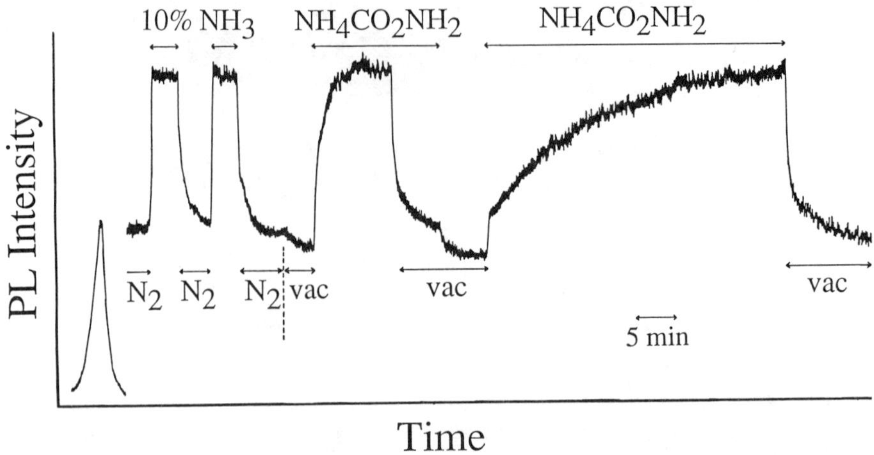

Figure 6. Results for CdSe/ammonium carbamate headspace experiment. After scanning PL intensity vs. wavelength in a nitrogen ambient, we monitored PL intensity at the band maximum (720 nm) vs. time. To the left of the dashed line are the results for exposure to 10% NH_3 in nitrogen (1 atm total pressure; flow rate 100 ml min^{-1}) relative to a nitrogen reference. To the right of the dashed line is the PL response to ammonia produced from ammonium carbamate decomposition relative to a vacuum reference (see text for further details).

between the gas and the semiconductor surface are quite rapid. In the second part of the experiment, CdSe is used to monitor the production of ammonia as the decomposition reaches equilibrium. After evacuation of both the ammonium carbamate bulb and CdSe chamber, we observe a gradual increase in PL as the ammonia reaches its equilibrium concentration once again. This time the adduct formation depends on the ammonium carbamate decomposition rate; the PL gradually increases as the partial pressure of ammonia increases.

Detection in Organic Solution

We have constructed a simple and portable apparatus for detecting adduct-forming compounds in organic solution (Figure 4). This compact system can be used to detect the presence of weakly acidic or basic compounds dissolved in an organic solvent. The experiment uses highly emissive CdS:Te because the Te doping produces a very broad, highly intense subband gap emission centered at about 600 nm that is readily detected with either a Si photodiode or a CdS photocell. The subband gap emission involves holes trapped at Te electronic levels that lie energetically within the band gap (Karas, et al., 1981). A benefit of using CdS:Te is that the semiconductor can serve as both PL source and excitation wavelength cutoff filter. Impinging light with energy greater than or equal to the band gap energy is absorbed by the semicon-

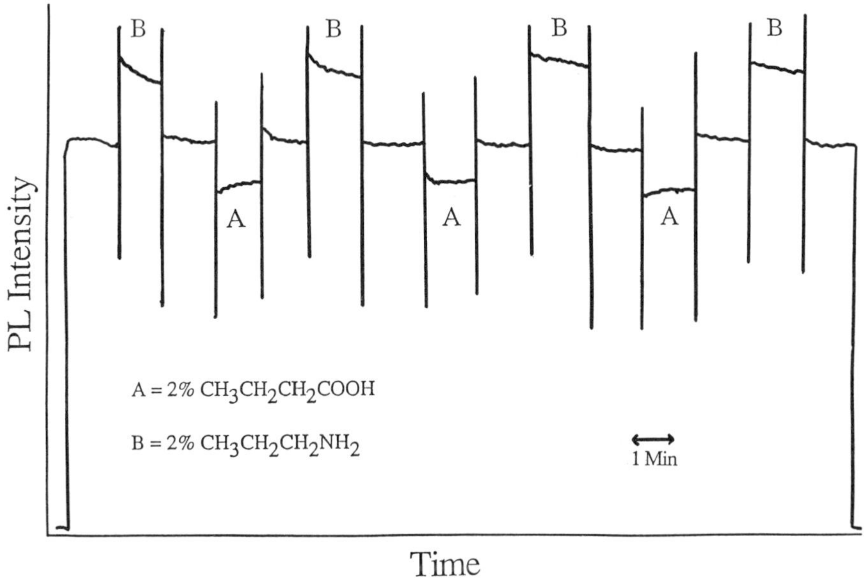

Figure 7. Results for CdS:Te in a toluene reference solution using a portable PL detection apparatus. The PL intensity was monitored at 600 nm, the band maximum for the Te based PL. The unlabeled intermediate responses are obtained in the reference solvent, toluene. After draining the toluene, refilling the cell with 2% propylamine in toluene produces a higher PL intensity. By draining the propylamine solution, rinsing the cell several times with toluene, and refilling the cell with toluene, the original PL intensity is restored. The same procedure using 2% butanoic acid in toluene decreases the PL intensity.

ductor while the lower energy Te-based emission will pass through the semiconductor to the detector.

The portable system reproduces the results obtained in the non-portable laboratory setting: Lewis acids decrease the PL intensity while Lewis bases increase the PL intensity. Using toluene as the reference solvent, we were able to demonstrate that low concentrations (2% by volume) of butanoic acid and propylamine have opposing effects on the PL, decreasing or increasing the PL relative to toluene, respectively (Figure 7). Data obtained with this simple system can be fit to the Langmuir adsorption isotherm model.

Detection in Aqueous Solution

Given that II-VI semiconductors respond to amines in gaseous and organic solution ambients, the next logical experiment was investigation of amines in aqueous solution. Our choice of semiconductor was limited; the instability of common II-VI semiconductors (CdS, CdS:Te, and CdSe) to irradiation in aqueous solution restricts their use as sensors to non-aqueous systems (Ellis, et al., 1976). By extending our studies to GaAs, a III-V semiconductor, we were able to conduct experiments in aqueous solution.

Recent advances in inorganic sulfide-passivation processes are prompting new investigations of GaAs (see for example Carpenter, et al., 1989; Sandroff, et al., 1989; and Spindt, et al., 1989). The high density of surface states on non-passivated n-GaAs limits its electronic and photoluminescent applications; the surface states effectively "pin" the electric field in the semiconductor. Passivation decreases the number of these oxide-based surface states and not only improves the electrical properties of GaAs for device applications but also enhances the PL intensity. Wet chemical surface passivation with aqueous Na_2S or $(NH_4)_2S$ solutions increases the PL dramatically. Although Na_2S creates a higher PL surface than the $(NH_4)_2S$ treatment, we restricted our studies to the latter because Na_2S is less durable in aqueous solution. The effects of Na_2S passivation are reported to last for only 18–48 hr and are lost by rinsing the surface with deionized H_2O (Carpenter, et al., 1988). In contrast, the ammonium sulfide-treated GaAs is substantially more durable. The ammonia is highly volatile, resulting in a sulfide passivation that is less readily reversed by water.

We found that $(NH_4)_2S$-passivated GaAs, used in water as a PL-based sensor, forms a reversible adduct with aniline. A decrease in the PL intensity of $(NH_4)_2S$-passivated GaAs is observed in the presence of aqueous aniline solutions relative to a water reference and can be quantitatively fit to the Langmuir adsorption isotherm model. The PL quenching increases with concentration and saturates by ~ 15 mM, as shown in Figure 8. The double reciprocal plot yields an adduct formation constant of $450 \pm 35\ M^{-1}$.

The $(NH_4)_2S$-passivated GaAs PL is also perturbed by the presence of wastewater influent or effluent, relative to a distilled water reference. Although these PL intensity changes could indicate the presence of a multitude of dissolved chemicals present in the wastewater, we have evidence that ammonia-derived species contribute to the response. Analysis of raw influent and final effluent samples using the Kjeldahl method (American Public Health Association, et al., 1985) revealed typical concentrations of 4–30 mg l^{-1} for ammonia and < 3 mg l^{-1} for organic nitrogen. In exploring the role of ammonia in wastewater solution adduct formation, we investigated the effects of two ammonium salts, ammonium nitrate and ammonium sulfate, dissolved in distilled water. These slightly acidic solutions were chosen because comparable ammonium ion concentrations from ammonium hydroxide in distilled water would have resulted in highly basic solutions (pH $\sim$10) that possibly could have etched the GaAs; the influent and effluent consistently had neutral to slightly basic pH values (between 7 and 8). We found that exposure to the ammonium salt solutions gave similar but smaller PL responses than influent wastewater samples of comparable ammonia concentrations, relative to distilled water. We can conclude from these experiments that ammonia/ammonium ion contributes to the PL intensity changes observed when comparing wastewater to a distilled water reference.

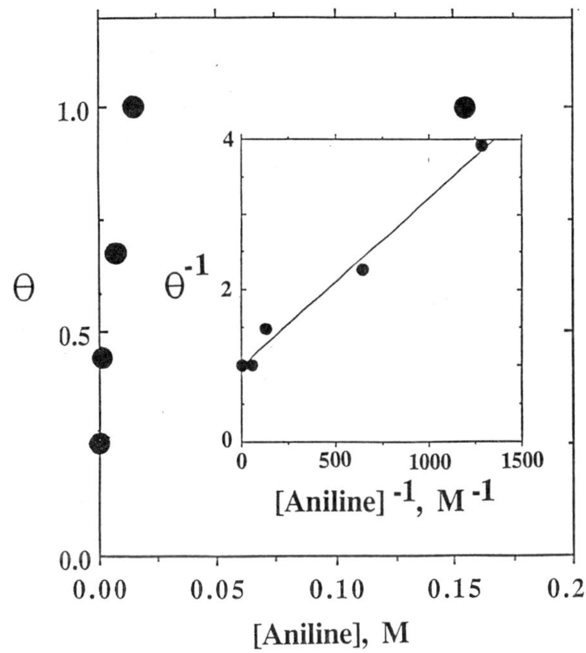

Figure 8. Fractional surface coverage θ, equated to the fractional change in ln (PL intensity), as a function of the concentration of aqueous aniline for $(NH_4)_2S$-passivated GaAs. The inset shows the same data as a double reciprocal plot; the line shown is a least-squares best fit, giving an adduct formation constant of 450 ± 35 M^{-1} using the equation given in the text.

Conclusions

In a headspace analysis of solid ammonium carbamate, we are able to detect the $NH_{3(g)}$ produced as the solid decomposes to its gaseous products. Using a closed system, we can observe the kinetically-controlled production of ammonia as the solid-gas equilibrium is attained. At equilibrium, the PL measurements indicate that the adduct-forming sites on the CdSe surface are saturated.

A simple and economical apparatus has been constructed for detection of weak Lewis acids and bases in organic solution using CdS:Te. Results from this system can be fit to the Langmuir adsorption isotherm model to calibrate the device for individual compound detection.

We have demonstrated that $(NH_4)_2S$-passivated GaAs is sufficiently stable in water to be used as an aqueous PL-based chemical sensor. This stability enables $(NH_4)_2S$-passivated GaAs to be used in wastewater applications, and a portable system has been constructed for on-line use at a wastewater treatment facility.

The three types of experiments discussed in this paper offer a glimpse at the potentially vast applications of emissive semiconductors in chemical sensing. There are a variety of designs available for compact detection equipment, and these can be varied with the semiconductor. One can envision coating the semiconductor to increase the selectivity for other classes of compounds, such as aromatic and halocarbon compounds. Because the semiconductor surface area is small (typically 0.25 cm^2) and the peripheral electronics needed for detection can be scaled down, emissive semiconductors could be unobtrusively positioned to be continuous on-line monitors. Adduct formation at the semiconductor surface is a contactless, *in situ* sensing method that can be effective in technological applications and useful in providing fundamental information about the semiconductor surface.

Acknowledgments

We thank the 3M Company and the Office of Naval Research for generous support of the headspace analysis and organic solution detection research. We thank University Technology Transfer, Inc. for generous support of experiments involving aqueous solution detection. We thank Mr. David Retzinger and the technical staff at the Racine, Wisconsin wastewater treatment facility for providing and analyzing wastewater samples.

References

American Public Health Association, American Water Works Association and Water Pollution Control Federation. 1985. Standard methods for the examination of water and wastewater. 16th ed. American Public Health Association, Washington, DC. pp. 406–412.

Burk, A.A., Jr., Johnson, P.B., Hobson, W.S. and Ellis, A.B. 1986. Photoluminescent properties of n-GaAs electrodes: Simultaneous determination of depletion widths and surface hole-capture velocities in photoelectrochemical cells. J. Appl. Phys. 59: 1621–1626.

Carpenter, M.S., Melloch, M.R. and Dungan, T.E. 1988. Schottky barrier formation on $(NH_4)_2S$-treated n- and p-type (100)GaAs. Appl. Phys. Lett. 53: 66–68.

Carpenter, M.S., Melloch, M.R., Cowans, B.A., Dardas, Z. and Delgass, W.N. 1989. Investigations of ammonium sulfide surface treatments on GaAs. J. Vac. Sci. Technol. B, 7: 845–850.

Ellis, A.B., Kaiser, S.W. and Wrighton, M.S. 1976. Visible light to electrical energy conversion. Stable cadmium sulfide and cadmium selenide photoelectrodes in aqueous electrolytes. J. Am. Chem. Soc. 98: 1635–1637.

Karas, B.R., Streckert, H.H., Schreiner, R. and Ellis, A.B. 1981. Luminescent photoelectrochemical cells. 5. Multiple emission from tellurium-doped cadmium sulfide photoelectrodes and implications regarding excited-state communication. J. Am. Chem. Soc. 103: 1648–1651.

Leung, L.K., Meyer, G.J., Lisensky, G.C. and Ellis, A.B. 1990. Modulation of the time-resolved photoluminescence of cadmium selenide by adduct formation with gaseous amines. J. Phys. Chem. 94: 1214–1216.

Lisensky, G.C., Meyer, G.J. and Ellis, A.B. 1988. Selective detector for gas chromatography based on adduct-modulated semiconductor photoluminescence. Anal. Chem. 60: 2531–2534.

Lisensky, G.C., Penn, R.L., Murphy, C.J. and Ellis, A.B. 1990. Electro-optical evidence for the chelate effect at semiconductor surfaces. Science. 248: 840–843.

Meyer, G.J., Leung, L.K., Yu, J.C., Lisensky, G.C. and Ellis, A.B. 1989a. Semiconductor-

olefin adducts. Photoluminescent properties of cadmium sulfide and cadmium selenide in the presence of butenes. J. Am. Chem. Soc. 111: 5146–5148.
Meyer, G.J., Lisensky, G.C. and Ellis, A.B. 1987. A new class of chemical sensors for gases based on photoluminescence from semiconductor-derived interfaces. Proc. Electrochem. Soc. 87–89, 438–448.
Meyer, G.J., Lisensky, G.C. and Ellis, A.B. 1988. Evidence for adduct formation at the semiconductor-gas interface. Photoluminescent properties of cadmium selenide in the presence of amines. J. Am. Chem. Soc. 110: 4914–4918.
Meyer, G.J., Luebker, E.R.M., Lisensky, G.C. and Ellis, A.B. 1989b. Photoluminescent properties of cadmium sulfide contacted with gaseous Lewis acids and bases. Photochemistry on Solid Surfaces. Elsevier, Amsterdam. pp. 388–403.
Murphy, C.J. and Ellis, A.B. 1990a. The coordination of mono- and diphosphines to the surface of cadmium selenide. Polyhedron. 9: 1913–1918.
Murphy, C.J. and Ellis, A.B. 1990b. Evidence for adduct formation at the semiconductor-solution interface. Photoluminescent properties of cadmium selenide in the presence of lanthanide β-diketonate complexes. J. Phys. Chem. 94: 3082–3085.
Sandroff, C.J., Hegde, M.S. and Chang, C.C. 1989. Structure and stability of passivating arsenic sulfide phases on GaAs surfaces. J. Vac. Sci. Technol. B, 7: 841–844.
Smith, R.A. 1978. Semiconductors. 2nd ed. Cambridge University Press, Cambridge.
Spindt, C.J., Liu, D., Miyano, K., Meissner, P.L., Chiang, T.T., Kendelewicz, T., Lindau, I. and Spicer, W.E. 1989. Vacuum ultraviolet photoelectron spectroscopy of $(NH_4)_2S$-treated GaAs (100) surfaces. Appl. Phys. Lett. 55: 861–863.

Discussion (DR. ELLIS)

Er-El: I would like to ask you two questions. First, you have shown response with 10% in the gas. What is the lowest level that you are able to detect?

Ellis: With ammonia we start to see a response at about a 0.1 mm pressure, about 10^{-4} atmospheres. That is the most sensitivity that we have seen with cadmium sulfide. There are a number of games you could play if you wanted to improve that. For example, you can take cadmium sulfide or cadmium selenide, and if you just dip them for a short time in aqueous solutions containing silver ions, you pull out the surface cadmium ions and replace them with silver ions. Then you can do the same luminescence experiment, but you have the silver coordination chemistry instead. It turns out that is a stronger binding constant for coordination of ammonia, and we can detect this gas at lower concentration than with the unexchanged solid. That is the kind of proof-of-concept experiment to show that you can actually tune the response to these various kinds of molecules based on coordination chemistry.

Er-El: Second, is it feasible to detect oxygen on-line, and is it feasible to sterilize a probe like that? Can it stand 120° or pressure of 1.5 atmospheres for 20 min?

Ellis: We have not made a point of detecting oxygen on-line, but we have seen responses to oxygen in some of our systems. We have done a little work on temperature and pressure in connection with using semiconductors as detectors for gas chromatography. With elevated temperature we can, in some cases, rejuvenate a response that has tailed off over time. We have not done a lot of work in this area.

Kornguth: I would like to compliment you on a very excellent demonstration of the potential use of small chips with multiple different receptor molecules for detection of chemicals. Two questions come up. During the on-line reading, what is the effect of fouling, pH, and salinity on the detection system for the amines or other sensed molecules?

Ellis: The closest we have come to addressing these issues are the waste water experiments, and there are so many other compounds that are present there that it is really hard to say. In general what you would expect is that if you have some other species that can competitively bind, you are obviously going to affect the response and you have to worry about that.

Kornguth: The peak seemed constant on sequential traces.

Ellis: That is right. We did see that; these were batch experiments where we took a little influent and a little effluent, but you can also just hook up a pump and do it on-line continuously.

Kornguth: The second question concerns the potential utility of this method for detecting specific biological molecules. Can one clad the surface of the crystal with polyacetylenes, polycinnamic acids and so on? Will such modifications permit the detection of compounds in the environment other than amines?

Ellis: Yes, there is no reason why that cannot be done. We have done a little along those lines using aromatics. We have been able to adsorb some aromatics to the surface that form charge transfer complexes. You can actually see differences; if you adsorb, for example, an acceptor molecule, you can see different donors because they give different PL signatures. So I think what you are saying is quite doable.

Kornguth: So in fact, could one potentially use either the porphyrins or other light-sensitive materials as part of your sensing component for a charge-transfer complex as well. Would that have a utility here?

Ellis: Yes, it could. Although with something like porphyrins or in fact, any highly-colored species, you have to correct for optical effects, and there may be some electronic interaction between the molecule and the semiconductor that you would have to investigate.

Smith: I wondered if there is any information in the wavelength of the emitted light. Can you get some more specific information about what the bound molecule is?

Ellis: We have not seen any evidence for that. All of our experiments have been done with low resolution, and if you go to higher resolution, in fact, you can often pick up what are called self-absorption effects: some of the emitted light is re-absorbed. In principle, if you go to high resolution you can detect that, but it is probably not going to tell you anything more about the molecule. It would help to confirm the dead-layer model.

Liron: Selectivity is a critical demand in the case of detection systems. I wonder what would be the case if you have in your system not just one analyte, but a mixture of analytes with each having a specific interaction with your system. What would you get?

Ellis: That is a good question. If you have a mixture of analytes, then obviously you are looking at a competition for the sites, and there are strategies for exploiting this. You could actually build the topography or landscape of the surface to impart additional selectivity. For example, you may find that a very bulky molecule cannot gain access to certain sites. In general, what we are finding is that our sensor really works best at looking at classes of compounds. You would be able to determine, for example, if there were amines present, depending on what else is in the mixture. The advantage of this is that, in principle, you might be able to rule out the presence of an entire class of compounds based on the lack of response; or to infer the presence of a member of a class based on a positive response. In the latter case you would then do some other analytical tests. If there were a particular molecule you wanted to sense, I am sure we could design a surface for it that would be selective, but that is a lot more work than just using the device as a sensor of a class of compounds. I should add that one could also have an array of sensors and use pattern recognition schemes to identify a specific compound.

Key Words: CdS, CdSe, GaAs, Amines, Adducts, Semiconductor, Sensors, Wastewater, Photoluminescence, Langmuir adsorption isotherm model, Dead-layer model, Aniline

High Speed DNA Sequencing and the Analysis of the Human Genome

Lloyd M. Smith, Jon D'Cunha and John A. Luckey
Department of Chemistry, University of Wisconsin, Madison, WI 53706 USA

The Human Genome Initiative is a massive international effort now underway to decode the total genetic content of *Homo sapiens*. The resultant information will provide a tool for fundamental biological and medical research of unprecedented power. The genome is encoded on 23 pairs of chromosomes and contains over three billion units of DNA sequence. The characterization of this immense structure is a project four orders of magnitude larger than any other such project ever undertaken. Accordingly, the major emphasis of the project is in the development of the new technologies required to make this undertaking feasible.

The most critical of these technologies is the sequence analysis of DNA. Conventional techniques for DNA sequencing, although powerful and effective, are extremely time-consuming, repetitive, and laborious procedures that employ hazardous and unstable radioisotopes and are not well-suited to very large-scale sequencing projects. For these reasons, we have developed an automated DNA sequencing method which uses fluorescent tags in place of radioisotopes for detection. The fluorescent tags are detected using a high-sensitivity laser-based detector, and the resultant data are analyzed by microcomputer to automatically yield the DNA sequence. A commercial version of this instrument has been developed which has a theoretical capability of sequencing 10,000 bases of DNA per day.

However, even though these automated fluorescence-based instruments are now widely available, they are not yet sufficiently accurate, rapid, and powerful to undertake the analysis of the human genome. There are two major reasons for this. First, the instruments available currently address primarily only one aspect of sequencing, the separation and analysis of DNA fragments. To realize the potential of automated sequencing, the other major aspects of sequencing must also be addressed, i.e., the pre-separation steps such as subcloning, DNA preparation, and fragmentation (Front End), and the post-separation steps of data analysis, assembly, and sequence (Back End). Second, the throughput of these instruments needs to be increased substantially.

We have developed two new tools to help address these limitations of

D. Kamely et al. (eds), Biotechnology: Bridging Research and Applications, 333–340.

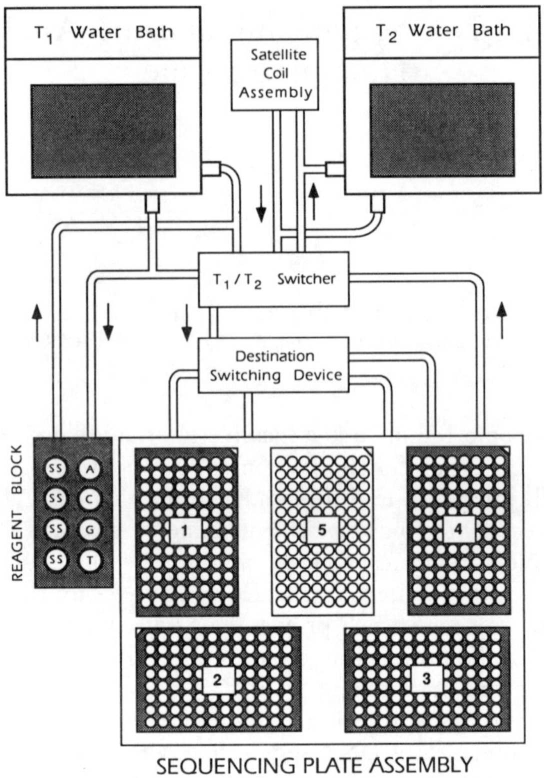

Figure 1. Block diagram of robotic system.

sequencing methodology, a robotic system for the performance of enzymatic sequencing reactions, and an instrument for high-speed DNA sequencing by capillary electrophoresis (CE). The robotic system is a Gilson Model 222 HPLC Autosampler modified to include two temperature-controlled assemblies (see Figure 1). One assembly holds the reagents used for the reactions and the second holds five 96–well microtiter plates in which the DNA templates to be analyzed are contained and the sequencing reactions are performed. The autosampler adds reagents to each of the templates at approximately 0°C, and then the temperature is raised to permit the enzymatic extensions to take place. Reactions have been optimized for conventional sequencing and automated fluorescence-based sequencing. Approximately 3 hours are required to perform 96 sets of sequencing reactions, and the accuracy of the resultant sequence data is equivalent to that obtained manually. Representative data obtained in conventional sequencing are shown in Figure 2.

The throughput of automated sequencers is limited in part by the separation speed, which is limited in slab gel electrophoresis by the magnitude of the electric field which can be applied to the gel without excessive heat

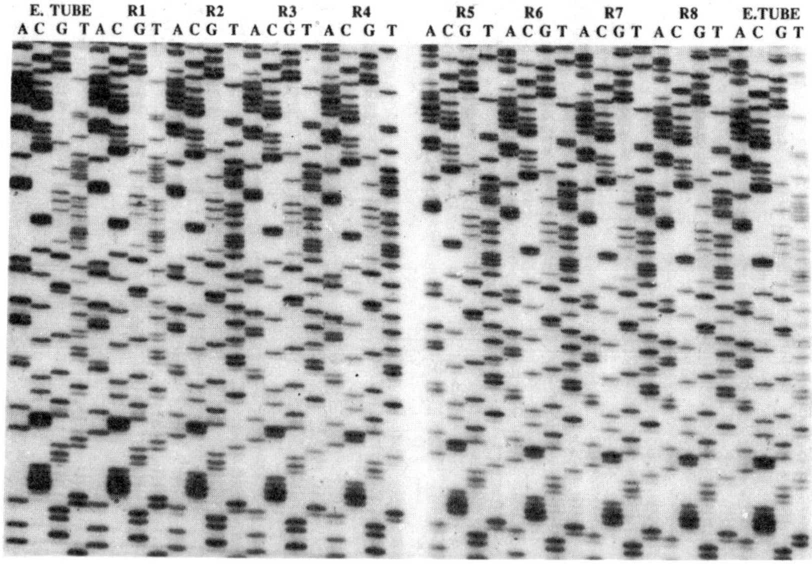

Figure 2. Data comparing manual (E. Tube) and robotic sequencing at eight positions (R1–R8) in the sequencing plate assembly.

production. Electrophoretic separations performed in very small diameter capillaries (typically 50–100 μm i.d.) permit a much greater field to be applied. CE separations are characterized by extremely high speeds and resolutions. We have developed an instrument for automated DNA sequencing using capillary gel electrophoresis, in which speed is increased by up to 14–fold over conventional methodology.

Figure 3 shows a block diagram of the optical configuration employed for such multiple wavelength fluorescence detection in gel-filled capillaries. Briefly, a 10 milliwatt beam from an argon ion laser operated in multiline mode is focused onto the capillary in a 50 μm spot to fill the inner diameter of the tube. The emitted fluorescence is collected at right angles with a microscope objective, spatially filtered to remove scattered excitation light, and split into four parts of approximately equal intensity using beamsplitters. Each of the four parts is passed through a different 10 nm bandpass interference filter to select a wavelength region of interest, and the fluorescence is measured with a photomultiplier tube. The signal is digitized and processed after acquisition by use of a digital filtering algorithm and multicomponent analysis similar to that described previously.

The performance of this system was tested by sequence analysis of the bacteriophage vector M13mp19. Mixtures of the four DNA sequencing reactions labeled with different fluorophores were applied to a 50 μm i.d., 375 μ o.d. denaturing polyacrylamide gel-filled capillary by electrokinetic

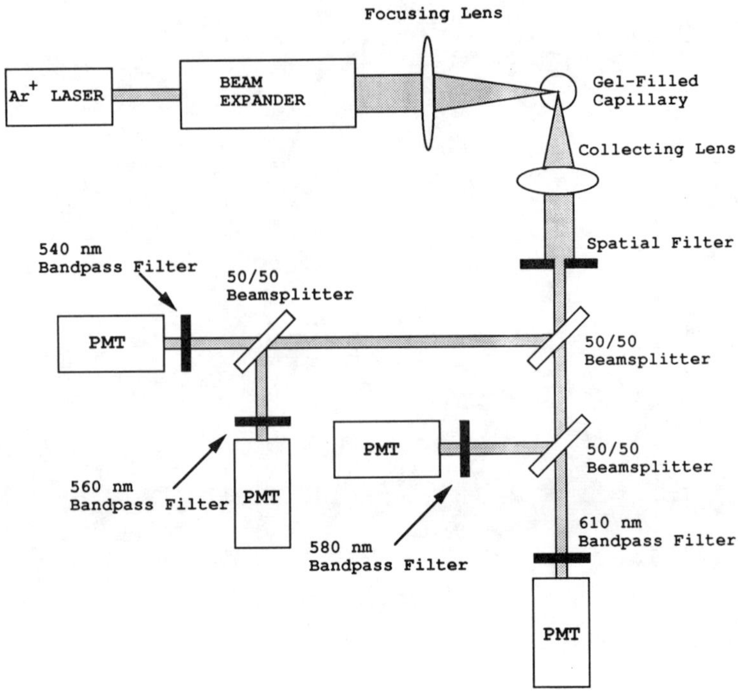

Figure 3. Block diagram of optical system employed in the capillary electrophoretic DNA sequencer.

injection and separated by electrophoresis. Representative data are presented in Figure 4.

Table 1 presents a summary of data obtained from multiple capillary and slab gel analyses performed under a variety of conditions. The retention time, resolution, and column efficiency obtained for three different length DNA fragments was measured. In each case, the measurements were made for the center T of a group of three T bases.

The most significant effect evident in these data is the high speed of the CE-based sequence analysis. In the most rapid analysis presented (70/40 cm capillary at 428 V/cm), speed was increased more than 14-fold relative to slab gel electrophoresis (31 minutes versus 452 minutes for DNA fragments 317 bases in length). This high speed is obtained at no cost in separation efficiency, as measured either by resolution or by plates obtained per meter (Table 1).

Discussion (DR. SMITH)

Er-El: You should stop doing research because the cheapest commercial sequencing I have seen is about \$0.95 per base. So if you can do 600,000

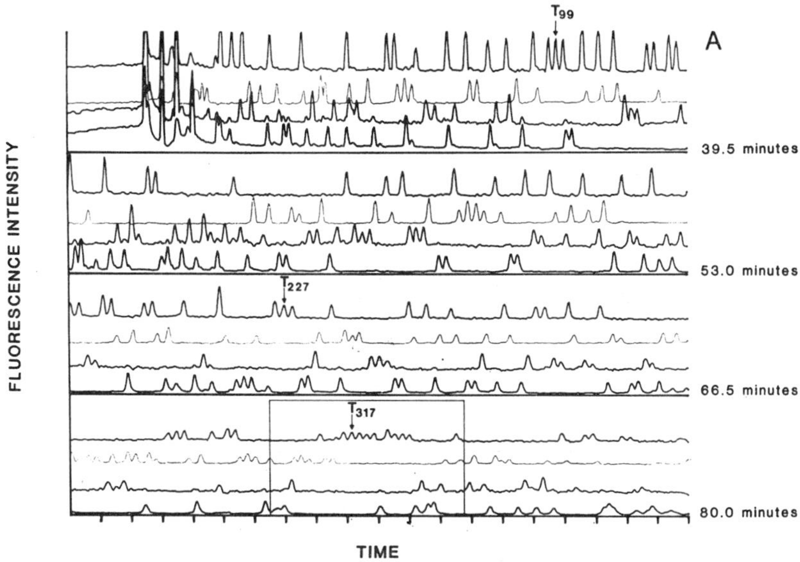

Figure 4. Representative data obtained from the system shown in Figure 3.

per day, you can be in a very good business. Is this system simple and rugged enough to be handled by an individual in the field laboratory?

Smith: Certainly not right now. I have a very simple acid test for when I think sequencing technology is "there," that will be when I am willing to do it. I am not willing to do it yet. So our basic goal is to make sequencing simple enough that it is straightforward, like running a pH meter or spectrophotometer. That is certainly a long way away. For instance, in the front end I showed you one little thing, a little robot that mixes things, but there are a lot of other things to do beside mixing things. This is a long way from being a mature technology.

If I were going to do sequencing right now and I were really in a hurry, a life or death situation, I do not know if I would use any of these fancy tricks. It does not mean that I am not interested in developing them but they are not really necessarily the best thing in the world yet, though, I hope they will be and I hope soon.

Szybalski: We tried a few runs making primers by ligation, and it works nicely – from hexamers we made 18–mers with reasonable efficiency. So it looks like it would be a very useful technique, which we named SPEL-6 (Gene 90 (1990): 177–178).

Smith: May I explain this idea to the audience? Dr. Szybalski recently attended a "Human Genome Workshop" where Bill Studier came up with a neat strategy for sequencing. Instead of sequencing many different templates, sequence a certain length, say 500 bases, along one template. Then synthesize a new primer from the end of that sequence and sequence from that point on. So you sequence sequentially. It is a nice strategy but has not been terribly practical yet because of the expense and time of making pieces of

Table 1. Summary of electrophoresis results[1]

		Capillary Electrophoresis[6]			Slab[7]
Length[2] (cm) Field		100/70 300 V/cm	70/40 300 V/cm	70/40 428 V/cm	41.5/26 30–40 V/cm
Base[5] 99	Retention time (minutes), t_R	39.2 ± 2.3	23.0 ± 2.9	17.7 ± 1.1	175.7 ± 3.5
	Resolution,[3] R	2.47 ± 0.53	1.92 ± 0.24	1.99 ± 0.10	1.13 ± 0.26
	Theoretical plates[4] (per meter), N	$9.1 \pm 5.3 \times 10^6$	$7.2 \pm 3.0 \times 10^6$	$13.2 \pm 1.5 \times 10^6$	$1.7 \pm 1.0 \times 10^6$
Base 227	t_R	59.4 ± 1.8	33.8 ± 2.9	25.7 ± 1.7	336.2 ± 5.2
	R	1.93 ± 0.52	1.34 ± 0.45	1.23 ± 0.30	1.02 ± 0.26
	N	$8.9 \pm 4.6 \times 10^6$	$7.8 \pm 2.0 \times 10^6$	$7.2 \pm 3.5 \times 10^6$	$3.9 \pm 1.9 \times 10^6$
Base 317	t_R	73.8 ± 2.0	41.4 ± 2.7	31.3 ± 3.4	452 ± 8.7
	R	1.27 ± 0.22	0.94 ± 0.33	0.70 ± 0.16	0.73 ± 0.11
	N	$6.7 \pm 1.4 \times 10^6$	$7.0 \pm 2.5 \times 10^6$	$3.8 \pm 1.0 \times 10^6$	$3.8 \pm 1.0 \times 10^6$

[1] Data are reported as the mean and standard deviation of three separate runs under identical conditions (employing different gel columns) for the CE results, and of five lanes on two gels for the slab results.
[2] x/y, where x is the overall length and y is the distance from the beginning of the column to the detector, i.e. the effective column length.
[3] Resolution R computed according to the formula $R = 1.18 \frac{[(t_R)_B - (t_R)_A]}{[(W_{1/2})_A + (W_{1/2})_B]}$.
[4] Theoretical plates N computed according to $N = 5.54\ (t_R/W_{1/2})^2$.
[5] Bases 99,227, and 317 are each the middle T of three consecutive T's and are underlined in Figure 2B.
[6] Performed on 3.2% total acrylamide/2.7% bis acrylamide crosslinker gels containing 7 M urea (7).
[7] Performed on 0.4 mm thick, 6% total acrylamide/5% bis acrylamide crosslinker gels containing 7 M urea (7) on an ABI DNA sequencer.

DNA over and over again. Dr. Szybalski is working on a technique in which you have all 4,096 possible 6–mers and a robot like the Gilson instrument. If you wish to synthesize an 18–mer primer, you simply mix the appropriate three 6–mers, hybridize them to the template, ligate them together, and go. If that does work, it could change the nature of the sequencing game considerably, and if you could combine that with a technique where you could sequence really long sequences at a time, that would be even more important.

Salemme: What are the desirable properties for a support matrix other than polyacrylamide?

Smith: One issue is chemical stability; our standard gels are urea-polyacrylamide gels run at pH 8.5. However, neither urea nor polyacrylamide are stable at pH 8.5; they hydrolyze. When they hydrolyze, you generate a variety of other molecules in the gel, some of which are charged; and you get breakdown of the gel, charge effects,etc. Another key issue is mechanical stability. Right now everybody in the world who runs gels has to make them in their own lab within three days of their use. That is a lot of extra work which is very time consuming. Improved stability could allow commercialization of pre-produced gels if we had better control of those properties; that would be really important. Also, a matrix should provide good resolution, at least as good or better than acrylamide. Another desireable feature would be a robust, simple polymerization reaction that is not too sensitive to oxygen and temperature. The list goes on and on.

Salemme: Is there some ultimate asymptotic limit to performance in sequencing?

Smith: I do not really know. If you have high enough resolution you should be able to do anything, potentially. We are looking for fun at using mass spectrometry for sequencing, not using gels at all. The limitations there would just be the resolution of the mass spec, just like you are limited in normal electrophoresis by the resolution of a gel. The fundamental limit depends on the properties of the gels, and we do not really understand very well how gels work and certainly do not have a good theory for predicting their resolution so I could not say. It is pretty empirical.

Kamely: Could you discuss the progress of the human genome, current time frames, and where the whole project stands right now?

Smith: Last year the NIH had $57 million to spend and DOE had about $26 million. They are succeeding in spending that money. They want to achieve a steady state level for funding from all agencies of about $200 million per year. The project seems to be moving along pretty well. A human genome center has been established which is separate from NIH-GMS, which required administrative changes to make it a separate entity within NIH. Funding mechanisms have been established, including regular R01 NIH grants, program project grants, training grants, and also a new Centers Program. The thrust is five-fold – doing things we know how to do just to get the job done (physical and genetic mapping), some sequencing, technology development, informatics (how to handle these large amounts of data and extract useful information), and ethics.

Mevarech: Why do people use urea instead of predenaturing the DNA chemically with glyoxyl and running gels without urea?

Smith: That would probably change its electrophoretic mobility, but maybe that is a good idea. You should try it.

Key Words: Capillary electrophoresis, DNA sequencing, Fluorescent nucleotide tags, Human genome

Interacting Light Energy with Biocatalytic Assemblies: Biocatalysed Photosynthesis and Photoregulation of Enzymes

Itamar Willner, Noa Lapidot, Shai Rubin, Azalia Riklin and Bilha Willner
The Institute of Chemistry, The Hebrew University of Jerusalem, Jerusalem 91904, Israel

Light energy can be coupled to biocatalytic assemblies by two basic features. First, solar light could provide the energy input for biocatalysed syntheses, similar to the performance of the photosynthetic cycle in nature. Thereby, light-induced biocatalysed processes mimicking photosynthesis could provide a novel biotechnological approach. Second, interaction of light energy with enzymes could eventually trigger their activities "on" or "off". As a result, enzymes could be photoregulated by external light signals and the application of biocatalytic materials as biochemical switches in sensor devices or information storage systems seems feasible. This article will describe the progress of our laboratory in harnessing light energy in the performance of biocatalytic assemblies.

Biocatalysed Artificial Photosynthetic Systems by Photochemical Regeneration of NAD(P)H Cofactors

Extensive efforts have been directed in recent years towards the development of artificial photosynthetic systems as solar energy conversion and storage devices (Grätzel, 1983; Harriman, 1983a; Bard, 1980; Meyer, 1989). The basic configuration of an artificial photosynthetic system is displayed in Figure 1. The system includes a light harnessing component, S, (photosensitizer) that upon excitation results in an electron-transfer (ET) process to the electron acceptor, A. Regeneration of the photosensitizer through oxidation of an electron donor, D, effects the net conversion of light energy to chemical potential stored in the products A^- and D^+. In the presence of proper catalysts, and provided that the photoproducts, A^- and D^+, exhibit adequate thermodynamic potentials, endoergic chemical transformations, by the subsequent utilization of the intermediate photoproducts, can be envisaged, i.e., CO_2-fixation or water photolysis. Substantial progress in developing the

D. Kamely et al. (eds), Biotechnology: Bridging Research and Applications, 341–369.

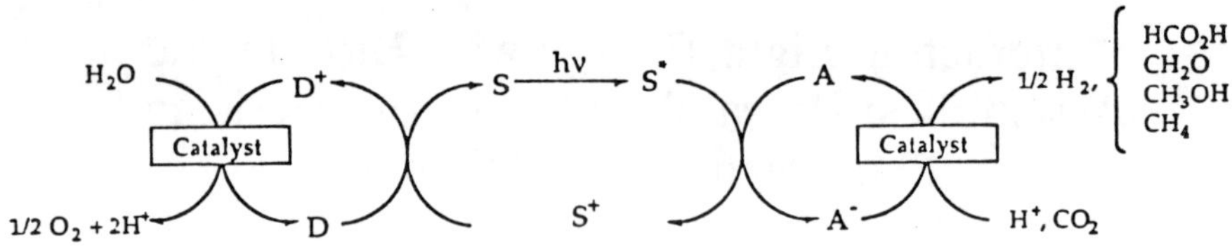

Figure 1. Photolysis of water and CO_2-fixation in an artificial photosynthetic system.

components of such artificial photosynthetic devices has been realized in recent years. Transition metal complexes, i.e., Ru(II)-tris-bipyridine, $Ru(bpy)_3^{2+}$ (1), Zn(II)-meso-tetra-(N-methylpyridinium)porphyrin, Zn-TMPyP^{4+} (2), or organic dyes, i.e., acridine orange (3), have been applied as photosensitizers that induce ET reactions (Harriman, 1980; Kalyanasundaram, 1982; Balzani, 1978; Juris, 1988). Electron acceptors, i.e., N,N′-dime-

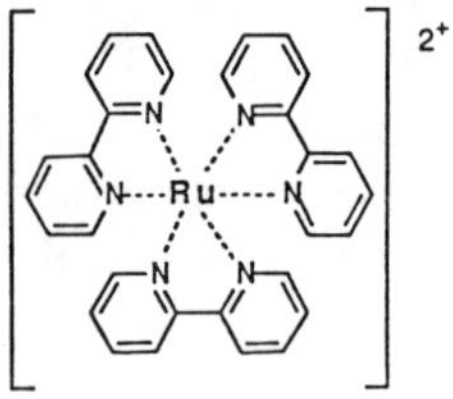

(1)

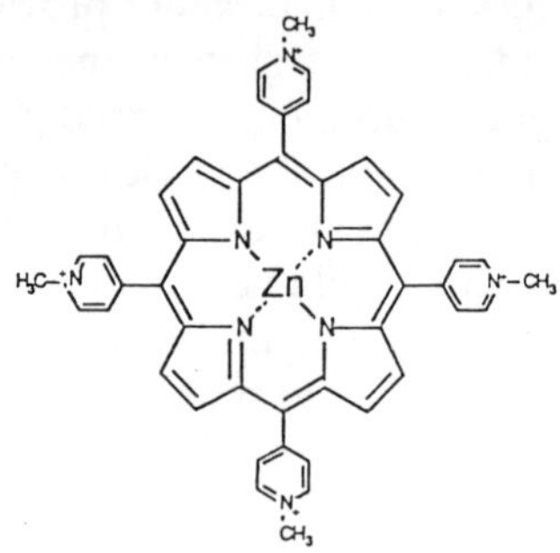

(2)

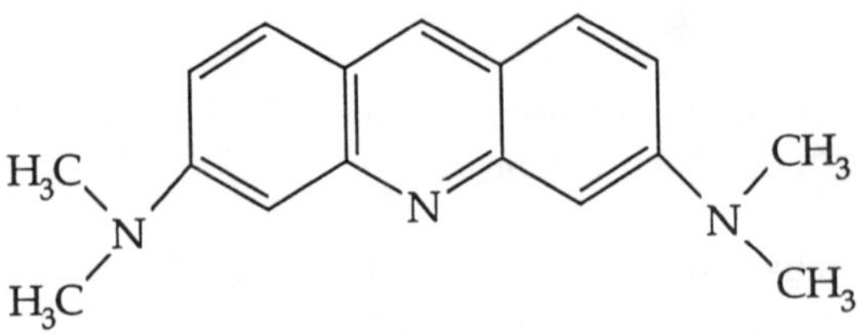

(3)

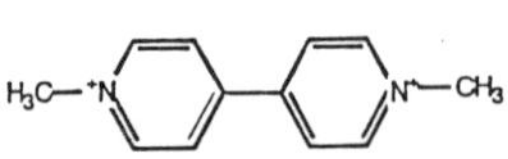

(4)

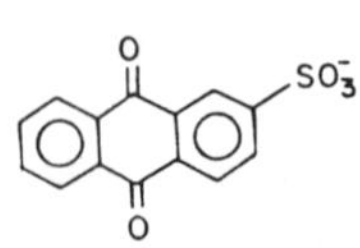

(5)

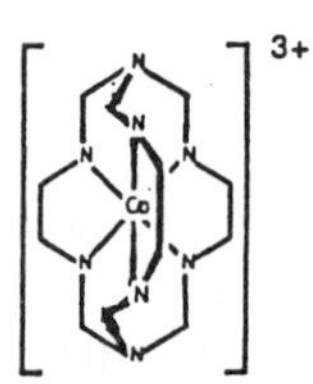

(6)

Schemes 1–6.

thyl-4,4′-bipyridinium (4), anthraquinone sulfonate (5), or metallo-complexes such as Co(III)-sepulchurate (6), and electron donors such as triethanolamine, thiols, i.e., cysteine or ethylenediaminetetraacetic acid have been broadly applied in such artificial photosynthetic assemblies (Kiwi, 1979; Keller, 1980; Miller, 1981; Richoux, 1980; Lay, 1983). Heterogeneous and homogeneous catalysts have been coupled to the photogenerated redox products, and H_2-evolution or CO_2-fixation processes have been effected (Willner, 1987, 1988, 1989a, 1990a; Harriman, 1983b).

Regeneration of 1,4–dihydronicotinamide adenine dinucleotide (NADH) and 1,4–dihydronicotinamide adenine dinucleotide phosphate (NADPH) by light-induced electron transfer processes provides a general route for developing biocatalysed photosynthetic systems. Since a variety of oxidoreductase enzymes are NAD(P)H cofactor-dependent, their coupling to the photoregenerated cofactors enables a variety of biocatalysed transformations. In fact, substantial activities are directed toward the development of NAD(P)H regeneration routes (chemical and electrochemical) (Chenault, 1987; Simon, 1985; Steckhan, 1987). Light-induced regeneration of NAD(P)H exemplifies an alternative concept in this area (Willner, 1989b).

Light-induced regeneration of NADH has been accomplished (Mandler, 1986) in a photosystem that includes Zn-TMPyP^{4+} (2), as photosensitizer, N,N′-dimethyl-4,4′-bipyridinium, MV^{2+} (4), mercaptoethanol as electron donor and lipoamide dehydrogenase, LipDH, as biocatalyst. Illumination of this photosystem results in oxidative ET quenching of the excited porphyrin, and photogenerated $MV^{+\cdot}$ acts as electron carrier for reduction of NAD^+ to NADH in the presence of LipDH, Figure 2(a). The oxidized photosensitizer, Zn-TMPyP$^{5+\cdot}$, is reduced by the thiol, and thereby the light harnessing compound is recycled. A closely related photosystem for the regeneration of NADPH has also been designed (Mandler, 1984, 1986). In this assembly, $Ru(bpy)_3^{2+}$ (1), and MV^{2+} act as photosensitizer and electron acceptor, respectively, and triammonium ethylenediaminetetraacetic acid, $(NH_4)_3$EDTA, as electron donor. The biocatalyst, ferredoxin-NADP-reductase, FDR, communicates the photosystem with the regeneration process of NADPH, Figure 2(b). The photophysical events leading to the regeneration of NADPH include oxidative ET quenching of excited $Ru(bpy)_3^{2+}$, and subsequent reduction of $NADP^+$ by the reduced relay, $MV^{+\cdot}$, in the presence of FDR. The quantum efficiencies for NADH and NADPH regeneration are $\phi = 9 \times 10^{-3}$ and $\phi = 1.9 \times 10^{-2}$ respectively. The net photoinduced biocatalysed reduction of NAD(P)$^+$ to NAD(P)H cofactors by mercaptoethanol (Equation 1) and $(NH_4)_3$EDTA (Equation 2) correspond to endoergic uphill chemical transformations.

The light-induced regeneration systems of NAD(P)H have been coupled (Mandler, 1986) to subsequent cofactor-dependent biocatalysed transformations, Figure 3. Reduction of pyruvic acid to lactic acid, and of acetoacetic acid to β-hydroxybutyric acid has been conducted using the NADH-dependent enzymes lactate dehydrogenase, LacDH, and β-hydroxybutyrate de-

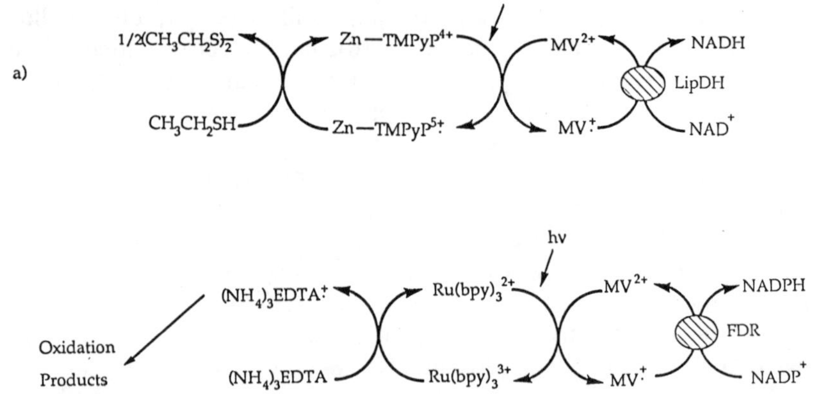

Figure 2. Photosensitized regeneration systems of NAD(P)H cofactors: a) Photochemical regeneration of NADH using Zn-TMPyP^{4+} as photosensitizer and lipoamide dehydrogenase, LipDH, as biocatalyst. b) Light-induced regeneration of NADPH using $Ru(bpy)_3^{2+}$ as photosensitizer and ferredoxin reductase, FDR, as biocatalyst.

hydrogenase, β-HButDH, respectively. Similarly, reductive amination of pyruvic acid to alanine proceeds in the presence of alanine dehydrogenase, AlaDH. Light-induced regenerated NADPH has been coupled to the reduction of ketones, i.e., 2–butanone to 2–butanol, and to the reductive amination of ketoglutaric acid to glutamic acid using alcohol dehydrogenase, AlcDH, and glutamate dehydrogenase, GluDH, as biocatalysts, respectively.

The endoergicity involved in the light-induced regeneration of NAD(P)H cofactors allows us to tailor biocatalysed CO_2-fixation processes using visible light as the energy driving force (Mandler, 1988; Willner, 1986). Biocatalysed decarboxylation reactions are exoergic processes utilized by the living organism as a source of energy. For example, decarboxylation of isocitric acid in the heterotrophic cycle, in the presence of isocitrate dehydrogenase, IcitDH, or decarboxylation of malic acid to pyruvic acid, in the presence of the malic enzyme, MalE, represent such transformations. In these processes, the decarboxylation processes are accompanied by the reduction of $NADP^+$ to NADPH. The endoergicity characterizing the light-induced regeneration system of NADPH, suggests that the carboxylation of ketoglutaric acid or pyruvic acid could be driven in the photochemical assembly, provided that the enzymes exhibit reversible activities. Namely, the light-induced regeneration of NADPH cofactors could act as the primary step for CO_2-fixation

1. $NAD^+ + 2HOCH_2CH_2SH \rightarrow NADH + (HOCH_2CH_2S)_2$
   $\Delta G° = 14.5\ \text{Kcal mole}^{-1}$

2. $NADP^+ + R{-}N(CH_2CO_2^-)_2 + 2H_2O \rightarrow NADPH + RNHCH_2CO_2^- + CH_2O + HCO_3^- + H^+$
   $\Delta G° = 28\ \text{Kcal mole}^{-1}$

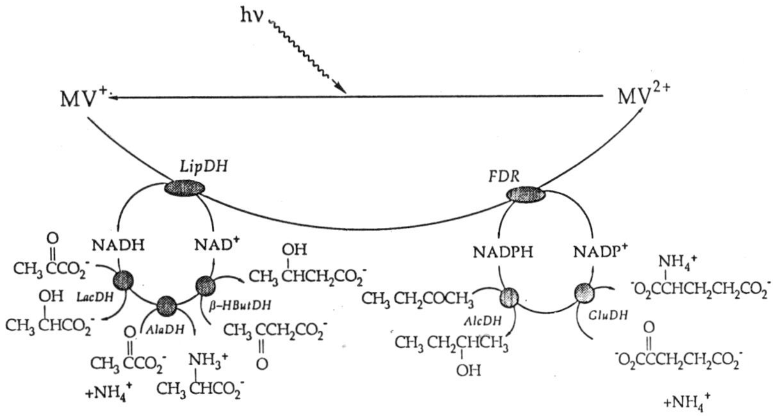

Figure 3. Light induced biocatalysed transformations utilizing photochemically regenerated NAD(P)H cofactors.

(carboxylation) of keto-acids. Indeed, illumination of the photochemical assembly composed of $Ru(bpy)_3^{2+}$ as photosensitizer, MV^{2+} as primary electron acceptor, mercaptoethanol or dithiothreitol as electron donors, $NADP^+$ and FDR, results in carboxylation of ketoglutaric acid to isocitric acid and of pyruvic acid to malic acid in the presence of IcitDH and MalE, respectively (Mandler, 1988). Figure 4(A) and Figure 4(B) show the rates of malic acid and isocitric acid formation as a function of illumination time. The sequence of reactions involved in the carboxylation processes, Figure 5, involves the photoinduced regeneration of NADPH by the primary formation of $MV^{+\cdot}$ in the photochemical ET process. The reduced cofactor, NADPH, mediates the carboxylation reactions in the presence of IcitDH and MalE, respectively. The net photochemical CO_2-fixation processes driven in this photosynthetic assemblies are summarized in Equation 3 and in Equation 4.

It can be seen that photosynthesis of malic acid through carboxylation of pyruvic acid is endoergic by ca. $\Delta G° = 11.5$ Kcal mole^{-1}. The carboxylated products could act as synthones for further biocatalysed transformations. For example, into the biocatalytic assembly for the photosynthesis of malic acid two additional enzymes, fumarase, Fum, and aspartase, Asp, and ammonium ions were added. Illumination of this photosystem results in the primary photoinduced formation of malic acid through regeneration of NADPH. This photoproduct undergoes dehydration to fumaric acid in the presence of Fum, and subsequent biocatalysed amination of fumarate to aspartic acid by Asp, Figure 5. The rates of malic acid and of aspartic acid production in the photosystem, are displayed in Figure 4(C). Thus, we realize that the four enzymes (FDR, MalE, Fum and Asp) operate in a synchronous feature, and the sequential light-driven synthesis of aspartic acid occurs through the primary ET process induced in the photosystem. Table 1 summarizes the turnover numbers of the different components included in the photochemical

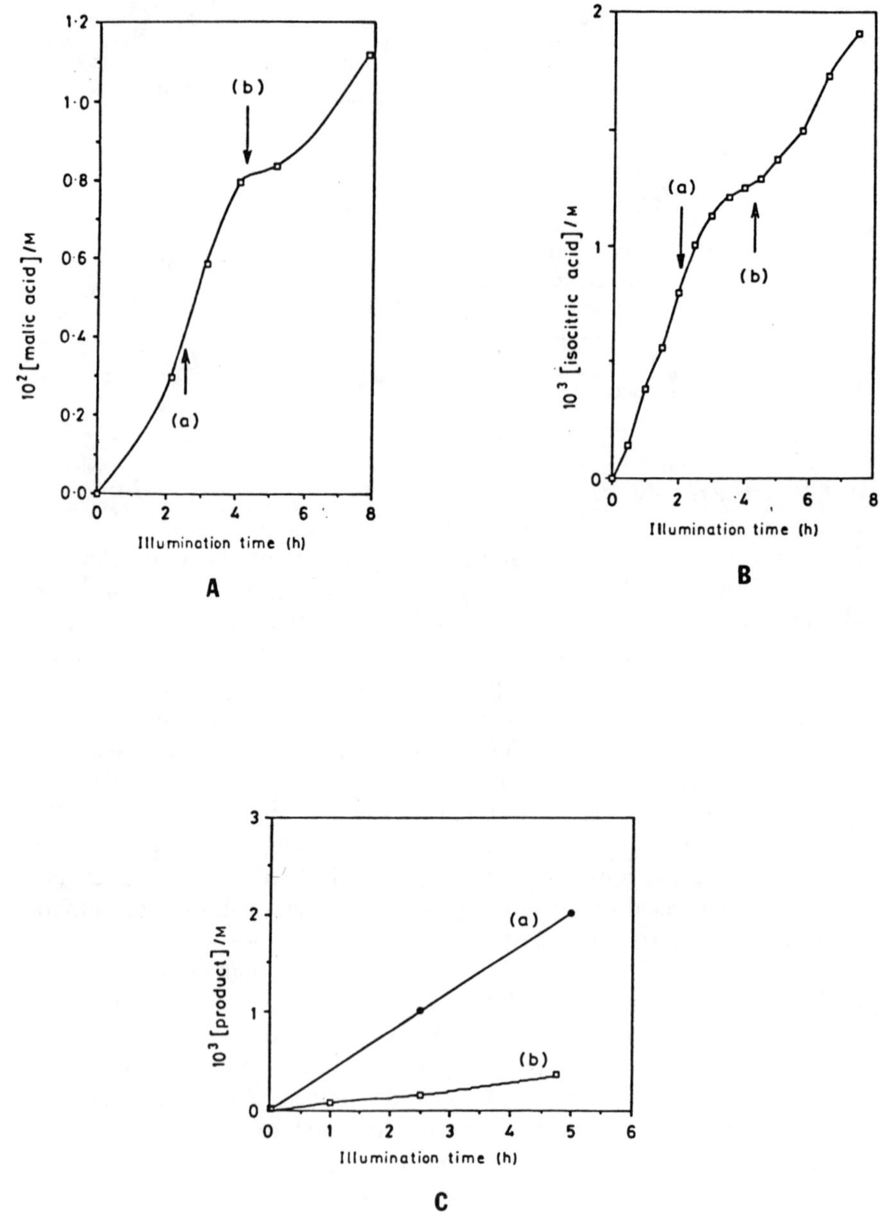

Figure 4. Photoinduced carboxylation reactions and photosynthesis through CO_2-fixation: A) Rate of malic acid formation as a function of illumination time with addition of mercaptoethanol at (a) 2.2 x 10^{-2} M, and (b) 2.7 x 10^{-2} M. B) Rate of isocitric acid formation as a function of illumination time with addition of dithiothreitol (5 x 10^{-3} M) at (a) and (b). C) Rate of (a) aspartic, and (b) malic acid formation as a function of illumination time.

3. $$CH_3C(=O)CO_2H + CO_2 + 2HOCH_2CH_2SH \rightarrow HO_2C - CH(OH) - CH_2CO_2H + (HOCH_2CH_2S)_2$$

$\Delta G° = 11.5\ \text{Kcal mole}^{-1}$

4. 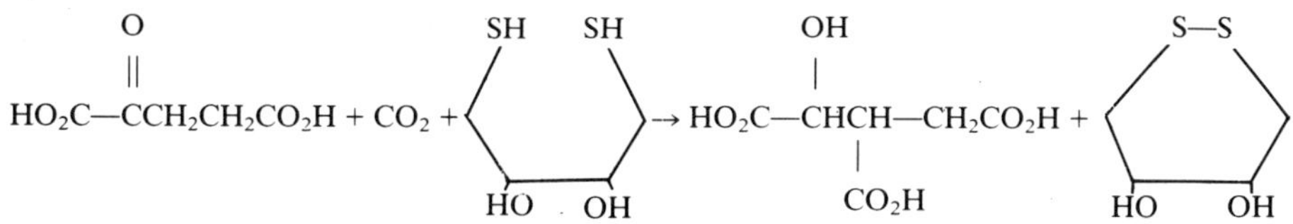

$\Delta G° = 0$

assemblies producing malic acid, isocitric acid and aspartic acid by CO_2-fixation. It can be seen that the bioactive materials exhibit high turnover numbers, implying their high stabilities in the artificial environment of the photosystems.

The previous systems have applied a biocatalyst (FDR or LipDH) for communicating the primary electron carrier, $MV^{+\cdot}$ with the reduction of NAD(P)H cofactors. The fact that NAD(P)H cofactors act as two electron relays (hydride equivalents), suggests that direct electron transfer communication, exclusive of enzymic participation, between $NAD(P)^+$ and hydrido species could be established. Indeed, a photosystem for the production of the hydrido-rhodium species derived from chloro-tris-(diphenylphosphinobenzene-m-sulfonate)rhodium(I), $RhCl(dpm)_3^{3-}$ (7), is capable of regenerating NAD(P)H. The photosystem is composed of $Ru(bpy)_3^{2+}$ as photosensitizer, ascorbate as electron donor and $RhCl(dpm)_3^{3-}$ as hydride carrier (Willner, 1990b). Illumination of this photosystem in the presence of $NAD(P)^+$ results

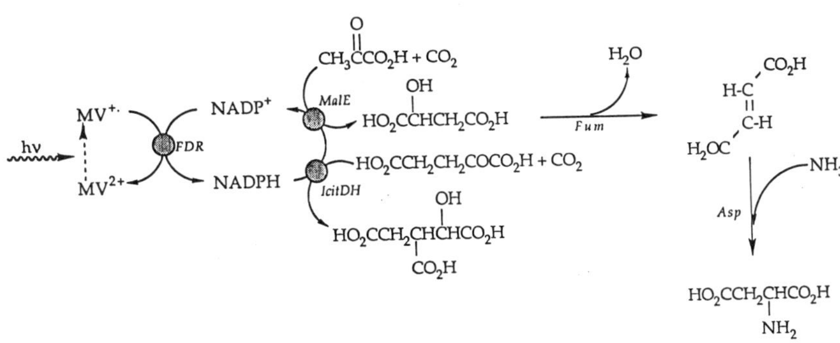

Figure 5. CO_2-fixation by coupling carboxylation biocatalysis to the photochemical NADPH regeneration system. CO_2-fixation products are subsequently coupled to the biocatalysed photosynthesis of aspartic acid.

Table 1. Turnover numbers (TN) of components included in the CO_2-fixation photosynthetic assemblies[a].

TN System	$Ru(bpy)_3^{2+}$	MV^{2+}	$NADP^+$	FDR[b]	MalE[c]	IcitDH[d]	Fum[e]	Asp[f]
Malic acid	1074	117	62.2	2.3×10^5	7.4×10^5			
Aspartic acid	174	25	6.3	1.6×10^3	6.3×10^4		8.5×10^2	5.2×10^2
Isocitric acid[g]	272	23	11.4	2.5×10^3		5.5×10^4		

[a] Turnover number is defined: TN = moles of product/moles of biocatalyst.
[b] F.W. ≈ 40,000; Shin, M. 1971. Methods Enzymol. 23: 441.
[c] F.W. ≈ 280,000; Hsu, R.Y. and Lardy, H.A. 1967. J. Biol. Chem. 242: 520.
[d] F.W. ≈ 58,000; Colman, R.F. 1968. J. Biol. Chem. 243: 2454.
[e] F.W. ≈ 48,500; Beekmans, S., and Kanarek, L. 1977. Eur. J. Biochem. 78: 437.
[f] F.W. ≈ 48,500; Suzuki, S., Yamaguchi, J., and Tokushige, M. 1973. Biochem. Biophys. Acta 321: 369.
[g] IcitDH is anchored to a cross-linked acrylamide polymer.

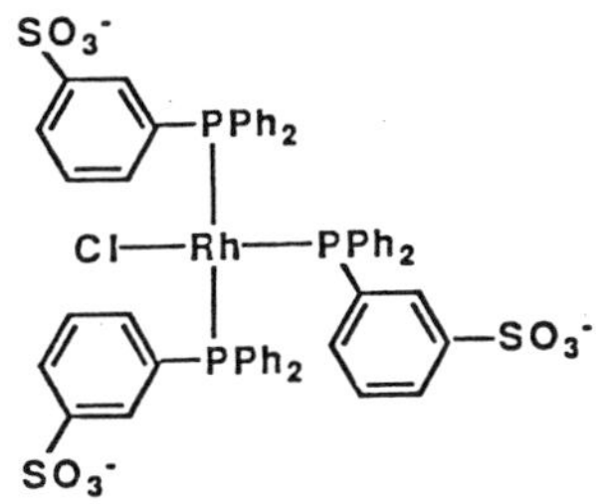

(7)

in the formation of NADH and NADPH. The quantum efficiencies for the formation of the two reduced cofactors is similar and corresponds to $\phi = 5 \times 10^{-3}$. Figure 6 shows the spectral characteristics of NADH formed at time intervals of illumination. The mechanism leading to the regeneration of NAD(P)H is schematically outlined in Figure 7. The primary step involves reductive ET quenching of $Ru(bpy)_3^{2+}$ by ascorbate (Equation 5). The resulting reduced photoproduct $Ru(bpy)_3^{+}$ mediates the reduction of $RhCl(dpm)_3^{3-}$ that generates a hydrido-rhodium species, (Equations 6–7), that subsequently effects the reduction of $NAD(P)^+$ to NAD(P)H, (Equation 8). Reduction of $NAD(P)^+$ by the hydrido-rhodium species proceeds with high regioselectivity, and the formation of the biologically active 1,4–dihydronicotinamide cofactors is substantially favored (selectivity > 90%). The fact

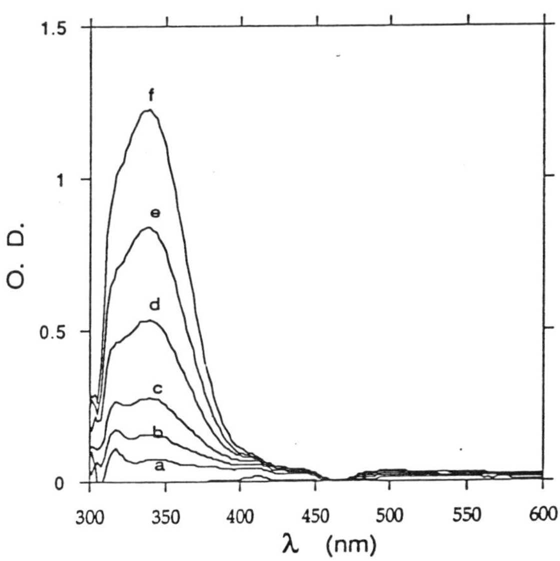

Figure 6. Photoinduced NADH formation using $RhCl(dpm)_3^{3-}$ as catalyst as a function of illumination time after a) 1 min, b) 2 min, c) 4 min, d) 7 min, e) 10 min, f) 15 min.

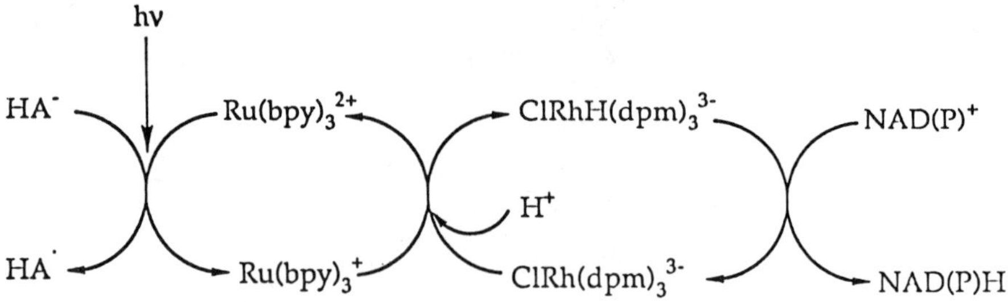

Figure 7. Photochemical regeneration of NAD(P)H cofactors using the homogeneous catalyst $RhCl(dpm)_3^{3-}$.

that NADH and NADPH are formed in the photosystem at similar quantum efficiencies suggests that the processes are controlled by the effectiveness of formation of the hydrido-rhodium complex in the photosystem.

Although the regeneration of NAD(P)H cofactors by the artificial hydride carrier is scientifically interesting, its practical application in NAD(P)H biocatalysed transformations is limited. We find that the ligand diphenylphosphinobenzene-3–sulfonate, dpm, acts as an inhibitor to most of the biocatalysts examined in our study, i.e., AlaDH, LacDH and GluDH. Upon formation of the hydrido-rhodium intermediate, $Rh(I)HCl(dpm)_3^{4-}$, the ligand, dpm, dissociates and inhibits the coupled NAD(P)H dependent enzymes. The only enzyme that exhibits moderate resistance towards inhibition by the dpm ligand is alcohol dehydrogenase, AlcDH. Accordingly, reduction of acetaldehyde to ethanol using AlcDH as biocatalyst proceeds in the photosystem by photoregenerated NADH at a rate corresponding to 1.5×10^{-3} M h^{-1}.

Photosynthetic Biocatalysed Transformations Mediated by Artificial Electron Carriers

The photosynthetic biocatalysed transformations discussed in Section I are mediated by the biologically active NAD(P)H cofactors. An interesting approach for developing photosynthetic biocatalysed systems could involve

5. $^*Ru(bpy)_3^{2+} + HA^- \rightarrow Ru(bpy)_3^+ + HA$
6. $Ru(bpy)_3^+ + Rh(I)Cl(dpm)_3^{3+} + H^+ \rightarrow Ru(bpy)_3^{2+}$
   $+ Rh(II)HCl(dpm)_3^{3-}$
7. $2\ Rh(II)HCl(dpm)_3^{3+} \rightarrow Rh(I)HCl(dpm)_3^{4-} + Rh(I)Cl(dpm)_3^{3-}$
8. $Rh(I)HCl(dpm)_3^{4-} + NAD(P)^+ \rightarrow Rh(I)Cl(dpm)_3^{3-} + NAD(P)H$

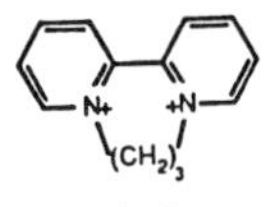

(9)

direct ET communication between redox enzymes and artificial photogenerated electron carriers. Coupling of enzymes, regularly activated by native NAD(P)H cofactors, to artificial electron carriers, requires mutual recognition of the electron carrier by the biocatalytic active sites.

We find that different enzymes, particularly biocatalysts that include iron-sulfur clusters or metal porphyrins at their active sites, can communicate with N,N′-dialkyl-bipyridinium radical cations acting as electron carriers. Systems for the photoreduction of bipyridinium salts using $Ru(bpy)_3^{2+}$, metalloporphyrins or organic dyes as photosensitizers in the presence of different electron donors, i.e., thiols, EDTA or triethanolamine, are well established (Kiwi, 1979; Keller, 1980; Miller, 1981; Krasna, 1980). Illumination of a photosystem composed of $Ru(bpy)_3^{2+}$ as photosensitizer, one of the electron relays, N,N′-dimethyl-4,4′-bipyridinium (4), MV^{2+}, N,N′-trimethylene-2,2′-bipyridinium (9), DT^{2+}, N,N′-tetramethylene-2,2′-bipyridinium (10), DQ^{2+}, or N,N′-dimethyl-2,2′-bipyridinium (11), DM^{2+}, in the presence of cysteine as electron donor, and under CO_2 in the presence of formate dehydrogenase, ForDH, results in the formation of formate (Mandler, 1988). The highest quantum efficiencies for formate production are observed in the photosystems that include MV^{2+} and DT^{2+} as electron carriers, $\phi = 1.6 \times 10^{-2}$. The sequential process leading to the formation of formate is schematically

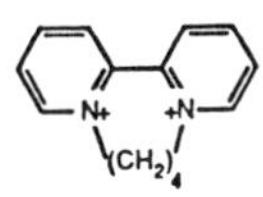

(10)

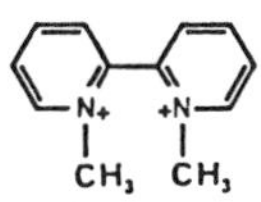

(11)

9. $$^*Ru(bpy)_3^{2+} + A^{2+} \xrightarrow{kq} [Ru(bpy)_3^{3+} \ldots A^{+\cdot}] \rightarrow Ru(bpy)_3^{3+} + A^{+\cdot}$$

10. $$2\ {}^-O_2C\underset{\substack{|\\ N^+H_3}}{C}HCH_2CH_2SH + CO_2 \rightarrow HCO_2H + ({}^-O_2C{-}\underset{\substack{|\\ N^+H_3}}{C}HCH_2CH_2S)_2$$

$$\Delta G° = 5.5 \text{ Kcal mole}^{-1}$$

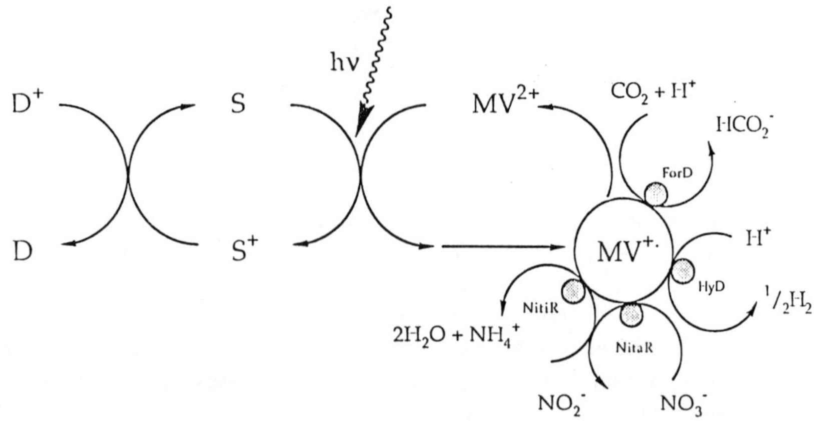

Figure 8. Biocatalysed photosynthetic systems mediated by an artificial electron carrier, N,N'-dimethyl-4,4'-bipyridinium radical cation $MV^{+\cdot}$ Photogenerated $MV^{+\cdot}$ mediates H_2-evolution, CO_2 reduction to formate and sequential reduction of NO_3^- to ammonia, using Hyd, ForDH and a mixture of NitraR and NitriR, respectively.

11. $NO_3^- + 8e^- + 10H^+ \rightarrow NH_4^+ + 3H_2O$

12. $NO_3^- + 2e^- + 2H^+ \rightarrow NO_2^- + H_2O$

13. $NO_2^- + 6e^- + 8H^+ \rightarrow NH_4^+ + 2H_2O$

presented in Figure 8. The primary step involves the oxidative ET quenching of excited $Ru(bpy)_3^{2+}$ by the bipyridinium electron relay (Equation 9). The reduced photoproduct, $A^{+\cdot}$, mediates the reduction of CO_2 to formate in the presence of ForDH. The light harnessing compound is concomitantly recycled through oxidation of cysteine. The net photosynthetic process driven in the system corresponds to the light induced reduction of CO_2 to formate by cysteine (Equation 10). This process is endoergic by ca. $\Delta G° = 5.5$ Kcal mole^{-1}.

Similarly, the light-induced multi-electron reduction of nitrate to ammonia (Equation 11) was conducted in a photosystem that includes the two enzymes, nitrate reductase, NitraR, and nitrite reductase, NitriR (Willner, 1989c). Both enzymes exhibit ET communication with the artificial electron carrier, N,N-dimethyl-4,4'-bipyridinium radical cation, $MV^{+\cdot}$. The photosystem is composed of $Ru(bpy)_3^{2+}$ as photosensitizer, MV^{2+}, as electron acceptor and Na_2EDTA as electron donor. Upon illumination of the photosystem in the presence of NO_3^- or NO_2^- and the biocatalysts NitraR or NitriR, respectively, NO_2^- and NH_4^+ are formed, respectively. The rates of NO_2^- and NH_4^+ formation, upon illumination, are displayed in Figure 9(A). The rate of ammonia formation in a photosystem that includes the two biocatalysts, NitraR and NitriR and the substrate NO_3^- is displayed in Figure 9(B). The mechanism

A

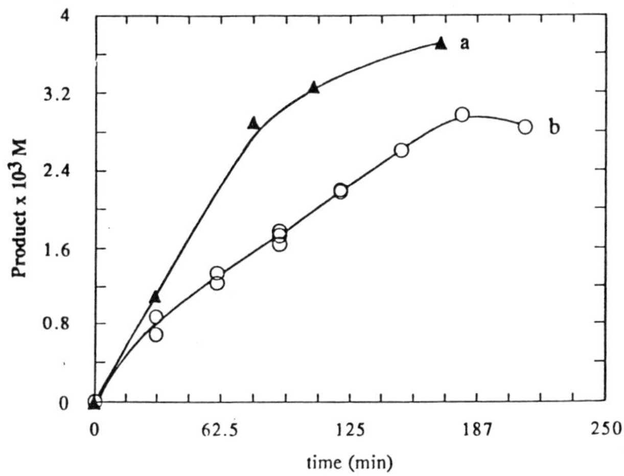

B

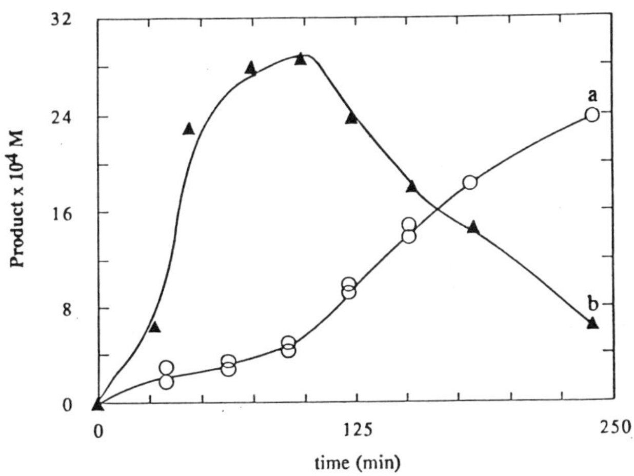

Figure 9. (A) Rates of products formation as a function of illumination time. In all systems $[Ru(bpy)_3^{2+}] = 7.4 \times 10^{-5}$ M, $[Na_2EDTA] = 0.02$ M. (a) (▲) NO_2^-formation, pH 7.0, Tris buffer 0.1 M, $[MV^{2+}] = 3.2 \times 10^{-4}$ M, $[NO_3^-] = 9.9 \times 10^{-3}$ M, nitrate reductase 0.2 U. (b) (o) NH_4^+ formation, pH 8.0, Tris buffer 0.1 M, $[MV^{2+}] = 4.2 \times 10^{-4}$ M, $[NO_2^-] = 0.01$ M, nitrite reductase 0.06 U. (B) NO_2^- and NH_4^+ concentrations in the combined system, as a function of illumination time. (a) (o) NH_4^+. (b) (▲) NO_2^-. pH = 8.0, Tris buffer 0.1 M, $[Ru(bpy)_3^{2+}] = 7.4 \times 10^{-5}$ M, $[Na_2EDTA] = 0.02$ M, $[MV^{2+}] = 4.2 \times 10^{-4}$ M, $[NO_3^-] = 0.01$ M, nitrate reductase 1.0 U, nitrite reductase 0.35 U.

that leads to the formation of ammonia in the photosystem is included in Figure 8. The primary step involves the photoinduced formation of $MV^{+\cdot}$ through oxidative ET quenching of the excited sensitizer, $Ru(bpy)_3^{2+}$. The light active material, $Ru(bpy)_3^{2+}$, is recycled by oxidation of the electron

Table 2. Quantum yields and turnover numbers of components in photosystems for CO_2 and NO_3^- fixation.

Photosystem	ϕ	$Ru(bpy)_3^{2+}$	MV_2^+	Biocatalyst
CO_2 fixation to HCO_2^- [a]	0.016	67	2	2×10^3 (ForDH)
NO_3^- reduction to NO_2^- [b]	0.08	80	18.5	6.2×10^4 (NitraR)
NO_2^- reduction to NH_4^+ [c]	0.06	38.5	7	2.4×10^4 (NitriR)
NO_3^- fixation to NH_4^+ [d]	0.06	32	6	9×10^4 (NitraR) 2.1×10^4 (NitriR)

[a] Composition of photosystem: $[Ru(bpy)_3^{2+}] = 3.0 \times 10^{-5}$ M, $[MV_2^+] = 1 \times 10^{-3}$ M, [Cysteine] = 1×10^{-2} M, $[NaHCO_3] = 2 \times 10^{-1}$ M, ForDH (1.2.1.2) 0.89 unit. F.W. of ForDH = 300,000 (Hopner, T. and Trutwein, A. 1972. Z. Naturforsch. 27b: 1075).

[b] Composition of photosystem: $[Ru(bpy)_3^{2+}] = 7.4 \times 10^{-5}$ M, $[MV_2^+] = 3.2 \times 10^{-4}$ M, $[EDTA] = 2 \times 10^{-2}$ M, $[NO_3^-] = 9.9 \times 10^{-3}$ M, NitraR (E.C.1.9.6.1) 0.2 unit.

[c] Composition of photosystem: $[Ru(bpy)_3^{2+}] = 7.4 \times 10^{-5}$ M, $[MV_2^+] = 3.2 \times 10^{-4}$ M, $[EDTA] = 2 \times 10^{-2}$ M, $[NO_3^-] = 1 \times 10^{-2}$ M, NitriR (E.C.1.6.6.4) 0.06 unit.

[d] Composition of photosystem: $[Ru(bpy)_3^{2+}] = 7.4 \times 10^{-5}$ M, $[MV_2^+] = 4.2 \times 10^{-4}$ M, $[EDTA] = 2 \times 10^{-2}$ M, NitraR 1 unit and NitriR 0.35 unit. F.W. of NitraR and NitriR ≈ 200,000 (Adams, M.W.W. and Mortenston, L.E. 1985. In "Molybdenum Enzymes", T.S. Spiro, Ed., John Wiley).

donor by the oxidized photoproduct, $Ru(bpy)_3^{3+}$. The reduced photoproduct, $MV^{+\cdot}$, is recognized by the two enzymes NitraR and NitriR and acts as electron carrier for these biocatalysts. The primary enzyme, NitraR, mediates the two electron reduction of NO_3^- to nitrite (Equation 12) while the second biocatalyst, NitriR mediates the multi-electron reduction of photogenerated NO_2^- to ammonia (Equation 13). Table 2 provides a summary of the quantum efficiencies and turnover numbers of the different components included in the biocatalytic photosynthetic assemblies for the reduction of CO_2 to formate and reduction of NO_3^- and NO_2^- to ammonia. We realize that the photosystem for the fixation of NO_3^- to ammonia exhibits substantial stability and high quantum efficiency. Practical implications of such a photosystem could be solar production of fertilizers, eventually through decontamination of atmospheric nitrous oxides.

Other enzymes that can be communicated with photogenerated $MV^{+\cdot}$ include hydrogenase (Okura, 1985; Kinumi, 1989), Hyd, that mediates H_2-evolution (Figure 8) (Equation 14) and glutathione reductase (I. Willner, unpubl.), GlutR that mediates the production of glutathione (Equation 15).

Electron Transfer Communication between Redox Enzymes and Redox Polymers

Electron transfer communication between the redox enzymes and the photogenerated electron carrier, $MV^{+\cdot}$, described in the last section is accomplished through diffusion of the electron relay into the protein backbone. Consequently, a proper distance, ca. 10 Å between the electron carrier and

14.

$$2MV^{+\cdot} + 2H^{+} \xrightarrow{\text{Hyd}} 2MV^{2+} + 2H^{2}$$

15.

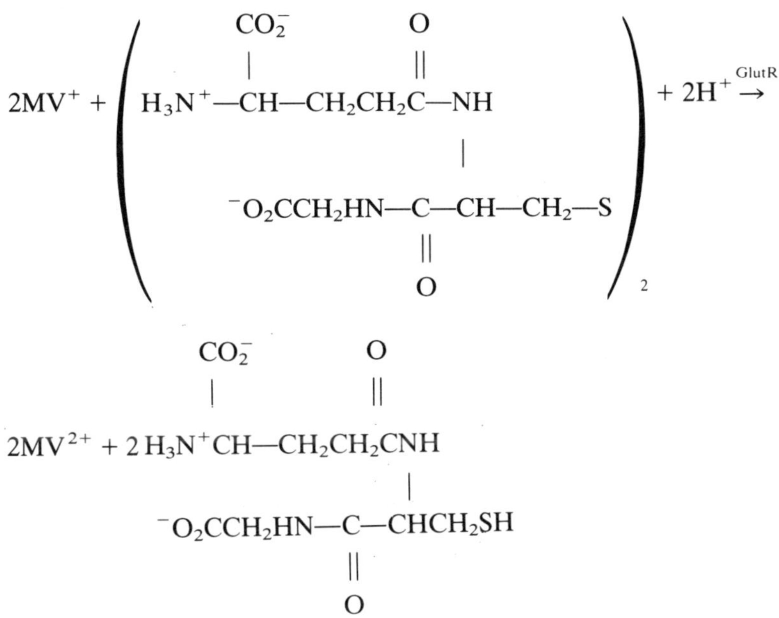

the active site is maintained and ET occurs. However, in a rigid assembly where diffusion is eliminated, the electron carrier and enzyme redox site must be forced into mutual proper distances that allow ET communication. The design of rigid organized systems, where electrical communication between redox sites of enzymes and their macroscopic environment is maintained, is of importance in developing biosensor devices, i.e., electrical communication between bulk electrode material and the redox active site (Degani, 1988, 1989; Heller, 1990; Armstrong, 1988).

We find that ET communication between redox enzymes and their macroscopic environment can be established by immobilization of the enzymes into a polymer matrix containing pendant redox groups (Willner, 1990c). The enzyme nitrate reductase (E.C. 1.9.6.1), NitraR, has been entrapped in an acrylamide copolymer composed of the monomers acrylamide and N-methyl-N′-propyl-4,4′-bipyridinium acrylamide (12). The average loading of the acrylamide copolymer by the bipyridinium redox groups corresponds to 2.7%. Introduction of the enzyme NitraR, immobilized in the redox polymer, into an aqueous photosystem that includes $Ru(bpy)_3^{2+}$ as photosensitizer, Na_2EDTA as electron donor, and nitrate as substrate results upon illumination, in the photoreduction of NO_3^- to nitrite. The rate of NO_2^- formation upon illumination of the system is shown in Figure 10(a). The reduction of NO_3^- to nitrite is also driven in the dark, by adding dithionite to the enzyme

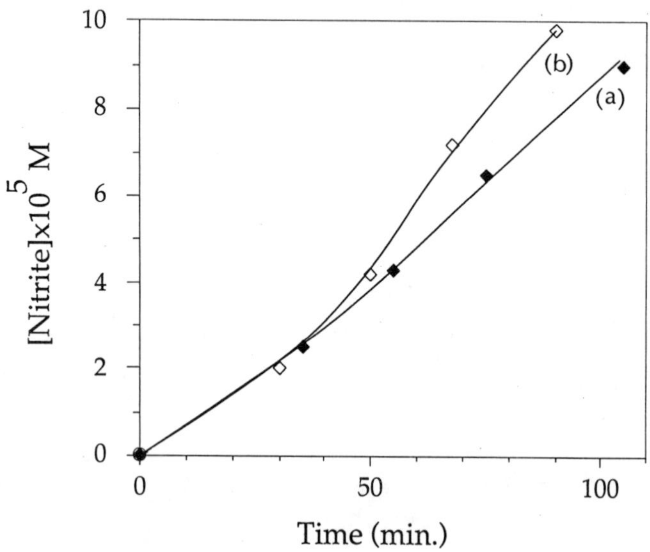

Figure 10. Rate of NO_3^- reduction in the assembly composed of nitrate reductase, 0.023 units (based on native enzyme) immobilized in the acrylamide: (12) copolymer. In all experiments, $[NO_3^-] = 4.4 \times 10^{-3}$ M in 2.3 ml Tris buffer, pH = 7.44: (a) Through illumination in the presence of $Ru(bpy)_3^{2+}$, 5.5×10^{-5} M, and Na_2EDTA, 4.4×10^{-3} M; (b) Dark reduction in the presence of sodium dithionite, 9.4×10^{-3} M.

being immobilized in the redox matrix. The rate of NO_2^- formation in the dark chemical reaction is displayed in Figure 10(b). It is evident that the rates of NO_2^- formation in the light-induced process and in the dark chemical reduction reaction are very similar, implying that the rate-limiting process of NO_3^- reduction is independent of the primary reducing component of the systems. Figure 11 outlines schematically the sequence of reactions that occurs in the photochemical assembly that includes the redox polymer. Excitation of the photosensitizer results in oxidative ET quenching by the bipyridinium polymer, and the formation of the reduced bipyridinium radical cation pendant polymer. The oxidized photoproduct, $Ru(bpy)_3^{3+}$, oxidizes Na_2EDTA, a process that recycles the photosensitizer. Electron exchange between polymer bound oxidized bipyridinium components and the reduced radical cation site occurs, a process leading to charge distribution in the polymer backbone. It is assumed that immobilization of the enzyme into the redox matrix forces some of the pendant bipyridinium groups into proper distance alignment with the biocatalyst active site, through which ET communication proceeds. Thus the bipyridinium polymer sites, in proximity to the enzyme active site, act as electron sinks for collecting polymer distributed electrons and the distribution source to the biocatalyst active site.

In fact, the rate of ET from the reduced redox polymer to the biocatalyst active site can be determined by means of time-resolved laser flash photolysis experiments (Figure 12). Upon flashing the redox polymer in the presence

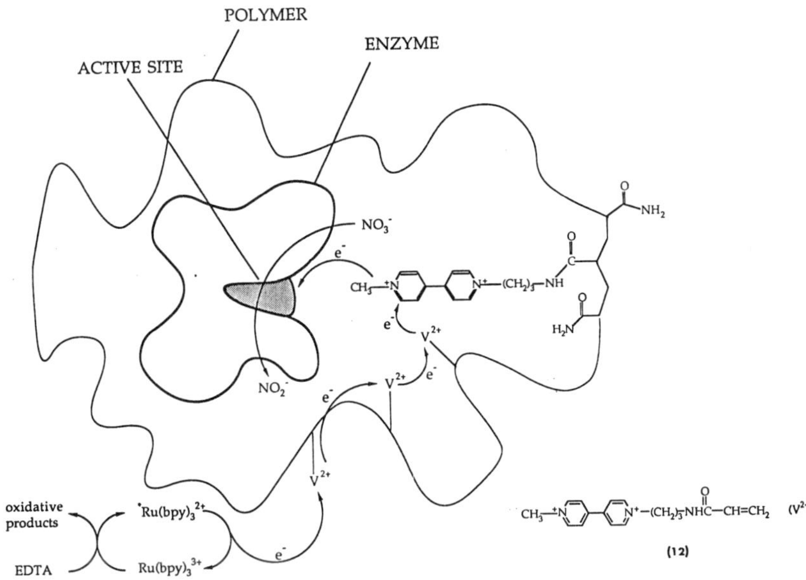

Figure 11. Photoinduced reduction of NO_3^- to NO_2^- in an organized assembly composed of NitraR immobilized in an acrylamide-bipyridinium redox polymer. ET communication between the redox polymer and the enzyme active site takes place in the redox polymer-enzyme assembly.

of the photosensitizer, $Ru(bpy)_3^{2+}$, but without including the NitraR enzyme, the reduced polymer is formed, and its decay (Figure 12a) through recombination with the oxidized photoproduct (Equation 16) can be followed. The recombination rate constant corresponds to $k_b = (9.5 \pm 2) \times 10^5\ M^{-1}\ s^{-1}$. A similar time-resolved experiment performed with the enzyme NitraR which is immobilized in the redox polymer shows the decay curve of the reduced polymer displayed in Figure 12(b). It is evident that the reduced polymer decays faster in the presence of the enzyme, implying that reduction of the enzyme provides a competitive route to the recombination rate (Equation 17). Deconvolution of the decay curve in the presence of the enzyme by the decay pattern observed in its absence allows the calculation of the ET rate constant, k_{et}, from the redox polymer to the enzyme active site. The value of k_{et} corresponds to $(9 \pm 3) \times 10^5\ s^{-1}$. We thus realize that ET communication in rigid assemblies can be tailored by the immobilization of catalysts in redox matrices.

Photoregulation of Enzymes by Chemical Modification of Proteins with Photochromic Components

Photoregulation of the activity of enzymes is an alternative route through which light energy could interact with biological materials. Light-triggered ON-OFF biocatalytic switching systems could be important as biocatalytic

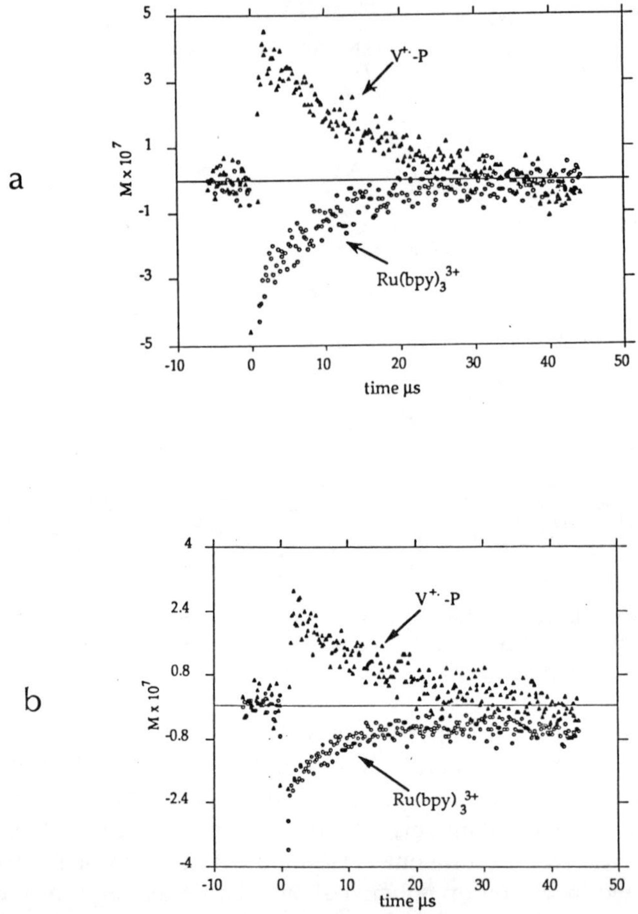

Figure 12. (a) Transient absorption spectra of photoproducts formed upon illumination of $Ru(bpy)_3^{2+}$ and bipyridinium-acrylamide redox copolymer. The system is illuminated with a laser flash, $\lambda = 430$ nm. Transient absorption corresponds to bipyridinium radical cation-copolymer (followed at $\lambda = 602$ nm). Bleach spectrum corresponds to formation of $Ru(bpy)_3^{3+}$ (followed at $\lambda = 465$ nm). Decay curves correspond to the recombination of electron transfer products. (b) Transient absorption spectra of redox products formed in a photosystem composed of $Ru(bpy)_3^{2+}$ and NitraR immobilized in the bipyridinium-acrylamide redox copolymer. Experimental conditions as in (a). Decay curves correspond to the back electron transfer reaction of intermediate photoproducts and competitive electron transfer process from the reduced polymer to the enzyme.

information storage devices, light signal amplifiers and sensoring assemblies (Martinek, 1979; Hug, 1978). The concept of designing a light-switchable biocatalyst is displayed in Figure 13. The enzyme is chemically modified by photochromic components. It is assumed that in the presence of the photochromic isomer I, the active site structure of the biocatalyst is maintained and the biocatalyst retains the activity of the native enzyme. Light-

16. $Ru(bpy)_3^{3+} + V^{+}\text{-Polymer} \xrightarrow{k_b} Ru(bpy)_3^{2+} + V^{2+}\text{-Polymer}$

17.
$$\begin{array}{ccc} & V^{+}\text{-polymer} & \\ \swarrow^{+Ru(bpy)_3^{3+}} & & \searrow^{+E(Enzyme)}_{k_{c+}} \\ Ru(bpy)_3^{2+} + V^{2+}\text{-Polymer} & & V^{2+}\text{-Polymer} + E^{-} \end{array}$$

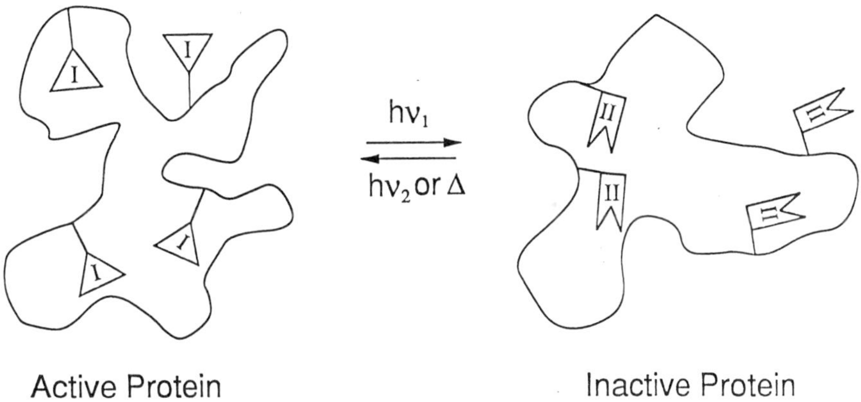

Figure 13. Photoregulation of biocatalysts through chemical modification of the protein by photochromic components.

induced isomerization of the photochromic component to isomer II, at wavelength, $h\nu_1$, perturbs the protein structure, and the active site configuration is destroyed. Consequently, the biocatalyst activity is switched off by the light signal. Subsequent reversible isomerization of the photochromic isomer II into structure I, by a light signal at a second wavelength, $h\nu_2$, or by a thermal process, restores the protein and its active site structures, and thereby the biocatalyst is switched on.

Photoregulation of papain was accomplished by following this conceptual approach (I. Willner, 1991). The enzyme was chemically modified by anchoring trans-azobenzene-4–carboxylic acid (13) on the lysine residues of the protein. The chemical derivatization process of the biocatalyst involves reaction of the N-hydroxysuccinimide ester of (13) with the enzyme in the presence of dicyclohexylcarbondiimide. The average extent of loading of papain by (13) corresponds to 4.6 molecules of dye per protein molecule. Since the protein backbone of papain includes ten lysine units (Arnon, 1970), the coverage by the photochromic components corresponds to 46% of the lysine groups. The trans-(13)-modified papain exhibits photochromic properties: upon illumination at $\lambda = 320$ nm, it undergoes isomerization into cis-(13)-

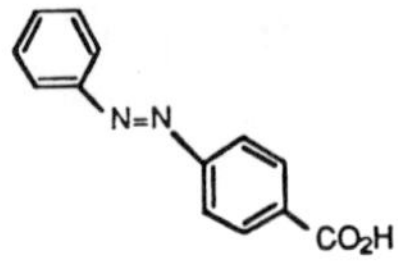

(13)

Table 3. Loading degrees(a) and activities(b) of cis- and trans-(13)-azo-benzene papain.

Enzyme	Loading	Activity (%)
papain	0	100
trans-(13)-papain	4.6	80
cis-(13)-papain	4.6	30
trans-(13)-papain	2.9	80
cis-(13)-papain	2.9	70

[a] Determined by (i) the dye content (mg dye mg^{-1} protein), ϵ(13) 322 nm = 26,600 M^{-1} cm^{-1}; (ii) staining the modified protein and native enzyme by fluorescamine and measuring the fluorescence intensity of the stained (13)-papain relative to the fluorescence intensity of stained native papain.
[b] Determined by following the hydrolysis rate of N-α-benzoyl-D,L-arginine-4-nitroanilide (14), Equation 19.

papain (Equation 18). Spectroscopic studies reveal that interconversion of trans-(13)-papain into cis-(13)-papain corresponds, in fact, to a photostationary equilibrium where [cis-(13)-papain]/[trans-(13)-papain] = 0.89. Thus, approximately 47% of the enzyme anchored trans-(13) groups are isomerized to cis-(13) moieties.

Illumination of photogenerated cis-(13)-papain with light, $\lambda > 400$ nm, restores entirely trans-(13)-papain. Thus the azobenzene papain exhibits reversible photochromic properties.

The activities of the photochromic papain in its trans-(13) and cis-(13) forms have been determined and compared to the native papain, Table 3. Activities are determined by following the hydrolysis of N-α-benzoyl-D,L-arginine-4–nitroanilide (14), BAPNA, (Equation 19). It is evident that trans-(13)-papain retains 80% of the activity of the native enzyme, while cis-(13)-papain exhibits only 30% of the activity of the native enzyme. Thus, the

18. 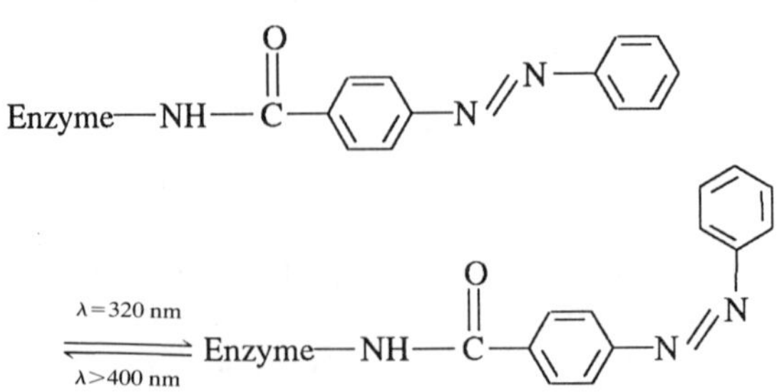

19.

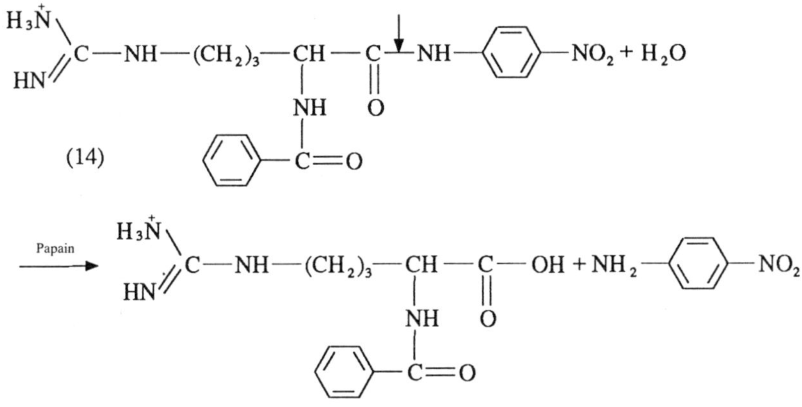

biocatalyst trans-(13)-papain is about 2.75–fold more active than the isomerized functionalized protein, cis-(13)-papain.

The differences in the biocatalytic activities of cis- and trans-(13)-papain allow us to photoregulate the enzyme by an external light source. Figure 14 shows the sequential "ON-OFF" and "OFF-ON" switching of (13)-modified papain. In Figure 14(a) hydrolysis of (14) is initiated by trans-(13)-papain, and after 15 min of reaction, the assembly is illuminated with light, $\lambda =$ 320 nm (arrow indication). Interconversion of trans-(13)-papain to cis-(13)-papain occurs, and simultaneously the hydrolysis rate of (14) is slowed down as compared to the starting rate. Similarly, the hydrolysis rate of (14) can be switched "ON" as shown in Figure 14(b). Here the reaction is initiated by photochemically pre-generated cis-(13)-papain, and a slow hydrolysis rate of (14) is observed for about 15 min. After this time, illumination of the system with light, $\lambda > 400$ nm, results in interconversion of the biocatalyst into trans-(13)-papain. Isomerization of the biocatalyst is accompanied by an acceleration of the hydrolysis rate of (14). We realize that chemical modification of papain by the photochromic material (13) allows the photoregulation of the biocatalyst. We note, however, that the photochromic papain does not exhibit complete switching properties. Yet, with the specific biocatalyst, a photostationary cis:trans composition is established and the cis-(13)-papain contains, in fact, 53% of non-isomerized trans azo components. Thus, selecting photochromic biocatalysts, where complete interconversion between active and inactive photochromic configurations occurs, is anticipated to improve the switching effectiveness.

Some insight into the origin of the difference in the activities of the photochromic papain is obtained from complementary studies. Decrease of the loading degree of papain by (13) to 2.9 photochromic units on the protein backbone results in almost similar activities of trans and cis-(13)-papain (see Table 3). Thus, the extent of loading of the protein backbone by the photochromic components strongly affects its photoregulative properties.

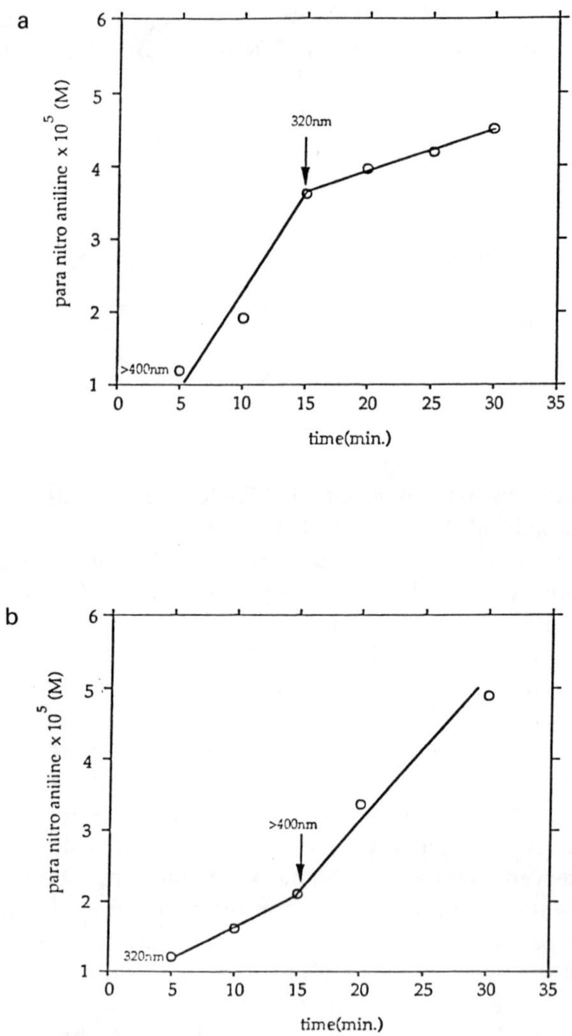

Figure 14. (a) Photoregulated hydrolysis of BAPNA (14). Hydrolysis is initiated by trans-(13)-papain and switched off by cis-(13)-papain. Arrow indicates photochemical conversion of trans-(13)-papain to cis-(13)-papain. The system is composed of an aqueous solution (3.61 ml, pH = 7.5), that includes (14), 8.3×10^{-4} M, and the photochromic enzyme, 0.5 mg, (dye loading 0.05 mg dye mg^{-1} protein). (b) Photoregulated hydrolysis of BAPNA (14). Hydrolysis is initiated by cis-(13)-papain and then accelerated by trans-(13)-papain. Arrow indicates photoinduced conversion of cis-(13)-papain to trans-(13)-papain. The system is composed of an aqueous solution (3.6 ml, pH = 7.5) that includes (14), 8.3×10^{-4} M, and the photochromic enzyme, 0.5 mg (dye loading 0.05 mg dye mg^{-1} protein).

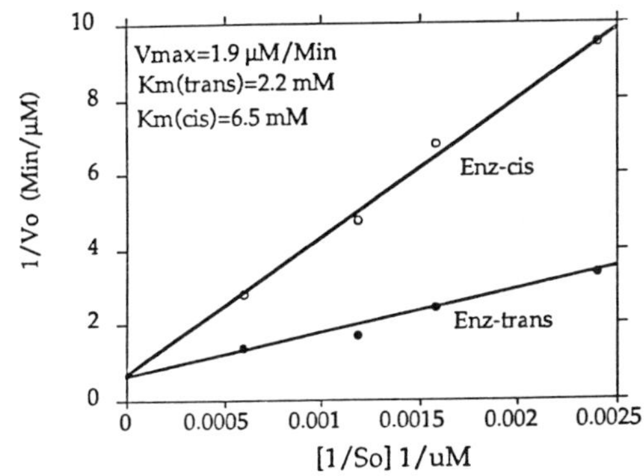

Figure 15. Lineweaver-Burke plots for (●) trans-(13)-papain and (○) cis-(13)-papain.

Namely, structural changes in the protein backbone induced by loaded photochromic components can account for its photoregulative behavior. Deactivation of cis-(13)-papain towards hydrolysis of (14) could originate from poorer binding properties of the biocatalyst towards the substrate, from perturbations in the catalytic functions of the active site, or from both effects operating together. One can distinguish between the modes of deactivation of cis-(13)-papain by comparing the kinetic behavior of cis and trans-(13)-papain. The Lineweaver-Burke plots of trans-(13)-papain and cis-(13)-papain are shown in Figure 15. It is evident that the two biocatalysts exhibit a similar V_{max} value (for an enzyme concentration of 2μM, $V_{max} = 1.9 \pm 0.2$ μM min^{-1}), but differ substantially in their K_m values: K_m(trans) $= 2.2 \pm 0.2$ mM; K_m(cis) $= 6.5 \pm 0.6$ mM. This kinetic behavior implies that cis-(13)-papain is deactivated as compared to trans-(13)-papain due to decreased binding properties of the inhibited biocatalyst towards the substrate, rather than from a difference in catalytic performance of the two enzymes. We thus interpret the photoregulative properties of (13)-papain in terms of structural changes in the protein backbone as outlined schematically in Figure 16. In trans-(13)-papain the protein structure is retained and the biocatalyst performance is maintained. Light-induced isomerization of the photochromic components to cis-(13)-papain effects structural perturbations in the protein backbone that decrease the binding features of the enzyme towards the substrate. As a result, the catalytic performance of the enzyme declines, and its activity is switched off. We thus conclude that structural perturbations in the protein backbone by means of photoisomerizable photochromic components generate photoregulative biocatalytic materials.

Conclusions, Outlook and Perspectives

Biocatalytic materials can interact with light energy by two modes. Light energy could be a continuous source of energy that drives biocatalysed

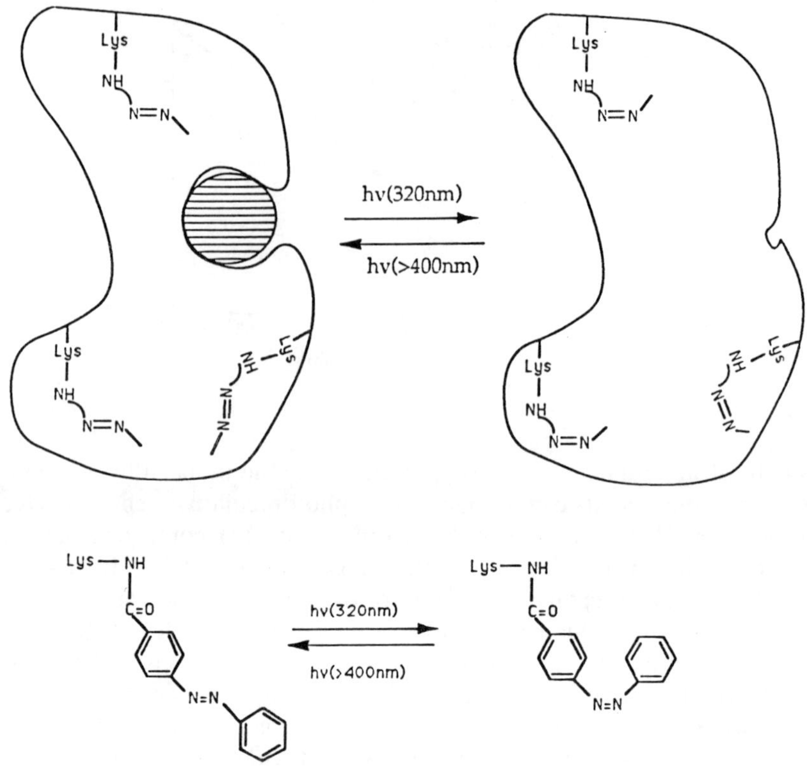

Figure 16. Schematic structural changes of the protein backbone through photochemical isomerization of anchored groups exerting a remote effect on the binding properties of the biocatalyst towards the substrate.

transformations. Alternatively, light energy could act as an external signal that triggers biocatalysed processes. Light-driven biocatalytic processes operate through primary conversion of the radiative energy into chemical energy by means of electron transfer reactions. The redox products are then utilized in the regeneration of biologically active cofactors or, alternatively, directly coupled to redox sites in enzymes. Supply of redox equivalents to the enzymes allows the occurrence of subsequent biocatalysed transformations, i.e., amino acid synthesis, carbon dioxide fixation. Thus, the preparation of feedstock compounds and fuel products by light driven biocatalysed processes is envisaged. The report highlights the organization of integrated polymer-enzyme assemblies where electrical communication between the photogenerated redox sites and enzymes occurs. Such organized assemblies exhibiting electrical communication between the redox site and external macroscopic environment are of general applicability in the development of biosensor devices.

Light-triggered biocatalytic materials could also find broad practical applicabilities. "ON-OFF" triggered biocatalysts could be useful materials in targeted therapy where enzymes are activated/deactivated at a localized site by guided light. Future applications of photo-switchable biological compounds could also involve their application as biosensors exhibiting improved sensitivities, and utilization as information storage catalysts.

Acknowledgments

The different topics discussed in the report are supported by the Kernforschung Anlage, Jülich, FRG, the Israel Academy of Sciences and Humanities and the Niedersachsen Land Foundation, FRG.

References

Armstrong, F.A., Hill, H.A.O. and Walton, N.J. 1988. Direct electrochemistry of redox proteins. Acc. Chem. Res. 21: 407–413.

Arnon, R. 1970. In: Methods in Enzymology. Perlman, G.E. and Lorand, L., eds. Academic Press, New York and London. Vol. 19.

Balzani, V., Bolletta, F., Gandolfi, M.T. and Maestri, M. 1978. Bimolecular electron transfer reactions of the excited states of transition metal complexes. Top. Curr. Chem. 75: 1–64.

Bard, A.J. 1980. Photoelectrochemistry. Science. 207: 139–144.

Chenault, H.K. and Whitesides, G.M. 1987. Regeneration of nicotinamide cofactors for use in organic synthesis. Appl. Biochem. Biotechnol. 14: 147–197.

Degani, Y. and Heller, A. 1988. Direct electrical communication between chemically modified enzymes and metal electrodes. 2 Methods for bonding electron-transfer relays to glucose oxidase and D-amino-acid oxidase. J. Am. Chem. Soc. 111: 2357–2358.

Grätzel, M., ed. 1983. Energy Resources through Photochemistry and Catalysis. Academic Press, New York.

Harriman, A., Porter, G. and Richoux, M.-C. 1980. Attempted photoreduction of hydrogen using sulphophthalocyanines as chromophores for three-component system. J. Chem. Soc. Faraday Trans. 2, 76: 1618–1626.

Harriman, A. and West, M.E., eds. 1983a. Photogeneration of Hydrogen. Academic Press, London.

Harriman, A. 1983b. The photodissociation of water. Platinum Metals Rev. 102–107.

Heller, A. 1990. Electrical wiring of redox enzymes. Acc. Chem. Res. 23: 128–134.

Hug, D. 1978. The activation of enzymes with light. Photochem. Photobiol. Rev. 3: 1–33.

Juris, A., Balzani, V., Barigelletti, F., Campagna, S., Belser, P. and von Zelewsky, A. 1988. Ru(II) polypyridine complexes: Photophysics, photochemistry, electrochemistry, and chemiluminescence. Coord. Chem. Rev. 84: 85–277.

Kalyanasundaram, K. 1982. Photophysics, photochemistry and solar energy conversion with tris(bipyridyl)ruthenium(II) and its analogues. Coord. Chem. Rev. 46: 159–244.

Keller, P. and Moradpour, A. 1980. Is there a particle-size dependence for the mediation by colloidal redox catalysts of the light-induced hydrogen evolution from water? J. Am. Chem. Soc. 102: 7193–7196.

Kinumi, Y. and Okura, I. 1989. Photoinduced hydrogen evolution with viologen-linked anionic zinc porphyrins. J. Mol. Catal. 52: L33–L36.

Kiwi, J. and Grätzel, M. 1979. Hydrogen evolution from water induced by visible light mediated by redox catalysis. Nature. 281: 657–658.

Krasna, A.I. 1980. Acridines, deazaflavins, and tris(2,2′-bi- pyridine)ruthenium as catalysts for photoreduction of hydrogen from organic compounds. Photochem. Photobiol. 31: 75–82.

Lay, P.A., Mau, A.W.H., Sasse, W.H.F., Creaser, I.I., Gahan, L.R. and Sargeson, A.M. 1983. Cobalt cage complexes as electron-transfer agents in the photoreduction of water. Inorg. Chem. 22: 2347–2349.

Mandler, D. and Willner, I. 1984. Solar light-induced formation of chiral 2–butanol in an enzyme catalyzed chemical system. J. Am. Chem. Soc. 106: 5352–5353.

Mandler, D. and Willner, I. 1986. Photosensitized NAD(P)H regeneration systems: Application in the reduction of 2–butanone, pyruvic and acetoacetic acids and in the reductive amination of pyruvic and ketoglutaric acids to amino acids. J. Chem. Soc. Perkin Trans. 2: 805–811.

Mandler, D. and Willner, I. 1988. Photochemical fixation of carbon dioxide by enzymatic photosynthesis of malic, aspartic, isocitric and formic acids in artificial media. J. Chem. Soc. Perkin Trans. 2: 997–1003.

Martinek, K. and Berezin, I.V. 1979. Artificial light-sensitive enzymatic systems as chemical amplifiers of weak light signals. Photochem. Photobiol. 29: 637–649.

Meyer, T.J. 1989. Chemical approaches to artificial photosynthesis. Acc. Chem. Res. 22: 163–170.

Miller, D. and McLendon, G. 1981. Model systems for photocatalytic water reduction: Role of pH and metal colloid catalysts. Inorg. Chem. 20: 950–953.

Okura, I. 1985. Hydrogenase and its application for photoinduced hydrogen evolution. Coord. Chem. Rev. 68: 53–99.

Richoux, M.-C. and Hall, D.O. 1980. Photoproduction of hydrogen activated by a colloidal platinum polyvinyl alcohol catalyst. Photobiochem. Photobiophys. 1: 375–383.

Simon, H., Bader, J., Günther, H., Newmann, S. and Thanos, J. 1985. Chiral compounds synthesized by biocatalytic reductions. Angew. Chem. Int. Ed. Engl. 24: 539–553.

Steckhan, E. 1987. Organic syntheses with electrochemically regenerable redox systems. Top. Curr. Chem. 142: 1–69.

Willner, I., Mandler, D. and Riklin, A. 1986. Photoinduced CO_2- fixation forming malic and isocitric acids. J. Chem. Soc. Chem. Commun. 1022–1024.

Willner, I., Maidan, R., Mandler, D., Dürr, H., Dörr, G. and Zengerle, K. 1987. Photosensitized reduction of CO_2 to CH_4 and H_2-evolution in the presence of Ru and Os-colloids: Strategies to design selectivity of products distribution. J. Am. Chem. Soc. 109: 6080–6086.

Willner, I. and Steinberger-Willner, B. 1988. Solar hydrogen production through photobiological, photochemical and photoelectrochemical assemblies. Int. J. Hydrogen Energy. 13: 593–604.

Willner, I. and Mandler, D. 1989a. Characterization of Pd-β cyclodextrin colloids as catalysts in the photosensitized reduction of bicarbonate to formate. J. Am. Chem. Soc. 111: 1330–1336.

Willner, I. and Mandler, D. 1989b. Enzyme catalysed biotransformations through photochemical regeneration of nicotinamide cofactors. Enzyme Microb. Technol. 11: 467–483.

Willner, I., Lapidot, N. and Riklin, A. 1989c. Photoinduced enzyme catalyzed reduction of nitrate (NO_3^-) and nitrite (NO_2^-) to ammonia (NH_3). J. Am. Chem. Soc. 111: 1883–1884.

Willner, I. and Willner, B. 1990a. Artificial photosynthetic model systems using light-induced electron transfer reactions in catalytic and biocatalytic assemblies. Top. Curr. Chem. In press.

Willner, I., Maidan, R. and Shapira, M. 1990b. Thermal and photochemical regeneration of nicotinamide cofactors and a nicotinamide model compound using a water-soluble rhodium phosphine catalyst. J. Chem. Soc. Perkin Trans. 2: 559–564.

Willner, I., Lapidot, N. and Riklin, A. 1990c. Electron transfer communication between a redox polymer matrix and immobilized enzyme. Activity of nitrate reductase in a viologen-acrylamide copolymer. J. Am. Chem. Soc. 112: 6438–6439.

Willner, I., Rubin, S. and Riklin, A. 1991. Photoregulation of papain activity through anchoring photochronic azo groups to the enzyme backbone. J. Am. Cham. Soc. In press.

Discussion (DR. WILLNER)

Kornguth: That was a very nice demonstration of the potential uses of electron conducting systems and biopolymers. I have several questions regarding these systems. First, is there any magnetic field effect on the efficiencies of the electron transfer process?

Willner: Magnetic fields can indeed affect photoinduced electron transfer reactions. External magnetic fields can influence conversion of singlet ion pairs into triplet ion pairs. Consequently, improved charge separation can be envisaged under such conditions. In addition, back electron transfer reactions of the primary photoproducts can be affected by an external magnetic field by T-S intersystem crossing. The systems I have described generate a primary triplet ion-pair and thus a magnetic field effect is only envisaged for the back electron transfer of the primary photoproducts. However, we did not check magnetic field effects on the photosystems that have been discussed.

Kornguth: I understand that electric initiation or photon initiation can be applied in your systems to get the same properties.

Willner: That is a correct statement. The systems operate by generating an intermediate electron carrier that mediates subsequent biotransformations. This electron carrier can be generated by the two alternative routes, i.e., electrochemically or photochemically.

Kornguth: You have described Ru(II)-tris-bipyridine and a Zn(II)-porphyrin as light active compounds that initiate the electron transfer process. What was the specific reason for introducing the Zn-porphyrin in the NADH regeneration system, and why is it not included in other photosystems?

Willner: The two photosensitizers, $Ru(bpy)_3^{2+}$ and Zn-porphyrin, Zn-$TMPyP^{4+}$, can be applied as photosensitizers for the production of $MV^{+\cdot}$, and therefore both can be applied to drive these biocatalyzed transformations. In fact, other compounds such as xanthene dyes or semiconductor materials (Z. Goren, N. Lapidot and I. Willner, J. Mol. Catal. 47 (1988): 21–32) can be applied for the same purpose. In the NADH regeneration system Ru(II)-tris-bipyridine cannot act as photosensitizer since photogenerated NADH acts as electron donor for the excited Ru(bpy)32 +. Thus the system is short-circuited, and photogenerated NADH is reoxidized by this pathway (I. Willner, R. Maidan and B. Willner, Isr. J. Chem. 29 (1989): 289–301). Hence we had to switch to the Zn-porphyrin, where photogenerated NADH does not act as an electron donor for the excited photosensitizer. Zn-porphyrins show photodegradability, while Ru(bpy)32 + exhibits superior photostability. Therefore, in most of the systems, Ru(bpy)32 + is applied as photosensitizer and only in the photochemical NADH regeneration system is Zn-TMPyP4 + the light active compound applied.

Kornguth: You have applied N,N′-dimethyl-4,4′-bipyridinium as the electron acceptor. In this compound free rotation between the two pyridinium units exists. Have you looked at bipyridinium salts where rotation is restricted to determine the structural effects on the electron transfer properties of the relay compound?

Willner: Indeed, the structural properties of the bipyridinium salt strongly affect its redox properties. For example, the reduction potential of N,N′-dimethyl-4–4′-bipyridinium, MV^{2+}, corresponds to $E^o = -0.42$ Volt vs. NHE. On the other hand, the reduction potential of the N,N′-dimethyl-(3,3′-dimethyl)-4,4′-bipyridinium, where the two methyl groups eliminate rotation across the pyridinium units (and the two rings are almost perpendicular), exhibits a reduction potential of $E^o = -0.70$ Volt vs. NHE. The bipyridinium radical cations generated in the electron transfer processes are in a planar structure to maintain effective $\pi - \pi$-overlap and resonance delocalization. The configuration of the oxidized bipyridinium salts is determined by the substituents in the ortho positions of the compounds. With bulky substituents in these positions, the pyridinium units are distorted, the reduction of the bipyridinium compound (and its enforcement into a planar structure) is thermodynamically more difficult, and its reduction potential is shifted to more negative values. In fact, we have been able to correlate structural features of bipyridinium salts (dihedral angles between the pyridinium units) and reduction potentials (I, Willner, A. Ayalon and M. Rabinovitz, New J. Chem. 14: 685–688). Thus, the redox potential of bipyridinium salts can be tuned by structural modification of the electron acceptor.

The redox potential of the electron acceptor affects the effectiveness of the photoinduced electron transfer process. The electron transfer rate from the excited species is determined, among other factors, by the thermodynamic driving force. Thus lowering the electron acceptor redox potential is accompanied by a decrease in the electron transfer rate. On the other hand, the reduced electron carrier must exhibit the proper redox potential to reduce the enzyme active site. Therefore, structural modifications of the bipyridinium electron acceptor allow us to tune electrical interactions with the excited species and the biocatalyst active site.

Ellis: In the azobenzene switching experiment is there any thermal control of the cis to trans isomerization process? Is the photostationary equilibrium constant measured in the dark?

Willner: In this system no thermal isomerization of cis-azobenzene papain to trans-azobenzene papain is observed. The photostationary [cis]:[trans] equilibrium is thus measured in the dark.

Ellis: Is that because it is in the protein matrix or would the free photochromic compound show this property in solution too?

Willner: The thermal isomerization of trans-azobenzene to cis-azobenzene is controlled by the substituents of azobenzene compound and the reaction solvent. 4–Carboxy-trans-azobenzene forms in water a similar photostationary equilibrium and no thermal cis to trans isomerization is detected. Thus, the thermal behavior of the photochromic enzyme probably does not originate from a matrix effect.

Rishpon: In the redox polymer electrical communication system, do you need the Ru(II)-tris-bipyridine component or can you get direct electron transfer to the polymer?

Willner: The $Ru(bpy)_3^{2+}$ component is only included to initiate a photochemical electron transfer process. The polymer can be immobilized onto an electrode. In such configuration direct electron transfer to the polymer backbone occurs.

B. Green: Back to the azobenzene substituted papain. What is the loading of the protein by the photochromic material? Could you improve the switching efficiency of the photochromic enzyme by initiating the chemical modification of the enzyme with 10% cis- carboxyazobenzene?

Willner: The results that have been described include an enzyme where 4.6 lysine residues out of the 10 lysine residues of papain are chemically modified. In fact, the extent of loading strongly affects the switching efficiency of the biocatalyst. Decrease of the loading of the biocatalyst by the photochromic material to an average value of 2.9 results in trans and cis-papains of similar activities.

Your suggestion to initiate the chemical modification of the biocatalyst by cis-carboxyazobenzene is very good. It should certainly be tried, and improved switching efficiencies could be anticipated, provided that a similar extent of loading can be achieved. Yet one should remember that such improved switching can be maintained only for one cycle. Subsequent cycles will generate the photostationary equilibrium.

Key Words: Photobiocatalysis, Photochemical regeneration of cofactors, Photosynthesis, CO_2-fixation, Nitrate reduction, Ammonia biocatalysed photosynthesis, Electrical communication in proteins, Redox polymers, Photoregulative enzymes, Photoresponsive papain.

Modifying Specificities of Restriction Enzymes

Waclaw Szybalski
McArdle Laboratory for Cancer Research, University of Wisconsin, Madison, WI 53706

Introduction

For the last five years, our laboratory has been studying various aspects of controlled cleavage of DNA, both for (1) the 50 bp-10 kb range of single-stranded (ss) and double-stranded (ds) DNA fragments, plasmids or phages, and (2) for very large genomes of bacteria, yeast and higher eukaryotes. We are using commercial restriction enzymes (ENases) either (i) of the class-IIS variety (ENase-IIS) (Szybalski, 1985), where the recognition site on the DNA for the ENase is separated by several base pairs from the cleavage site, together with an adapter oligodeoxynucleotide (oligo), or (ii) various combinations of a particular ENase, its cognate methyltransferase (MTase) and DNA-binding protein whose binding site contains the corresponding ENase/MTase recognition site. The ENase-IIS enzymes were used either for precise cutting (Podhajska and Szybalski, 1985) or for precise trimming of DNA (Hasan et al., 1986), whereas the MTase/ENase/DNA-binding protein combination was used for very rare cuts in large genomes (Koob et al., 1968; Grimes et al., 1990; Koob and Szybalski, 1990).

ENase-IIS/Adapter Combination: A Universal ENase

We have described a method to transfer the ENase-IIS specificity to any specially designed sequence of a synthetic oligo adapter (Podhajska and Szybalski, 1985; Kim et al., 1988). The adapter consists of the hairpin ds domain, which contains the ENase-IIS recognition site (e.g., 5′-GGATG for *Fok*I), flanked by at least 2 bp on both sides, and an ss domain which is complementary to the ss target DNA that is to be cleaved. For *Fok*I, the ENase binds to its ds recognition site and cleaves the duplex between the adapter and the target ss DNA, at 9 and 13 nucleotides (nt) away from the recognition site. By proper design of the nt sequence of the adapter, one could cleave any ss DNA target between any two nt (Podhajska and Szybalski, 1985).

D. Kamely et al. (eds), Biotechnology: Bridging Research and Applications, 371–376.

The oligo adapter-*Fok*I method was primarily designed to cleave ss DNA, but could be adapted to cut one or even both strands of a ds DNA target by hybridizing adapter within partially denatured regions of ds DNA and at the same time fixing it by a variety of chemical reactions, involving functional crosslinking groups present in the ss domain of the adapter. It was shown also that the product of the *Fok*I-adapter-primer cleavage could be a ds DNA fragment with precisely defined ends (Kim et al., 1988). The cleavage is precise not only on the level of target DNA but also on the level of adapter, which leads to another kind of application, namely, development of probes for assay of mutations or detection of targeted DNA molecules.

Probes

Hairpin Oligo Adapters

We designed several kinds of oligoprobes that permit one to detect point mutations. For example, adapters have been designed which would hybridize with target DNAs in such a manner that the mutations to be determined would fall into the 4–bp *Fok*I cleavage site. When using denatured wild-type (wt) target DNA and a properly designed wt probe, labeled mono- (or di- or tri-) nucleotides would be released. The same would happen with the mutant probe and denatured mutant DNA target. However, no cutting and no release of labeled nucleotides would be observed with mutant versus wt combinations, when the probes are properly designed with nonhomologies varying between one and two base pairings (Kim, S.C. and Szybalski, W., in preparation). This approach is the basis of the mutant assay. One should stress that it is quite easy to automate the assay for labeled nucleotides released from rather large probes (adapters), which makes the method well adapted to clinical applications.

End-filling and Amplification by the "Printing" Reaction

Another variant of this method is based on our observation that the 4–nt cohesive end formed by *Fok*I cleavage, when filled in with dNTPs by DNA polymerase(s), becomes again a target of one-strand cutting by *Fok*I, with release of the "filled-in" tetranucleotide. Actually, S.C. Kim and W. Szybalski (in preparation) have found that it is enough to fill in the cohesive end with only two or three nt (one is not enough) to permit cutting off a di- or trinucleotide, respectively. These findings permit two kinds of mutant detection: (1) a modification of the method just described, where the adapter could be labeled either before or after the first cutting, and (2) introduction of a powerful amplification step (signal amplification), based on our so-called "printing reaction," where simultaneous addition of the *Fok*I enzyme (or other ENases IIS) and special DNA polymerases (which does not digest the product) with the required nucleoside triphosphates, results in many rounds of re-synthesis and re-cutting, while releasing copious amounts of tetranucleotide (or di- or trinucleotide, depending on design), which could easily be labeled and detected, distinguishing between mutant and wt DNA target.

The original version of our "printing reaction" was designed to produce only di-, tri-, or tetranucleotides, but it should be possible to produce much longer DNA chains by using a variety of designs, including development of a *Fok*I mutant enzyme which would cut only one particular DNA strand (carrying the 5′-GGATG sequence). Since the two cutting reactions could be dissociated by using the intercalating agents, it should be possible to separate the cleavage of one or another DNA strand by mutations in cutting domain of the ENase-IIS. Moreover, I predict that many ENases, which cut precisely only one strand, might exist in nature, but were missed in various ENase screening procedures which were designed to detect ds cuts. Actually, there exist endonucleases of such kind, e.g., the product of gene *A* of phage ϕX174 (or analogous product of phages f1 or M13) which cuts only one specific strand between precisely defined nt.

Precise Trimming of DNA

By incorporating a unique ENase(s) IIS-recognition site into a cloning vector, and by using the corresponding ENase(s), mung bean nuclease, and ligase, we were able to precisely trim cloned DNA inserts in a pre-programmed fashion, as to remove an exact number of bp, usually 1–40. Each cycle of cleavage removes 1, 4 or 12 nt, in various combinations (Hasan et al., 1966; Hasan and Szybalski, 1989). For these experiments, we developed also a double-methylation reaction, which permits increasing the *Fok*I recognition site from 5 to 7 bp (Pósfai and Szybalski, 1988). Other variants of the trimming reaction permits trimming DNA in both directions allowing precise fusion of two genes. In this way we succeeded in construction of hybrid MTases which displayed enzymatic activity (Kim et al., 1990). We are working on further application of such trimming procedures.

Amplification of DNA Fragments

The asymmetric nature of the ss ends produced by several ENase-IIS permitted us to construct special amplifying vectors, as described by Kim and Szybalski (1988). New amplifying plasmids are under development.

Achilles' Heel Cleavage (AC): Cleavage of Large Genomes at a Single Site

This novel method combines the recognition specificity of DNA-binding proteins, e.g., *lac* or phage λ repressors or various transcriptional factors, with that of an ENase (and cognant MTase) whose recognition specificity, in the case of the *lac* repressor, whose recognition sequence, *lacO*$_s$ (5′-AATTGTGAGCGCTCACAATT) contains sites for the *Hha*I (overlined) and *Hae*II (underlined) ENases. Thus, the *lac* repressor protects these sites from M·*Hha*I MTase, which does methylate all other *Hha*I/*Hae*II sites, rendering them uncleavable by these two ENases (Koob et al., 1988; Grimes et al., 1990).

Using this approach, we have cleaved the *E. coli* 4.7–Mb genome in one place (*att*λ) or in two defined places (*att*λ and Tn7), and also completely cleaved a single chromosome (V) in the 15–Mb *Saccharomyces cerevisiae* genome at a unique predetermined site, using the LacI-*lacO*$_s$-M·*Hha*I-*Hae*II AC protocol. In the case of yeasts, the symmetric *lacO*$_s$ operator was introduced into the *ura*3 gene on chromosome V. To protect DNA from shear, cells were embedded in agarose microbeads, DNA was released, deproteinized, and methylated with M·*Hha*I in the presence of the LacI repressor. All *Hae*II sites (5′-PuGCGCPy), with the exception of the one in the introduced *lacO*$_s$ that was protected by LacI, were methylated. After heat inactivation of the M·*Hha*I and LacI proteins, the 580–bp chromosome V was completely and specifically cleaved by *Hae*II into 450–kb and 130–kb fragments, as identified by pulsed-field gel electrophoresis (CHEF) (Koob and Szybalski, 1990). The success of this methodology was aided by many technical improvements, especially in preparing genome-carrying agarose microbeads (by vortexing or atomizing), their coating with non-ionic detergents, rapid and efficient methylation, and other experimental details (Koob and Szybalski, 1990; and in preparation).

The AC methodology has many applications, including (1) initiation of sequencing at unique sites, using the SPEL-6 approach (Szybalski, 1990); (2) mapping of protein binding sites on large genomes ("inverse footprinting," since the binding sites result in unique cuts); (3) function-specific cleavage, when using transcriptional factors for protecting from MTases; (4) cloning very large DNA fragments excised by cleaving two AC sites (within special, e.g., transposable elements with other enabling modules), either as YACs or by direct insertion into bacterial genomes (BACs) using *att*P modules and Int function, and many other applications.

Conclusions

Using ENases, MTases, ss-specific nucleases, ligases, oligo hairpin adapters, and DNA binding proteins, we have designed various strategies for very specific cleavage of small, medium, and very large DNA molecules, for detecting point mutations or other genetic changes, for a very efficient "printing reaction" to produce tetranucleotides, as a highly amplified detection signal, or to synthesize copious amounts of longer sequences, for development of trimming, gene-fusion, or amplification plasmids, and for various other applications, some already tested and some in various stages of development.

Acknowledgments

This outline summarizes studies of several collaborators specified in text and references, and was supported by NIH Grant GM39715 to W. Szybalski, and by NCI Core Grant 5–P30–CA-07175.

References

Grimes, E., Koob, M. and Szybalski, W. 1990. Achilles' heel cleavage: creation of rare restriction sites in λ phage genomes and evaluation of additional operators, repressors and restriction/modification systems. Gene 91: 1–7.

Hasan, N., Kim, S.C., Podhajska, A.J. and Szybalski, W. 1986. A novel multi-step method for generating precise unidirectional deletions using *Bsp*MI, a class-IIS restriction enzyme. Gene 50: 55–62.

Hasan, N., Kur, J. and Szybalski, W. 1989. An *Mbo*II/*Fok*I trimming plasmid allowing consecutive cycles of precise 1- to 12–base-pair deletions in cloned DNA. Gene 82: 305–311.

Kim, S.C., Podhajska, A.J. and Szybalski, W. 1988. Cleaving DNA at any predetermined site with adapter-primers and class-IIS restriction enzymes. Science 240: 504–506.

Kim, S.C., Pósfai, G. and Szybalski, W. 1990. A novel gene fusion vector: construction of a GGmCC-specific chimeric methyltransferase. Second New England Biolabs Workshop on Biological DNA Modification. Berlin, September 2–7, 1990, p. 99.

Kim, S.C. and Szybalski, W. 1988. Amplification of cloned DNA as tandem multimers using *Bsp*MI-generated asymmetric cohesive ends. Gene 71: 1–8.

Koob, M., Grimes, E. and Szybalski, W. 1988. Conferring operator specificity on restriction endonucleases. Science 241: 1084–1086.

Koob, M. and Szybalski, W. 1990. Cleaving yeast and *Escherichia coli* genomes at a single site. Science 250: 271–273.

Podhajska, A.J. and Szybalski, W. 1985. Conversion of the *Fok*I endonuclease to a universal restriction enzyme: Cleavage of phage M13mp7 DNA at predetermined sites. Gene 40: 175–181. [Errata: 1986. Gene 43: 325].

Pósfai, G. and Szybalski, W. 1988. A simple method for locating methylated bases in DNA using class-IIS restriction enzymes. Gene 74: 179–181.

Szybalski, W. 1985. Universal restriction endonucleases: designing novel cleavage specificities by combining adapter oligodeoxynucleotide and enzyme moieties. Gene 40: 169–173.

Szybalski, W. 1990. Proposal for sequencing DNA using ligation of hexamers to generate sequential elongation primers (SPEL-6). Gene 90: 177–178.

Discussion (DR. SZYBALSKI)

Smith: Is *lac* repressor hard to come by in the purity that you need. Szybalski: We prepared *lac* repressor in large quantities, both wild-type *lac* repressor and a special mutant which has ten times higher affinity to DNA. Isolation was done by our Biotechnology Center or by Mr. Piotr Skowron. So we might have the world's largest supply of the X86 mutant repressor. There is also available a commercial version of *lac* repressor fusion protein (Promega, Madison, WI, USA) which works fine.

Smith: Does the fact that DNA is methylated at all the *Hha*I sites have any downstream repercussions? Does that interfere at all?

Szybalski: All sites must be methylated perfectly, otherwise the method would not work because wherever a site is not methylated, it will be cut. We might not see cutting if only 1–2% of molecules are not methylated; they would melt into the background. But fortunately, in all of our experiments we must have had perfect methylation at all non-protected sites.

Smith: Now you have this methylated DNA and you want to do something with it. Do the methyl groups get in the way of any subsequent things that you would like to do?

Szybalski: Frequently, DNA is methylated naturally and researchers still can work with such DNA, so it probably does not really make much difference. I have not really given much thought to it, because I know that DNA is methylated only at the GCGC sequences. Maybe you could even take advantage of the C^5 methylated 5′-GCGC sequences in some cleverly designed experiments.

Key Words: Achilles' heel cleavage, DNA-binding protein, DNA cleavage, Methyltransferase, Oligodeoxynucleotide adapter, Printing reaction, Restriction enzymes, Amplification, Detection of point mutants, Universal restriction endonuclease

Pulsed Laser Fluorometry for Environmental Monitoring

George C. Saunders, John C. Martin, James H. Jett, Mark E. Wilder, Adelmo Martinez, Bill F. Bentley, John Lopez, and Lara Hutson
Information Technologies Group and Cell and Molecular Biology Group, Los Alamos National Laboratory, Los Alamos, NM 87545

Introduction

Fluorescence chemistry offers the potential for extremely sensitive detection analyses. At the Los Alamos National Laboratory, we have been developing for several years both instrumentation and analytical procedures to detect and measure fluorescence at the molecular level (reviewed in Jett et al., 1990). Recently, single chromophores of Rhodamine 6 G have been reliably detected in a flowing sample stream using laser-induced fluorescence (E.B. Sherer, et al., submitted for publication). While this level of sensitivity requires very sophisticated instrumentation, moderately priced ($5,000–10,000) commercially available instrumentation can detect fluorescent compounds in the nanomolar to picomolar range. One such instrument is a pulsed laser fluorometer, a nitrogen pumped dye laser with the capability to excite and measure fluorescence from the near-ultraviolet to the near-infrared spectral regions. The authors were tasked to develop a point detector capable of detecting both acetylcholinesterase inhibitors and biological molecules (amino acids, peptides, and proteins) in water or air. As part of this system, we incorporated a pulsed laser fluorometer into a continuous flow monitor for each class of compounds above. The instrument should prove useful for detecting and monitoring any compound or class of compounds for which fluorescent chemistry exists.

Fluorescence detection of the organophosphorous nerve agent, sarin, was reported by Gehauf and Goldenson (1957). Their method was based on the formation of highly fluorescent indoxyl by the direct reaction of sarin with indole and sodium perborate. Sensitivity of the method approached 1 ng of sarin ml^{-1}. A problem with this procedure is that the indoxyl is a short-lived (less than 1 min) chemical intermediate in the reaction sequence between indole (the starting compound) and indigo (the end compound).

Guilbault and Kramer (1965a) combined immobilized acetylcholinesterase

D. Kamely et al. (eds), Biotechnology: Bridging Research and Applications, 377–397.

(AChE) with a fluorogenic substrate to monitor the reaction of the enzyme with anticholinesterase compounds. AChE was immobilized within a starch gel placed on a polyurethane foam pad. A solution of 2–naphthyl acetate (non-fluorescent) was slowly passed through the enzyme pad, where it was enzymatically converted to the highly fluorescent 2–naphthol (ex 320 nm, em 410 nm). A drop in the baseline fluorescence indicated the presence of an inhibitor, and the rate of fall provided some idea of the activity of the inhibitor. The reported system easily detected 1 μg sarin ml^{-1} water.

Other fluorogenic substrates for AChE include resorufin butyrate and indoxyl acetate (Guilbault and Kramer, 1965b), N-methylindoxyl acetate (Guilbault, et al., 1968), and 1–methyl-7–acetoxyquinolinium iodide (Prince, 1966). The most sensitive of these substrates for AChE detection is reported to be N-methylindoxyl acetate (Guilbault, 1973). More recently, Parvari (Parvari, et al., 1983) described a very sensitive two-step fluorometric reaction for detecting AChE activity. Acetylthiocholine is first converted by the enzyme to thiocholine, which is then reacted with coumarinylphenylmaleimide (CPM) to form a fluorescent product.

We explored both N-methylindoxyl acetate and CPM chemistries for use in our monitoring system.

Many methods are currently available for sensitive detection of amino acids, peptides, and proteins. For a fluorescent chemical process to be practical in a continuous flow water monitor, it must be sensitive and quick, and the starting reactants should not be fluorescent. The reactants and products should also be compatible with an aqueous environment. For our experiments, we chose to examine both CPM, which reacts with thiols, and fluorescamine, which reacts with primary amines. CPM initially was of great interest to us because a single laser dye and set of detection optics would allow us to monitor both AChE activity and proteinaceous material. CPM was initially synthesized and used for the histochemical staining of protein thiol groups on tissue sections (Sippel, 1981a, 1981b). Fluorescamine (4–phenylspiro [furan-2(3H) 1′-phthalan]-3, 3′dione) was introduced in 1972 (Weigele, et al., 1972a; Udenfriend, et al., 1972). This reagent reacts with primary amines to form the same fluorophors as are produced in the ninhydrin-phenylacetaldehyde reaction (Weigele, et al., 1972b). At room temperature the fluorescamine reaction with primary amines is extremely rapid, proceeding with a half-time of less than a second. CPM reacts more slowly, but significant reaction does occur within one minute.

As stated earlier we were tasked to develop a monitoring system suitable to continuously sample both water and air. Because the monitoring system was designed to perform continuous aqueous phase wet chemistry, we required an air sampler that would scrub contaminants from the air into an aqueous phase suitable for injection into the detector. We could not obtain a compact, reasonably priced device from commercial sources to perform this task. Therefore, we built and tested a compact air sampler/scrubber based on our own unique design (described later). A prototype system, sampling at the rate of 50 L of air per minute, concentrated protein aerosols at least 3000 times. We believe that because of its effectiveness, simplicity of design, and projected low cost, this device will have broad application for environmental sampling and monitoring.

Materials and Methods

Chemicals

N-methyl indoxyl acetate (NMIA), purified electric eel acetylcholinesterase (AChE), fluorescamine, acetylthiocholine, all buffer chemicals, amino acids, (BSA) bovine serum albumin (fraction V), and solvents were obtained from Sigma Chemical Company, St. Louis, MO, 63178. 7–diethylamino-3–(4′-maleimdylphenyl)-4–methylcoumarine (CPM) was obtained from Molecular Probes, Inc., Eugene, OR 97402. CNBr-activated Sepharose 4B was obtained from Pharmacia AB, Uppsala, Sweden. Isopropyl p-nitrophenyl methylphosphonate (INP) was a gift from Dr. Thomas Whaley, Biochemistry/Biophysics Group, Life Sciences Division, Los Alamos National Laboratory, Los Alamos, NM 87545.

Immobilization of Acetylcholinesterase

One unit (about 1 μg) of AChE in 5 ml of 0.1 M carbonate buffer, pH 9.7 was added to 1 gm of CNBr-activated Sepharose 4B and gently rocked for 4 h at room temperature. The mixture was refrigerated overnight to hydrolyze remaining active sites. The Sepharose-AChE was washed with 300 ml of PBS, resuspended in 10 ml of PBS containing 5 mg of BSA and 0.1 mg ml^{-1} of sodium azide. This reagent was stable for at least 6 months at 4°C.

Stock and Working Reagent Solutions

NMIA: stock, 0.62 M in ethylene glycol monomethyl ether (stable at least 3 months); working, 75 μl of stock in 50 ml 0.1 M PO_4 buffer, pH 7.5 (stable for at least 24 h). CPM: stock 2 mg in 10.0 ml of acetonitrile; working, small amounts of stock were added directly to the sample to be analyzed. Aqueous CPM solutions were not stable because of spontaneous hydrolysis. INP: stock, 100 μg ml^{-1} in acetonitrile (stable at least 3 months); working dilutions of INP were made fresh daily in acetonitrile from stock. Sepharose-AChE: 50 μl (0.005 unit AChE per test). Acetylthiocholine: Stock (prepared weekly), 2 mg ml^{-1} 0.1 M PO_4 buffer, pH 7.5; working dilutions varied. CPM: stock, 2 mg in 10 ml of acetonitrile; working dilutions varied but were prepared fresh daily from stock. Fluorescamine: stock/working, 200–400 μg

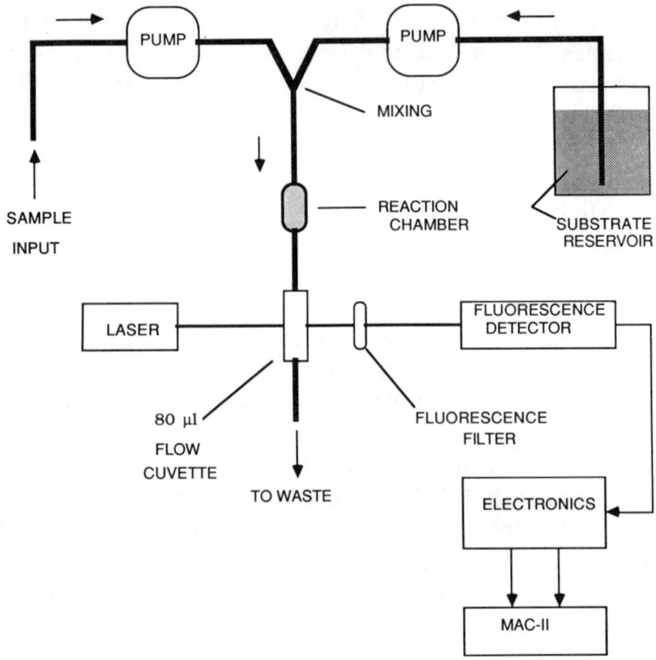

Figure 1. Los Alamos chemical detector. Schematic diagram of system components. See text for narrative.

ml^{-1} in acetonitrile (stable at least 3 mo). Fluorescamine rapidly hydrolyzes in the presence of water. The buffer used for amino acid/protein reactions with fluorescamine was 0.05 M sodium borate, pH 10.0. Small amounts of stock fluorescamine were added directly to the buffered sample to be analyzed.

Continuous Flow Monitor for AChE Inhibitors

The sample and NMIA substrate are individually pumped (Peristaltic Pump P-3, Pharmacia, Uppsala, Sweden) into a mixing chamber and then flow through a reaction chamber that contains sepharose-AChE (Figures 1 and 2). Silicon tubing is used throughout. The reaction chamber consists of a small disposable pipette tip fitted with a large glass bead. The large bead is sequentially layered with small glass beads and Sepharose-AChE. A stopper with a 20 gauge needle seals the chamber. The fluorescent product flows into the measurement chamber, a low volume flow cuvette (Hellma Cells Inc., Box 544, Jamaica, NY 11424; flow cell Model No. 176.751–QS Z-8.5). Typically, about 2 min are required for the sample to arrive at the measurement chamber. Then, the beam of a nitrogen-pumped, pulsed dye laser (Laser

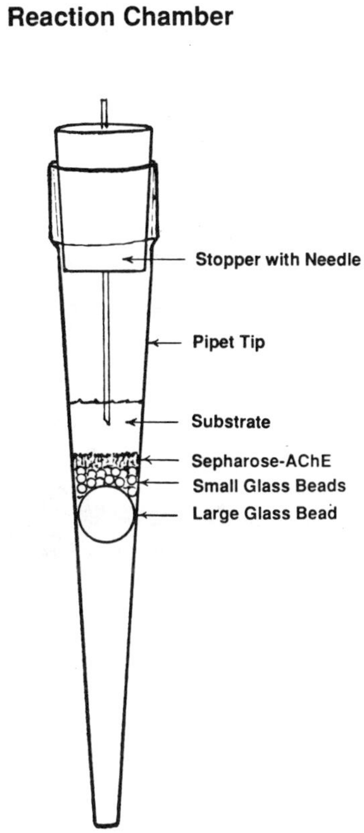

Figure 2. Drawing illustrating the components of the reaction chamber used for the generation of fluorescent reaction product which results from the interaction of immobilized AChE and a fluorogenic substrate.

Sciences Inc., 80 Prospect St., Cambridge, MA 02139; Model 337500) impinges on the flow channel of the cuvette. The fluorescence signals are electronically processed (described in detail later) and then acquired and stored in a MAC-II computer. Presence of AChE inhibitors is indicated by a reduction of fluorescence intensity.

Continuous Flow Monitor for Proteins, Peptides, Amino Acids

The sample is pumped by a peristaltic pump and is mixed with both fluorescamine and borate buffer (pumped with a Harvard 22 Syringe Pump, Harvard Apparatus, South Natick, MA 01760, USA) (Figure 3). The fluorescent reaction products pass through the flow cuvette, where the fluorescence intensity is determined by the pulsed laser fluorometer. The fluorescence

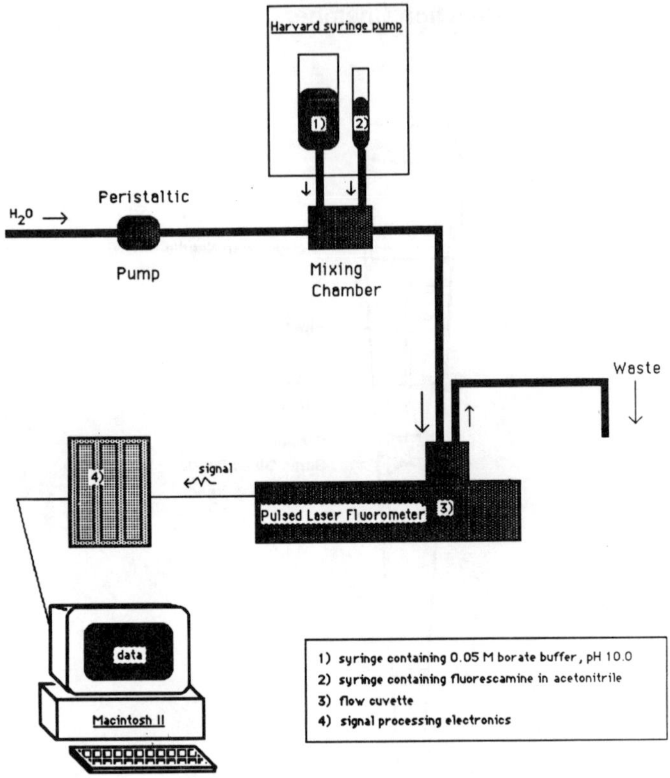

Figure 3. Apparatus for detecting protein using fluorescamine. Schematic diagram of system components. See text for narrative.

intensity increases with increasing levels of compounds containing primary amines. Signals are processed and stored as described below.

Pulsed Laser Fluorometer

A nitrogen-pumped, pulsed dye laser (Laser Sciences Inc., 80 Prospect St., Cambridge, MA 02139; Model 337500) impinges on the flow channel of the cuvette. To detect primary amines and thiols we use the laser dye BBQ (Laser Sciences Inc. Cat. No. 337876) to obtain output at wavelength ~395 nm wavelength for the excitation of the fluorescent products of fluorescamine and CPM. To detect AChE inhibitors, we use the laser dye Coumarin 450 (Photochemical Research Associates, 45 MEG. Drive, London, Ontario, Canada NGE2U2; Cat. No. 2A446) to obtain output at 436 nm wavelength for the excitation of the fluorescent product n-methylindoxyl. The pulsed laser is externally triggered, usually at a rate of about one pulse sec^{-1}. We chose this slow rate for two reasons: first, it maximizes the laser life time (the nitrogen laser tube will last for about 20 million shots or about 8 months at this rate); second, this slow rate minimizes the amount of data stored by the computer-based data acquisition system.

The fluorescence detection system is contained in a light-tight housing. The dye laser beam enters the housing after passing through a band-pass filter. This filter blocks spontaneous fluorescence emission from the laser dye; this emission gives an undesired background for the fluorescence measurement. Although the flow cuvette is centered in the laser beam, most of the laser energy is not absorbed by the sample, but passes on through the cuvette and is absorbed by a beam dump. Sample fluorescence, detected at right angles to the laser beam, passes through a long-pass (and/or band-pass) filter to block any laser light that might have been scattered by the sample or the cuvette. The sample fluorescence is detected by a small (1.125 inch) diameter photomultiplier tube (PMT, Thorn EMI Gencom Inc., 23 Madison Rd., Fairfield, NJ 07006; Model 9924B and Model QL/30 PMT housing) attached to the detector housing. The PMT high voltage is operated in the range of 500 – 700 V DC. Fluorescence emission from the sample during each laser pulse gives rise to a current pulse from the PMT. The PMT signals are converted to voltage pulses by a transimpedance amplifier (Hiebert, 1990). Without further amplification these signals are routed to a Gated Peak-Sense and Hold circuit (GPS&H) (Steinkamp and Hiebert, 1982). This circuit is gated on only during each laser pulse; thus the detection electronics processes signals only for each laser pulse. The signal amplitude is held for about 20 μs, then cleared after the signal is digitized.

The signal amplitude is digitized by a 12–bit Analog-to-Digital Converter (ADC) contained on a data acquisition board from G.W. Instruments (35 Medford St., Somerville, MA 02143; Model No. GWI-625). The board is interfaced to a Macintosh II (Apple Computers Inc., 20525 Mariana Ave., Cupertino, CA 95014) and has several useful features. Its ADC can digitize 4 to 12 signals using an expansion daughter board. The G.W. board can generate up to four logic output signals, with programable control of timing and pulse width. These logic signals are used for external triggering of the pulse laser and for triggering the GPS&H circuit. The board also contains two Digital-to-Analog Converters (DAC) under programmable control. The DACs can be used to control the PMT high-voltage power supply output or to generate special purpose pulses with programmable waveforms.

The fluorescence data is recorded and stored by a Mac II program based on GW Instruments' control function library for the Model GWI-625 data acquisition board. The data record, displayed in chart recorder format, gives a history of the fluorescence intensity. The program also has a threshold alarm that analyzes the data for significant excursions from the background fluorescence as it is recorded. The usual alarm analysis tests the running average of the ten most recent measurements and looks for a more than a 10% change from the long term average. An alarm is read out in several ways: 1) as a message on the computer terminal, 2) as a DAC output used to generate an audio alarm and to turn on a flashing light, and 3) as a trigger for a remote alarm either by a direct wire hookup or by a microwave link. The response time of the computer alarm discriminator is close to real time, 10 sec of data inspection. The total system response time (about 2.16 min) is the sum of the sample delay time and the software discriminator time.

Data Acquisition and Analysis

Data acquisition and analysis are accomplished using a Macintosh II (Apple Computers, Sunnyvale, CA) computer with an internal analog-to-digital converter (ADC) board (GW Instruments, Inc., Somerville, MA). The computer, ADC board, and laser are controlled by a program written in binary basic (Microsoft, Redmond, WA) and a library of ADC control functions supplied by GW Instruments.

The program triggers the laser at a user-controlled frequency. At a specific time after each trigger, the ADC converts the voltage available at the analog input to the board into integers with 12–bit precision (10 volts/4096). The data are plotted on the screen of the computer and (optionally) written into a file. Two averages are calculated after each conversion. The short average is the average of the current and previous relatively few (on the order of 10) conversions. The long average is the average of the current and previous relatively many (on the order of 200) conversions. The number of conversions in each average are user selectable. If the short average is greater than the long average by a user-selectable factor, a 5–volt level is set at the analog output of the board (suitable for triggering an external alarm), and a bell is sounded by the computer. The factor could be related to the statistics of the measurements, for example, two standard deviations of the long average. The program can be set to run for any practical amount of time.

Ultrasonic Micromist Air Scrubber

Figure 4 schematically illustrates the principles used in the Los Alamos designed air sampler/scrubber. A small vacuum cleaner continuously draws air through the unit. A piezoelectric crystal oscillating at 2 million cycles per sec generates a micromist (most of the mist will pass through a 0.5 μm pore-size filter) whose droplets scrub chemical vapors from the sampled air. A series of baffles (two are illustrated in the figure) condense and precipitate the contaminant containing droplets. The resulting fluid is pumped through the detector where it is analyzed for components of interest.

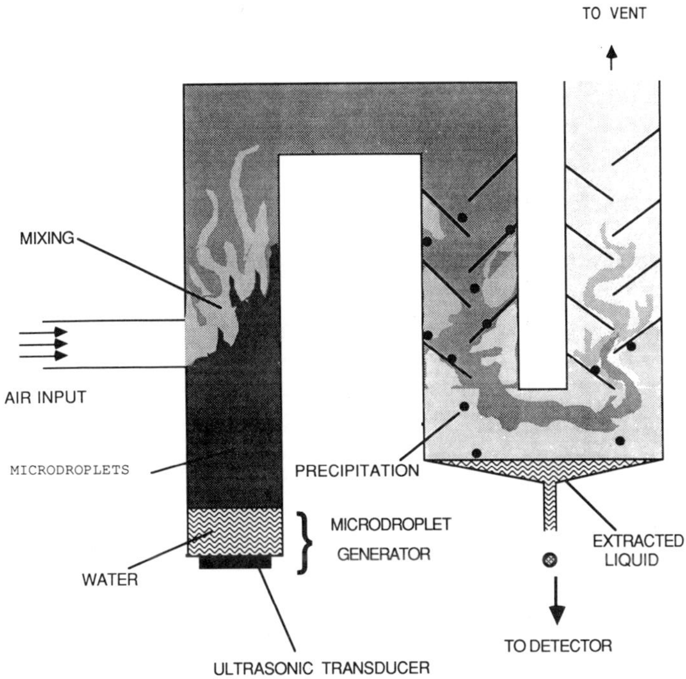

Figure 4. Ultrasonic micromist air sampler. Schematic diagram which illustrates the principles of the device. See text for narrative.

Results

Monitoring for Cholinesterase Inhibitors with NMIA

In the presence of AChE, the non-fluorescent NMIA is rapidly hydrolyzed to intensely fluorescent (ex 430 nm, em 501 nm) N-methylindoxyl (NMI) and acetic acid. NMIA stock solutions in ethylene glycol monomethyl ether were stable for at least 1 month. Because working dilutions in 0.1 M phosphate buffer undergo slow spontaneous hydrolysis, the working solution is prepared just before analysis. INP was used as the AChE inhibitor in all of our experiments. The AChE inhibitory activity of this compound is estimated to be about 100 X less (per molecule) than the nerve agent sarin (T. Whaley, Los Alamos National Laboratory, personal communication). Stock and working dilutions were prepared in a biosafety cabinet. Rubber gloves, labcoat, and safety glasses should be used when working with this compound. Spills can be rapidly neutralized with molar sodium hydroxide solutions.

Preliminary Experiments in a Static System

Varying amounts of INP were placed into a cuvette containing 10^{-4} M NMIA. At time t = 0, either 10^{-3} or 2.5×10^{-3} units of AChE were added

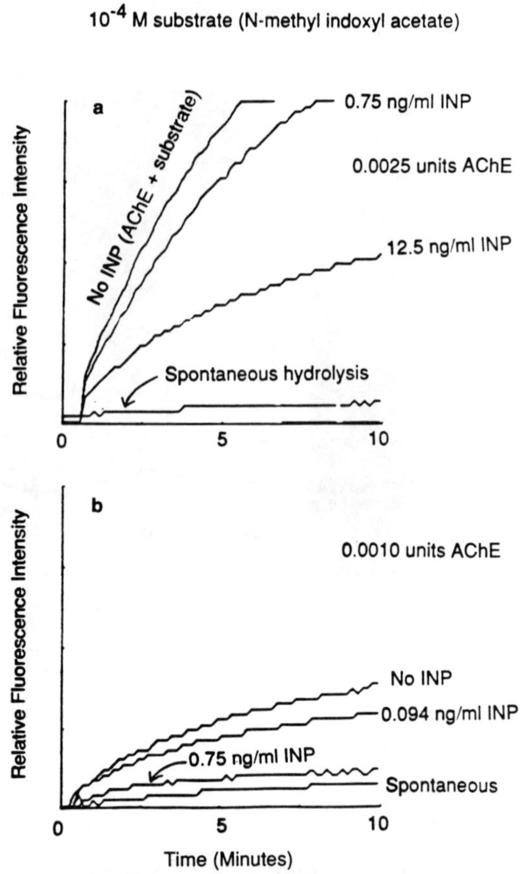

Figure 5. Kinetics of the inhibition of AChE by INP. (a) Using 2.5×10^{-3} units of enzyme; (b) using 1×10^{-3} units of enzyme.

with rapid mixing. Hydrolysis kinetics of the conversion of NMIA to NMI were monitored for 10 min. Figure 5 shows the type of results obtained in this experiment. Under limiting AChE levels (10^{-3} units), significant inhibition of hydrolysis was seen at about 100 pg ml^{-1} of INP. In data not shown, we established that the INP solvent, acetonitrile, in the volume used, did not affect AChE-induced NMIA hydrolysis. The results obtained in the static system were very reproducible (Figure 6). The data illustrate the overlay results of three separate experiments comparing spontaneous, slightly inhibited (130 pg INP ml^{-1}), and non-inhibited hydrolysis of 10^{-4} M NMIA. The amount of AChE used in Figure 6b-c was 0.002 units ml^{-1}.

Measurements in Flow

The instrumental setup, as shown in Figure 1, had a flow rate through the reaction chamber of approximately 3 ml hr^{-1}. Figure 7a shows a 6–hr run with

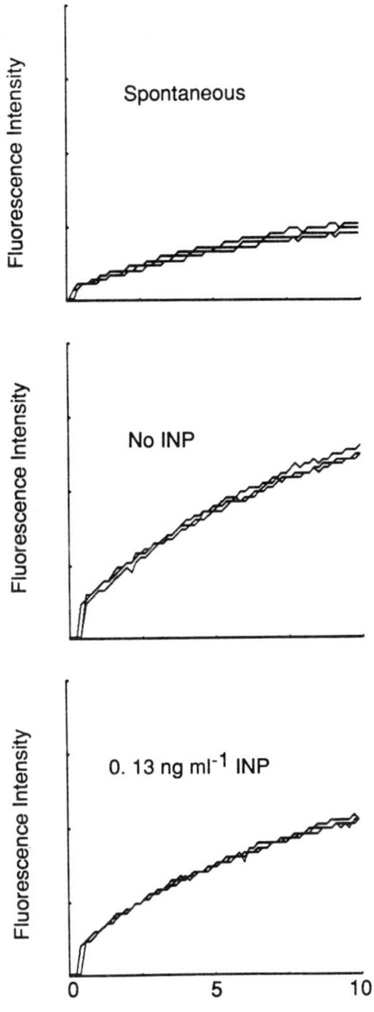

Figure 6. Results of three replicate kinetic experiments which illustrate the reproducibility of data obtained with the pulsed laser fluorometer. (a) spontaneous hydrolysis of 10^{-4} M NMIA; (b) hydrolysis of NMIA in the presence of 2×10^{-3} units AChE; (c) hydrolysis of NMIA in the presence of AChE and 130 pg INP.

NMIA (10^{-4} M) in the absence of AChE to illustrate the very low spontaneous hydrolysis rate observed. For the 6–hr experiment shown in Figure 7b the reaction chamber contained 0.005 unit of Sepharose-AChE. The fluorescence intensity of the AChE-catalyzed NMI remained remarkably constant throughout the run. Figure 8 illustrates the effect on NMI generation when either 10 ng (8a) or 1 ng (8b) of INP contained in 1 ml of NMIA solution passed through the reaction chamber. The gradual decrease in fluorescence intensity was caused by the poisoning of the immobilized AChE as the INP passed through the sys-

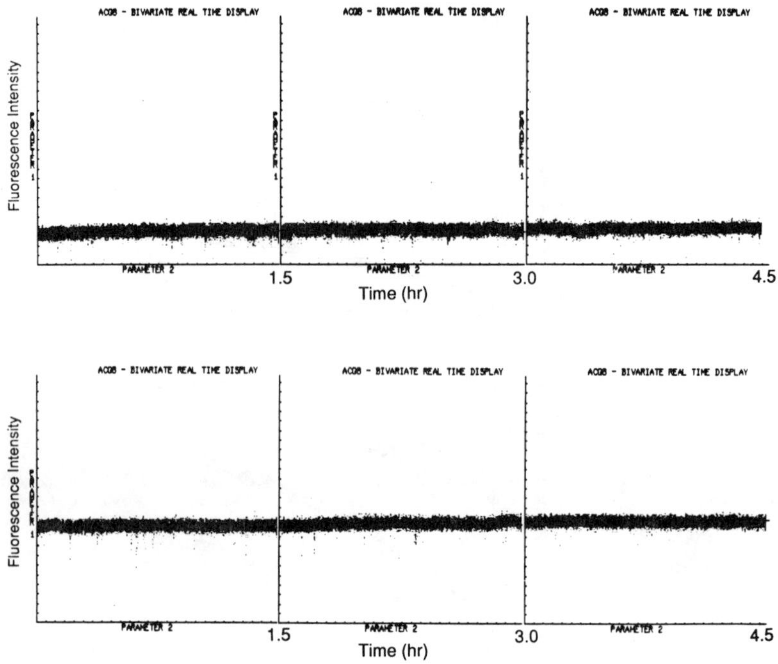

Figure 7. Continuous flow analysis of the hydrolysis of 10^{-4} M NMIA in the absence (a) and presence (b) of 5×10^{-3} units of immobilized AChE.

tem. Once the bolus of INP has passed through the system, no furthur decrease in fluorescence occurred, thus indicating a new steady state directly proportional to the amount of active enzyme remaining. A reaction chamber could be used for experiments for at least 5 days without apparent loss of enzyme activity. However, once the chamber was challenged with INP, we did not attempt to reverse the poisoning; therefore, a reaction chamber was used only once in a challenge experiment.

The generation of NMI reaction product was demonstrably effected by changes in the ambient temperature. Figure 9 illustrates this effect during a 48–hr run. Especially significant changes in reaction rate occurred during afternoons when the black shade in our laboratory absorbed and released significant amounts of solar energy into the room. The data acquisition program (described in the Materials and Methods section) compensated for these gradual changes in reaction rate. Alternatively the reaction chamber could be placed in a temperature controlled compartment.

Monitoring AChE Activity with Acetylthiocholine and CPM

AChE catalyzes acetylthiocholine to thiocholine and acetic acid. The thiocholine can then react with CPM to yield an intensely fluorescent product

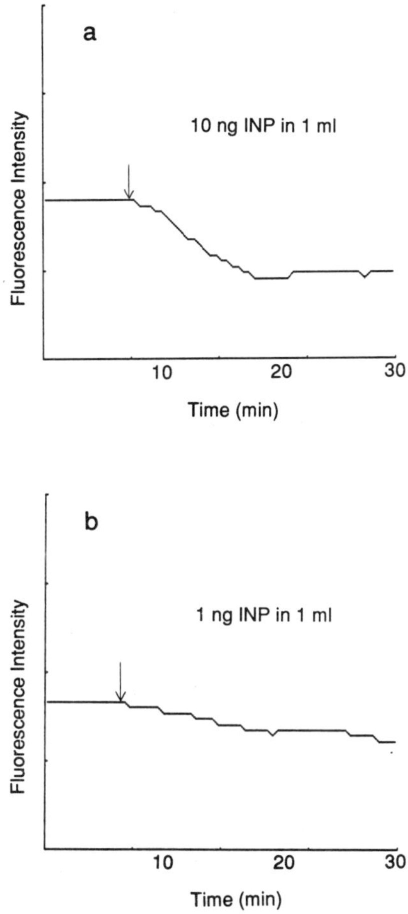

Figure 8. Sensitivity of detection of AChE inhibition by INP in flow. (a) injection of 1 ml of a sample containing 10 ng INP; (b) injection of 1 ml of a sample containing 1 ng INP.

(Parvari, et al., 1983). We explored the use of this reaction sequence as a scheme to monitor AChE activity. Stock solutions of CPM were stable in acetonitrile for at least 2 weeks. However, working solutions of CPM in 0.1 M PO_4 buffer, pH 7.5 were not suitably stable for use in a continuous-flow long term monitoring system because of rapid (within 2 hr) spontaneous conversion of CPM (in the abscence of acetylthiocholine) to a fluorescent product (Figure 10). No improvement in stability of CPM was seen in tris buffer. Additionally, preliminary kinetic experiments in a non-flowing system indicated that the CPM scheme was significantly less sensitive than the NMIA scheme (Figure 11). As shown previously (Figures 5, 8) using NMIA, we could detect less than 1 ng of INP. Therefore, we decided not to pursue CPM as a fluorogenic reagent for a continuous monitoring system for AChE inhibitors.

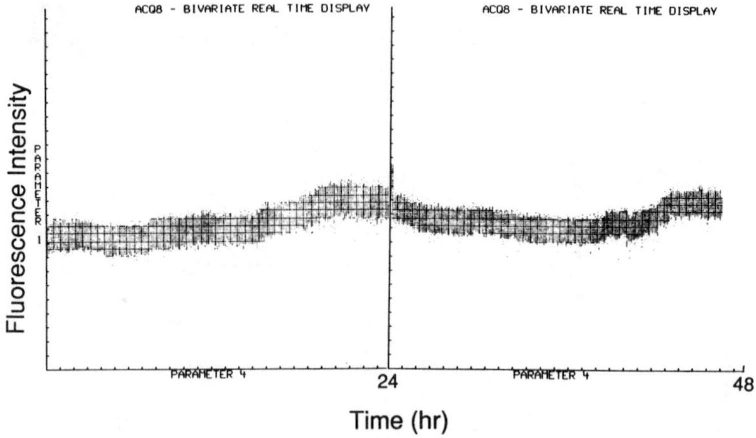

Figure 9. Effect of changes in ambient temperature (see text) on the rate of enzymatic hydrolysis of NMIA during a 48 hr continuous run in flow.

Use of CPM to Monitor Proteins, Peptides, and Amino Acids

We chose not to pursue using CPM because of both the limitations discussed above and initial experimental observations that, while CPM rapidly reacts with proteins (thiol groups), it only slowly reacts with the primary amines of amino acids (data not shown).

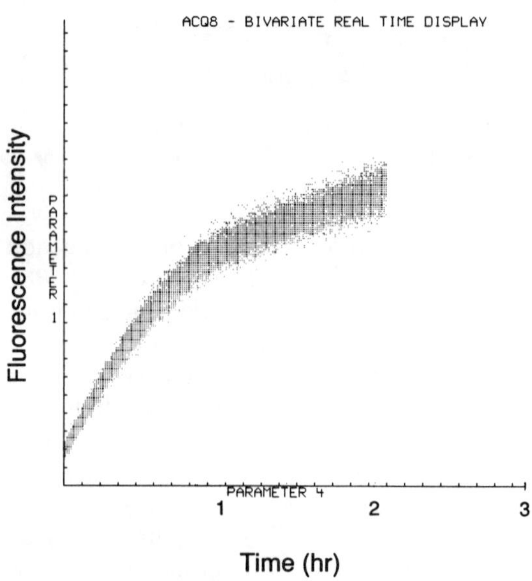

Figure 10. Kinetics of the spontaneous hydrolysis of 50 nM CPM in phosphate buffer.

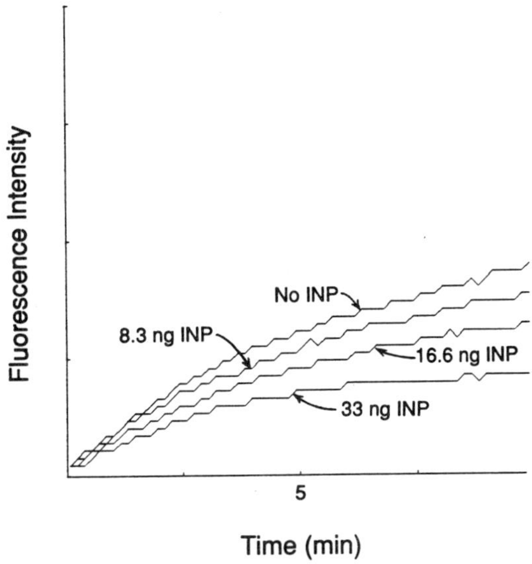

Figure 11. Kinetics of acetylthiocholine (8×10^{-5} M) hydrolysis (as measured by CPM chemistry) by AChE (2.5×10^{-3} units) in the presence of various amounts of INP. CPM was at 50 nM. A comparison of these results with those obtained with NMIA (Figure 5) indicates signifcantly less INP detection sensitivity when using the CPM scheme.

Fluorescamine as a Fluorogenic Reagent for Proteins, Peptides, and Amino Acids

Fluorescamine rapidly reacts with proteins, peptides, and amino acids to form stable fluorescent products. Because fluorescamine rapidly hydrolyzes in aqueous media, it must be dissolved in water-miscible solvents and added rapidly to test solutions. All tests were performed in 0.05 M borate buffer, pH 10.0. Fluorescamine was in acetontrile at 0.2 mg ml^{-1}.

Static Experiments

The ability of fluorescamine to react with a number of amino acids was established by rapidly adding 100 ul of fluorescamine (200 $\mu g\ ml^{-1}$) to 2 ml of buffer containing 0.5 ng of the amino acid. The pulsed laser fluorometer was used to measure the relative fluorescence intensity of each sample. The results of this experiment are shown in Table 1. All the amino acids tested except proline, which contains a secondary rather than a primary amine, yielded detectable fluorescent products.

In similar experiments we examined the reaction between fluorescamine and the purified protein BSA. Table 2 shows the results of this type of experiment. The reproducible detection limit was found to be 5×10^{-10} M.

Table 1. Detection of amino acid-fluorescamine reaction products

Amino acid[1]	Relative fluorescence intensity[2]
Buffer only	5
Buffer + Fluorescamine	42
Methionine	75
Cysteine	135
Lysine	108
Cystine	100
Phenylalanine	111
Tyrosine	60
L-glutamine	110
Arginine	147
Isoleucine	96
Proline	40
Asparagine	120
L-valine	102
L-serine	113
Threonine	167

[1] 0.5 ng/tube.
[2] Arbitrary units; mean of replicate determinations.

Experiments in Flow

Figure 3 shows the instrumental setup. In contrast to the AChE-NMIA reaction, the reaction of fluorescamine with amino acids, peptides, and proteins is not visibly temperature-dependent. Therefore, baseline fluorescence, representing ambient levels of these compounds, is relatively stable for long times.

Figure 12 shows examples of detecting short injected pulses of BSA. The borate buffer flow rate was 0.4 ml min^{-1}; the fluorescamine (400 $\mu g\ ml^{-1}$) flow rate was 0.5 ml hr^{-1}. A pulse containing BSA at a concentration of 5×10^{-9} M can routinely be detected (Figure 12b). The decrease in the sensitivity seen in the flow system (compared to the static system) is largely due to the lower mixing efficiency of the reactants. To help rectify this situation, we are in the process of redesigning the mixing chamber.

Table 2. Detection of BSA-fluorescamine reaction products

BSA concentration[1]	Relative fluorescence intensity[2]
Buffer	50
Buffer + fluorescamine	80
10^{-10}	95
5×10^{-10}	110
10^{-9}	135
5×10^{-9}	235
10^{-8}	>500 (off scale)

[1] Mol. l^{-1}
[2] Arbitrary units; mean of 3 determinations.

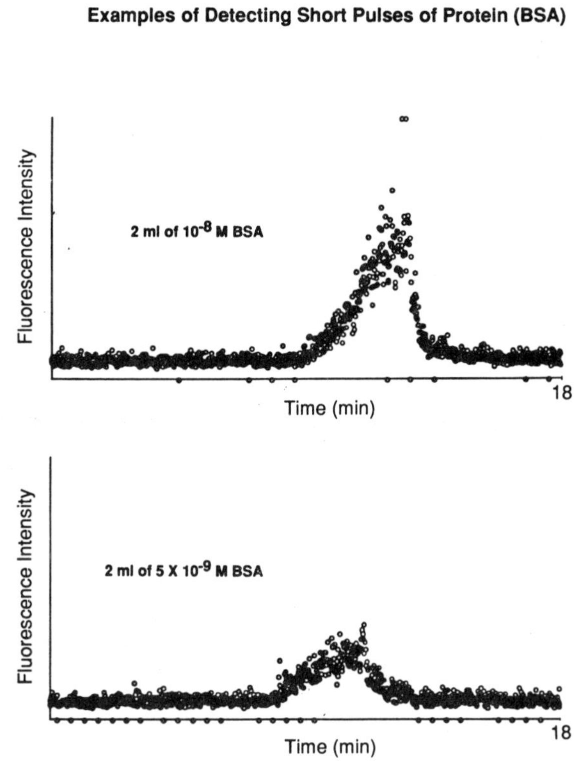

Figure 12. Examples of detecting short injection pulses of BSA by fluorescamine chemistry in flow. (a) injection of 2 ml × 10^{-8} M BSA; (b) injection of 2 ml of 5 × 10^{-9} M BSA.

For long runs, the borate buffer is pumped from a reservoir using a peristaltic pump instead of the syringe drive. The fluorescamine is pumped using the syringe drive. Figure 13 illustrates continuous flow performance over two 4–day periods. Once each day, the system was challenged with 1 ml of a 5 × 10^{-8} M BSA solution.

Ultrasonic Micromist Air Sampler

Efficiency of Aerosol Collection

We put 3 ml of a 2.5 × 10^{-6} M BSA solution into a small Hudson disposable nebulizer (obtained from our local hospital). The nebulizer was attached to the intake stack of the air sampler and was driven at about 12 psi with the lab air line. At this pressure, approximately 0.45 ml of BSA solution was nebulized per min. The nebulized sample was introduced into the sampled air stream (flowing at 50 l min^{-1}) and mixed with the micromist, which was

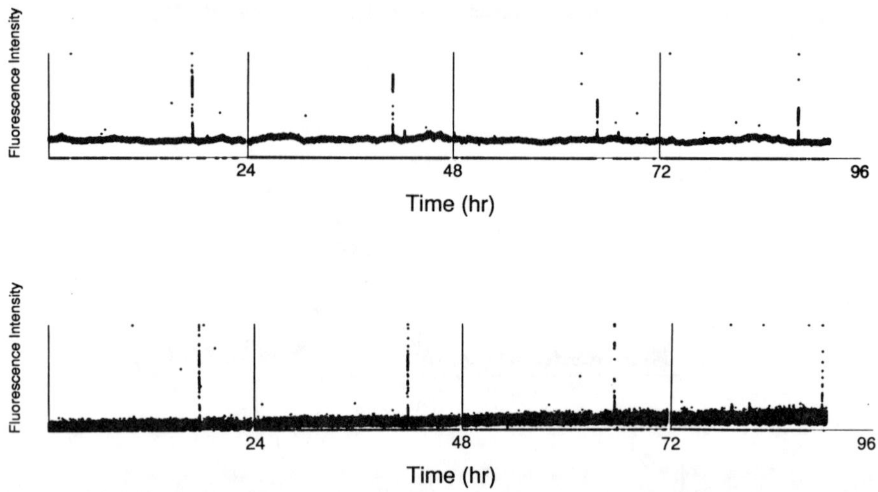

Figure 13. Results obtained in two 4-day water monitoring experiments using fluorescamine chemistry. Each day the system was challenged with an injection of 1 ml of a 5×10^{-8} M BSA (spikes).

ultrasonically generated from the borate buffer containing reservoir. The micromist was condensed by the baffles and precipitated. Precipitated samples were collected at various time intervals. Collected samples were analyzed for protein content by a Bio-Rad Protein Assay (Bio-Rad Laboratories, Richmond CA). The amount of protein collected was compared to the amount introduced, thus allowing us to determine the scrubbing efficiency of the device for BSA aerosols. In one experiment (see Table 3 for results), 2.3 ml of the BSA solution was delivered into the sampler over a 5–min period. As determined in the protein assay, the 402.4 μg BSA delivered contained 9844 optical density (OD) units. Precipitated fluid samples were collected continuously for four 5–min intervals. About 2 ml of fluid per min was collected from the device. The number of OD units contained in each 5–min aliquot were determined and totaled. A total of 2803 OD units were obtained in 20 min, indicating a collecting efficiency of about 28%. Approxi-

Table 3. Collection efficiency of the ultrasonic micromist air sampler[1]

Collection period	Optical density units collected
0–5 min	1260
5–10 min	1160
10–15 min	287
15–20 min	96

[1] 402.5 μg BSA containing 9844 Bio Rad OD units were delivered into 250 l of air. Scrubbed samples were collected for four 5-min intervals. About 10 ml of sample was collected during each interval. The protein content, expressed as OD units, was determined for each sample. Approximately 24% of the delivered BSA was scrubbed from 250 l air and collected into 20 ml liquid within 10 min.

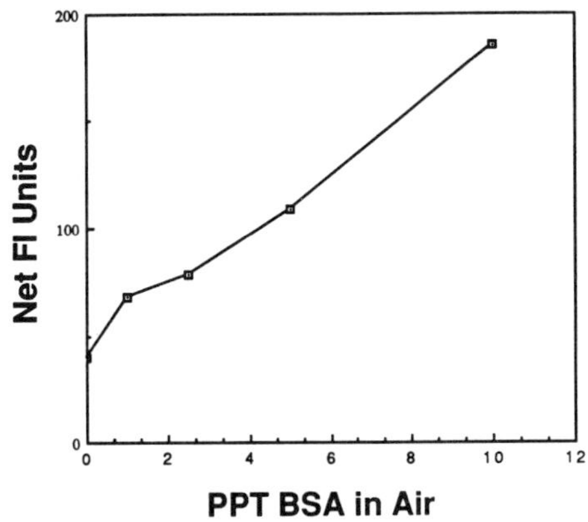

Figure 14. BSA aerosol detection response curve. Aerosols were collected by the air sampler. The precipitated samples were reacted with fluorescamine and subsequently analyzed for fluorescence in the pulsed laser fluorometer. See text for details.

mately 45% of the collected OD units were obtained in the first 5–min sample, and more than 86% of the scrubbed sample was cleared within 10 min (5 min after collection ceased). Therefore, about 24% of the entering sample was scrubbed from 250 l of air and delivered into 20 ml liquid within 10 min. This translates into a concentration factor of about 3,000 (402.5 μg BSA delivered in 250 l air = 1.61 μg l^{-1}or 1.61 ng ml^{-1}). Of this, 97 μg BSA was trapped and collected in 20 ml of liquid. Each ml of liquid contained 4.8 μg BSA, or about 3000 × as much BSA (4800/1.61) than was contained in 1 ml of original air.)

Aerosolized BSA Detection in Conjunction with Fluorescamine Chemistry

Three ml of each BSA concentration was placed in the nebulizer. The solution,was nebulized for 2 min at 10 psi into the flowing air sample (50 l min^{-1}). The reservoir contained borate buffer, pH 10. The condensed liquid precipitate was collected for 2 min for each BSA concentration nebulized. Fluorescamine (20 μl at 400 μg ml^{-1}) was added to the collected precipitate, and the samples were read in the pulsed laser fluorimeter. The results were plotted as a function of the BSA concentration within the flowing air stream (Figure 14). We can readily detect 1 ppt of BSA in air when concentrated by the air sampler and coupled to the fluorescamine detection chemistry.

Conclusions

This chapter describes the development and application of a pulsed laser fluorometer that has been incorporated into a continuous flow system for use in environmental monitoring. The system provides a high degree of sensitivity in a compact, relatively inexpensive device, and should be capable of monitoring for any class or individual compound for which a fluorescent chemistry exists. By replacing the flow cuvette with a stationary cuvette, the pulsed laser fluorometer component can also be used as an instrument to measure fluorescence of large numbers of individual samples.

The compact ultrasonic micromist air sampler described herein offers the environmental monitoring community a powerful and economical new tool for continuous air sampling and scrubbing of ambient air for an almost limitless number of contaminants. The device can serve as a "front end" to a variety of analytical systems.

Acknowledgments

This work was supported by the U.S. Department of State, the U.S. Department of Energy, and the Los Alamos National Laboratory.

References

De Bernardo, S., Weigele, M., Toome, V., Manhart, K., Leimgruber, W., Bohlen, P., Stein, S. and Udenfriend, S. 1974. Studies on the reaction of fluorescamine with primary amines. Arch. Biochem. Biophys. 163: 390–399.

Gehauf, B. and Goldenson, J. 1957. Detection and estimation of nerve gases by fluorescence reaction. Anal. Chem. 29: 276–278.

Guilbault, G.G. 1973. Fluorescence in enzymology. In: Practical Fluorescence. Marcel Dekker, Inc. New York. pp. 354–395.

Guilbault, G.G. and Kramer, D.N. 1965a. Resorufin butyrate and indoxyl acetate as fluorogenic substrates for cholinesterase. Anal. Chem. 37: 120–123.

Guilbault, G.G. and Kramer, D.N. 1965b. Fluorometric system employing immobilized cholinesterase for assaying anticholinesterase compounds. Anal. Chem. 37: 1675–1680.

Guilbault, G.G., Sadar, M.H., Glazer, R. and Skou, C. 1968. N-methylindoxyl esters as substrates for cholinesterase. Anal. Letters. 1: 365–379.

Hiebert, R.D. 1990. Electronics and signal processing. In: Flow Cytometry and Sorting, 2nd ed. Melamed, M.R., Lindmo, T. and Mendelsohn, M.L., eds. Wiley-Liss, NY. pp. 127–144.

Jett, J.H., Keller, R.A., Martin, J.C., Nguyen, D.C. and Saunders, G.C. 1990. Ultrasensitive molecular-level flow cytometry. In: Flow Cytometry and Sorting, 2nd ed. Melamed, M.R., Lindmo, T. and Mendelsohn, M.L., eds. Wiley-Liss, NY. pp. 381–396.

Parvari, R., Pecht, I. and Soreq, H. 1983. A microfluorometric assay for cholinesterases, suitable for multiple kinetic determinations of picomoles of released thiocholine. Anal. Biochem. 133: 450–456.

Prince, A.K. 1966. A sensitive fluorometric procedure for the determination of small quantities of acetylcholinesterase. Biochem. Pharmacol. 15: 411–417.

Sippel, T.O. 1981. New fluorochromes for thiols: Maleimide and iodoacetamide derivatives of a 3–phenylcoumarin fluorophore. J. Histochem. Cytochem. 29: 314–316.

Sippel, T.O. 1981. Microfluorometric analysis of protein thiol groups with a coumarinylphenylmaleimide. J. Histochem. Cytochem. 29: 1377–1381.

Steinkamp, J.A. and Hiebert, R.D. 1982. Signal processing electronics for multiple electronic and optical measurements on cells. Cytometry 2: 232–237.

Udenfriend, S., Stein, S., Bohlen, P., Dairman, W., Leimgruber, W. and Weigele, M. 1972.

Fluorescamine: A reagent for assay of amino acids, peptides, proteins, and primary amines in the picomole range. Science. 178: 871–872.
Weigele, M., De Bernardo, S.L., Tengi, J.P. and Leimgruber, W. 1972. A novel reagent for the fluorometric assay of primary amines. J. Am. Chem. Soc. 94: 5927–5928.

Discussion (DR. SAUNDERS)

Kornguth: First, is it possible to look at the effect of droplet size on detection capability using your air sampling system? Second, is there a problem of possible contamination of the flow system in the solution detector with the fluorescence?

Saunders: We have measured the droplet size; almost all of them pass through a 0.3 micron filter, so they are very small. They are fairly uniform; just like a cloud. We have been very lucky as far as contamination of the water. We have used good water from several different cities. If the water were contaminated, you would have to pre-filter it to get out the large particles. However, a large particulate load (e.g., silt) could be a problem.

Key Words: Fluorescence, Laser induced fluorescence, Laser fluorometry, Acetylcholinesterase , Fluorescamine, N-methylindoxyl acetate, Biosensor, Flow injection analysis, Cholinesterase inhibitors, Environmental monitoring, Air sampling

Session V
Protein Engineering

Recognition of the Human CD4 Receptor by the AIDS Virus: A Molecular Target for AIDS Therapeutics

Raymond Sweet and Martin Rosenberg
Smith Kline Beecham Biopharmaceutical R&D, King of Prussia, PA 19406

Introduction

The human immunodeficiency virus (HIV) is the causative agent of acquired immune deficiency syndrome (AIDS) and the related disorder, AIDS related complex (ARC). Infection by the virus leads to the loss of immune function through depletion of the helper/inducer CD4$^+$ T-cell population. Research on HIV is directed towards both prophylactic and therapeutic intervention. An effective vaccine for the prevention of HIV infection has not been developed and azidothymidine (AZT) is the only drug that has received regulatory approval for therapy. AZT has only limited efficacy and causes toxic side-effects in chronic dosing. Thus, less toxic, more effective therapeutic agents are needed for the treatment of AIDS, ARC and early viral infection.

One target for antiviral intervention is the interaction of HIV with CD4, the cell surface protein which is used by HIV as its receptor (Sattentau and Weiss, 1988). The human CD4 receptor is a 55,000–dalton glycoprotein found primarily on the surface of a subset of human lymphocytes and certain other human cells. The extracellular surface of CD4 is comprised of four tandem domains showing limited homology to immunoglobulin variable (V) regions (See Figure 1), which suggests that CD4 is a member of the immunoglobulin supergene family (Maddon, et al., 1985; Clark, et al., 1987). The presence of CD4 on certain T-cell subsets (T4$^+$ lymphocytes) is thought to promote antigen-specific T-cell activation through the association of the extracellular domain of CD4 with class II major histocompatibility (MHC) molecules on the surface of antigen presenting cells (Gay, et al., 1987; Doyle and Strominger, 1987; Sleckman, et al., 1987). The function of CD4 on other human cell types is not known but may also involve association with MHC class II proteins.

There is now substantial evidence that CD4 serves as the human receptor for HIV (Dalgleish, et al., 1984; Klatzman, et al., 1984; McDougal, et al., 1986; Maddon, et al., 1986). The interaction between virus and CD4 is mediated by the viral envelope protein, gp120. Moreover, cell-to-cell spread of the virus is also dependent on CD4/viral envelope interaction. The gp120 expressed on the surface of infected cells is known to promote fusion with CD4 positive uninfected cells (that is, syncytia formation). The resulting

D. Kamely et al. (eds), Biotechnology: Bridging Research and Applications, 401–411.

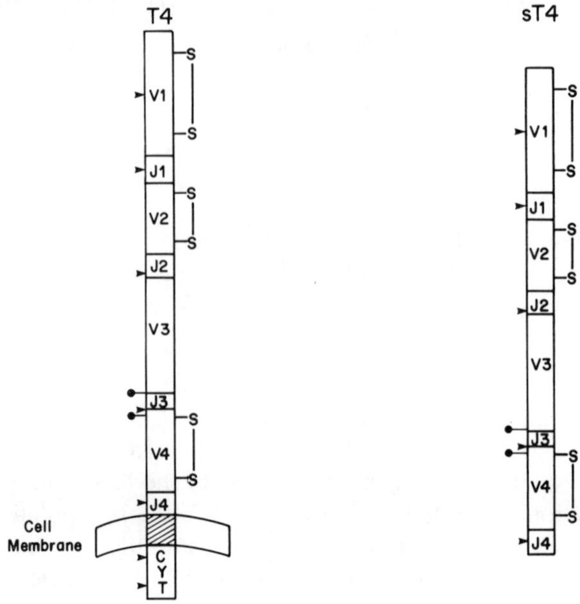

Figure 1. Schematic representation of the human CD4 receptor shown both in its cell-associated, membrane-bound form (T4) and in its recombinant soluble form (sT4). Domain structural homology to the immunoglobulin variable domain regions is indicated (V1, V2, and so forth). Disulfide-linked cysteine residues are represented by S-S. Tm is transmembrane domain; cyt is cytoplasmic domain. Numbering system starts with lysine residue found at NH2–terminus of mature recombinant form of receptor.

giant, multinucleated syncytia continue to support viral infection and concomitantly eliminate normal $CD4^+$ lymphocytes. In addition, free viral gp120 released from infected cells may bind to $CD4^+$ uninfected cells and thereby target these cells for destruction and elimination by the host immune system (Lyerly, et al., 1987; Siliciano, et al., 1988). It remains unclear as to which of these mechanisms occur and predominate *in vivo*; nonetheless, the interaction of gp120 with CD4 is a critical step in viral infection, spread, and elimination of $CD4^+$ lymphocyte populations. Consequently, inhibitors of this interaction might slow or even halt the progression of the infection in man.

In this paper, we describe the expression and characterization of one such inhibitor which consists of the entire extracellular domain of CD4, and describe studies on the localization and structure of the gp120 binding region on CD4.

A Soluble Form of the CD4 Receptor

The mature human CD4 protein consists of a 372 amino acid extracellular segment, a 23 amino acid transmembrane domain, and a 38 amino acid

Properties of sT4

- **sT4 is a conformational mimic of human CD4**
- **sT4 binds to gp120 with high affinity**
- **sT4 blocks HIV binding to $CD4^+$ human cells**
- **sT4 inhibits HIV infectivity of human lymphocytes and monocytes**
- **sT4 inhibits multiple HIV isolates**
- **sT4 inhibits HIV-induced cell fusion and subsequent infection of non-infected cells**

Figure 2. Properties of the soluble CD4 receptor (sT4).

cytoplasmic domain (Landau, et al., 1988) (Figure 1). We sought to produce a recombinant form of the human CD4 receptor to use as a molecular target for identifying selective inhibitors of the interaction between CD4 and gp120.

Reasoning that a CD4 protein lacking the transmembrane and cytoplasmic domains might be transported to the cell surface and then secreted from the cell, we truncated the CD4 coding sequence at the boundary between the extracellular and transmembrane regions. Upon introduction and amplification of this sequence in a Chinese hamster ovary cell line (CHO), the "soluble" CD4 protein sT4 (Figure 1) accumulated to high levels in the culture supernatant. The expression and characterization of this protein have been described (Deen, et al., 1988), and its properties are summarized in Figure 2. The sT4 protein was purified to near homogeneity and was recognized by a panel of anti-CD4 monoclonal antibodies developed against the native receptor. Most of these antibodies recognize conformational epitopes within the amino-terminal domain, which indicates that this portion of sT4 mimics the structure of native CD4.

The sT4 protein binds to HIV-1 gp120 with an affinity of approximately 3nM, which is in agreement with that reported for native CD4 (Stein, et al., 1987) and other soluble CD4 proteins (Fisher, et al., 1988). As a consequence of this high-affinity association with gp120, the sT4 protein dramatically inhibited virus binding and infection of $CD4^+$ cells *in vitro* (Figures 2 and 3). At concentrations of approximately 3 mg ml^{-1}, sT4 completely blocked the binding of HIV-1 (Figure 3A). At 8 mg ml^{-1}, sT4 reduced viral infectivity of human lymphocytes 10^4-fold, as measured by the diminished appearance of viral antigen in the culture supernatant (Figure 3B). The sT4 protein was also shown to inhibit virus-mediated cell fusion as well as virus infection for several different isolates of HIV-1, HIV-2, and simian immunodeficiency virus (SIV) (Clapham, et al., 1988). Interestingly, inhibition of the HIV-2 and SIV strains required higher concentrations of sT4. The basis for these differences in inhibition are currently under study. Preliminary data (M.

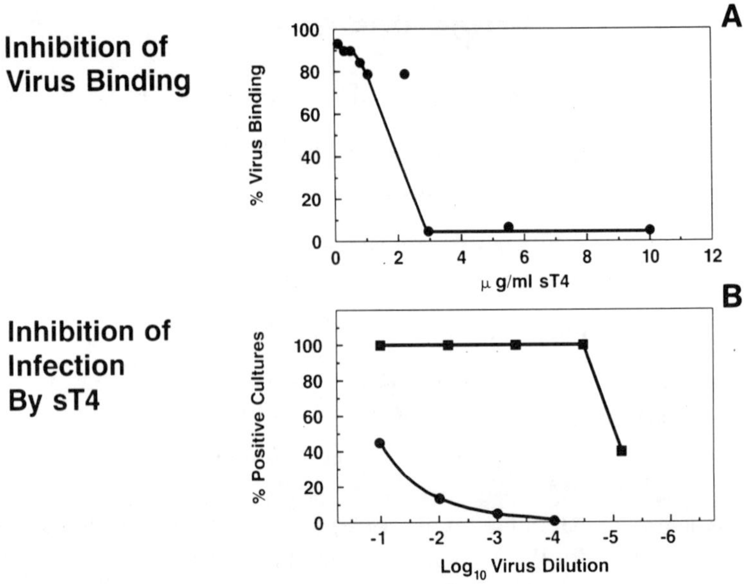

Figure 3. Inhibition of virus binding and infection. (A) Inhibition of HIV-1 binding to CD4+ CEM cells at the indicated concentrations of sT4 protein. (B) Inhibition of HIV-1 infection of PHA-stimulated human lymphocytes for the indicated virus dose in the presence of 0.8 μg/ml of sT4.

Hoyle and M. Rosenberg, unpublished) indicates that there are significant differences in the inherent affinity of the envelope proteins of these virus isolates for the CD4 receptor.

In contrast to the potent viral inhibitory activity of sT4, the molecule had essentially no effect on human T-cell function *in vitro*. At concentrations as high as 1.0 μM, no effect was observed on tetanus toxoid-induced lymphocyte proliferation or interleukin 2 secretion. Similarly, sT4 did not inhibit mitogen-induced or mixed-lymphocyte responses (Sweet, et al., 1990).

These studies demonstrate that *in vitro*, sT4 can antagonize HIV infection of $CD4^+$ cells without affecting normal T-cell function. These combined properties suggested that recombinant sT4 could serve as a potential therapeutic agent for the treatment of AIDS, AIDS-related complex (ARC), or early infection. Similar results have been reported for other variations of soluble CD4 protein produced in other laboratories (Fisher, et al., 1988; Smith, et al., 1987; Hussey, et al., 1988; Traunecker, et al., 1988; Berger, et al., 1988).

The Binding Site for gp120 is in the V_1 Domain

A variety of analogues of sT4 have been examined both for their ability to retain gp120 recognition and their viral inhibitory activity. A series of

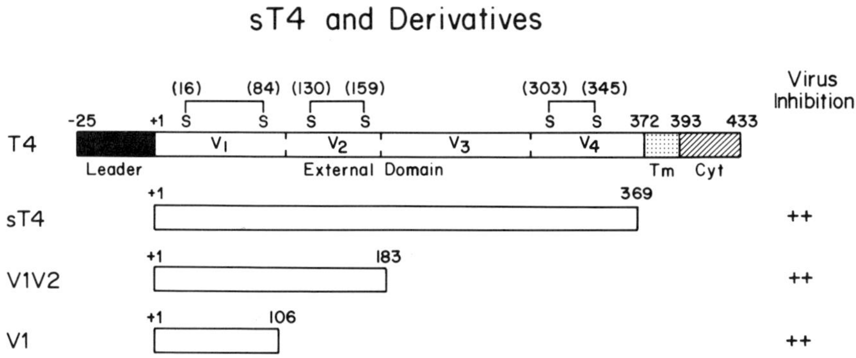

Figure 4. Schematic representation of native CD4 protein (T4), the recombinant truncated soluble form of CD4 comprising the entire extracellular domain (sT4), and two other recombinant truncated derivatives containing the first 183 (V1V2) and 106 (V1) amino acid residues of CD4, respectively (see text for details). Also indicated is the relative ability (++) of each protein to inhibit human immunodeficiency virus infection. All other designations are as shown in Figure 1.

truncated soluble CD4 derivatives were produced and assayed (Arthos, et al., 1989). These proteins (two of which are shown schematically in Figure 4) demonstrated that the amino terminal domain of CD4, the V_1 region, contained the high-affinity binding site for viral gp120. These recombinant proteins were expressed in either mammalian cell culture or in bacteria, then purified, and assayed for their ability to compete with sT4 for binding to purified gp120. These proteins all compete equally well with sT4. Moreover, both the V_1 and V_1V_2 proteins inhibited HIV infection and virus-induced cell fusion at similar molar concentrations (Arthos, et al., 1989). We conclude that the high-affinity binding site for gp120 resides solely within the V_1 domain.

V_1 is Structurally Related to IgG V_k

Crystallographic analysis of CD4 is in progress. Meanwhile, epitope analysis using a panel of anti-CD4 mABs has provided evidence supporting a folding of the V_1 domain in a manner analogous to an immunoglobulin light-chain variable region (V_k) (Sattentau, et al., 1988).

Among immunoglobulin proteins of known structure, the V_1 domain of CD4 aligns most closely with the V_k of the Bence-Jones protein REI. An α-carbon trace of one variable region from the crystal structure of REI reported by Epp, et al. (1975) is shown in schematic form in Figure 5. Based on the sequence alignment, we have superimposed the amino acid residue numbers of the V_1 domain of CD4 on this structure. Mapping data for a series of conformationally dependent mABs indicate that with rare exception, the epitopes lie in turns and exposed loops. Also, analysis of a variety of CD4 mutational derivatives (see following section and Figure 6) which disrupt the

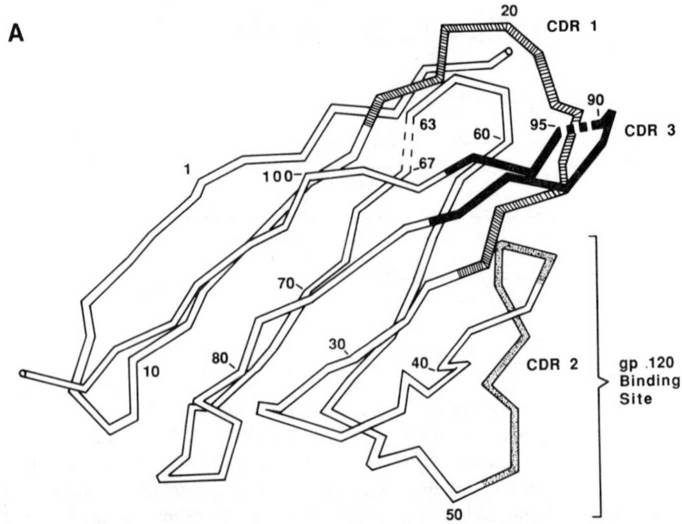

Figure 5. Structural model of the V1 domain. Schematic α-carbon trace of one variable region of REI. Numbers correspond to residues in V1 of CD4. Dashes indicate two positions of amino acid insertions in V1 relative to REI.

binding of antibodies such as Leu3a predict that the altered positions lie in close proximity. This predicted spatial proximity of nonlinear epitopes and the coincidence of epitopes with predicted loops and turns supports the idea that the V_1 domain bears strong structural similarity to an immunoglobulin V_k region.

Localization of the Binding Site within V_1

To further define the binding site within the V_1 domain, a series of single and multiple substitution mutants were constructed using the mouse CD4 protein (L3T4) as a template for this mutagenesis (Arthos, et al., 1989 and 1990). L3T4 shares approximately 50 percent amino acid sequence homology with human CD4 but does not bind the HIV virus. By this approach, gross structural distortions resulting from mutagenesis were minimized. The positions of these mutations are shown in Figure 6. Each of these mutants was expressed and quantitatively analyzed for its ability to bind gp120 (Arthos, et al., 1989). Only two closely spaced clusters of mutants affected the binding to gp120 (Figure 6). One site includes residues 41–43, and the second site involves residues 51–55. These mutations reduce the affinity for gp120 by at least 20-fold. In addition, two other mutations within this same region moderately disrupt the interaction with gp120. All of the disruptive mutants lie within the region between residues 41–55, but certain alterations within this region had no effect. Recently, ten additional mutants within the V_1

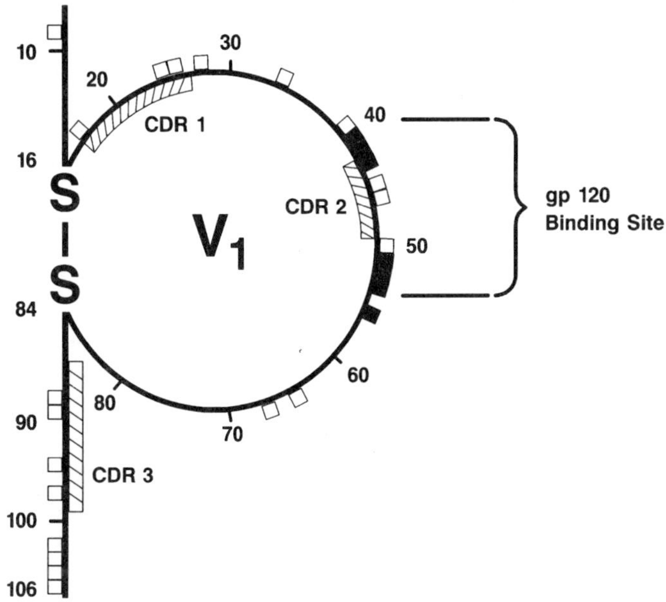

Figure 6. Schematic diagram depicting V1 domain of human CD4 extending from amino acid residue 1–106. Location of various substitution mutants is indicated by squares: open squares represent alterations at positions that have no effect on gp120 binding; stippled squares represent alterations at positions which have partial effects on gp170 binding; filled squares represent alterations at positions that eliminate gp120 binding. Also denoted is the disulfide linkage occurring between cys16 and cys84 and the proposed positions of the complementarity determining regions (CDRS) in CD4 determined by aligning CD4 homology to the light chain variable region of the Bence-Jones immunoglobulin protein REI.

domain have been examined (J. Arthos, et al., unpublished), and these studies have yielded essentially identical results. The data strongly implicate this 15 amino acid region as the high-affinity recognition site for gp120 (Arthos, et al., 1989 and 1990).

Other genetic analyses (Landau, et al., 1988; Clayton, et al., 1988; Mizukami, et al., 1988; Jameson, et al., 1988; Rossman, et al., 1985) have localized the binding site to the same region. The fact that the soluble CD4 proteins containing the mutations that disrupt binding to gp120 fail to inhibit virus attachment, virus infection, and cell fusion indicates that virus inhibition is strictly dependent upon the association of gp120 with the high-affinity binding site.

Based on the homology of the V_1 domain with the immunoglobulin light chain variable region, the gp120 binding site encompasses the second complementarity determining region (CDR2), one of the three hypervariable sites in immunoglobulins that participate in antigen recognition.

It is important to note that mutations in CD4 that might have been expected to dramatically alter the CDR1- and CDR3-like regions had no effect on gp120 binding. If gp120 interacted with CD4 in a manner similar to that thought to occur for antigen recognition by immunoglobulin (Ig), then these mutations should have affected the interaction. The results suggest that although CD4 mimics the Ig variable region structure, only a small segment of this structure (that is, the CDR2–like loop regions) is being "seen" by the viral envelope protein.

Conclusions

sT4, a soluble CD4 protein consisting of the entire extracellular domain of the mature receptor (amino acids 1–369), was expressed at high levels in CHO cells. In vitro, this protein binds to gp120 with high affinity resulting in the efficient inhibition of virus binding and infection and of virus-induced cell fusion. However, sT4 does not affect normal T-cell function *in vitro*. These properties have led to the use of sT4 as a potential therapeutic agent for the treatment of AIDS.

Through quantitative binding assays of CD4 mutants, a single high affinity site for gp120 was identified. Our results localized this site to amino acid residues 41–55 in the V_1 domain. The locations of specific epitopes for a panel of CD4 monoclonal antibodies are consistent with the folding of the V_1 domain in a manner similar to that of an IgG K-light chain variable region.

In the context of a V_k structure, the high affinity binding site for gp120 appears restricted to a region of about 12 amino acids which encompass the CDR2 loop in V_1. We suggest that this small loop may be accommodated by a compact binding pocket in gp120 which by its size excludes recognition by host antibodies, a proposal initially forwarded by Rossman, et al., 1985, for the receptor binding site on rhinoviruses. Protection of the receptor binding site on gp120 in this manner may allow the virus to evade the humoral immune response. The observation that the V_1 domain of CD4 alone retains high affinity for gp120 and the localization of the gp120 binding site to a region of about 12 amino acids may lead to the design or discovery of other novel inhibitors of HIV infection.

References

Arthos, J., Deen, K.C., Chaikin, M.A., Fornwald, J., Sathe, G.M., Clapham, P., Weiss, R., McDougal, J.S., Pietropaolo, C., Maddon, P.J., Truneh, A., Axel, R. and Sweet, R.W. 1990. Analysis of the HIV envelope binding domain on CD4 through site-directed mutagenesis. Human Retroviruses, UCLA Symposia on Molecular and Cellular Biology, New Series. Alan R. Liss, Inc., New York. 119: 171–179.

Arthos, J., Deen, K.C., Chaikin, M.A., Fornwald, J.A., Sattentau, Q.J., Clapman, P.R., Weiss, R.A., McDougal, J.S., Pietropaolo, C., Axel, R., Truneh, A., Maddon, P.J. and

Sweet, R.W. 1989. Identification of the residues in human CD4 critical for binding of HIV. Cell 57: 469–481.

Berger, E.A., Fuerst, T.R. and Moss, B. 1988. A soluble recombinant polypeptide comprising the amino-terminal half of the extracellular region of the CD4 molecule contains an active binding site for human immunodeficiency virus. Proc. Natl. Acad. Sci. 85: 2357–2361.

Clapham, P.R., Weber, J.N., Whitby, D., McIntosh, K., Dalgleish, A.G., Maddon, P.J., Deen, K.C., Sweet, R.W. and Weiss, R.A. 1989. Soluble CD4 blocks the infectivity of diverse strains of HIV and SIV for T cells and monocytes but not for brain and muscle cells. Nature 337: 368–370.

Clark, S.J., Jefferies, W.A., Barclay, A.N., Gagnon, J. and Williams, A.F. 1987. Peptide and nucleotide sequences of rat CD4 (W3/25) antigen: Evidence for derivation from a structure with four immunoglobulin-related domains. Proc. Natl. Acad. Sci. 84: 1649–1653.

Clayton, L.K., Hussey, R.E., Steinbrich, R., Ramachandran, H., Husain, Y. and Reinherz, E.L. 1988. Substitution of murine for human CD4 residues identifies amino acids critical for HIV-gp120 binding. Nature 335: 363–366.

Dalgleish, A.G., Beverly, P.C.L., Clapham, P.R., Crawford, D.H., Greaves, M.F. and Weiss, R.A. 1984. The CD4 (T4) antigen is an essential component of the receptor for the AIDS retrovirus. Nature 312: 763–766.

Deen, K.C., McDougal, J.S., Inacker, R., Folena-Wasserman, G., Arthos, J., Rosenberg, J., Maddon, P.J., Axel, R. and Sweet, R.W. 1988. A soluble form of CD4 (T4) protein inhibits AIDS virus infection. Nature 331: 82–84.

Doyle, C. and Strominger, J.L. 1987. Interaction between CD4 and class II MHC molecules mediates cell adhesion. Nature 330: 256–259.

Epp, O., Lattman, E.E., Schiffer, M., Huber, R. and Palm, W. 1975. The molecular structure of a dimer composed of the variable portions of the Bence-Jones protein REI refined at 2.0 Å resolution. Biochem. 14: 4943–4952.

Fisher, R.A., Bertonis, J.M., Meier, W., Johnson, V.A., Costopoulos, D.S., Liu, T., Tizard, R., Walker, B.D., Hirsch, M.S., Schooley, R.T., Flavell, R.A. 1988. HIV infection is blocked *in vitro* by recombinant soluble CD4. Nature 331: 76–78.

Gay, D., Maddon, P., Sekaly, R., Talle, M.A., Godfrey, M., Long, E., Goldstein, G., Chess, L., Axel, R., Kappler, J. and Marrack, P. 1987. Functional interaction between human T-cell protein CD4 and the major histocompatibility complex HLA-DR antigen. Nature 328: 626–629.

Hussey, R.E., Richardson, N.E., Kowalski, M., Brown, N.R., Chang, H-C., Silician, R.F., Dorfman, T., Walker, B., Sodroski, J. and Reinherz, E.L. 1988. A soluble CD4 protein selectively inhibits HIV replication and syncytium formation. Nature 331: 78–81.

Jameson, B.A., Rao, P.E., Kong, L.I., Hahn, B.H., Shaw, G.M., Hood, L.E. and Kent, S.B.H. 1988. Location and chemical synthesis of a binding site for HIV-1 on the CD4 protein. Science 240: 1335–1338.

Klatzman, D., Champagne, E., Chamaret, S., Guest, J., Guetard, D., Hercend, T., Gluckman, J-C. and Montagnier, L. 1984. T-lymphocyte T4 molecule behaves as the receptor for human retrovirus LAV. Nature 312: 767–768.

Landau, N.R., Warton, M., Littman, D.R. 1988. The envelope glycoprotein of the human immunodeficiency virus binds to the immunoglobulin-like domain of CD4. Nature 334: 159–162.

Lyerly, H.K., Matthews T.J., Langlois, A.J., Bolognesi, D.P. and Weinhold, K.J. 1987. Human T-cell lymphotropic virus III b glycoprotein (gp120) bound to CD4 determinants on normal lymphocytes and expressed by infected cells serves as target for immune attack. Proc. Natl. Acad. Sci. 84: 4601–4605.

Maddon, P.J., Dalgleish, A.G., McDougal, J.S., Clapham, P.R., Weiss, R.A. and Axel, R. 1986. The T4 gene encodes the AIDS virus receptor and is expressed in the immune system and the brain. Cell 47: 333–348.

Maddon, P.J., Littman, D.R., Godfrey, M., Maddon, D.E., Chess, L. and Axel, R. 1985. The isolation and nucleotide sequence of a cDNA encoding the T cell surface protein T4; a new member of the immunoglobulin gene family. Cell 42: 93–104.

McDougal, J.S., Kennedy, M.S., Sligh, J.M., Cort, S.P., Mawie, A. and Nicholson, J.K.A. 1986. Binding of HTLV-III/LAV to T4 + T cells by a complex of the 110K viral protein and the T4 molecule. Science 231: 382–385.

Mizukami, T., Fuerst, T.R., Berger, E.A. and Moss, B. 1988. Binding region for human immunodeficiency virus (HIV) and epitopes for HIV-blocking monoclonal antibodies of the CD4 molecule defined by site-directed mutagenesis. Proc. Natl. Acad. Sci. 85: 9273–9277.

Rossman, M.G., Arnold, E., Erickson, J.W., Frankenberger, E.A., Griffith, J.P., Hecht, H-J., Johnson, J.E., Kramer, G., Luo, M., Mosser, A.G., Rueckert, R.R., Sherry B. and Vriend, G. 1985. Structure of a human common cold virus and functional relationship to other picornaviruses. Nature 317: 145–152.

Sattentau, Q.J., Arthos, J., Deen, K.C., Hanna, N., Healy, D., Beverly, C.L.P., Sweet, R.W. and Truneh, A. 1988. Structural analysis of the human immunodeficiency virus-binding domain of CD4 epitope mapping with site-directed mutants and anti-idiotypes. J. Exptl. Med. 170: 1319–1334.

Sattentau, Q.J. and Weiss, R.A. 1988. The CD4 antigen: Physiological ligand and HIV receptor. Cell 52: 631–633.

Siliciano, R.F., Lawton, T., Knall, C., Karr, R.W., Berman, P., Gregory, T. and Reinherz, E.L. 1988. Analysis of host-virus interactions in AIDS with anti-gp120 T cell clones: Effect of HIV sequence variation and a mechanism for CD4 + cell depletion. Cell 54: 561–575.

Sleckman, B.P., Peterson, A., Jones, W.K., Foran, J.A., Greenstein, J.L., Seed, B. and Burakoff, S.J. 1987. Expression and function of CD4 in a murine T-cell hybridoma. Nature 328: 351–353.

Smith, D.H., Bryn, R.A., Marsters, S.A., Gregory, T., Groopman, J.E. and Capon, D.J. 1987. Blocking of HIV-1 infectivity by a soluble secreted form of the CD4 antigen. Science 238: 1704– 1707.

Stein, B.S., Gowda, S.D., Lifson, J.D., Penhallow, R.C., Bensch, K.G. and Engleman, E.G. 1987. pH-independent HIV entry into CD4–positive T cells via virus envelope fusion to the plasma membrane. Cell 49: 659–668.

Sweet, R., Arthos, J., Deen, K., Chaikin, M., Fornwald, J., Sathe, G., Truneh, A., Rosenberg, M., Sattentau, Q., Clapham, P., Weiss, R., McDougal, J.S., Pietropaolo, C., Maddon, P.J. and Axel, R. 1990. Recognition of CD4 by the AIDS virus: Inhibition by soluble CD4 and analysis of the virus binding site. Frontiers in Chemistry: Biotechnology, R. E. Stobaugh, ed. Chemical Abstracts Service, Columbus, Ohio. pp. 151–167.

Traunecker, A., Luke, W. and Karjalainen, K. 1988. Soluble CD4 molecules neutralize human immunodeficiency virus type 1, Nature 331: 84–86.

Discussion (DR. M. ROSENBERG)

Fish: It is beautiful work. I was wondering if you could increase avidity instead of affinity by presenting multimers or multiple molecules on a single surface instead of presenting the molecules individually.

M. Rosenberg: That is being tried. Once the V1 or the V1V2 regions were found to bind to the envelope, people have taken those regions and attached them to everything you can imagine in order to study affinity as well as avidity. In addition to the Genentech work of attaching CD4 to immunoglobulin molecules, researchers also have fused them to various toxins in an attempt to bring toxins to the virus and virus infected cells. The problem with increasing avidity is to make sure that you are not creating an antigen which the person will respond to as an immunogen. One has to balance putting things in multiple copies with increasing size and the concern for presentation to the human immune system such that it becomes an immunogen. The last thing you want to do is create an antibody response against your own CD4 receptor.

Szybalski: You have presented data on the mutants which have lost affinity. Do you have any system for selecting the opposite kind of mutants with a much higher affinity, as was done a long time ago for the lac repressor?

M. Rosenberg: You must know what is going on in my laboratory. Yes, because we can express the molecule in bacteria, we can start with one affinity and try to create a window in which we can select for a higher affinity. By using classical bacterial genetics, we can mutagenize and select up mutations to get mutations in this V1 or V1V2 domain which increase affinity. We are in the middle of those experiments, but I have nothing to report as of yet.

Gershoni: Regarding the CDR3 domain and gp120 binding, there have been numerous reports, especially from Dr. Lee Eiden's laboratory, indicating that there may be some contribution from this region. Have your mutants been specifically directed towards looking at the significant residues that Dr. Eiden has pointed out?

With respect to the negative role of gp120, it is interesting to note, that whereas it could recruit the immune system to selectively kill uninfected cells, there are other alternative effects. Savita Pawa at North Shore Hospital and Gus Dahlgleish from London have illustrated that gp120 binding to T4 cells down regulates the IL-2 autocrine response thus preventing the proliferation of these cells.

One last comment to Dr. Fish, we have made polymers, such as V1V2 and extensive V1 polymers. These do have unique characteristics for gp120 binding but they have not been fully tested with regard to virus neutralization.

M. Rosenberg: Dr. Eiden made peptides to the CDR3 region and showed that they blocked virus infection. When he looked at the peptides, the only peptides that worked were peptides in which he did not remove various phenoxy ring blocking groups. In fact, if you look at the peptides themselves, without the phenoxy rings, they have absolutely no effect. Given the mutations that have been made in the CDR3, I think there are some mutants that prove that certain residues found in the peptides are not important. More importantly, Eiden has demonstrated that you must have the phenolic rings on these molecules to get them to block. It is not clear what specificity the system has because of this complication.

Key Words: AIDS, HIV, CD4 receptor, gp120 binding, Therapeutic agent

Antibody-Catalyzed Concerted Chemical Reactions

Donald Hilvert
Departments of Chemistry and Molecular Biology, Research Institute of Scripps Clinic, 10666 North Torrey Pines Road, La Jolla, CA 92037

Biocatalysis plays a key role in the field of biotechnology. Enzymes are essential components in biodetection systems, bioremediation strategies, and approaches toward development and synthesis of new biomedical and industrial materials. For many applications of potential practical interest, however, natural enzymes may be unavailable, unstable, or otherwise unsuitable. Natural enzymes may not even exist at all for specific chemical reactions. For these reasons, the development of general strategies for creating enzyme-like catalysts from scratch in the laboratory is a significant scientific goal. One recent advance in the area of enzyme engineering exploits the diversity and specificity of the immune system to create catalytic antibodies with tailored activities and specificities. This emerging field is briefly reviewed in this chapter with emphasis placed on work carried out in the author's laboratory to generate antibody catalysts for concerted carbon-carbon bond-forming reactions.

The Immune System and Catalytic Antibodies

The mammalian immune system is certainly the most prolific source of specific receptor molecules known (Kabat, 1976; Pressman and Grossberg, 1968; Nisonoff, et al., 1975). When it encounters a foreign material, the immune system is able to fashion an almost limitless supply of high affinity antibody molecules that bind the invader tightly and specifically, targeting it for elimination from the body. It is estimated that this protein-generating system can produce up to 10^{12} unique receptors by a process of genetic recombination and somatic mutation at the level of DNA.

For the purposes of catalyst design what is important is that scientists have learned how to exploit this diversity system to produce, on demand, high affinity antibodies against virtually any substance (Goding, 1986). Suitable antigens can be large or small, charged or neutral, natural or man-made. If a compound can be synthesized by the organic chemist in the lab, it is

D. Kamely et al. (eds), Biotechnology: Bridging Research and Applications, 413–427.

more than likely that an antibody can be produced that binds it tightly and specifically. It is also important to note that the immune system is a microevolutionary system. That is, it continually refines its response to a given immunogen, producing receptor antibodies of ever increasing affinity (and hence tighter binding) for the target antigen. Typical dissociation constants for antibody-antigen complexes span the range 10^{-4} to 10^{-14} M. At the low end of this range, there is certainly enough binding energy to do chemical work. The key question therefore becomes to what extent the chemist can exploit this directable binding energy for catalysis.

The similarity between enzymes and antibodies was recognized nearly forty years ago (Pauling, 1948). Both sets of molecules are proteins that bind ligands tightly and specifically. The basic sizes and shapes of their respective active (combining) sites are similar. A typical antibody can hold five or six amino acids or sugar residues, much like a typical proteolytic or glycolytic enzyme (Kabat, 1976; Pressman and Grossberg, 1968; Nisonoff, et al., 1975). Moreover, antibodies and enzymes exploit the same hydrogen bonding, hydrophobic, electrostatic, and dispersion forces to hold on to their ligands (Kabat, 1976; Pressman and Grossberg, 1968; Nisonoff, et al., 1975). However, whereas an enzyme's active site is complementary in shape and charge to the high energy, ephemeral transition state structure of the reaction it catalyzes, an antibody's active site matches the ground state. This fundamental difference explains why enzymes, but not antibodies, catalyze reactions. By binding and stabilizing the transition state relative to the ground state, an enzyme lowers the activation energy needed for reaction.

In 1969, Jencks proposed an ingenious strategy for preparing antibody molecules with catalytic activity based on this simple paradigm for enzymatic catalysis. He suggested that stable immunogens could be synthesized that resemble the high energy transition state of individual chemical transformations in their overall shape and charge. If the transition state analog design is a good one, some fraction of the resulting antibodies should possess the desired catalytic activity.

In the last few years, this approach has been reduced to practice and catalytic antibodies (or abzymes) have been produced for a number of different chemical transformations. Three factors were crucial for the success of these experiments. First, it was imperative to develop effective transition state analogs for specific chemical reactions. Over the last twenty years chemists and biologists have designed many such molecules as tight binding inhibitors to a variety of enzymes (Wolfenden, 1976; Lienhard, 1973). Many of these same molecules are now being used successfully to create catalytic antibodies. Second, the importance of monoclonal antibody technology cannot be underestimated. In any immunization a large number of immunoglobulins is generated, each with a slightly different shape and affinity for the immunizing antigen. Monoclonal antibody technology allows these individual proteins to be separated, making it possible to work with homogeneous preparations of single, chemically well-defined antibody molecules (Goding,

1986). Finally, because only a fraction of the polyclonal response to a given antigen will possess the desired properties, efficient screening techniques are essential to identify the most effective catalysts for the target reaction. Screening is likely to be particularly important in the identification of abzymes for reactions that are energetically more demanding than those studied to date.

The first antibodies to exhibit substantial turnover catalysis promoted the hydrolysis of activated esters and carbonates. They were prepared independently by the Lerner group at Scripps (Tramontano, et al. 1986; Tramontano, et al. 1988) and the Schultz group at Berkeley (Pollack, et al., 1986) by immunizing mice with aryl phosphonate esters. The tetrahedral, anionic phosphonate is a good mimic of the tetrahedral, anionic transition states and intermediates that occur along the reaction coordinate for ester hydrolysis. The induced antibodies exhibited substantial rate accelerations, with k_{cat}/k_{un} approaching 10^6-fold in the best cases. More importantly, the substrate specificity of the antibodies that were elicited correlated precisely with the structure of the homologous hapten. The fact that catalyst specificity can be programmed by the researcher at the stage of hapten design confers enormous versatility to this technology and, for practical applications, is likely to be the most valuable property of abzymes.

Recently, this work has been extended to the catalysis of other hydrolytic reactions, including cleavage of p-nitroanilides (Janda, et al., 1988) and unactivated esters (Janda, et al., 1989). In the presence of a metal cofactor it has even been possible to hydrolyze an unactivated amide bond in a peptide using an antibody (Iverson and Lerner, 1989). The latter result is exciting, as it demonstrates the feasibility of promoting particularly difficult transformations.

Concerted Carbon-Carbon Bond Forming Reactions

Although considerable success has been achieved in catalyzing hydrolytic reactions, such processes are not necessarily the best for uncovering enzyme-like activities in antibodies. Enzymes typically rely on precisely positioned constellations of multiple catalytic groups, including nucleophiles, general acids and general bases, to achieve enormous accelerations of such reactions. However, introduction of catalytic functionality into an antibody combining site during the immunization event is difficult. Particularly when several catalytic groups are needed to act in concert, the probability of obtaining an effective catalytic antibody is expected to be low.

In addition to chemical catalysis, enzymes can utilize binding energy directly to do chemical work (Page and Jencks, 1971; Jencks, 1975). For example, an enzyme can exploit specific binding interactions to induce strain or distortion into a ligand and thereby help it over an enthalpic barrier for reaction. Alternately, an enzyme can use binding energy to bring two molecules together in the proper orientation for reaction and thereby lower the entropic barrier for reaction. Enzymes can also use desolvation to destabilize a charged ground state molecule relative to a transition state in which the charge is substantially delocalized. These factors are exactly the catalytic effects an antibody should be able to impart effectively. It has been known for years that antibodies can distort the ligands that they bind into unusual conformations and that they can bring two molecules together within their combining site.

My research group has set out to exploit each of these mechanisms of action – induced strain, proximity and desolvation – in designing abzymes for novel chemical transformations (Hilvert, et al., 1988, 1989). Concerted reactions in which carbon-carbon bonds are made or broken are particularly attractive candidates for catalysis. These include Diels-Alder cycloadditions, sigmatropic rearrangements, and other pericyclic processes. They generally do not require chemical catalysis by nucleophiles, general acids or general bases, but should be sensitive to strain and proximity effects. Efforts to catalyze such reactions with antibodies, as outlined below, have the potential for elucidating the relative importance of these factors in catalysis by naturally occurring enzymes. In addition, abzymes for concerted chemical reactions would have great promise as selective catalysts for a variety of practical applications.

Diels-Alder Reactions

To study proximity effects we have chosen a bimolecular reaction of considerable importance in organic synthesis. The Diels-Alder cycloaddition involves concerted addition of a conjugated diene to an olefin to give a cyclohexene derivative (Wassermann, 1965; Sauer, 1967). As such, it represents one of the most versatile and valuable methods available to the organic chemist for constructing carbon-carbon bonds in complex natural products, therapeutic agents and synthetic materials of all kinds (Sauer, 1966). Although enzyme-catalyzed Diels-Alder reactions are unknown in nature, we felt that antibodies were ideally suited for accelerating these processes.

Bimolecular Diels-Alder reactions are entropically very unfavorable. Activation entropies are typically between -30 and -40 cal K^{-1} mol^{-1} due to the large loss of translational and rotational entropy that occurs when the two substrates are brought together in the transition state (Wassermann, 1965; Sauer, 1966). In principle, it should be possible to pay for this enormous loss in entropy by binding the substrates together in an antibody combining site in the proper orientation for reaction. Theoretical calculations suggest that conversion of a bimolecular reaction into a unimolecular process on the surface of protein in this way can lead to rate accelerations as high as 10^8 M for 1 M standard states (Page and Jencks, 1971).

The transition state for the Diels-Alder reaction has a highly ordered,

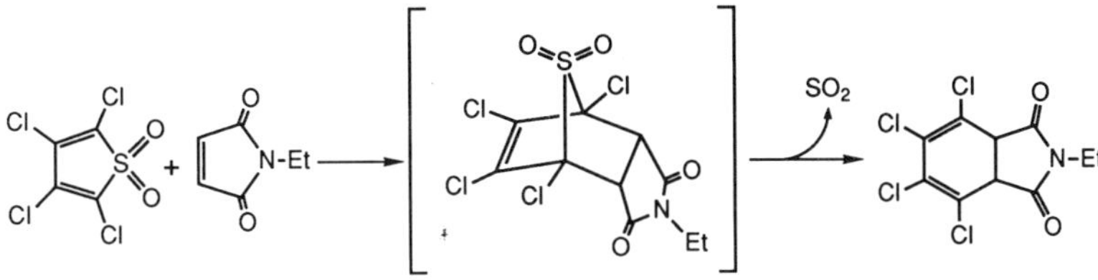

1

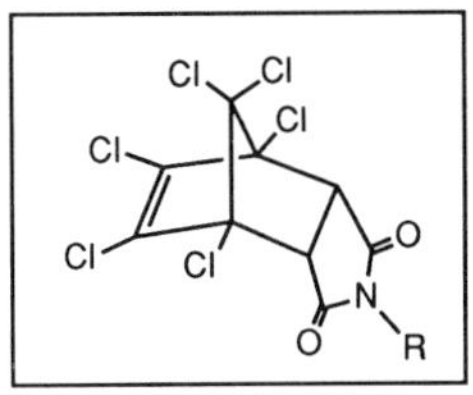

2a $-(CH_2)_5COOH$

2b $-CH_2CH_3$

Figure 1.

boat-like cyclohexene conformation (Sauer, 1967; Brown and Houk, 1994). It is this structure that must be mimicked in the design of haptens for the production of catalytic antibodies. Because the transition state resembles the product of the reaction more closely than the starting materials, one might use the product itself as a hapten. Such a strategy is, however, unlikely to yield an effective *turnover* catalyst. Combining sites elicited against a product molecule will be complementary in shape and charge to this species, and severe product inhibition will likely limit the induced "catalysts" to a single turnover. In order to produce a multiturnover antibody catalyst a transition state analog must therefore be devised that not only has the necessary boat-like geometry, but that does not resemble the final product of the reaction too closely.

One solution to this problem is illustrated in Figure 1. A number of cyclic dienes undergo Diels-Alder reactions with olefins to give bicyclic adducts which are intrinsically unstable and decompose by cycloreversion, for example by chelotropic elimination of nitrogen, carbon dioxide, sulfur dioxide, nitriles, etc. Tetrachlorothiophene dioxide, for instance, reacts with N-alkylmaleimides to give the bicyclic intermediate **1** (Raasch, 1980). This material is unstable and, in turn, eliminates SO_2 to give an N-alkyl-tetrachlorodihydrophthalimide. Under the reaction conditions, the latter is subsequently oxidized to the N-alkyl-tetrachlorophthalimide. The transition states for both the cycloaddition and the subsequent cycloreversion steps of this reaction resemble the high energy intermediate **1**. Antibodies raised against a stable analog of this material should possess the proper shape for bringing the two substrates together. In addition, because **1** has a much different structure than the planar starting materials and reaction products (tetrachlorophthalimide and SO_2), the induced combining site would not be expected to be subject to severe product inhibition.

We have prepared the hexachloronorbornene derivative **2** as a potential mimic of the transition states of the reaction depicted in Figure 1 (Hilvert, et al., 1989). It is a close analog of **1**, containing a Cl-C-Cl moiety in place of the O = S = O bridgehead. We prepared eight monoclonal antibodies with a KLH-**2a** protein conjugate according to standard immunization and fusion protocols (Goding, 1986). Hybridomas secreting high affinity immunoglobulins were identified by an enzyme-linked immunosorbent binding assay (ELISA). They were propagated individually in ascites and the resulting antibodies were purified to homogeneity by affinity chromatography on protein A and by FPLC MonoQ ion-exchange chromatography.

The purified antibodies were tested individually for their ability to catalyze the cycloaddition between tetrachlorothiophene dioxide and N-ethylmaleimide (Hilvert, et al., 1989). Because the thiophene dioxide undergoes a rapid and undesired side reaction with lysine residues of the antibody, it was necessary to protect the ε–amino groups by reductive methylation with formaldehyde and sodium cyanoborohydride (Jentoft and Dearborn, 1979). We found that the permethylated antibodies retained their ability to bind **2** but no longer reacted with the substrate. Chemical modification of antibodies in this way is likely to be generally useful, as it makes it possible to employ reactive substances as substrates for catalytic antibodies.

Preliminary screening of the methylated antibodies was accomplished by spectroscopic and HPLC assays. Authentic product was isolated from the reaction mixture and characterized by NMR, mass spectrometry, and by comparison with authentic material. Six of the eight antibodies that bound **2** also accelerated the reaction between N-ethylmaleimide and tetrachlorothiophene dioxide. One of these, 1E9, was substantially faster than the others and was characterized in some detail (Hilvert, et al., 1989).

Three pieces of evidence indicate that catalysis by 1E9 is not artifactual. First, the reaction is first-order in 1E9. Second, the antibody-catalyzed reaction is strongly inhibited by the transition state analog. When 1E9 is mixed with an equimolar amount of **2b**, no catalysis above background is observed for the bimolecular cycloaddition, even in the presence of a 10^3-fold molar excess of substrate. This result supports the notion that the reaction occurs within the induced binding pocket. Finally, the substrate specificity of 1E9 correlates with expectations based on the immunogen structure (K.W. Hill and D. Hilvert, unpubl.). Hapten **2** was linked to the carrier protein KLH via a hexanoic acid spacer unit, and N-alkylmaleimides with long alkyl tails are substantially better substrates for the antibody than maleimides with short alkyl groups. Thus, the rate of the catalyzed reaction increases in the order: H $\ll$ methyl $<$ ethyl $<$ propyl $\approx$ butyl. In contrast, the uncatalyzed thermal process shows little dependence on the length of the alkyl group.

To gauge the chemical efficiency of 1E9, a series of kinetic experiments was undertaken (Hilvert, et al., 1989). Holding the concentration of tetrachlorothiophene dioxide constant, initial rates of reaction were determined as a function of N-ethylmaleimide concentration. Hyperbolic kinetics were observed, indicating the formation of a Michaelis-complex between the substrates and the antibody. Analysis of the rate data by standard methods allowed us to calculate apparent values of k_{cat} and $(K_m)_{NEM}$ of 4.3 min^{-1} and 21 mM, respectively, at 0.61 mM tetrachlorothiophene dioxide. Comparison of $(k_{cat})_{app}$ obtained under these conditions with the second-order rate constant for the uncatalyzed reaction ($k_{un} = 0.040\ M^{-1}\ min^{-1}$) gives an apparent effective molarity of 110 M for the Diels-Alderase antibody. The effective molarity is the concentration of substrate that would be needed in the uncatalyzed reaction to achieve the same rate as in the antibody ternary complex (Jencks, 1975). Clearly, the binding site of 1E9 confers a significant kinetic advantage to this bimolecular Diels-Alder reaction: the observed effective molarity is several orders of magnitude larger than the maximum physically accessible concentration of substrate in aqueous buffer. However, the measured value of 110 M underestimates the actual chemical efficiency of 1E9. The true effective molarity could not be measured because of the low solubility of tetrachlorothiophene dioxide in the reaction medium, but it is likely to be at least 10– to 100–times larger than the measured value.

Our results demonstrate the feasibility of exploiting proximity effects in antibody catalysis of bimolecular reactions. Significantly, we explicitly incorporated a mechanism for avoiding product inhibition into our hapten design. Available data suggest that our approach to this problem works very well. The antibody catalyst 1E9 undergoes multiple turnovers (>50), and we were able to show by competition ELISA that the final reaction product, N-ethyl-tetrachlorophthalimide, binds roughly 10^3-times less tightly to the combining site than does the hapten derivative **2b**. We are currently extending these ideas to the catalysis of other cycloaddition reactions. Ultimately, we anticipate that it will be possible to apply our strategy for obviating product inhibition to other bimolecular reactions, as well.

Catalysis of a bimolecular Diels-Alder cycloaddition by an antibody is notable because this reaction has no physiological counterpart. Our success at promoting this process therefore underscores both the power and versatility of the catalytic antibody approach to enzyme engineering. Because cycloadditions are the prototype for a broad class of concerted pericyclic processes that are important in organic synthesis, it should now be possible to generalize our findings to some of these other reactions. In the near future we consequently expect to be able to prepare abzymes for additional Diels-Alder reactions, dipolar cycloadditions, and the like. The ability to carry out these transformations with the high rates and exacting selectivities of enzymes will have considerable practical value in organic synthesis.

Claisen Rearrangements

Sigmatropic processes, particularly Cope and Claisen rearrangements, represent another group of pericyclic reactions that are useful to synthetic chemists

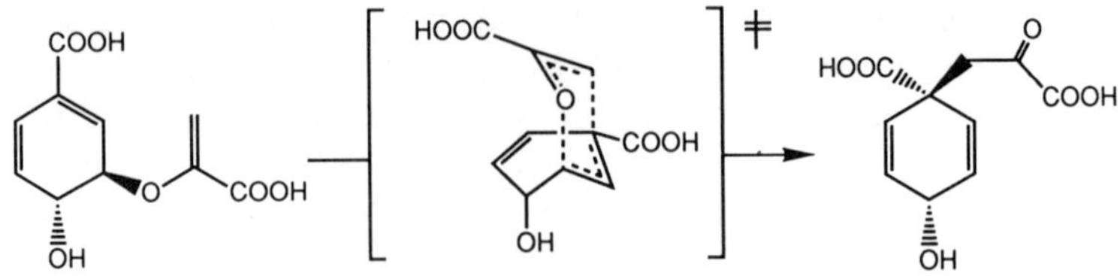

3

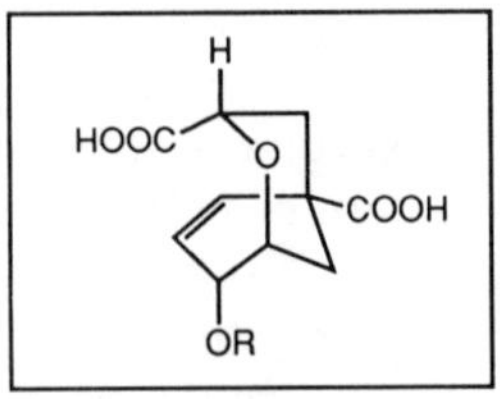

4a -H

4b -CO$(CH_2)_3$COOH

Figure 2.

for constructing carbon-carbon bonds (Lutz, 1984; Ziegler, 1988). We expected that they, like Diels-Alder reactions, would be susceptible to strain and proximity effects and hence to antibody catalysis.

A biologically important example of a Claisen rearrangement is the conversion of the metabolite (−)-chorismate into prephenate as shown in Figure 2 (Weiss and Edwards, 1980). The enzyme chorismate mutase accelerates this reaction by a factor of two million fold over the spontaneous thermal reaction. Although the mechanism of action of chorismate mutase is still under investigation, extensive inhibitor studies (Andrews, et al., 1977; Chao and Berchtold, 1982) and elegant stereochemical investigations (Sogo, et al., 1984) have suggested that the transition state has the diaxial chair-like geometry illustrated by structure **3**.

Bartlett and coworkers have prepared the oxabicyclic molecule **4a** as a mimic of the putative high energy species **3** (Bartlett and Johnson, 1985; Bartlett, et al., 1988). Although it has normal carbon-carbon and carbon-oxygen bonds in place of the elongated making and breaking bonds in the true transition state, and although the bridging element that mimics the migrating enolpyruvate moiety contains a carboxylate attached to an sp^3 rather than to an sp^2 center, compound **4a** is the best inhibitor known for chorismate mutase. It binds to the enzyme about 200–times more tightly than does chorismate itself. Consequently, it is a good candidate for use as a hapten for eliciting chorismate mutase antibodies.

We synthesized **4a** according to Bartlett's procedure and linked it through its pendant hydroxyl group to the carrier protein KLH with a glutaric acid spacer (Hilvert, et al., 1988). The hydroxyl group is known to be unimportant in the rearrangement chemistry (Pawlak, et al., 1989). We prepared 45 high affinity monoclonal antibodies against the KLH-**4b** protein conjugate using standard techniques. The purified immunoglobulins were assayed individually for their ability to catalyze the rearrangement of (−)-chorismate. Two of the antibodies, 1F7 and 27G5, were found to be catalytic. The similarity in their kinetic properties (Hilvert, et al., 1988) and recent partial sequence data (K. Bowdish, J.B. Hicks and D. Hilvert, unpubl.) suggest that these two molecules originated from the same daughter clone. Differences distant from their active sites cannot, however, be ruled out at this time.

Control experiments with 1F7 have established that prephenate is the authentic product of the antibody promoted reaction and that its rate of formation equals the rate of disappearance of chorismate. The catalyzed reaction is first-order in antibody and can also be specifically inhibited by the transition state analog. We determined the K_i for racemic **4a** to be ca. 0.6 μM. Saturation kinetics were observed when the initial rate of reaction was monitored as a function of chorismate concentration. The observation of hyperbolic kinetics in this case is consistent with a kinetic mechanism in which the substrate and antibody form a Michaelis-complex prior to the rearrangement event. Fitting the data obtained at 14°C to the standard Michaelis-Menten equation gave values of k_{cat} and K_m of 0.0234 min^{-1} and 49 μM, respectively (Hilvert, et al., 1988). In this case, the k_{cat} parameter is the first-order rate constant for conversion of the bound starting material into the bound product. It can be compared directly to the first-order rate constant for the uncatalyzed thermal reaction to provide an estimate of the chemical efficiency of the abzyme. Making this comparison, we find that 1F7 accelerates the chorismate mutase reaction by a factor of ca. 250–fold.

Working independently, Schultz, Bartlett and their colleagues have also identified a chorismate mutase antibody from a mouse hyperimmunized with hapten 4 (Jackson, et al., 1988). Their antibody exhibits a rate acceleration of roughly 10^4-fold, or 50– to 100–times more efficient than ours. This result is important for several reasons. First, it demonstrates that an imperfect transition state analog can be employed to obtain a catalytic antibody that is remarkably efficient. The Berkeley antibody has about 1% of the activity of the natural enzyme. Second, this result underscores the importance of screening the immune system widely in order to identify those molecules that possess the desired specific activity. In a typical fusion we screen less than 50 hybridomas. This represents but a tiny fraction of the estimated 10^{12} binding specificities potentially available from the diversity of the immune system. Better access to the large numbers of immunoglobulins in a given immune response, together with improved screening techniques, would consequently have a major impact on our ability to identify new antibody catalysts. In this regard, the production of very large combinatorial libraries of Fab proteins in *E. coli* using the polymerase chain reaction and lambda phage technology has recently generated considerable excitement (Huse, et al., 1989).

Rate accelerations are only one aspect of enzyme-catalyzed reactions. In practical terms, the ability of enzymes to distinguish between closely related substances or to steer a reaction down a single regio- and stereochemical pathway is far more important. Because antibodies are inherently chiral objects, it was not unreasonable to expect that they might be able to exert considerable control over the reactions that they promote. Not surprisingly, experiments with racemic chorismate have established that the chorismate mutase antibody 1F7 does exhibit high enantioselectivity (Hilvert and Nared, 1988).

Under conditions in which all of (−)-chorismate rearranges, only half of the racemic substrate is converted to prephenate by 1F7. This is exactly what is expected if the (+)-isomer is not a substrate for the antibody. We took advantage of this finding to prepare an authentic sample of (+)-chorismate by kinetic resolution of racemic chorismate with 1F7 (Hilvert and Nared, 1988). We used the pure (+)-isomer that we isolated to obtain a more accurate estimate of the enantioselectivity of the catalyst. Rate measurements with the individual isomers in the presence of the antibody showed that (-)-chorismate is favored over (+)-chorismate by a factor of at least 90 to 1 at low substrate concentrations. These results demonstrate the hand-in-glove relationship between the induced immunoglobulin binding pocket and its preferred substrate. The high selectivity of this and other antibodies is likely to be one of the most valuable properties of these inducible catalysts, particularly as the specificity can be set by the researcher through appropriate design of the transition state analog.

Since the immunizing antigen **4** in our system was racemic, catalytic antibodies specific for rearrangement of both (−)- and (+)-chorismate might have been elicited. Nevertheless, assays with racemic chorismate of the 45 monoclonals specific for **4** did not reveal any antibodies with reversed substrate preference, i.e., that selectively rearranged the (+)-isomer (Hilvert, 1989). By extensively screening larger immunoglobulin libraries (Huse, et al., 1989), however, it should be possible to find such catalysts. Indeed, in another system, a racemic hapten was employed to produce individual antibody catalysts for specific hydrolysis of both enantiomers of the homologous ester substrate (Janda, et al., 1989).

Future Directions

Although it is now possible to create new enzyme active sites with the immune system, the chemical efficiency of these catalysts is typically considerably lower than that of naturally occurring enzymes. The chorismate mutase antibody 1F7, for instance, is four orders of magnitude less effective than chorismate mutase from *Aerobacter aerogenes* (Hilvert, et al., 1988). Consequently, general strategies are needed to improve the catalytic potential of first generation abzymes.

Naturally occurring enzymes have been brought to peak efficiency through billions of years of evolution, and it should be possible to mimic this molecular selection process in the laboratory with catalytic antibodies. To that end we recently cloned and sequenced the genes for 1F7 and expressed them as assembled and functional Fab protein in the cytoplasm of the yeast *Saccharomyces cerevisiae* (Y. Tang, K. Bowdish, J.B. Hicks and D. Hilvert, unpubl.). Because the rearrangement of chorismate into prephenate is the committed step in the biosynthesis of phenylalanine and tyrosine in yeast (Weiss and Edwards, 1980), our expression system provides a powerful assay for identifying changes that result in increased chemical efficiency of the antibody. The host cells that çontain the abzyme have all the enzymes needed for aromatic amino acid biosynthesis *except* for the essential chorismate mutase.

Experiments are now underway to determine whether the intracellular antibody can repair the metabolic deficiency of its yeast host. Random mutagenesis of the immunoglobulin genes, followed by genetic selection, will then allow us to identify improved versions of 1F7. In essence, we are attempting to mimic in the microorganism what normally happens during maturation of the immune response in vertebrate cells. However, rather than select for tighter binding to an imperfect transition state analog, we will select directly for *function*. Since sensitive selection assays can be devised for virtually any metabolic reaction as well as for reactions in which drugs or hormones are synthesized or in which pesticides or other toxins are destroyed, genetic selection is likely to be a powerful and general tool for investigating and extending the properties of a wide range of antibody catalysts.

Concluding Remarks

Catalytic antibody technology is exciting because it provides researchers with the means to create entirely new protein catalysts for use in research, industry and medicine. In the last few years, a large number of different chemical reactions have been catalyzed by abzymes, including ester and amide hydrolysis (Tramontano, et al., 1986; Pollack, et al., 1986; Janda, et al., 1988; Janda, et al., 1989; Iverson and Lerner, 1989), acyl transfer to amines and alcohols (Benkovic, et al., 1988), photochemical processes (Balan, et al., 1988; Cochran, et al., 1988), a β-elimination (Shokat, et al., 1989), various redox reactions (Shokat, et al., 1988; Janjić and Tramontano, 1989), and the pericyclic processes discussed above (Hilvert, et al., 1988; Jackson, et al., 1988; Hilvert, et al., 1989). In the next few years, abzymes are likely to be produced that can site-specifically cleave peptides and proteins, modify complex oligosaccharides regio- and stereospecifically, carry out selective aldol condensations, polyene cyclizations and a host of other redox processes. So long as a suitable transition state analog can be devised, and the target reaction is compatible with an aqueous milieu and a protein active site, it should be possible to create tailored antibody catalysts for virtually any chemical transformation. This technology is already inherently multidisciplinary, drawing on chemistry, immunology and enzymology. By bringing the powerful tools of molecular biology and genetics to bear on the problem of improving first-generation abzyme molecules, we anticipate that it will be possible to produce novel catalysts within the next few years that rival the chemical efficiency of naturally occurring enzymes.

Acknowledgments

This work was supported in part by a grant from the National Institutes of Health (GM38273), the National Science Foundation (DMB-8912068), and a Junior Faculty Research Award from the American Cancer Society (JFRA-182).

References

Andrew, P.R., Cain, E.N., Rizzardo, E. and Smith, G.D. 1977. Use of chorismate mutase inhibitors to define the transition state structure. Biochemistry 16: 4848.

Balan, A., Doctor, B.P., Green, B.S., Torten, M. and Ziffer, H. 1988. Antibody combining sites as templates for selective organic chemical reactions. J. Chem. Soc., Chem. Commun. 106.

Bartlett, P.A. and Johnson, C.R. 1985. An inhibitor of chorismate mutase resembling the transition-state conformation. J. Am. Chem. Soc. 107: 7792.

Bartlett, P.A., Nakagawa, Y., Johnson, C.R., Reich, S.H. and Luis, A. 1988. Chorismate mutase inhibitors: Synthesis and evaluation of some potential transition-state analogues. J. Org. Chem. 53: 3195.

Benkovic, S.J., Napper, A.D. and Lerner, R.A. 1988. Catalysis of a stereospecific bimolecular amide synthesis by an antibody. Proc. Natl. Acad. Sci. USA. 107: 5355.

Brown, F.K. and Houk, K.N. 1984. The STO-3G transition structure of the Diels-Alder reaction. Tetrahedron Lett. 25: 4609.

Chao, H.S.-I. and Berchtold, G.A. 1982. Inhibition of chorismate mutase activity of chorismate mutase-prephenate dehydrogenase from *Aerobacter aerogenes*, Biochemistry 21: 2778.

Cochran, A.G., Sugasawara, R. and Schultz, P.G. 1988. Photosensitized cleavage of a thymine dimer by an antibody. J. Am. Chem. Soc. 110: 7888.

Goding, J.W. 1986. Monoclonal antibodies: Principles and practice. Academic Press, New York.

Hilvert, D. 1989. Design of enzyme enzymatic catalysts. In: ACS Symposium Series 389: Biocatalysts in Agricultural Biotechnology. Whitaker, J.R. and Sonnet, P., eds. American Chemical Society, Washington, D.C. pp. 14–23.

Hilvert, D.H., Carpenter, S.H., Nared, K.D. and Auditor, M.-T.M. 1988. Catalysis of concerted reactions by antibodies: The Claisen rearrangement. Proc. Natl. Acad. Sci. USA. 85: 4953.

Hilvert, D.H. and Nared, K.D. 1988. Stereospecific Claisen rearrangement catalyzed by an antibody. J. Am. Chem. Soc. 110: 5593.

Hilvert, D., Hill, K.W., Nared, K.D. and Auditor, M.-T.M. 1989. Antibody catalysis of a Diels-Alder reaction. J. Am. Chem. Soc. 111: 9261.

Huse, W.D., Sastry, L., Iverson, S.A., Kang, A.S., Alting-Mees, Burton, D.R., Benkovic, S.J. and Lerner, R.A. 1989. Generation of a large combinatorial library of the immunoglobulin repertoire in phage lambda. Science. 246: 1275.

Iverson, B.L. and Lerner, R.A. 1989. Sequence-specific hydrolysis of an amide bond. Science. 243: 1184.

Jackson, D.Y., Jacobs, J.W., Sugasawara, R., Reich, S.H., Bartlett, P.A. and Schultz, P.G. 1988. An antibody-catalyzed Claisen rearrangement. J. Am. Chem. Soc. 110: 4841.

Janjić, N. and Tramontano, A. 1989. Antibody-catalyzed redox reaction. J. Am. Chem. Soc. 111: 9109.
Janda, K.D., Benkovic, S.J. and Lerner, R.A. 1989. Catalytic antibodies with lipase activity and R and S substrate selectivity. Science. 244: 437.
Janda, K.D., Schloeder, D., Benkovic, S.J. and Lerner, R.A. 1988. Induction of an antibody that catalyzes the hydrolysis of an amide bond. Science. 241: 1188.
Jencks, W.P. 1975. Binding energy, specificity, and enzymic catalysis: The Circe effect. Adv. Enzymol. 43: 219.
Jencks, W. P. 1969. Catalysis in Chemistry and Enzymology. McGraw Hill, New York.
Jentoft, N. and Dearborn, D.G. 1979. Labeling of proteins by reductive methylation using sodium cyanoborohydride. J. Biol. Chem. 254: 4359.
Kabat, E. A. 1976. Structural concepts in immunology and immunochemistry. Holt, Reinhart and Winston, New York.
Lienhard, G.E. 1973. Enzymatic catalysis and transition-state theory. Science. 180: 149.
Lutz, R.P. 1984. Catalysis of the Cope and Claisen rearrangements. Chem. Rev. 84: 205.
Nisonoff, A., Hopper, J. and Spring, S. 1975. The antibody molecule, Academic Press, New York.
Pauling, L. (1948). Chemical achievement and hope for the future. Amer. Sci. 36: 51.
Page, M.I. and Jencks, W.P. 1971. Entropic contributions to rate accelerations in enzymic and intramolecular reactions and the chelate effect. Proc. Nat. Acad. Sci. USA 68: 1678.
Pawlak, J.L., Padykula, P.E., Kronis, J.D., Aleksejczyk, R.A. and Berchtold, G.A. 1989. Structural requirements for catalysis by chorismate mutase. J. Am. Chem. Soc. 111: 3374.
Pollack, S.J., Jacobs, J.W. and Schultz, P.G. 1986. Selective chemical catalysis by an antibody. Science. 243: 1570.
Pressman, D. and Grossberg, A. 1968. The structural basis of antibody specificity. Benjamin, New York.
Raasch, M.S. 1980. Annelations with tetrachlorothiophene 1,1–dioxide. J. Org. Chem. 45: 856.
Sauer, J. 1966. Diels-Alder reactions: New preparative aspects. Angew. Chem., Int. Ed. Engl. 5: 211.
Sauer, J. 1967. Diels-Alder reactions: The reaction mechanism. Angew. Chem., Int. Ed. Engl. 6: 16.
Shokat, K.M., Leumann, C.J., Sugasawara, R. and Schultz, P.G. 1989. A new strategy for the generation of catalytic antibodies. Nature. 338: 269.
Shokat, K.M., Leumann, C.J., Sugasawara, R. and Schultz, P.G. 1988. An antibody-mediated redox reaction. Angew. Chem., Int. Ed. Engl. 27: 1172.
Sogo, S.G., Widlanski, T.S., Hoare, J.H., Grimshaw, C.E., Berchtold, G.A. and Knowles, J. R. 1984. Stereochemistry of the rearrangement of the chorismate to prephenate: Chorismate mutase involves a chair transition state. J. Am. Chem. Soc. 106: 2701.
Tramontano, A., Janda, K.D. and Lerner, R.A. 1986. Catalytic antibodies. Science. 234: 1566.
Tramontano, A., Ammann, A.A and Lerner, R.A. 1988. Antibody catalysis approaching the activity of enzymes. J. Am. Chem. Soc. 110: 2282.
Wassermann, A. 1965. Diels-Alder reactions. Elsevier Publ. Co., Amsterdam.
Weiss, U. and Edwards, J.M. 1980. The biosynthesis of aromatic amino compounds. Wiley, New York. pp. 134–184.
Wolfenden, R. 1976. Transition state analog inhibitors and enzyme catalysis. Ann. Rev. Biophys. Bioeng. 5: 271.
Ziegler, F.E. 1988. The thermal aliphatic Claisen rearrangement. Chem. Rev. 88: 1423.

Discussion (DR. HILVERT)

B. Green: Regarding the two reactions, you reported that in the Diels-Alder reaction there are approximately 50 turnovers.

Hilvert: Yes, we see at least 50 turnovers. The antibody is not dead after 50 turnovers, but we do not know what the upper limit for turnovers is.

B. Green: In the chorismate mutase reaction was turnover observed there as well?

Hilvert: Yes, again we see multiple turnovers and no irreversible inhibition of the antibody.

Salemme: Many studies show that protein structures recur with different functions. If the antibody structural motif is a good one for making enzyme catalytic sites, why is it that it does not seem to have been used before?

Hilvert: Enzymes play a variety of roles. They have to be regulated by allosteric mechanisms, etc. Perhaps these roles are reflected in the varied structures of existing enzymes. Of course, not every ezyme structure has been determined, and it is conceivable that some may take advantage of the immunoglobulin fold. Also, a number of naturally occurring antibodies have been identified that do catalyze reactions – hydrolysis of the vasoactive intestinal peptide by a human autoantibody, for example – although I do not think that this will turn out to be a general phenomenon.

Fish: You isolated about 60 hybridomas producing antibodies to the transition state. Is that a successful fusion?

Hilvert: In a typical fusion one can get as many as 50 antibodies, but it is very expensive and time consuming to maintain multiple cell lines. Of course 50 monoclonal antibodies represent a tiny fraction of the diversity that is potentially available in the immune system. One recent exciting development at Scripps is the use of a bacteriophage system for expressing large libraries of Fab's in *E. coli*. This allows one to access a much larger fraction of the immune response than is possible with conventional hybridoma techniques.

Fish: Did you try to correlate the affinity of the antibodies to their success as enzymes?

Hilvert: Basically we had two sets of molecules, those that worked and those that did not. Those that worked bound the transition state analogue several orders of magnitude more tightly than they bound the substrates. However, the non-catalytic antibodies also bound the transition state analogue tightly.

M. Rosenberg: You said something that absolutely amazed me. You have been able to obtain expression of functional antibodies inside of yeast?

Hilvert: That is correct. It has been done before by Dr. Carlson at Yale who has been able to express full-length, functional IgG molecules in yeast cytoplasm. We have repeated that work with our chorismate mutase antibody, although we only used the Fab portion of the molecule rather than the full-length IgG.

M. Rosenberg: You have demonstrated full folding of molecules that normally have to pass through membranes and the Golgi system to achieve those types of foldings?

Hilvert: When we break the cells up we obtain functional antibody. We think it unlikely that the disruption process is promoting folding, so we presume that the intracellular material is also assembled. Carlson has evidence that his antibody is able to bind and block cytoplasmic alcohol dehydrogenase.

M. Rosenberg: They got a binding activity, but do they know that they formed an antibody?

Hilvert: Although indirect, the evidence for formation of a functional antibody seems convincing to me.

Lotan: It is certainly excellent work and a beautiful presentation. Just a brief comment. You mentioned the 34 entropy units involved in such a reaction; I would suggest that, in your list of mechanisms of catalysis, you may also add the orientation factor, not just the proximity.

Hilvert: Orientation and proximity are related.

Key Words: Biocatalysis, Abzymes, Catalytic antibody, Diels-Alder reactions, Claisen rearrangements

Engineering Altered Substrate Specificity into Subtilisin BPN′

Richard Bott, Thomas Graycar and David Estell
Genencor International, 180 Kimball Way, So. San Francisco CA 94080

Introduction

The discovery of restriction enzymes has allowed researchers to undertake far more exact analysis of DNA fragments. It has provided a means of extracting and manipulating specific segments of DNA. Segments of DNA coding for a particular protein can now be incorporated into the genome of a prokaryotic or eukaryotic host in a manner that provides for overexpression of that protein.

Recombinant DNA technology allows researchers to take an additional step beyond the cloning of a gene. Once the DNA segment coding for a desirable protein has been isolated and its sequence determined, it is also possible to make selective alterations in the sequence of DNA, thereby altering the amino acid sequence at one or more positions and resulting in the production of a variant protein. The underlying principle in this process is that the amino acid sequence coded by the DNA determines the final tertiary structure of the protein. In the tertiary structure of the protein, sidechains of amino acids that are separated in the linear amino acid sequence are brought together to form the reactive center. In serine proteases, for example, an aspartic acid, histidine and serine are brought together to form a "catalytic triad" (Blow, et al., 1969). The same catalytic triad is formed in proteins where the sequential positions of the residues forming it are interposed in the linear amino acid sequence. In chymotrypsin His57, Asp102 and Ser195 form the catalytic triad (Blow, et al., 1969), while in subtilisin Asp32, His64 and Ser221 are brought into an identical juxtaposition (Wright, et al., 1969). The tertiary folding of these two serine proteases is substantially different, and thus these proteases represent two families of proteases that have convergently evolved from different ancestral genes. The recent finding that two lipases share a similar catalytic triad (Brady, et al., 1990; Winkler, et al., 1990) points to the importance of this particular juxtaposition of aspartic acid, histidine and serine sidechains and the importance of the three-dimensional folding of a protein for its function.

Although the exact nature of how a protein adopts its particular tertiary

D. Kamely et al. (eds), Biotechnology: Bridging Research and Applications, 429–440.

structure is far from being well understood, it is now expected that variant enzymes involving a few substitutions in the amino acid sequence of a protein consisting of several hundred residues will often have the same tertiary structure as the native enzyme. This expectation comes from two general observations. The first observation is that in nature a number of related proteins share a common tertiary folding pattern, even though their amino acids sequences have diverged. For example, two subtilisins, BPN′ (from *Bacillus amyloliquefaciens*) and Carlsberg (from *B. licheniformis*), that differ at 88 out of a possible 275 residues, have virtually the same folding pattern (Bott, et al., 1988; Neidhart and Petsko, 1988; McPhalen and James, 1988). The second observation is that there are now a number of crystallographic studies involving variants enzyme created by introducing one or a few site-specific changes where only subtle conformational changes have been detected (Bott and Ultsch, 1986; Alber, et al., 1987). In this paper we will contrast two related strategies that use the potential of recombinant DNA technology to study the enzyme subtilisin BPN′.

The Model System

Subtilisin BPN′ has proved to be an excellent model system for studying the feasibility of using site-directed mutagenesis for altering the functional properties of an enzyme.

The three-dimensional structure of subtilisin BPN’ has now been determined independently in four laboratories, each using a different crystal form (Wright, et al., 1969; Bott, et al., 1988; McPhalen and James, 1988; Drenth, et al., 1972). The sequences of a number of related subtilisins from different species of *Bacillus* were known (Markland and Smith, 1967; Smith, et al., 1968; Nedkov, et al., 1983; Kurihara, et al., 1972). Subtilisin is a serine protease having a well characterized mechanism of action. The structural gene for *B. amyloliquefaciens* subtilisin had already been cloned and expressed from *B. subtilis* (Wells, et al., 1983). The mature subtilisin BPN’ consists of a single polypeptide chain of 275 residues.

A number of the functional properties of subtilisin BPN’ have been modified. The overall enzyme activity has been increased by single and multiple site-specific mutations, the selectivity for different substrates has been increased by as much as 1000–fold, the pH activity profile has been shifted, and the stability to chemical oxidants and thermal stability have also been increased (Wells and Estell, 1988).

The alteration of substrate specificity is one of the most straightforward objectives to “engineer” into an enzyme. If an enzyme with good reactivity on a particular substrate already exists, it should be possible to tailor its specificity towards another, related substrate with little or no loss in the catalytic activity. This has been achieved following two different strategies. One approach has been to identify the relevant structural differences between

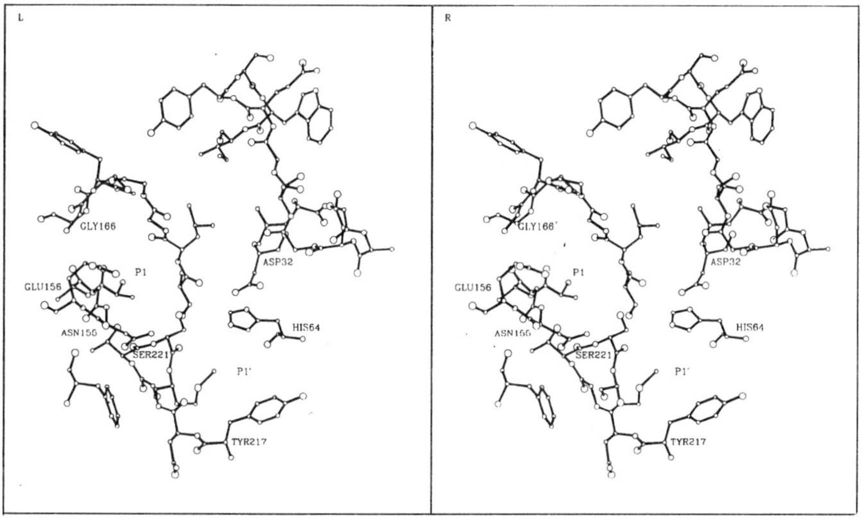

Figure 1. Stereographic view of the substrate binding region. The subsites S1 and S1′, binding P1 and P1', are indicated.

two enzymes such as subtilisins BPN′ and Carlsberg, which manifest different substrate specificities (Svensen, 1976), and then attempt to "recruit" the desirable specificity of one (Carlsberg) into the other (BPN′) (Wells, et al., 1987a). The second, more general approach is to identify sites for mutagenesis in the x-ray crystal structure of subtilisin BPN′ that should affect substrate specificity. We report here a comparison of some of the successes that have been realized from both approaches.

The substrate binding region for a polypeptide substrate is usually defined in terms of subsites. The definition of these subsites follows the nomenclature suggested for a polypeptide substrate bound to the protease papain (Berger and Schecter, 1970). The different sidechains of the polypeptide are designated with respect to the scissle peptide bond, so that the sidechain of the residue preceding the cleaved bond is designated P1, the next preceding residue is designated P2, and so forth. The sidechain of the residue following the cleaved bond is designated P1′, the next following P2′, and so forth. The subsite on the surface of the protein that interacts with the P1 sidechain is designated S1. The subsites interacting with the P2, P1′ and P2′ sidechains are designated S2, S1′ and S2′, respectively, and so forth.

In subtilisin BPN′, the S1 subsite is formed by three segments: residues 126–129, 152–156 and 166 (Figure 1). Unlike the trypsin/chymotrypsin class of serine proteases, where the S1 subsite is a well defined pocket, the subtilisin S1 subsite is an open trough bounded by the catalytic site including Asn155 along with Ser221, His64 and Asp32 and open at the other end. The glycine residue found at position 166 is situated below this open end.

The size, shape and electrostatic nature of this subsite determines in large part the specificity of the enzyme for the P1 sidechain of the substrate. The specificity of subtilisin is reported in the literature to be chymotrypsin-like, favoring large hydrophobic residues at P1. The specificity of subtilisin's S1 subsite has been examined with a series of synthetic substrates having the form succinyl-Ala-Ala-Pro-X-para-nitroanilide (sAAPXpNA), where the amino acid sidechain at P1 (X) is varied (Estell, et al., 1986, Wells, et al., 1987b). The results from these studies indicate that subtilisin has a broad specificity for substrate P1 sidechains. While the optimum is for residues such as tyrosine and phenylalanine at P1, consistent with the earlier studies, substrates having small or hydrophilic sidechains are also hydrolyzed at rates comparable to substrates having tyrosine or phenylalanine.

Results

Two of the residues that form the S1 subsite have been modified by site-specific mutagenesis, at positions 156 and 166. A glutamic acid is found at position 156 in the native subtilisin BPN′ sequence, and we have postulated that this residue may provide some interference to the binding of proteinaceous substrates (Bott, et al., 1987). Glycine is found at position 166 in BPN′ and it has been demonstrated that substitution of different amino acids at this site can dramatically alter the activity and specificity of subtilisin (Estell, et al., 1986, Wells, et al., 1987b).

At position 166, all of the remaining nineteen animo acids have been substituted for glycine. All of these variants retained enzymatic activity ranging from 18–126% of the activity found for the native enzyme on casein (Bott, et al., 1987). The variant having asparagine at position 166 (N166) was more active on casein than the native enzyme. When the kinetic parameters were determined for synthetic peptide substrates, it was evident that the substrate specificity for a number of variants had changed dramatically. The parameter kcat/Km is taken as a measure of the efficiency of the enzyme for a particular substrate. The N166 variant, having improved activity on proteinaceous substrates, has a specificity pattern that is virtually identical to the native enzyme except that the specificity of the N166 is 100–fold lower for the substrate having tyrosine at P1 (Table 1). This is one example of the changes in specificity seen for position 166 variants (Estell, et al., 1986; Wells, et al., 1987b). Another example involves substituting hydrophobic sidechains such as leucine and isoleucine at position 166 (L166 and I166), these variants show a 100–fold discrimination for substrates having phenylalanine at P1 (Estell, et al., 1986). The effect of substituting lysine at position 166 (K166) is even more dramatic for a substrate having glutamic acid in the P1 position where a 330–fold improvement is seen in kcat/Km (Table 1).

The three-dimensional structures of these position 166 variants have been determined. One striking observation is the effect that flexibility in the conformation of a particular sidechain at position 166 can have on the func-

Table 1. Kinetic parameters for site-specific variants at positions 156 and 166 and native subtilisn BPN'

	Substrate											
	sAAPY-pNA			sAAPF-pNA			sAAPK-pNA			sAAPE-pNA		
	k_{cat}[1]	Km[2]	k_{cat}/Km[3]	k_{cat}[1]	Km[2]	k_{cat}/Km[3]	k_{cat}[1]	Km[2]	k_{cat}/Km[3]	k_{cat}[1]	Km[2]	k_{cat}/Km[3]
Enzyme												
BPN'	25.0	0.18	140	50.0	1.4	35.0	28.0	7.5	4.0	0.18	52.0	0.0035
S156	18.0	0.086	210	6.8	0.35	100.0	4.7	20.0	0.23	0.47	12.0	0.039
Q156	36.0	0.12	300	48.0	0.45	87.0	7.2	15.0	0.47	0.86	33.0	0.027
K166	20.0	3.0	6.7	21.0	0.37	57.0	20.0	12.0	1.6	0.69	0.56	1.2
R166	24.0	3.9	6.1	27.0	0.64	42.0	1.4	8.7	0.16	0.40	5.0	0.081
N166	9.5	7.7	1.2	29.0	2.2	13.0	15.2	8.4	1.8	0.25	60.3	0.0042
D166	2.6	19.0	0.14	9.5	18.0	0.5	1.28	21.5	0.059	n.a.	n.a.	n.a.

[1] sec^{-1}.
[2] $\times 10^{-4}$ M.
[3] $\times 10^{4}$.

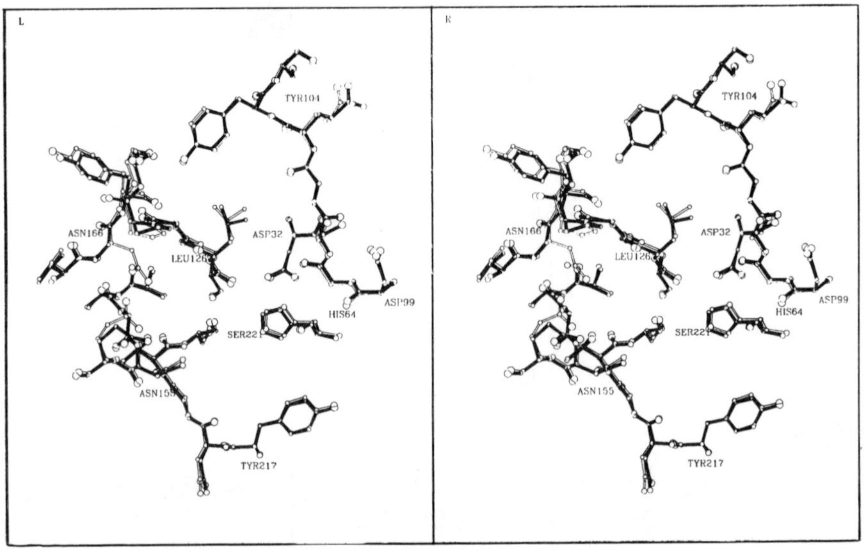

Figure 2. Stereographic view comparing the structure of a variant having asparagine in place of glycine at position 166 (N166) is compared in open bonds with the structure of the native enzyme, in solid bonds. The result of the interaction between the Asn166 and Glu156 sidechains to form a hydrogen bond can be seen with respect to the position of Glu156 sidechain in the native enzyme.

tion of subtilisin BPN′(22). In the three-dimensional structure of the N166 variant (Figure 2), the sidechain of Asn166 forms a hydrogen bond with the sidechain of Glu156. In the native enzyme, the Glu156 sidechain is quite flexible as indicated by a high crystallographic temperature factor. The increased activity of this variant on protein substrates such as casein is postulated to come from the tethering of the charged Glu156 side away from the catalytic site. The three-dimensional structure of the K166 variant (having lysine at position 166) shows a benefit that seems related to the flexibility of the Lys166 sidechain. Unlike the case of the N166 variant, where we propose that stability of a sidechain may be an important factor, the flexibility of the Lys166 sidechain seems to enable lysine to adopt multiple conformations so that the substrate specificity of the K166 variant for the negatively charged glutamic acid sidechain at P1 is increased more than 300–fold, but the specificity for some hydrophobic sidechains such as leucine and phenylalanine at P1 are also increased by 10–fold. These effects are larger than the changes seen in Table 1 for other substitutions that have similar net changes in the overall net charge of residues in the S1 subsite. The R166 (substituting arginine for Gly166) and S156 (replacing serine for Glu156) both have the effect of altering the overall charge in the S1 subsite by +1 either by removing a negative charge contributed by Glu156 or adding a positive charged arginine. Both variants show increased activity on sAAPE-pNA (having glutamic acid at P1) and lower activity on sAAPK-pNA (having lysine at P1). The effect of charge is not always uniform, the data for the D166 (having aspartic acid at position 166) on sAAPK-pNA shows that even though an additional negative charge has been introduced into the S1 subsite, the activity on a substrate with a positively charged sidechain at P1 is reduced. Likewise, the data for the variant having glutamine in place of glutamic acid (Q156) which also alters the net charge at S1 by + 1, shows decreased activity on sAAPKpNA but no increase in its activity relative to the native enzyme on sAAPEpNA. Taken together these results show that while the charge can be important factors the nature of the sidechain and the site of substitution can be determining factors.

We have seen that changes in the S1 subsite in subtilisin BPN′ have successfully modified substrate specificity. As might be expected from the above results, attempts to recruit the substrate specificity of subtilisin Carlsberg into BPN′ has also been successful (Wells, et al., 1987a). Three of the 88 sequence differences in the amino acid sequences of subtilisins BPN′ and Carlsberg occur at positions 156, 169 and 217. At these positions serine replaced Glu156, alanine replaced Gly169 and leucine replaced Tyr217. A triply substituted variant of subtilisin BPN′ having only these three changes was constructed (S156/A169/L217). The activity of this variant for a substrate having glutamic acid at P1 was intermediate between the activity of the parent subtilisin BPN′ and the target subtilisin Carlsberg (Table 2). It was also possible to demonstrate that for a number of different P1 sidechains, the introduction of these three substitutions was additive. Here we will focus on the alteration of specificity for substrates having glutamic acid at P1. The efficiency of BPN′ for the substrate sAAPE-pNA is poor (kcat/Km = 34.5), while the efficiency of Carlsberg is about 70–fold better (kcat/Km = 2160). The efficiency of the triple variant is 35–fold greater than the native BPN' enzyme or half of Carlsberg (kcat/Km = 1050). The additivity can easily be seen in comparing the efficiency of the singly substituted variants S156, A169 and L217 having serine at position 156, alanine at position 169, or leucine at position 217 respectively. These variants have increased efficiency on the SAAPE-PNA substrate: kcat/km = 390 (S156), kcat/Km = 100 (A169), and kcat/Km = 150 (L217) compared to 35 for the native enzyme. The doubly substituted variant S156/L217, having both serine at position 156 and leucine at position 217, has a synergistic increase greater than the combined kcat/Km values of each single variant relative to the native enzyme (**kcat/Km** = 580) for the substrate s-AAPE-PNA. The triple variant S156/A169/L217 shows even greater synergy for SAAPE-PNA (kcat/Km = 1100).

It is also worth noting that the increase in kcat seen for most substrates in Table 2 for subtilisin Carlsberg relative to BPN′ is likely to be a consequence of the substitution of leucine for Tyr217. The increase in kcat is approximately 10–fold on three of the four substrates (having tyrosine, phenylalanine and glutamic acid at P1). This effect on kcat is offset in large part by a comparable increase in Km for some substrates such as sAAPY-pNA.

Table 2. Kinetic parameters for subtiloisn BPN', Carlsberg and variants of BPN' designed to recruit Carlsberg activity

	Substrate											
	sAAPY-pNA			sAAPF-pNA			sAAPK-pNA			sAAPE-pNA		
	k_{cat} [1]	Km [2]	k_{cat}/Km [3]	k_{cat} [1]	Km [2]	k_{cat}/Km [3]	k_{cat} [1]	Km [2]	k_{cat}/Km [3]	k_{cat} [1]	Km [2]	k_{cat}/Km [3]
Enzyme												
BPN'	25.0	0.18	140.0	50.0	1.40	35.0	30.0	7.5	4.0	0.18	52.0	0.0035
S156	18.0	0.086	210.0	6.8	0.35	100.0	4.7	20.0	0.23	0.47	12.0	0.039
L217	230.0	1.4	170.0	71.0	7.7	59.0	88.0	19.0	4.6	0.68	44.0	0.015
A169	230.0	0.08	280.0	52.0	0.5	110.0	26.0	40.0	0.66	0.22	21.0	0.01
S156/L217	180.0	0.78	240.0	300.0	2.0	150.0	14.0	33.0	0.42	1.2	20.0	0.058
S/L/A169	140.0	0.36	380.0	250.0	0.94	260.0	15.0	16.0	0.92	1.3	13.0	0.11
Carlsberg	230.0	0.78	290.0	510.0	2.0	250.0	680.0	420.0	1.6	3.7	17.0	0.22

[1] sec^{-1}.
[2] $\times 10^{-4}$ M.
[3] $\times 10^{4}$.

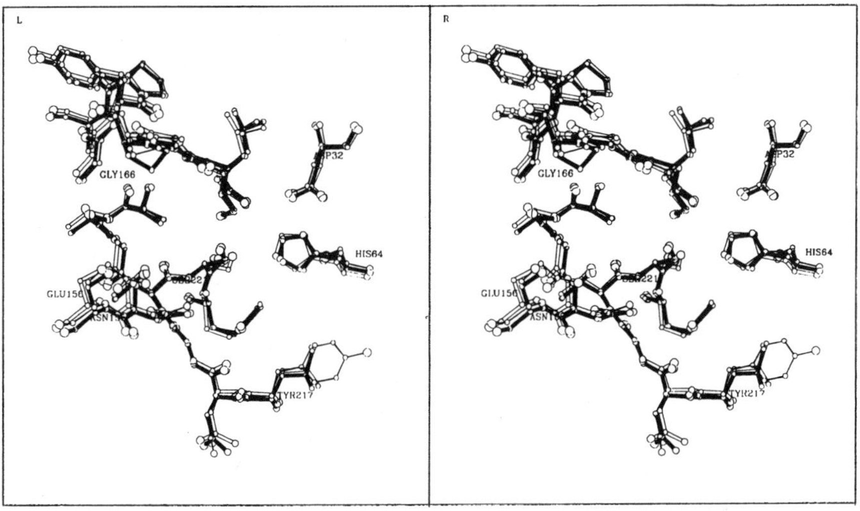

Figure 3. Stereo view comparing native subtilisin BPN' (thin bonds), subtilisin Carlsberg (open bonds) with a quadruple variant FASL (Phe50/Ser156/Ala169/L217) (solid bonds). The conformations of the substituted sidechains in the FASL variant are identical to those seen for the same sidechains in subtilisin Carlsberg.

Perhaps nature has found in the two subtilisins a compromise in different combinations of residues at the active site that result in enzyme having the same level of catalytic efficiency for most substrates.

The three-dimensional structures of the L217 variant and a quadruple variant having S156, A169 L217 along with a substitution of phenylalanine for Met50 (FASL) have been determined at 2.0 å resolution. The structure of the FASL variant has been compared with the structures of native subtilisin BPN' and subtilisin Carlsberg (Figure 3). In the region of position 217 we find that the L217 and F50/S156/A169/L217 variants have the same tertiary folding pattern that is shared by subtilisins BPN' and Carlsberg. The conformation of Leu217 in both the L217 and the FASL variant structures is identical to the same residue in subtilisin Carlsberg. Similar conservation of sidechain conformation is seen for the substitutions at positions 50, 156 and 169; all adopt the same conformations as those seen in subtilisin Carlsberg. These results confirm the expectations that recruitment of substrate specificity can be achieved with the substitutions of a select few residues identified by close inspection of the crystallographic structures of related enzymes when they are available.

In many cases the three-dimensional structures of related proteins will not be immediately available. In those cases the more general approach indicated earlier giving rise to the K166 variant also shows increased activity on S-AAPE-pNA having a kcat/Km = 12300. In this instance the increase was even greater than the variant designed to recruit the Carlsberg activity and indeed greater than the kcat/Km of Carlsberg. It is encouraging that in this case a variant selected solely on the basis of an analysis of the tertiary structure of subtilisin BPN', along with the available information on specificity and substrate binding, exceeds the activity found in two naturally occurring enzymes.

Conclusions

We have presented a few examples illustrating two related approaches that have resulted in variants of an enzyme that have altered specificities. Importantly, the improvement can be manifested as a significant increase in catalytic activity that can often be achieved without loss of the previously existing enzymatic activity. The substitutions of one or a few sidechains at positions that are separated in the tertiary structure of the enzyme are accommodated without any major conformational rearrangement. In the case of the triply substituted variant (S156/A169/L217) sidechain conformations could be predicted on the basis of the existing crystallographic structures and act synergistically of the Kinetic parameters of the variant enzyme. All of this supports the development of predictive programs as an aid to protein engineering.

We have also found single substitutions can have equally dramatic effects on substrate specificity. In the case of K166, the Lys166 sidechain at position 166 can effect changes significantly larger than the similarly charged Arg166 in the R166 variant or the removal of a charge in the case of the S156 or Q156 variants. The effect of the Lys166 sidechain seems to exploit its tendency to adopt multiple conformations and its unique juxtaposition arising from the main chain conformation of residues 165–167 that is conserved in all position 166 variants (Bott, et al., 1987).

We anticipate that, as the data on site-specific variants accumulate, we will begin to recognize other sites similar to position 166 that will afford the opportunity to dramatically alter the specificity of subtilisin and other enzymes toward specific substrate targets.

References

Alber, T., Doa-Pin, S., Wilson, K., Wozniak, J.A., Cook, S.P. and Matthews, B.W. 1987. Contributions of hydrogen bonds of Thr157 to the thermodynamic stability of phage T4 lysozyme. Nature. 330: 41–47.

Berger, A. and Schecter, I. 1970. Mapping the active site of papain with the aid of peptide substrates and inhibitors. Phil. Trans. Roy. Soc. Lond. B 257: 249–264.

Blow, D.M., Birktoft, J.J. and Hartley, B.S. 1969. Role of a buried acid group in the mechanism of action of chymotrypsin. Nature. 221: 235–242.

Bott, R. and Ultsch, M. 1986. Structural consequences of site- specific mutagenesis. Fifth International Symposium on the Genetics of Industrial Microorganisms 1986. Alacevic, M., Hranueli, D. and Toman, Z., eds. Zagreb. pp. 375–385.

Bott, R.R., Ultsch, M., Wells, J., Powers, D. Burdick, D., Struble, M., Burnier, J. Estell, D., Miller, J., Graycar, T., Adams, R. and Power, S. 1987. Importance of conformational variability in protein engineering of subtilisin. ACS Symposium Series 334, Washington D.C. pp. 139–147.

Bott, R., Ultsch, M., Kossiakoff, A., Graycar, T., Katz, B. and Power, S. 1988. The three-dimensional structure of *Bacillus amyloliquefaciens* subtilisin at 1.8 å and an analysis of the structural consequences of peroxide inactivation. J. Biol. Chem. 263: 7895–7906.

Brady, L., Brzozowski, A.M., Derewenda, Z.S., Dodson, E., Dodson, G., Tolley, S., Turkenburg, J.P., Christiansen, L., Huge-Jensen, B., Norskov, L., Thim, L. and Menge, U. 1990. A serine protease triad forms the catalytic centre of a triaglycerol lipase. Nature. 343: 767–770.

Drenth, J., Hol, W.G.J., Jansonius, J.J. and Koekoek, R. 1972. Subtilisin novo: The three-dimensional structure and its comparison with subtilisin BPN'. Eur. J. Biochem. 26: 177–181.

Estell, D.A., Graycar, T.P., Miller, J.V., Powers, D.B., Burnier, J.P., Ng, P.G. and Wells, J.A. 1986. Probing steric and hydrophobic effects on enzyme-substrate interactions by protein engineering. Science. 233: 659–663.

Markland, F.S. and Smith E.L. 1967. Subtilisin BPN′ V11 isolation of cyanogen bromide peptides and the complete sequence. J. Biol. Chem. 242: 5198–5211.

McPhalen, C.A. and James, M.N.G. 1988. Structural comparison of two serine proteinase-protein inhibitor complexes: Eglin-C-subtilisin Carlsberg and CI-2–subtilisin Novo. Biochem. 27: 6582–6598.

Nedkov, P., Oberthur, W. and Braunitzer, G. 1983. Amino acid sequence of subtilisin DY. Hoppe-Seyer's Z. Physiol. Chem. 364: 1537–1540.

Neidhart, D.J. and Petsko, G.A. 1988. The refined crystal structure of subtilisin Carlsberg at 2.5 å. Protein Eng. 2: 271–276.

Kurihara, M., Markland, F.S., and Smith E.L. 1972. Subtilisin Amylosacchariticus III isolation and sequence of the chymotryptic peptides and the complete sequence. J. Biol. Chem. 247: 5619–5631.

Smith, E.L. DeLange, R.J., Evans, W.H., Landon, M. and Markland, F.S. 1968. Subtilisin Carlsberg V: The complete sequence. Comparison with subtilisin BPN′: Evolutionary relationships. J. Biol. Chem. 243: 2184–2191.

Svensen, I.B. 1976. Chemical modifications of the subtilisins with special reference to the binding of large substrates. Carlsberg Res. Commun. 41: 237–291.

Wells, J.A., Ferrari, E., Henner, D.J., Estell, D.A. and Chen, E.Y. 1983. Cloning, sequencing and secretion of *Bacillus amyloliquefaciens* subtilisin in *Bacillus subtilis*. Nuc. Acid Res. 11: 7911–7925.

Wells, J.A., Cunningham, B.C., Graycar, T.P. and Estell, D.A. 1987a. Recruitment of substrate-specificity properties from one enzyme into a related one by protein engineering. Proc. Nat. Acad. Sci. USA. 84: 5167–5171.

Wells, J.A., Powers, D.B., Bott, R.R., Graycar, T.P. and Estell, D.A. 1987. Designing substrate specificity by protein engineering of electrostatic interactions. Proc. Nat. Acad. Sci. USA. 84: 1219–1223.

Wells, J.A. and Estell, D.A. 1988. Subtilisin – an enzyme designed to be engineered. Trends in Biochem. Sci. 13: 291–297.

Winkler, F.K., D'Arcy, A. and Hunziker, W. 1990. Structure of human pancreatic lipase. Nature. 343: 771–774.

Wright, C.S., Alden, R.A., and Kraut, J. 1969. Structure of subtilisin BPN' at 2.5 Å resolution. Nature. 221: 337–340.

Discussion (DR. BOTT)

Mevarech: For clarification, when you showed us a model, does that mean you crystallized every mutant, or is it just a model?

Bott: All of the variant enzyme models I have presented any crystallographic structures, refined at high resolution (2 Å) to crystallographic R-factors below 16%.

Key Words: Enzyme modification, Subtilisin, Site-directed mutagenesis

Summary of Conference

These past three days of scientific exchanges between our U.S. and Israeli colleagues have informed us of the state of the art techniques and strategies used by the biotechnology community. The speakers are the primary innovators of new technologies (e.g., the rapid DNA sequencer, gene probe detectors and amplifiers using Q beta replicase, and nuclear magnetic resonance spectroscopic determination of protein conformation in the solid state on an atomic level).

A goal of these techniques is the preparation of naturally occurring or novel proteins, polysaccharides and nucleic acids in large scale and with highly reproducible stable activities. The proteins may have catalytic, receptor, hormonal, immunological, electron/photon transfer or pharmaceutical properties of interest. From the defense perspective these properties have applications in the detection and degradation of toxic materials, waste solvents, and polyhalogenated aromatics and hydrocarbons; in the construction of detector systems utilizing biomaterials; and in the development of electron-/photon switches for optical recording and information storage films and for visual display systems utilizing thin films.

The polysaccharides and lipopolysaccharides have rheological, emulsifier and receptor properties of interest. These polymers reduce frictional drag experienced by ships and the energy required for pulverizing minerals.

Nucleic acid polymers provide the genetic coding materials that can be used as probes or in the production of proteins of interest by genetic engineering. Genetic engineering provides a mechanism to produce protein biomaterials in a highly reproducible and cost effective manner. The large amount of gene sequence information gathered during the past decade has facilitated the development of polypeptides with desired catalytic, adhesive and receptor properties. From knowledge of the nucleic acid sequences in the human genome and therefore the primary sequence of proteins in the "standard" human, it is anticipated that new rules predicting functional activity from primary structure will emerge. These rules will depend upon the development of peptides/proteins with highly ordered conformations. The availability of polymers in a stable active form, with a shelf-life of months to years, will increase the utility of these materials in the defense, energy and environmental communities. One report discussed the stabilization of proteins by immobilization in glass beads. These studies provide one mechanism for obtaining stable, functionally active biomaterials.

D. Kamely et al. (eds), Biotechnology: Bridging Research and Applications, 441–444.

The technologies by which biopolymers with receptor, immunologic or electron/photon transducing properties are embedded on surfaces are important in the construction of sensors, catalysts, switching and optical storage devices. We have heard presentations on electrode adsorbed proteins which can function as sensors. Luminescent bacteria were shown to have utility in the detection of 100 ng levels of antibiotics and 0.001 ppm levels of parathion and nitrogen mustards. The photoluminescent properties of gallium arsenide and cadmium selenide crystals were shown to have utility in the detection of amines on-line; when these crystals are coated with biopolymers they will have the capability to detect additional chemical components in the liquid or gas phase. Immobilized bipyridinium complexes, when associated with redox enzyme systems, function as electron/photon transducers with applications in information storage films or molecular switches.

Because the reproducibility and sensitivity of the biomaterial systems depend on the structure of the polymer absorbed to surfaces, techniques that reveal such structure at an atomic level have importance. Fourier-Transformed Infra Red Spectral Analysis provides resolution of surface adsorbed proteins in the hundreds of Angstroms to micron level. Our Conference has heard an excellent presentation of the utility of NMR spectroscopy to reveal atomic resolution (2–3 Å) of proteins in the aqueous state or when the polymers are adsorbed to surfaces in the solid phase. The three-dimensional structure of the membrane bound PF1 protein was demonstrated by NMR spectroscopy and the three-dimensional structure of acetylcholinesterase, a membrane bound ion channel, is being determined by this technique.

The NMR technology also provides the capability to analyze the structure of normal and tumor tissue in vivo in small animals. The method is non-invasive and provides resolution at a level of 90×90 μm in the X,Y plane and 125 μm in the Z plane (Figure 1). This image was obtained by our group at the University of Wisconsin, Madison. The collaborators on this project included Mark Anderson, William Milo Westler and John Markley, the Director of the NMR National Facility.

Methodologies which permit the characterization of the polymers which have potential utility as materials with sensor, lubricating, diagnostic, adhesive, catalytic properties are shown in Table 1.

Table 2 illustrates the application of optical detection systems to analytical procedures of biotechnological importance.

The integration of biopolymers into solid and gel matrices for applications as detectors is shown in Table 3.

The reclamation of contaminated soils, equipment and ground water is a high priority for the 1990's as indicated by the introductory remarks of W. Schultz, J. Cornette and D. Kamely. The reclamation process involves the conversion of materials, solvents and waste stream effluents to non-toxic products, most preferably in situ. Among the steps in reclamation may be the energy efficient concentration of solvents, the partial degradation of

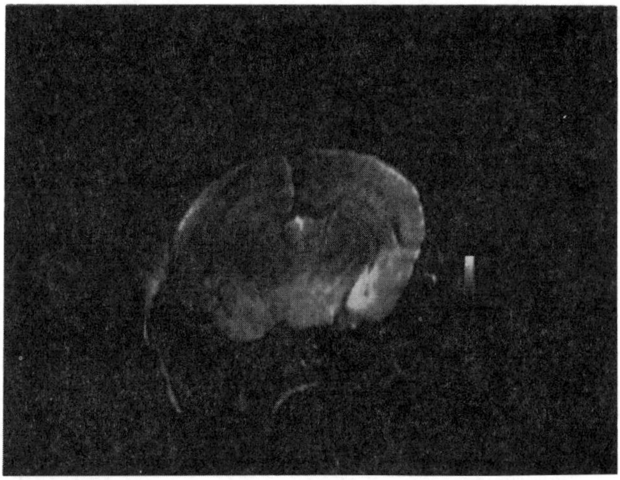

Figure 1. Magnetic resonance image, acquired at 9.4 tesla of brain from a 38 day old mouse. The magnification is 6x.

Table 1. New technologies for structure and surface analysis

Crystallographic Applications in Protein Engineering – Salemme atomic resolution 2–3 Å need crystal and site substitutions
Nuclear Magnetic Resonance Spectroscopy - Opella atomic resolution 2–3 Å aqueous or solid phase proteins are characterized
Rapid DNA Sequencer – Smith 730 bases/hr; 60,000 molecules or 10^{-19} moles with S/N = 2
Magnetic Resonance Imaging – Markley and Kornguth 9.5 tesla; *in vivo* study of small animals resolution is 90 × 90 × 125 Å

Table 2. Optical detection systems

Bacterial Luminescence and Analytic Measures – Ulitzur sensitive at the 100 ng ml^{-1} level for antibiotics
Optically Coupled Amine Sensors – Ellis μM to mM sensitivity for detection of amines on-line
Photocatalysis and Electron Transfer with Modified Proteins – Willner
Pulsed Laser Fluorometry in Environmental Monitoring – Saunders sensitive at ppb

Table 3. Redox detection systems

Biosensors Using Bilayer Electrodes – Gitler
Immobilized Enzymes in Sol/Gel Glass Matrices – Braun

Table 4. Environmental decontamination

Bacterial Degradation – Chakrabarty bacterial selection and genetic engineering
Catalytic Antibodies – Green and Hilvert
Remediation of Heavy Metal Wastes – Silver
Human Cholinesterase Targets for Organophosphorus Poisons – Soreq
Devices for Toxicology – Fish
Antibodies in the Degradation of Toxic Agents – Green

materials to less toxic or more readily degraded intermediates, and the final conversion of the intermediates to metabolites which meet the standardized limits (EPA guidelines). Technologies including sequential composting, photolytic/ catalytic cleavage of carbon-carbon bonds in aromatic systems, and metabolic oxidation of open ring structures are examples of a multi-stage strategy which may be necessary for in situ remediation.

Table 4 illustrates the utilization of cells, subcellular components and biopolymers in bioremediation efforts. These strategies have direct applications for in situ remediation of sites contaminated with petrochemicals, nitrocelluloses, nitroaromatics, polyhalogenated aromatics and hydrocarbons.

This conference demonstrates the great benefits that cooperative ventures may provide to the scientific, commercial and cultural communities of the United States and Israel. The unique strengths available in each of our nations have been illustrated by the lectures and discussions held during the past three days. It is the hope of the Organizing Committee, the supporting agencies and participants that future collaborative efforts will emerge from the Conference. I would anticipate that the result of such collaborative efforts will provide the basis for scientific exchanges at future meetings. I want to thank our Israeli hosts from the Weizmann Institute, Hebrew University, Technion, the Israeli Ministry of Defense, the multiple Agencies of the United States Department of Defense that provided support for this Conference, the presenters, discussants and participants.

Steven E. Kornguth

Concluding Discussion

Kornguth: We have representatives from the Israeli and U.S. private sectors from whom we would like input on how the identified capabilities can be applied – topics of interest include cost effectiveness and the improved properties of novel materials. We are also interested in identifying basic research components which will have application in 5–10 years.

Fish: The biological industry can help by moving new technologies into the marketplace. For example, there are very useful methods developed by Dr. Ulitzur for use in the laboratory for detecting toxic materials. Two years ago we had a kit which required less machinery, took 1–2 h, and required a laboratory; now we have chemistry which takes 12 min, is more sensitive, can be used on a larger scale, and is completely portable.

Er-El: Industry is geared to do development, but ideas must come from the university. Industry can serve as a link between the scientific ideas and what the community requires. Academia creates the idea and industry can best develop, validate and package it in the right form. That is the contribution of industry. The pharmaceutical industry has been asked to present techniques which are more accurate, valid, and sensitive. There are no strict guidelines for registering a new biopharmaceutical, as a result each industry sets the limit higher. For example, five years ago you could register a drug with 95% purity, two years ago with 99% purity, and now we are at a level of 99.99%. As a result we are required to present routine on-line production techniques to detect very low levels of viruses, foreign proteins, metals. Leaching from columns at the low level of contaminants becomes a very severe problem for which we need fully validated techniques. The FDA requires that and Europe after 1992 will be even more strict. The difference between the military and the pharmaceutical industry is that when you are dealing with a biopharmaceutical, you can identify the possible contaminants. Many years ago there was an attempt to look into a detection system that could clearly identify a biological assault in a non-specific way. Most of the current techniques fit biopharmaceutical industry needs since they identify the source. But we have not developed any tools to detect contaminants from unknown sources.

C. Green: In the last year I have had the opportunity to work with a team to look at both research and production issues associated with military questions in biotechnology from an industrialist's perspective. After the last three days no one can question that there are obviously a number of excep-

D. Kamely et al. (eds), Biotechnology: Bridging Research and Applications, 445–449.

tionally important potential military applications in biotechnology. That obvious fact is not entirely clear in the upper echelons of military or industrial America. In some way or another, the arcane world that has been discussed here, largely explicable and understandable only by people such as us in this room, must be translated into industrialist's language, or progress will only come through shallow and meager steps in terms of funding in the next decade. Paradoxically, the discovery rate in applied biotechnology is absolutely astonishing. Biotechnology in the next decade will make physics of the 1920's, electronics of the 1930's and 1940's, and computing sciences of the 1950's and 1960's pale by comparison in terms of the rate of change, discovery, and degree of extensive development of the science. That makes the challenge doubly important for us. What is often referred to in biotechnology laboratories as "secret" or "proprietary" research is actually just emerging technology, severely limited by non-bioscientists who place limits because we have not adequately explained ourselves.

In the next ten years the choices that must be made for our military will not be choices relating to application but choices relating to research-intensive questions. What should we work on? The issue is in finding some strategy to determine which of the hundreds of ideas are important, and no longer finding out whether biotechnology has application. At this time, our domain essentially is that of biophysics. In the next decade, however, we will be dealing with technology choices and the limitations will be largely that of the genome. Thirty years from now both military and industry systems will largely be driven by regulatory, environmental and ethical choices and will not have much to do with science and engineering.

We should now, therefore, begin to pay more attention to the engineering-intensive questions. Where are the process technologies and mathematics going to come from in order to produce the sorts of things that our science can make possible?

Kornguth: This essentially sets the direction for our discussion. It is the integration of each of the individual components identified that will determine the status of things to come in the year 2020.

Bott: There is a different time frame for biotech when compared with physical or chemical engineering. From initiation to fruition, projects in biotechnology take five to ten years. You have to be prepared for that time frame.

Ellis: It is essential for my academic colleagues to get an opportunity to network with those of you in the military and in the private sector to learn your concerns and interests. There is a very imperfect communications system between us. There is tremendous expertise in the universities and many of us have a choice of problems. It is frustrating to realize too late our missed opportunities.

B. Green: I agree, it is very important to publicize the opportunities and needs in biotechnology. Some of our best students are very interested in working on applied problems. I would like this group to discuss mechanisms for facilitating joint efforts, for instance collaboration between crystallographers and chemists.

Juris: I would like to address the time line issue. We need to move very rapidly to develop these technologies for implementation in the field because, from the Army's perspective, the environmental restoration of hazardous waste contaminated sites is in high gear, and expected to continue for about 20 years. The DoD predicts that the bulk of restoration will take place within the next five years. We are marching along with conventional technologies, for instance incineration, which will be utilized to a substantial degree because it is available, tested, tried and true. Biotechnological solutions will have to happen very quickly to be of any use. We do not have the luxury to wait for biotechnological developments to get here in order to clean up these sites. Also there will be a long lead time built time built into getting public acceptance of approaches that biotech may take.

Kornguth: Can you give us a perspective on the cost issue?

Juris: For incineration of contaminated soils programs the cost is in the vicinity of $250–300 per ton. In a recent project we incinerated 100,000 tons of soil, so we're talking about substantial amounts of money. If new technology can reduce that by an order of magnitude, there is a big incentive, but it must move rapidly.

Kornguth: This emphasizes why the university, private and federal sectors have to work together. The university in the United States is not a liability assuming entity. We have excellent scientists, eagerly developing technologies, but if we are going to incorporate biotechnology in remediation strategy we must couple it to an underwriting of the liability. The technology must be demonstrated within an acceptable time frame and in conformance with U.S. and E.C. standards and policies, and that is where the private sector has its strength and its participation is absolutely mandatory. It is essential to point out that the incinerated effluents are monitored, and that the procedure produces inactive materials, not volatile toxic fumes.

Juris: We use a high temperature incineration process, monitor the emissions, and there are no components released that would be of any concern, as far as we can detect. We use this process for explosive components – TNT, RDX, HMX, etc. With respect to the question of funding, our agency is involved in R&D, and we are always looking for promising potential projects to support.

Salant: The Natick Research, Development and Engineering Center is somewhat atypical. We have two organizational lines for R&D – a series of laboratories (research institutions) and research, development and engineering centers which have bottom-line responsibilities for producing end items and systems. Both do basic research and exploratory development. Our center is trying to develop a biotechnology program; we have been identified internally as a future growth area. We are looking at research investment strategies, a long and laborious process. At the same time we must deal with the "Peace Dividend" which is a very real thing. When the Army looks at substantially reducing its forces, there are real implications in terms of new technology, procurement, and industrial base which becomes the ultimate consideration, certainly from an R,D&E Center point of view. Trying to plan in this environment, we must realize that we are looking at a finite set of resources (both financial and intellectual) which are going to become even more finite in the future. We must look for collaborative opportunities to leverage those resources. Fortunately for such actions as the 1986 Amendment to the Technology Transfer Act, this is a much easier task today. We have at this moment our first consortium – with the University of Lowell and three industrial companies – working on biodegradable polymers. We are looking for a similar collaboration to begin scale up as the result of the successful cloning of the spider silk gene. We recognize that there are capabilities that are required that are not indigenous to our organization that are spread throughout university and industrial areas. We must use this collaborative approach to exploit the opportunities that exist today.

Grossman: I would like to offer a criticism of this meeting. Many scientists have become so caught up in these new techniques at the molecular level that are technically sweet, that they have forgotten some basic principles. There are established procedures for detecting environmental changes that have been used for a long time – simple things like miners taking canaries down into mines. We are dealing with whole animals that must be protected from their environment. All military organizations are responsible for environmental hazards that are affecting people now and in the future. Not everyone who has been exposed to a toxic environment is necessarily going to be affected by that environment. It would help if we could identify those that are vulnerable to advise them to avoid environmental conditions that are hazardous to them. Prospective studies on populations that are going to exposed to environmental agents should be done.

Er-El: A historical perspective of biotechnology in the 20th century reveals that biotechnology always advanced in great steps according to military needs. World War I saw the Weizmann process applied to large scale production of acetone because the British Navy needed acetone. That was the first large scale anaerobic fermentation to produce an industrial chemical. The second major development in biotechnology came in World War II with the penicillin project as the first large scale aerobic fermentation system. The third major advance in fermentation technology came during the oil crisis, when tremendous government support was made available to support research into cellulose degradation, etc. The only revolution I find difficult to link to direct military assistance is genetic engineering. Industry is not going to explore crazy ideas and novel technology. We now have good tools to genetically engineer genes, transfer them to organisms and mammalian cells, express them, and make a nice development program. We need some sensitive assays. But we cannot dream of technology in the year 2000 without the thrust of government financial support to enable novel development. The civilized industrial world is under constant assault, in a war not between east and west, which may dictate the future of mankind on the earth. We see proliferation of arms into third world countries which may take advantage of their new "toys" to exploit for instance the food resources of the industrial world. The premise that we are approaching a world of eternal peace where no army will be needed and no money will be available for future technology is very shallow.

Shain: The only parallel I can think of is when President Reagan asked for help on the Star Wars program, we were the first to volunteer. The MoD served as the funnel for all research group funding. We should be looking at how we can develop the MoD as a focal point to get these efforts together to represent the research efforts of groups in Israel in establishing partnerships with groups in the U.S. The advantage we have in the MoD is that we can talk with our colleagues in the U.S. and translate some of these projects into research efforts and funnel funding for international cooperation.

Kornguth: Thank you all for participating in the meeting. It has been an exciting scientific, cultural and social experience. I hope it will lead to mutual interactions between Israeli and U.S. university, private sector and federal laboratories. Good wishes for success to all of you.

List of Participants

Dr. Gabriel Amitai
Israel Institute for Biological Research
Division of Chemistry
P.O. Box 19
Ness-Ziona 70450
Israel

Prof. Shoshana Arad
Ben-Gurion University of the Negev
The Institutes for Applied Research
Beer-Sheva 84110
Israel

Dr. Ruth Arnon
Vice President
Weizmann Institute of Science
Rehovot 76100
Israel

Mr. Avi Avital
Ministry of Defence
Directorate of Defence R&D
Mafat
Hakirya, Tel-Aviv 61909
Israel

Mr. Joseph Bartov
Ministry of Defence
Directorate of Defence R&D
Mafat
Hakirya, Tel-Aviv 61909
Israel

Dr. Richard Bott
Genencor International
180 Kimball Way
South San Francisco, CA 94080

Dr. Sergei Braun
The Hebrew University
Department of Biological Chemistry
Institute of Life Sciences
Jerusalem 91904
Israel

Mr. Bruce Brodman
Chief, Advance Concept Team
U.S. Army Research, Development, and Engineering Center
Attn: SMCAR-AEE-BR,
Building 455T
Picatiny Arsenal, NJ 07806–5000

Dr. Ananda Chakrabarty
University of Illinois College of Medicine
Department of Microbiology
835 South Wolcott Avenue
Chicago, IL 60612

Ms. Leah Chapnik
Ministry of Defence
Directorate of Defence R&D
Mafat
Hakirya, Tel-Aviv 61909
Israel

Dr. Asher Cohen
Ministry of Defence
Directorate of Defence R&D
Mafat
Hakirya, Tel-Aviv 61909
Israel

Capt. Thomas J. Contreras
Office of Naval Technology
Code 22D
800 North Quincy Street
Arlington, VA 22217

Dr. Jimmy Cornette
Senior Scientist
U.S. Air Force
HQ AFESC/RDV
Tyndall Air Force Base,
FL 32403–6001

Dr. Itzhak Duer
Ministry of Defence
Directorate of Defence R&D
Mafat
Hakirya, Tel-Aviv 61909
Israel

Dr. Eric Eisenstadt
Office of Naval Research
Attn: Code 1141MB
800 North Quincy Street
Arlington, VA 22091

Dr. Mary Pat Eisman
Naval Civil Engineering Lab
Code L71
Port Hueneme, CA 93043

Dr. Arthur Ellis
University of Wisconsin
Department of Chemistry
1101 University Avenue Madison,
WI 53706

Dr. Zvi Erel
Interpharm Laboratories Ltd.
Science Based Industrial Park
Kiryat Weizmann
Ness-Ziona 76110
Israel

Dr. Avshalom Falk
Ministry of Defence
Directorate of Defence R&D
Mafat
Hakirya, Tel-Aviv 61909
Israel

Dr. Falk Fish
Orgenics Ltd.
Post Office Box 360
Industrial Zone
Yavne 70650
Israel

Dr. David Friedman
Ministry of Defence
Directorate of Defence R&D
Mafat
Hakirya, Tel-Aviv 61909
Israel

Dr. Jonathan Gershoni
National Institutes of Health
National Cancer Institute
Building 37, Room 6A09
Bethesda, MD 20892

Dr. David Gillespie
Hahnemann University
Department of Chemistry
Mail Stop 412
Broad Street and Vine
Philadelphia, PA 19102

Prof. Carlos Gitler
Weizmann Inst of Science
Rehovot 76100
Israel

Dr. Moshe Goldberg
Ministry of Defence
Directorate of Defence R&D
Mafat
Hakirya, Tel-Aviv 61909
Israel

Dr. Marian Gorecki
Kiryat Weizmann
Biotechnology General
Rehovot 76326
Israel

Dr. Bernard S. Green
The Hebrew University
Department of Pharmaceutical Chemistry
Faculty of Medicine
Post Office Box 12065
Jerusalem, 91120
Israel

Dr. Christopher Canfield Green
Head, Biomedical Science
General Motors Research Laboratory
30500 Mound Road
Warren, MI 48090–9055

Ms. Julia Grinshpon
Ministry of Defence
Directorate of Defence R&D
Mafat
Hakirya, Tel-Aviv 61909
Israel

Dr. Lawrence Grossman
Johns Hopkins University
Department of Biochemistry
615 North Wolfe Street
Baltimore, MD 21205

Prof. David Gutnick
Tel-Aviv University
Department of Microbiology
Tel-Aviv 69978
Israel

Dr. David Haselkorn
Kiryat Weizmann
Biotechnology General
Rehovot 76326
Israel

Dr. Donald Hilvert
Research Institute of Scripps Clinic
Molecular Biology MB-15
10666 North Torrey Pines Road
La Jolla, CA 92037

Dr. Joav Jechezkely
Ministry of Defence
Directorate of Defence R&D
Mafat
Hakirya, Tel-Aviv 61909
Israel

Dr. Kenneth Juris
U.S. Army Toxic and Hazardous Materials Agency
Attn: CETHA-CO (Dr. Juris)
Aberdeen Proving Ground,
MD 21010–5401

Dr. Daphne Kamely
Director, Research and Laboratory Management
U.S. Army
ATTENTION: SARD-TR
The Pentagon
Washington, D.C. 20310–0103

Aryeh Kaplan
Ministry of Defence
Directorate of Defence R&D
Mafat
Hakirya, Tel-Aviv 61909
Israel

Dr. Fortune Kohen
Weizmann Institute of Science
Hormone Research
Rehovot 76100
Israel

Dr. Steven E. Kornguth
University of Wisconsin
Departments of Neurology and Physiological Chemistry
Room 659 Waisman Center
Madison, WI 53705

Ms. Esther Krasner
Ministry of Defence
Directorate of Defence R&D
Mafat
Hakirya, Tel-Aviv 61909
Israel

Dr. Harry Langbeheim
Biomakor-Sigma Immunochemicals
Biological Sciences
Kiryat Weizmann
Rehovot 76362
Israel

Dr. Zvi Liron
Israel Institute of Biological Research
P.O. Box 19
Ness-Ziona 70450
Israel

Prof. Noah Lotan
Technion-Israel Institute of Technology
Biomedical Engineering
Haifa 32000
Israel

Dr. Itzhak Mayer
Ministry of Defence
Directorate of Defence R&D
Mafat
Hakirya, Tel-Aviv 61909
Israel

Dr. Moshe Mevarech
Tel-Aviv University
Department of Microbiology
G.S. Wise Faculty of Life Science
Tel-Aviv 69978
Israel

A. Moontzaz
Ministry of Defence
Directorate of Defence R&D
Mafat
Hakirya, Tel-Aviv 61909
Israel

Dr. Jacob Negreanu
Ministry of Defence
Directorate of Defence R&D
Mafat
Hakirya, Tel-Aviv 61909
Israel

Dr. Stanley Opella
Department of Chemistry
University of Pennsylvania
Philadelphia, PA 19104

Mr. Arie Ramati
Ministry of Defence
Directorate of Defence R&D
Mafat
Hakirya, Tel-Aviv 61909
Israel

Dr. Shaul Reuveny
Israel Institute for Biological

Research
Post Office Box 19
Ness-Ziona 70450
Israel

Dr. Judith Rishpon
Tel-Aviv University
Department of Biotechnology
Ramat Aviv 69978
Israel

Prof. Eugene Rosenberg
Tel-Aviv University
Department of Microbiology
Ramat Aviv 69978
Israel

Dr. Martin Rosenberg
Smith Kline & French
Biopharmaceutical R&D
P.O. Box 1539, MC L-28
King of Prussia, PA 19406–0939

Dr. Eli Sahar
Tel-Aviv University
Department of Biotechnology
Tel-Aviv 69978
Israel

Dr. Abner Salant
Director, Soldier Science
U.S. Army Natick Research,
Development and Engineering
Attn: STRNC-Y, Kansas Street
Natick, MA 01760–5020

Dr. F. Raymond Salemme
E.I. Du Pont de Nemours
Post Office Box 80228
Wilmington, DE 19880–0228

Dr. George Saunders
Los Alamos National Labs
Life Sciences Division
Mail Stop M-888
Los Alamos, NM 87545

Dr. Warren W. Schultz
Naval Research Laboratory
Code 6100.1
4555 Overlook Avenue
Washington, DC 20375–5000

Dr. Ralph Shain
Ministry of Defence
Directorate of Defence R&D
Mafat
Hakirya, Tel-Aviv 61909
Israel

Dr. Simon Silver
University of Illinois College of Medicine
Department of Microbiology
835 South Wolcott
Chicago, IL 60612

Dr. Lloyd Smith
University of Wisconsin
Department of Chemistry
1101 University Avenue
Madison, WI 53706

Dr. Hermona Soreq
The Hebrew University
Department of Biological Chemistry
The Life Sciences Institute
Jerusalem 91904
Israel

Dr. Paul T. Strickland
Johns Hopkins School of Public Health
Environmental Health Sciences
615 North Wolfe Street
Baltimore, MD 21205

Dr. Waclaw Szybalski
University of Wisconsin Medical School
Oncology

1400 University Avenue
Madison, WI 53706

Dr. Eliana Tapuhi
Tel-Aviv University
ICTAF
Tel-Aviv
Israel

Prof. Shimon Ulitzur
Technion-Israel Institute of Technology
Biolume
Haifa 32000
Israel

Dr. James J. Valdes
Department of the Army
Biotechnology Division
Attn: SMCCR-RSB
Aberdeen Proving Ground,
MD 21010–5423

Dr. Mirian Waldman
Ministry of Science and Technology
Post Office Box 18195
Jerusalem
Israel

Dr. Keith B. Ward
Naval Research Laboratory
Code 6030
4555 Overlook Avenue S.W.
Washington, DC 20375–5000

Dr. M. Weinstein
Ministry of Defence
Directorate of Defence R&D
Mafat
Hakirya, Tel-Aviv 61909
Israel

Prof. Itamar Willner
The Hebrew University
Organic Chemistry
Jerusalem, 91904
Israel

Dr. Lemuel Wingard
University of Pittsburgh
Pharmacology
School of Medicine
Pittsburgh, PA 15213

Dr. Itzhak Yuli
Weizmann Institute of Science
Rehovot
Israel

Dr. Mark J. Zoller
Genentech, Inc.
460 Point San Bruno Boulevard
South San Francisco, CA 94080

Index

Abzymes, 427
Acetylcholine receptor, 294, 297
Acetylcholinesterase, 165, 166, 379, 397
Achilles' heel cleavage, 375, 376
Acid phosphatase, 218
Acinetobacter, 239, 248
Adducts, 130, 170, 331
Affinity capture, 146, 151
AIDS, 1, 203, 263, 401, 404, 408, 409, 410, 411
Air sampling, 397
Amines, 331
Ammonia biocatalysed photosynthesis, 369
Amplification, 93, 146, 148, 149, 153, 160, 161, 372, 373, 375, 376
Analytical applications of bacterial bioluminescence, 87
Aniline, 320, 331
Antibiotics, 204
Antibodies, 10, 12, 77, 167, 168, 177, 191, 249, 250, 251, 254, 255, 256, 258, 413, 417
Antibody electrode, 107
Anti-viral agents, 297
Arsenite oxidation, 287

Benzo(a)pyrene, 168, 177
Bioaccumulation, 266, 281, 287
Biocatalysis, 413, 424, 427
Biocatalyst, 264, 354
Biocomposites, 42
Biodegradation, 25, 42, 221, 224, 225, 228
Biodispersan, 232, 239, 240, 241, 242, 245
Bioenergy, 42
Bioluminescence applications, 87
Biomaterials, 30, 42
Biorecognition, 297
Bioremediation, 30, 42, 221, 265, 287
Biosensors, 42, 43, 61, 93, 107, 127, 397
Biotechnology, 15, 16, 17, 18, 19, 20, 28, 30, 33, 35, 36, 37, 40, 41, 42, 79, 95, 215, 221, 262, 265, 287, 313, 314, 376, 410, 424
Butyrylcholinesterase, 166

Capillary electrophoresis, 340
Catalase, 207, 214, 218
Catalytic antibody, 423, 427
CD4 receptor, 401, 402, 403, 404, 411
CdS, 317, 321, 322, 324, 325, 327, 331
CdSe, 317, 318, 319, 320, 321, 323, 324, 327, 331
Chaotropic salt, 144, 151
Chlamydia, 190, 191, 202, 203, 204
Cholera toxin, 4, 10, 12
Cholinesterase inhibitors, 397
Chronocoulometric, 107
Chromate reduction, 281, 287
Cis-diamine dichloroplatinum, 12
Claisen rearrangements, 419, 425, 427
CO_2-fixation, 341, 342, 343, 344, 345, 347, 369
Computerized electrode, 107
Conducting polymers, 93
Copolymer I (Cop I), 12

Dead-layer model, 331
Decoy, 297
Detection of point mutants, 376
Detoxification, 264, 281
Diagnostics, 58, 181, 204
Diels-Alder reactions, 252, 253, 416, 417, 419, 420, 425, 427
Directed evolution, 230
Dispersant, 244, 248
DNA adducts, 167, 170, 171, 174, 175, 177
DNA-binding protein, 223, 371, 374, 376
DNA cleavage, 376
DNA damage, 140, 166, 175, 198
DNA modification, 127, 140, 167
DNA repair enzymes, 140
DNA sequencing, 58, 333, 334, 335, 340
DNA transfection, 166
Drag-reduction, 244, 248
Drug design, 297
Drug targeting, 12

D. Kamely et al. (eds), Biotechnology: Bridging Research and Applications, 457–459.

Electrical communication in proteins, 369
Electropolymerization, 89, 90, 93
ELISA, 106, 169, 171, 177, 183, 191, 202, 233, 234, 256, 418, 419
Emulsan, 232, 233, 234, 235, 236, 237, 238, 245, 246, 247, 248, 246, 248
Emulsifier, 245, 246, 247, 248
Entrapment, 89, 209, 218
Environmental monitoring, 397
Enzymatic detoxification, 287
Enzyme electrode, 95, 107
Enzyme immobilization, 93, 218
Enzyme-like catalysis, 264
Enzyme modification, 440
Europium, 65, 68, 76, 78
Experimental allergic encephalomyelitis (EAE), 12

fd coat protein, 113, 115, 116, 117, 118, 119, 120, 121, 122, 123, 124
Field operation, 204
Filamentous bacteriophage coat protein, 124
Fish bacteria, 248
Flow injection analysis, 397
Fluorescamine, 378, 379, 380, 391, 391, 392, 395, 397
Fluorescence, 106, 191, 202, 215, 377, 383, 392, 396, 397
Fluorescent nucleotide tags, 340
Fluoroimmunoassay, 67, 78, 69

GaAs, 321, 323, 326, 327, 328, 329, 331
Gene amplification, 161, 164, 166
Gene expression, 230
Genetic organization, 230
Glass, 90, 205, 209, 210, 218, 261, 322
Glucose monitoring, 107
Glucose oxidase, 90, 93, 96
gp120 binding, 407, 408, 409, 411
Guanidine isothiocyanate, 151

Heavy metal resistance, 287
Hemagglutinin, 12
Hepatitis, 194, 204
HIV, 186, 191, 203, 275, 401, 403, 404, 405, 406, 408, 409, 410, 411
Host cell reactivation (HCR), 140
Human genome, 340
Hydrocarbon, 174, 233, 245, 248
Hydrophilic linker arms, 61

Immunoassay, 65, 66, 68, 69, 77, 78, 167, 175, 176, 180
Immunodiagnostics, 204
Immunoelectrochemical sensor, 107
Immunofluorometric assays, 78
Industry, 15, 18, 20, 35, 42, 314
Influenza virus, 12
Ion channels, 46, 61, 164
Isoluminol, 65, 78

Langmuir adsorption isotherm model, 319, 325, 326, 327, 331
Laser fluorometry, 397
Laser induced fluorescence, 378, 397
Ligand, 204, 294
Ligand binding site, 297
Lipid bilayer, 61
Luciferase, 87

Magainin, 114, 117, 120, 123, 124
Manufacturing, 15, 42
Melittin, 49, 50, 58, 60, 61
Membrane proteins, 109, 110, 111, 124
Mercury reduction, 287
Metallothioneins, 287
Methyltransferase, 376
Microalgae, 301, 302, 307, 314, 315, 316
Molecular hybridization, 150, 151, 201, 204
Monoclonal antibodies, 77, 174, 175, 251, 261, 264, 424
MS2 bacteriophage, 2, 12
Multiple sclerosis, 7, 12
Mutagenesis, 166, 255, 282, 283, 284

Natural dyes, 316
Nitrate reduction, 369
N-methylindoxyl acetate, 397
NMR spectroscopy, 109, 111, 112, 113, 115, 122, 123, 124, 243, 304
Nucleation complex, 151
Nucleic acid probe, 151

Oligodeoxynucleotide adapter, 376
Oligonucleotide, 143, 151
Organophosphate, 258, 264
Organophosphorous poisons, 166

Papilloma, 196, 204
Peptides, 8, 109, 124, 381, 390, 391
Peroxidase, 180, 207, 218
Photobiocatalysis, 369
Photochemical regeneration of cofactors, 369
Photoluminescence, 317, 331
Photoregulative enzymes, 369
Photoresponsive papain, 369
Photosynthesis, 341

Polycyclic aromatic hydrocarbons, 174, 177
Polyethylene sleeves, 307, 316
Polyglutamic acid, 12
Polymerase chain reaction, 151
Polysaccharides, 303, 314, 315, 316
Polythiol polypeptides, 287
Porphyridium, 301, 302, 303, 304, 305, 306, 307, 309, 311, 312, 313, 314, 315, 316
Poultry, 189, 204
Printing reaction, 376
Probe amplification, 148, 151

Receptor, 44, 53, 204, 401, 402
Redox conversion, 287
Redox polymers, 369
Restriction enzymes, 376
Rhodella, 316
Rhodophyta, 316
Riboprobe, 151

Selection advantage, 166
Semiconductor, 317, 321, 328, 329, 331
Sensors, 331
Shiga toxin, 4, 5, 6, 10, 12
Site-directed mutagenesis, 440
Sol-gel, 206, 209, 210, 211, 215, 218
Solid-state NMR, 112, 120, 124
Streptavidin-biotin, 78
Structure-function relationships, 166, 237
Subtilisin, 429, 430, 439, 440
Surfactant, 24, 248
Synthetic peptides, 10, 12
Synthetic vaccines, 12

Therapeutic agent, 411
Thin films, 93
Toxic chemicals, 230
Toxicology, 204
Toxin, 4, 289, 295, 297
Transcription amplification system, 151
Transition state, 264, 425
Trypsin, 207, 212, 213, 218
Two-dimensional NMR, 124

Universal restriction endonuclease, 375, 376
Unsaturated fatty acids , 316
UV photoproducts, 130, 167, 177

Variant alleles, 166
Volatilization, 269, 287

Wastewater, 27, 331

Xenopus oocyte microinjection, 166
Xerogel, 218